Renal and Urinary Proteomics

Edited by
Visith Thongboonkerd

Related Titles

von Hagen, J. (ed.)

Proteomics Sample Preparation

2008

ISBN: 978-3-527-31796-7

Westermeier, R., Naven, T., Höpker, H.-R.

Proteomics in Practice
A Guide to Successful Experimental Design

2008

ISBN: 978-3-527-31941-1

Van Eyk, J. E., Dunn, M. J. (eds.)

Clinical Proteomics
From Diagnosis to Therapy

2008

ISBN: 978-3-527-31637-3

Omenn, G. S. (ed.)

Exploring the Human Plasma Proteome

2006

ISBN: 978-3-527-31757-4

Desiderio, D. M., Nibbering, N. M.

Redox Proteomics
From Protein Modifications to Cellular Dysfunction and Diseases

2006

ISBN: 978-0-471-72345-5

Hamacher, M., Marcus, K., Stühler, K., van Hall, A., Warscheid, B., Meyer, H. E. (eds.)

Proteomics in Drug Research

2006

ISBN: 978-3-527-31226-9

Renal and Urinary Proteomics

Methods and Protocols

Edited by
Visith Thongboonkerd

WILEY-VCH Verlag GmbH & Co. KGaA

The Editor

Dr. Visith Thongboonkerd
Medical Proteomics Unit
Office for Research and Development
Siriraj Hospital, Mahidol University
12th Fl. Adulyadej Vikrom Bldg.
2 Prannok Rd., Bangkoknoi
Bangkok 10700
Thailand

Cover
Two-dimensional map of urine,
with kind permission from the American
Chemical Society.
For details see Fig. 15.3 in this book.

Library of Congress Card No.:
applied for

British Library Cataloguing-in-Publication Data
A catalogue record for this book is available from the British Library.

Bibliographic information published by the Deutsche Nationalbibliothek
The Deutsche Nationalbibliothek lists this publication in the Deutsche Nationalbibliografie; detailed bibliographic data are available on the Internet at http://dnb.d-nb.de.

Cover Design Adam Design, Weinheim
Typesetting Laserwords Private Limited, Chennai, India
Printing and Binding Strauss GmbH, Mörlenbach

Printed in the Federal Republic of Germany
Printed on acid-free paper

ISBN: 978-3-527-31974-9

Contents

Renal and Urinary Proteomics: Methods and Protocols. Edited by Visith Thongboonkerd

ISBN: 978-3-527-31974-9

Preface

Renal and urinary proteomics is one of the most rapidly growing subdisciplines in the clinical proteomics arena. Investigations of normal kidney and urine proteomes provide invaluable information to better understand the renal physiology and function, whereas analyses of the kidney and urine proteomes in various diseases would lead to unraveling the pathogenic mechanisms and pathophysiology of kidney diseases and related disorders. Additionally, comparative analyses of the kidney and urine proteomes in a specific disease with those of the normal state (normal control) and other diseases (disease controls) would lead to the discovery of disease-specific biomarkers. Because of the simplicity of sample collection and availability of urine samples in almost all patients, the urine is an ideal source of clinical sample for the discovery of novel noninvasive biomarkers for clinical diagnostics and prognostics. Approximately 30% of the identified proteins in the normal human urinary proteome are derived from the plasma (by filtrating through the renal microstructure, namely, glomerulus). Therefore, urinary proteomics offers opportunities for biomarker discovery not only in kidney diseases but also in nonkidney diseases.

With this great potential, a group of investigators (including nephrologists, proteomists, biochemists, bioinformaticians, biostatisticians, and others) have initiated an internationally collaborative project/network, namely, "Human Kidney and Urine Proteome Project" (HKUPP) (http://www.hkupp.org) to promote and facilitate analyses of the kidney and urine proteomes worldwide. Recently, the HKUPP has been approved by the Human Proteome Organisation (HUPO) (http://www.hupo.org) as one of its official initiatives. More recently, another network, namely, "European Kidney and Urine Proteomics" (EuroKUP) (http://eurokup.com) has been established by a group of proteomists and nephrologists from (mainly) European countries to enhance the HKUPP activities in Europe. The rapid growth of the *renal and urinary proteomics* field has been well documented also in a recent special issue of PROTEOMICS–Clinical Applications (Volume 2; Issue 7–8; July 2008) (http://www3.interscience.wiley.com/journal/120840688/issue), which contains many research articles and reviews detailing advancements in this field.

However, one of the major obstacles in the *renal and urinary proteomics* arena is the comparability of the results obtained from different laboratories using various

Renal and Urinary Proteomics: Methods and Protocols. Edited by Visith Thongboonkerd

ISBN: 978-3-527-31974-9

technologies/methodologies. The lack of universal standards in this field limits robust comparability and reduces the significance of the renal and urinary proteome data obtained previously. Marked variability has been observed in several aspects of methods and protocols used for renal and urinary proteome analyses. Therefore, the first priority (and most pressing) objective of both HKUPP and EuroKUP is to establish standard protocols and guidelines for analyses of the kidney and urine proteomes. This mission has been ongoing and will take some time before these standards and guidelines can be finalized and are ready for use.

The objective of this book is to provide a practical guide for analyses of the kidney and urine proteomes. In addition to methods and protocols, background and some examples of clinical applications of *renal and urinary proteomics* are provided. Written by the acclaimed experts in the field (most of these experts are the members of HKUPP and/or EuroKUP), this book will be an excellent resource of references for performing high-quality studies in *renal and urinary proteomics.*

Finally, I wish to thank all the authors for their great contributions to this book.

Bangkok,
August 2009

Visith Thongboonkerd, *MD, FRCPT*

List of Contributors

Günter Allmaier
Vienna University of Technology
Institute of Chemical
Technologies and Analytics
Getreidemarkt 9/164
1060 Vienna
Austria

Brian M. Balgley
Calibrant Biosystems, Inc.
Gaithersburg
MD
USA

Rosamonde E. Banks
St James's University Hospital
Clinical and Biomedical
Proteomics Group
Cancer Research UK Clinical
Centre
Beckett Street
Leeds LS9 7TF
UK

Karl-Friedrich Beck
Klinikum der Johann Wolfgang
Goethe-Universität
Pharmazentrum
Frankfurt/ZAFES
Theodor-Stern-Kai 7
60590 Frankfurt am Main
Germany

Rainer Bischoff
University of Groningen
University Centre for Pharmacy
Department of Analytical
Biochemistry
Antonius Deusinglaan 1
Groningen 9713 AV
The Netherlands

Egisto Boschetti
Bio Rad Laboratories
c/o CEA-Saclay
Bât 142
Gif-sur-Yvette 91191
France

Cecilia M. Canessa
Yale University School of
Medicine
Department of Cellular and
Molecular Physiology
New Haven
Connecticut 06520-8026
USA

Annalisa Castagna
University of Verona
Department of Clinical and
Experimental Medicine
P.le L.A.Scuro 10
37134 Verona
Italy

Renal and Urinary Proteomics: Methods and Protocols. Edited by Visith Thongboonkerd

ISBN: 978-3-527-31974-9

Helen P. Cathro
University of Virginia Health System
Department of Pathology
Charlottesville
VA 22908
USA

Daniela Cecconi
University of Verona
Department of Biotechnology
Strada le Grazie 15
37134 Verona
Italy

Shu-Hui Chen
National Chen Kung University
Institute of Nanotechnology and MicroSystem Engineering
Tainan 701
Taiwan
and
Department of Chemistry
National Cheng Kung University
Tainan 701
Taiwan

Rachel A. Craven
St James's University Hospital
Clinical and Biomedical Proteomics Group
Cancer Research UK Clinical Centre
Beckett Street
Leeds LS9 7TF
UK

Prasad Devarajan
Louise M. Williams Endowed Chair
Professor of Pediatrics and Developmental Biology
Cincinnati Children's Hospital Medical Center
3333 Burnet Avenue
Cincinnati
OH 54229
USA

Hassan Dihazi
Georg-August University Göttingen
Department of Nephrology and Rheumatology
Robert-Koch-Strasse 40
37075 Göttingen
Germany

Beverly J. Gabert
University of California
Department of Animal Science
Physiological Genomics Group
One Shield Avenue
Davis
CA 95616
USA

Youhe Gao
School of Basic Medicine
Peking Union Medical College
Institute of Basic Medical Sciences
Chinese Academy of Medical Sciences
Proteomics Research Center
National Key Laboratory of Medical Molecular Biology
Department of Physiology and Pathophysiology
No. 5 Dong Dan San Tiao
10005 Beijing
China

Alex German
University of Liverpool
Small Animal Hospital, Leahurst
Faculty of Veterinary Science
Chester High Road
Neston
Cheshire CH64 7TE
UK

Patricia A. Gonzales
National Institutes of Health
Laboratory of Kidney and
Electrolyte Metabolism, NHLBI
10 Center Drive
Bldg 10, Room 6N
260 Bethesda
MD 20892
USA

Thomas Gronemeyer
Ruhr-Universität Bochum
Medizinisches Proteom-Center
Zentrum für Klinische Forschung
Universitätsstraße 150
44801 Bochum
Germany

Harry Holthofer
Dublin City University
Center for BioAnalytical Sciences
Glashevin
D9 Dublin
Ireland

Peter L. Horvatovich
University of Groningen
University Centre for Pharmacy
Department of Analytical
Biochemistry
Antonius Deusinglaan 1
Groningen 9713 AV
The Netherlands

Isao Ishikawa
Kanazawa Medical University
Asanogawa General Hospital
Division of Nephrology
Department of Internal Medicine
83-KosakaNaka
Kanazawa
Ishikawa 920-8621
Japan

Ramses F.J. Kemperman
University of Groningen
University Centre for Pharmacy
Department of Analytical
Biochemistry
Antonius Deusinglaan 1
Groningen 9713 AV
The Netherlands

Dontscho Kerjaschki
Medical University of Vienna
Clinical Institute of Pathology
Währinger Gürtel 18–20
1090 Vienna
Austria

Urban A. Kiernan
Intrinsic Bioprobes, Inc.
2155 E. Conference Dr., Ste 104
Tempe
AZ 85284
USA

Mark A. Knepper
National Institutes of Health
Laboratory of Kidney and
Electrolyte Metabolism, NHLBI
10 Center Drive
Bldg 10, Room 6N
260 Bethesda
MD 20892
USA

Sigurd Krieger
Medical University of Vienna
Clinical Institute of Pathology
Währinger Gürtel 18–20
1090 Vienna
Austria

Dietmar Kültz
University of California
Department of Animal Science
Physiological Genomics Group
One Shields Avenue
Davis
CA 95616
USA

Cheng S. Lee
University of Maryland
Department of Chemistry and
Biochemistry
College Park
Maryland 20742-4454
USA

Rudolf Lichtenfels
Martin Luther University
Halle-Wittenberg
Institute of Medical Immunology
Magdeburger Straße 2
06112 Halle
Germany

Bi-Cheng Liu
Southeast University
Zhong Da Hospital
Institute of Nephrology
Nanjing 210009
China

Jie Li
National Institutes of Health
(NIH)
National Institute of Neurological
Disorders and Stroke (NINDS)
Surgical Neurology Branch
Bethesda
MD 20892
USA

Lin-Li Lv
Southeast University
Zhong Da Hospital
Institute of Nephrology
Nanjing 210009
China

David Marples
University of Leeds
Institute of Membrane and
Systems Biology
Garstang Building
Leeds LS2 9JT
UK

Pablo Martín-Vasallo
Universidad de La Laguna
Department of Bioquímica y
Biología
Molecular Laboratorio de Biología
del Desarrollo
Avda Astrofásico Sánchez s/n
La Laguna
Tenerife 38206
Spain

Corina Mayrhofer
Medical University of Vienna
Clinical Institute of Pathology
Währinger Gürtel 18–20
1090 Vienna
Austria
and
Vienna University of Technology
Institute of Chemical
Technologies and Analytics
Getreidemarkt 9/164
1060 Vienna
Austria

Helmut E. Meyer
Ruhr-Universität Bochum
Medizinisches Proteom-Center
Zentrum für Klinische Forschung
Universitätsstraße 150
44801 Bochum
Germany

Harald Mischak
Mosaiques Diagnostics &
Therapeutics AG
Mellendorfer Straße 7–9
30625 Hannover
Germany

Ali Mobasheri
University of Nottingham
School of Veterinary Science and
Medicine
Division of Comparative Medicine
Sutton Bonington Campus
Loughborough
Leicestershire LE12 5RD
UK

Harry Mushlin
George Washington University
District of Columbia
Washington D.C. 20852
USA

Gerhard A. Müller
Georg-August University
Göttingen
Department of Nephrology and
Rheumatology
Robert-Koch-Straße 40
37075 Göttingen
Germany

Dobrin Nedelkov
Intrinsic Bioprobes, Inc.
2155 E. Conference Dr., Ste 104
Tempe
AZ 85284
USA

Randall W. Nelson
Intrinsic Bioprobes, Inc.
2155 E. Conference Dr., Ste 104
Tempe
AZ 85284
USA

Eric E. Niederkofler
Intrinsic Bioprobes, Inc.
2155 E. Conference Dr., Ste 104
Tempe
AZ 85284
USA

Rob Ofman
University of Amsterdam
Lab Genetic Metabolic Diseases
Academic Medical Center
Melbergdreef 9
1105 AZ, Amsterdam
The Netherlands

Josef Pfeilschifter
Klinikum der Johann-
Wolfgang-Goethe-Universität
Pharmazentrum
Frankfurt/ZAFES
Theodor-Stern-Kai 7
60590 Frankfurt am Main
Germany

Trairak Pisitkun
National Institutes of Health
Laboratory of Kidney and
Electrolyte Metabolism, NHLBI
10 Center Drive
Bldg 10, Room 6N
260 Bethesda
MD 20892
USA

Satish P. RamachandraRao
University of California at San
Diego
Department of Medicine
Division of Nephrology
Translational Research in Kidney
Disease
California
USA

Maria Pia Rastaldi
Fondazione IRCCS Ospedale
Maggiore Policlinico &
Fondazione D'Amico per la
Ricerca sulle Malattie Renali
Department Renal Research
Laboratory
Via Pace 9
Milano 20122
Italy

Pier Giorgio Righetti
Politecnico di Milano
Department of Chemistry
Materials and Chemical
Engineering 'Giulio Natta'
Via Mancinelli 7
20131 Milano
Italy

Gary F. Ross
Bio-Rad Laboratories
1000 Alfred Nobel Drive
Hercules
CA 94547
USA

Lars-Christian Rump
Ruhr-Universität
Bochum
Medizinisches Proteom-Center
Universitätsstr. 150
44801 Bochum
Germany

Eric Schiffer
Mosaiques Diagnostics &
Therapeutics AG
Mellendorfer Straße 7–9
30625 Hannover
Germany

Barbara Seliger
Martin Luther University
Halle-Wittenberg
Institute of Medical Immunology
Magdeburger Straße 2
06112 Halle
Germany

Kumar Sharma
University of California at San
Diego
Department of Medicine
Division of Nephrology
Translational Research in Kidney
Disease
3350 La Jolla Village Drive, 91114
San Diego, CA 92161-9111
USA

Michael A. Shaw
Thomas Jefferson University
Jefferson Medical College
Kimmel Cancer Center
Proteomics & Mass Spectrometry
Core Facility
10th Street
Philadelphia 19107
USA

Barbara Sitek
Ruhr-Universität Bochum
Department of Nephrology
Universitätsstr. 150
44801 Bochum
Germany

Graeme Smith
AstraZeneca
Alderley Park
Macclesfield
Cheshire SK10 4TG
UK

Robert A. Star
Renal Diagnostics and
Therapeutics Unit, NIDDK/NIH
10 Center Drive, Bldg 10
Room 3N
108 Bethseda
MD 20892-1268
USA

Christian Stephan
Ruhr-Universität Bochum
Medizinisches Proteom-Center
Zentrum für Klinische Forschung
Universitätsstraße 150
44801 Bochum
Germany

Kai Stühler
Ruhr-Universität Bochum
Medizinisches Proteom-Center
Zentrum für Klinische Forschung
Universitätsstraße 150
44801 Bochum
Germany

Abbreviations

ABC	ATP-binding cassette
Abhd14b	abhydrolase containing protein 14b
AC	analytical column
ACN	acetonitrile
Acad11	acyl-Coenzyme A dehydrogenase family member 11
Acot	acyl-CoA thioesterase
AEBSF	4-(2-aminoethyl) benzenesulfonyl fluoride hydrochloride
a-GvHD	acute graft-versus-host disease
ALDP	adrenoleukodystrophy protein
ALDR	adrenoleukodystrophy related
ALIX	apoptosis-linked gene-2 interacting protein X
AMIDA	autoantibody-mediated identification of antigens
ANOVA	analysis of variance
AQP	aquaporin
AQP2	aquaporin-2
AR	aldose reductase
ARF	acute renal failure
AS	autosampler
ATAD1	AAA domain containing protein 1
AUC	area under the curve
BACAT	bile acid-CoA : amino acid N-acyltransferase
BCA	bicinchoninic acid
bFGF	basic fibroblast growth factor
BLAST	basic local alignment search tool
BRP	bioreactive probe
BSA	bovine serum albumin
BVA	biological variation analysis
CAIX	carbonic anhydrase IX
CALR	calreticulin
CaM	calmodulin
CAPD	continuous ambulatory peritoneal dialysis
CASK	calcium/calmodulin-dependent serine protein kinase
CCB	colloidal Coomassie Blue

Renal and Urinary Proteomics: Methods and Protocols. Edited by Visith Thongboonkerd

ISBN: 978-3-527-31974-9

CDA	cellulose diacetate
CE	capillary electrophoresis
CE-MS	capillary electrophoresis coupled to mass spectrometry
CHAPS	3-[3-cholamidopropyl dimethylammonio]-1-propanesulfonate
CHCA	alpha-cyano-4-hydroxy cinnamic acid
CID	collision-induced dissociation
CITP	capillary isotachophoresis
CK	cytokeratin
CKD	chronic kidney disease
CL	capillary lumen
CNTF	ciliary neurotrophic factor
Con A	Concanavalin A
CPB	cardiopulmonary bypass
CSF	cerebrospinal fluid
CVD	cardiovascular disease
CVVH	continuous venovenous hemofiltration
Cy	cyanine
Cyb5	cytochrome b5
CZE	capillary zone electrophoresis
1-D GE	one-dimensional gel electrophoresis
1-D LC/MS	one-dimensional liquid chromatography/mass spectrometry
2-D GE	two-dimensional gel electrophoresis
2-D	two-dimensional
2-DE	2-D electrophoresis
2-D DIGE	two-dimensional difference gel electrophoresis
2-D PAGE	two-dimensional polyacrylamide gel electrophoresis
3-D	three-dimensional
DAPI	4′,6-diamidino-2-phenylindole
DC	detergent compatible
ddAVP	1-deamino-8-D-arginine vasopressin (or desmopressin)
DDRT-PCR	differential display RT-PCR
DEAE	diethylaminoethyl
DETA-NO	1-[2-(2-aminoethyl)-N-(2-ammonioethyl)amino]-diazen-1-ium-1,2-diolate
DGC	density gradient centrifugation
dI	deionized
DIA	differential in-gel analysis
DIGE	difference gel electrophoresis
DME	Dulbecco's Modified Eagle's
DMEM	Dulbecco's Modified Eagle's medium
DMF	dimethylformamide
DN	diabetic nephropathy
DTE	dithioerythrit

DTT	dithiothreitol
EAM	energy absorbing molecule
ECM	extracellular matrix
EDM	expression difference mapping
EGTA	ethylene glycol tetraacetic acid
EIC	extracted ion chromatogram
EMBL	European Molecular Biology Laboratory
ENaC	epithelial Na channel
ENO	expression of enolase
ER	enhanced resolution
ER	endoplasmic reticulum
ESI-MS	electrospray ionization mass spectrometry
ESI-MS/MS	electrospray ionization coupled to tandem mass spectrometry
ESI-TOF	electrospray ionization time-of-flight
ESI-TOF-MS	electrospray ionization time-of-flight mass spectrometry
ESRD	end-stage renal disease
EVAL	ethylene vinyl alcohol
EWP	extra wide pore
FA	formic acid
FACS	fluorescence-activated cell sorting
FALDH	fatty aldehyde dehydrogenase
FBS	fetal bovine serum
FCS	fetal calf serum
FDA	Food and Drug Administration
FFPE	formalin-fixed and paraffin-embedded
FISH	fluorescence in situ hybridization
FT-ICR	Fourier transform-ion cyclotron resonance
FT-ICR-MS	Fourier-transform ion cyclotron resonance mass spectrometer
FSGS	focal segmental glomerulosclerosis
FWHM	full width at half maximum
GAPDH	glyceraldehyde-3-phosphate dehydrogenase
GBM	glomerular basement membrane
GC	gas chromatography
GeLC–MS	gel electrophoresis followed by liquid chromatography coupled to mass spectrometry
GlcNAc	N-acetyl-glucosamine
GLUT	glucose transporter
GPC	glycerophosphocholine (or glycerophosphorylcholine)
GPF	gas-phase fractionation
GRIM	gene associated with retinoic-interferon-induced mortality
GRP	glucose-regulated protein
GSK	glycogen synthase kinase

GST	glutathione S-transferase
GTP	guanosine triphosphate
HBS	Hanks' buffered solution
HBSS	Hanks' buffered salt solution
HDSS	high dimensionality low sample size
H&E	hematoxylin and eosin
HEPES	4-(2-hydroxyethyl)-1-piperazineethanesulfonic acid
HIC	hydrophobic interaction chromatography
HIF	hypoxia-inducible factor
HPC	hydroxypropyl cellulose
HPLC	high-performance liquid chromatography
HSC	heat shock cognate
HSS	Hank's balanced salt solution
hTERT	human telomerase reverse transcriptase
HUPO	Human Proteome Organization
HV	high voltage
IAA	iodoacetamide
IAM	iodoacetamide
ICAT	isotope-coded affinity tagging
IEF	isoelectric focusing
IFN	interferon
IGF	insulin-like growth factor
IGFBP	insulin-like growth factor binding protein
IHC	immunohistochemistry
IL-1β	interleukin-1β
IM	inner medulla
IMAC	immobilized metal affinity capture
IMCD	inner medullary collecting duct
IPG	immobilized pH gradient
IPI	International Protein Index
IQGAP1	IQ motif-containing GTPase-activating protein1
iTRAQ	isobaric tags for relative and absolute quantitation
IVICAT	in vacuo isotope-coded alkylation technique
jck	juvenile cystic kidney
JRAP	java random access parser
LC	liquid chromatography
LCA	Lens culinaris agglutinin
LCM	laser capture microdissection
LC-ESI-IT MS	liquid chromatography-electrospray ionization-ion trap mass spectrometry
LC-MS	liquid chromatography coupled to mass spectrometry
LC-MS/MS	liquid chromatography coupled to tandem mass spectrometry
LC-SRM	liquid chromatography-selected reaction monitoring
LDH	lactate dehydrogenase

LMP	low melting point
LMW	low-molecular-weight
LOD	limit of detection
LOOCV	leave-one-out cross-validation
LOQ	limit of quantitation
LTQ	linear trap quadrupole
LTQ-FT	linear trap quadrupole combined with FT-ICR
mAb	monoclonal antibody
MAGI-2/S-SCAM	membrane-associated guanylate kinase inverted2/ synaptic scaffolding molecule
MALDI	matrix-assisted laser desorption/ionization
MALDI-TOF	matrix-assisted laser desorption/ionization time-of-flight
MALDI-TOF MS	matrix-assisted laser desorption/ionization time-of-flight mass spectrometry
MALDI-TOF/TOF	matrix-assisted laser desorption/ionization time-of-flight tandem mass spectrometry
MCD	minimal change disease
MDH	malate dehydrogenase
MHC	major histocompatibility complex
MIAPE	minimum information about a proteomics experiment
mIMCD	mouse IMCD
MNGN	membranous glomerulonephritis
Mn-SOD	manganese superoxide dismutase
MOPS	morpholinopropanesulfonic acid
Mosc2	MOCO sulfurase C-terminal domain containing 2 protein
mPTS	peroxisomal membrane targeting signal
mRNA	messenger RNA
MSIA	mass spectrometric immunoassay
MS	mass spectrometry
MS/MS	tandem MS
MudPIT	multidimensional protein identification technology
MUP	major urinary protein
MVB	multivesicular body
MW	molecular weight
MWCO	molecular weight cut-off
m/z	mass to charge ratio
NCBI	National Center for Biotechnology Information
NDSB	non-detergent sulfobetaines
NHE-3	sodium-proton exchanger 3
NHS	N-hydroxysuccinamide
NIDDK	National Institute of Diabetes and Digestive and Kidney Diseases
NIEHS	National Institute of Environmental Health Sciences
NIH	National Institutes of Health

NO	nitric oxide
NOMIS	normalization using optimal selection of multiple internal standards
NSC	nearest shrunken centroid
NTA	nitrilotriacetic acid
Nudt7	nudix hydrolase 7
OMSSA	open mass spectrometry search algorithm
OSF	osteoblast-specific factor
OWL	web ontology language
PAGE	polyacrylamide gel electrophoresis
PAI	plasminogen activator inhibitor
PANTHER	protein analysis through evolutionary relationship
PBE	peroxisomal bifunctional enzyme
PBS	protein biology system
PBS	phosphate-buffered saline
PCA	principal component analysis
PCP	protein correlation profiling
PCR	polymerase chain reaction
PDGF	platelet-derived growth factor
PDI	protein disulfide isomerase
PDMS	polydimethylsiloxane
pI	isoelectric point
PKB/Akt	protein kinase B
PKD	polycystic kidney disease
PMMA	polymethyl methacrylate
PMP	peroxisomal membrane protein
PMSF	phenylmethylsulfonyl fluoride
PNA	peanut agglutinin
PPAR-γ	peroxisome proliferator activated receptor γ
PPBP	pro-platelet basic protein
PSD	postsource decay
Pte1c	peroxisomal acyl-CoA thioesterase Ic
PTM	posttranslational modification
PTS	peroxisomal targeting signal
PVDF	polyvinylidene fluoride
qRT-/RT-PCR	semiquantitative/real time polymerase chain reaction
(q)RT-PCR	semiquantitative/real time polymerase chain reaction
RAP-PCR	RNA arbitrarily primed PCR
RBP	retinol-binding protein
RCA	rolling circle amplification
RCC	renal cell carcinoma
RCF	relative centrifuge force
ROC	receiver-operating characteristic
ROS	reactive oxygen species
RP	reverse phase

RPLC	reverse-phase liquid chromatography
RPLC-ESI	RPLC coupled to ESI
RSD	relative standard deviation
RT	room temperature
RT-PCR	real time polymerase chain reaction
SA	sinapic acid
SAGE	serial analysis of gene expression
SAW	surface acoustic wave
SAX	strong anion exchange
SCX	strong cation exchange
SD	standard deviation
SDS	sodium dodecyl sulfate
SDS-PAGE	sodium dodecyl sulfate polyacrylamide gel electrophoresis
SELDI	surface-enhanced laser desorption/ionization
SELDI-MS	surface-enhanced laser desorption/ionization mass spectrometry
SELDI-TOF	surface-enhanced laser desorption/ionization time-of-flight
SELDI-TOF MS	surface-enhanced laser desorption/ionization time-of-flight mass spectrometry
SEPT	septin
SEREX	serological identification of antigens by recombinant cDNA expression cloning
SERPA	serological proteome analysis
SILAC	stable-isotope labeling by amino acids in cell culture
SISCAPA	stable isotope standards and capture by anti-peptide antibodies
SKL	serine-lysine-leucin
SLE	systemic lupus erythematosus
SMIT	sodium-dependent myo-inositol transporter
S/N	signal/noise
SPA	sinipinic acid
SPEAR	serological and proteomic evaluation of antibody response
SPE	solid-phase extraction
SSH	suppression subtractive hybridization
STD	standard
STY	serine/threonine/tyrosine
TAA	tumor-associated antigen
TALH	thick ascending limb of Henle's loop
TBP	tributylphosphine
TBS	Tris-buffered saline
TCA	trichloroacetic acid
TCEP	tris(2-carboxyethyl) phosphine

TEM	transmission electron microscopy
TEMED	*N*, *N*, *N'*, *N'*,tetramethyl-ethylenediamine
TFA	trifluoroacetic acid
TGF	transforming growth factor
THIO	thioredoxin
THP	Tamm–Horsfall protein
TIC	total ion chromatogram
TIFF	tag image file format
TMA	tissue microarray
TNF	tumor necrosis factor
TNF-α	tumor necrosis factor alpha
TOF	time-of-flight
TPM	tropomyosin
TSG	tumor susceptibility gene
TSA	tyramide signal amplification
TTR	transthyretin
TUBA	tubulin α-chain
UCHL1	ubiquitin carboxyl-terminal hydrolase L1
UFR	ultrafiltration rate
US	urinary space
VEGF	vascular endothelial growth factor
VHL	von Hippel-Lindau
VIM	vimentin
WB	Western blot
WGA	wheat germ agglutinin
wt	wild-type
WT	Wilms' tumor
Zadh2	zinc-binding alcohol dehydrogenase domain containing protein 2
ZADH	zinc-binding alcohol dehydrogenase
ZO	zonula occludens

Part One
Renal Proteomics

1
Isolation and Enrichment of Glomeruli Using Sieving Techniques

Tadashi Yamamoto

1.1
Introduction

The glomerulus is a globular structure, which is composed of capillary vessel networks. A primary function of the glomerulus is to filter small solutes from the blood plasma as raw urine to the Bowman's space through the glomerular capillary wall. The glomerulus also provides a primary site for pathologic lesions, where chronic kidney diseases initially occur in humans. Therefore, it is important to study the glomerulus for understanding the mechanisms of the ultrafiltration and kidney diseases. Recently, molecules involved in the glomerular filtration have been partially clarified. The slit diaphragm between foot processes of glomerular epithelial cells, podocytes, has been found to consist of nephrin and other proteins and to form a pivotal filtration barrier for large molecules in the plasma [1]. However, the pathogenesis and pathophysiological processes of chronic kidney disease have not been clarified yet at the molecular level.

Isolation of the glomeruli is essential to study the physiological characteristics and pathologic molecular events in the glomerulus. The isolated glomeruli are used to obtain glomerular cells for culture and to purify RNA or proteins for the analysis of gene or protein expression in the glomerulus. In this chapter, sieving protocols for isolation of the glomerulus from various mammals are introduced.

1.2
Size and Number of the Glomerulus in Mammalian Kidneys

The size and number of the glomerulus in a kidney of various mammals have been evaluated by morphometry, showing generally that they are proportional to the sizes of their kidneys and bodies, as summarized in Table 1.1 [2]. Glomeruli can be isolated from kidneys of various animals by methods based on their sizes, which differ from other kidney components. However, the glomerular size also increases with maturation of animals [3] and the size of individual glomeruli is variable even within a kidney. The size of the glomeruli at the jaxtamedullary cortex is generally

Renal and Urinary Proteomics: Methods and Protocols. Edited by Visith Thongboonkerd

ISBN: 978-3-527-31974-9

Table 1.1 Glomerular size and number in association with body and kidney weights in several mammalian species.

Animal	Glomerular diameter (μm)	Number of glomeruli in in one kidney	One kidney weight (g)	Body weight (g)
Mouse	73	12 430	0.123	20
Rat	122	30 800	0.746	241
Guinea pig	126	75 700	1.9	565
Rabbit	142	207 000	15.3	2320
Cat	132	184 000	8.0	2750
Dog	180	408 100	31.3	9300
Pig	166	1 193 000	76.7	46 650
Man	201	1 095 000	156.5	70 000
Ox	244	3 992 000	640.0	410 000
Elephant	338	7 510 000	3650	4 545 000

larger than those at the other cortical regions. The number of the glomeruli in a kidney is also proportional to the size of their kidneys and bodies (Table 1.1).

1.3 Methods and Protocols

1.3.1 Original Sieving Method

The glomerulus has been isolated from kidney cortex of several animal species in a size-dependent manner, using stainless steel mesh sieves. Krakower and Greenspon [4] first reported a technique to isolate glomeruli from bovine kidneys. Bovine kidney cortex slices (40 g) were forced through #115 (opening size: 130 μm) stainless steel mesh sieve with the bottom surface of a beaker. The material that emerged through the sieve with cold saline was collected in a receiving pan on ice. The sieved material was gently shaken, and then poured on top of a #80 (184 μm) mesh sieve that was positioned on top of a #150 (103 μm) mesh sieve, rested on a receiving pan. These sieves were washed in the same sequence with ice-cold saline and the glomeruli retained on the #150 mesh sieve were free of tubular fragments and cells. The glomeruli were then suspended in ice-cold saline or PBS, transferred to a 50-ml centrifuge tube, and left standing on ice for 10–15 minutes. The supernatant was removed gently using a Pasteur pipette. The pellet resuspended in the solution was rich in the glomeruli and a small drop of the suspension put on a slide was examined for the purity of the glomeruli under a phase-contrast microscope.

This technique was modified and applied also for isolation of glomeruli from other mammals [5]. To culture rat glomeruli, the cubes of rat renal cortical tissue were pressed through a stainless mesh sieve of #60 (pore size: 250 μm) with a spatula and rinsed with culture medium, and the filtered material was poured on a #300 mesh (pore size: 53 μm). The glomeruli were collected on this mesh screen. Meshes of different sizes were used to isolate the glomeruli from kidneys of different animal species [5].

1.3.2 Standard Sieving Method

Burlingtron and Cronkite [6] introduced a simplified protocol for isolation of the glomeruli from rats using a set of three different stainless steel mesh sieves placed in series (Figure 1.1). To avoid contamination of the blood, kidneys of rats weighing 200–250 g were perfused *in situ* via the aorta with Hank's balanced salt solution (HSS) or PBS at a pressure of 120 mmHg under anesthesia. The kidneys were excised and the cortices were dissected from each kidney and sliced with a razor blade. They were pressed with a spatula through a stainless steel screen of #60 mesh (pore size: 250 μm) and rinsed with HSS or PBS through successive screens of #150 mesh (pore size: 150 μm) and #200 mesh (pore size: 75 μm), respectively. The glomeruli were collected on the #200-mesh screen and suspended in HSS or PBS to transfer to a plastic tube. The tube was left on ice for about 10 minutes and the supernatant was removed to collect glomeruli at the bottom. About 3×10^4 glomeruli are collectable from two kidneys.

This protocol has been used widely to isolate the glomeruli from other animal spices (Table 1.2) as it is simple and practical. Human glomeruli are purified by using #40-, #80-, and #100-mesh sieves, respectively, as for the bovine glomeruli [7].

Although the purity of the glomeruli isolated by the sieving method is relatively high from large animals such as cows or humans, it is not so high enough from small animals such as rats, mice, or rabbits for precise studies on the glomerulus.

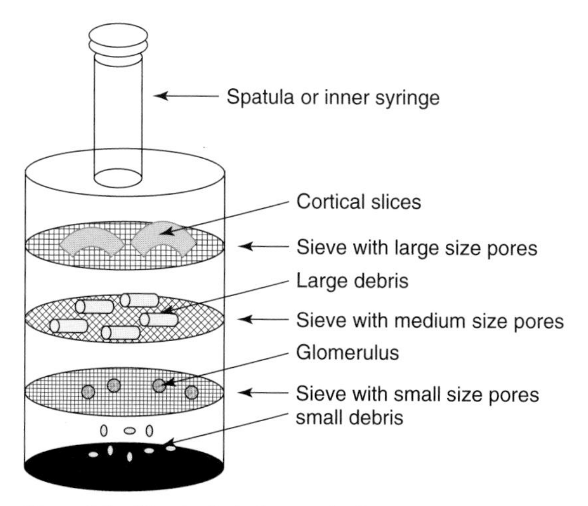

Figure 1.1 Standard sieving method.

Table 1.2 The numbers of sieving mesh in triple sets for glomerular isolation from several mammals.

	Rat	Mouse	Human	Cow	Rabbit
Top sieve	#60 (250 μm)[a]	#100 (150 μm)	#36 (425 μm)	#36 (425 μm)	#50 (300 μm)
Middle sieve	#100 (150 μm)	#200 (75 μm)	#60 (250 μm)	#60 (250 μm)	#83 (180 μm)
Bottom sieve	#200 (75 μm)	#282 (53 μm)	#100 (150 μm)	#100 (150 μm)	#166 (90 μm)
Glomerular size (μm)	122	73	201	250	142

[a] Opening size of mesh is indicated in parenthesis.

From our experience, the purity of the glomeruli isolated by this protocol is approximately 85% (80–90%) for adult rats, approximately 90% (85–95%) for adult humans and cows, but is much less, approximately 70% (ranging 65–80%), for mice, probably because the diameter of the glomerulus is similar to that of tubules in small animals.

1.3.3 Modified Method for High Purification of the Glomeruli

Several attempts have been made to increase the purity. A Ficoll-gradient centrifugation step was added after sieving of the glomeruli from rats, rabbits, and pigs to increase their purity to 97–98, 94–98, and 97–99%, respectively [8].

To increase the purity of the glomeruli, individual glomeruli may be picked up from the glomerulus-rich fractions obtained by the standard sieving method described above, with a pipette connected to a hand pipette pump and controlled by a micromanipulator under a phase-contrast microscope [9]. By this method, the

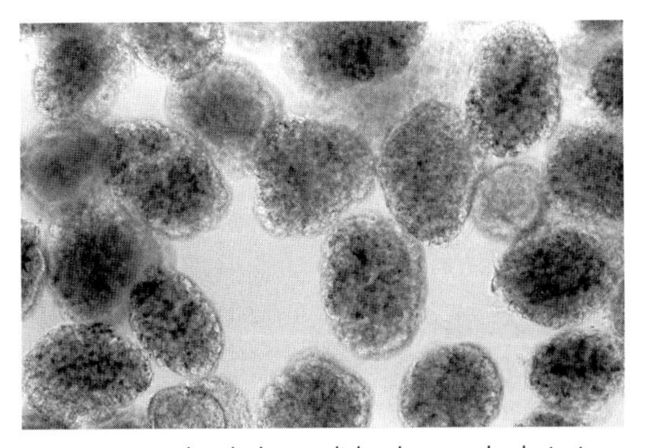

Figure 1.2 Isolated glomeruli by the standard sieving method and purified by collection with a suction pipette connected to a pipette pump under a phase-contrast microscope.

glomeruli were highly purified up to almost 100% (Figure 1.2). Since it takes time to collect the glomeruli (approximately 500–2000 glomeruli per hour) or even to remove contaminants from the glomerulus-rich fraction, this procedure may not be applicable to obtain a large number of the glomeruli in a short time. However, this method has an advantage to isolate the glomeruli without Bowman's capsule or those with Bowman's capsule from their mixture for investigation of difference between them in subsequent culture.

References

1. Tryggvason, K., Patrakka, J., and Wartiovaara, J. (2006) Mechanisms of disease: hereditary proteinuria syndromes and mechanisms of proteinuria. *New Engl J Med*, **354**, 1387–1401.
2. Smith, H.W. (1951) *XVII Comparative Physiology of the Kidney in The Kidney: Structure and Function in Health and Disease*, Oxford University Press, New York.
3. Fetterman, G.H., Shuplock, N.A., Philipp, F.J., and Gregg, H.S. (1965) The growth and maturation of human glomeruli and proximal convolutions from term to adulthood: studies by microdissection. *Pediatrics*, **35**, 601–619.
4. Krackower, C.A. and Greenspon, S.A. (1954) Factors leading to variation in concentration of "nephrotoxic" antigens of glomerular basement membranes. *Arch Pathol*, **58**, 401–432.
5. Holdworth, S.R., Glasgow, E.F., Atkins, R.C., and Thomson, N.H. (1978) Cell characteristics of cultured glomeruli from different animal species. *Nephron*, **22**, 454–459.
6. Burlington, H. and Cronkite, E.P. (1973) Characteristics of cell cultures derived from renal glomeruli. *Proc Soc Exp Biol Med*, **142**, 143–149.
7. Nagi, A.H. and Kirkwood, W. (1972) A quick method for the isolation of glomeruli from human kidney. *J Clin Pathol*, **25**, 361–362.
8. Norgaard, J.O. (1976) A new method for the isolation of ultrastructurally preserved glomeruli. *Kidney Int*, **9**, 278–285.
9. Yaoita, E., Kurihara, H., Sakai, T., Ohshiro, K., and Yamamoto, T. (2001) Phenotypic modulation of parietal epithelial cells of Bowman's capsule in culture. *Cell Tissue Res*, **304**, 339–349.

2
Isolation and Enrichment of Glomeruli Using Laser Microdissection and Magnetic Microbeads for Proteomic Analysis

Oliver Vonend, Barbara Sitek, Lars-Christian Rump, Helmut E. Meyer, and Kai Stühler

2.1
Introduction

The kidney is made up of various compartments with completely different functions. The major compartments defining renal function are called nephrons. They consist of glomeruli and draining tubular system. In humans, the glomeruli are able to produce up to 150-ml ultrafiltrate per minute and are the target of many renal diseases. However, the molecular mechanisms leading to progressive glomerular diseases are poorly understood [1]. Proteome analysis would be helpful to reveal the underlying mechanisms leading to disease progression. In particular, animal models are commonly used to find out more about the underlying mechanisms leading to disease progression [2]. Using knockout mice models, it becomes possible to study the effect of single genes on, for example, glomerulosclerosis *in vivo* [3, 4]. The main disadvantage of using mice, however, is the limited amount of sample size, in particular, when analyzing protein expression of the glomeruli. Even though the kidney cortex is enriched in the glomeruli, only a small percentage of glomerular proteins are expected in the kidney cortex lysate. Serial sieving technique has made it possible to separate the glomeruli from human kidney tissue in order to provide 100–500 μg of proteins that are needed for 2-D electrophoresis (2-DE) analysis [5]. However, when mice models are used (e.g., subtotal 5/6 nephrectomy), conventional 2-DE is not applicable since protein amounts in the above required range cannot be expected [5b]. The problem becomes even more pronounced when human samples are analyzed. In this case, small needle biopsy material of kidney cortex is the only material that is available and a maximum of 30–50 glomeruli can be expected in these kinds of samples.

Sample preparation for proteome analysis is a very critical step. This section discusses two different approaches that enable the separation of the glomeruli from mice and human samples. When working with murine animal models, embolization of ferromagnetic beads within the glomerular tuft has been shown to be a fast and effective method to purify the glomeruli [5b, 6], whereas laser capture microdissection (LCM) is the method of choice for isolating the glomeruli from

Renal and Urinary Proteomics: Methods and Protocols. Edited by Visith Thongboonkerd

ISBN: 978-3-527-31974-9

human kidneys since biopsy tissue is the only material available for diagnostic purpose [6].

Besides handling the sample preparation, the small amount of protein that can be expected after tissue lysis (0.5–3 μg) is the second challenge. A protocol will be given to facilitate the analysis of scarce sample amount resulting from mouse glomeruli preparation or human biopsy material for differential proteome analysis. The method of choice is the so-called difference gel electrophoresis (DIGE) saturation labeling. DIGE saturation labeling, with its more than 20-fold higher sensitivity compared to silver staining, enables a complete and quantitative 2-DE analysis of murine and human glomeruli micropreparations using ferromagnetic beads and LCM [5b, 6].

2.2 Methods and Protocols

2.2.1 Tissue Preparation

2.2.1.1 Isolation of Murine Glomeruli

Adult mice are anesthetized intraperitoneally with Ketanest and Xylazin (0.168 mg and 8 $\mu g\,g^{-1}$ bodyweight, respectively). After a medial laparatomy, the kidneys are perfused with phosphate-buffered saline (PBS) via abdominal aorta according to an amended method described previously [8]. The distal aorta is cannulated with polyethylene tubing (Portex, Germany; internal diameter 0.28 mm) and perfused *in situ* at a constant rate of 7.2 $ml\,min^{-1}\,g^{-1}$ kidney. The proximal aorta is ligated. To ensure venous drainage, a hole should be punched at the inferior vena cava. In the first place, kidneys are perfused with ice-cold PBS in order to free the blood vessels from any remaining blood (Figure 2.1).

After preparation, the kidneys need to be perfused with ferromagnetic beads with a diameter of 4.5 μm (Dynabeads Dynal Biotech, Invitrogen, Germany) in a concentration of 4×10^6 beads per milliliter PBS. The ferromagnetic beads are trapped within the vasa afferentes and glomerular capillaries and can be seen *in situ* with a magnification of 3× (Figure 2.2). The kidneys have to be removed and minced through a 100-μm cell mesh with intermittent PBS flushing (two to three times of flush with 5-ml ice-cold sterile PBS). After centrifugation at 4500 rpm for five minutes, the cell pellet is dissolved in 2-ml cooled PBS and transferred to a 2-ml tube. Using a magnet catcher and a three-step washing procedure, the ferromagnetic beads containing glomeruli are purified to approximately 95% homogeneity. The homogeneity can be ensured by microscopic inspection of one drop of the glomerular suspension (Figure 2.3).

The extracted glomeruli are lysed in 20 μl of 2-DE lysis buffer (30 mM Tris-HCl, 2 M thiourea, 7 M urea, 4% CHAPS; pH 8) and sonicated (6×10 second pulses on ice). To remove the beads, the lysates are centrifuged at 10 000 rpm for 10 minutes. Protein concentration can be determined using a protein assay. The

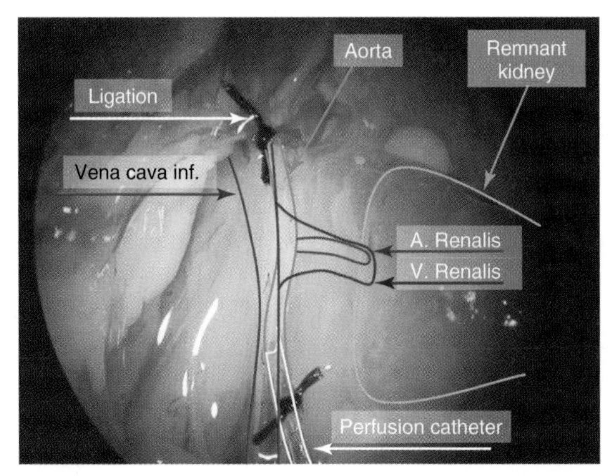

Figure 2.1 An intraoperative view of the mouse situs with highlighted anatomical structures is shown. The microbeads were applied through the perfusion catheter.

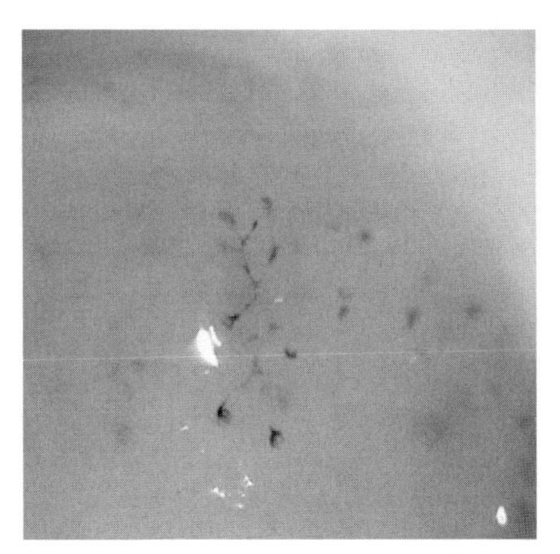

Figure 2.2 *In situ* picture (3× magnification) of kidney tissue, perfused with ferromagnetic beads. The trapped beads can bee seen within the vasa afferentes and glomerular capillaries.

samples need to be stored at $-80\,^{\circ}$C. When glomerular protein is compared to full cortex tissue, mice kidneys are removed and placed into ice-cold PBS. Cortical tissue can be manually separated from medulla. The cortex tissue is cut into small pieces, minced in 2-DE lysis buffer, sonicated, centrifuged, and stored, as explained earlier.

To study the differential glomerular proteome associated with chronic renal failure in mice, 5/6 of functional renal mass is surgically removed (uninephrectomy plus 2/3 resection of contralateral kidney). Subtotal nephrectomy is an accepted procedure that has been used to correlate molecular mechanisms with histopathological and clinical parameters found in chronic renal failure [9]. Samples with

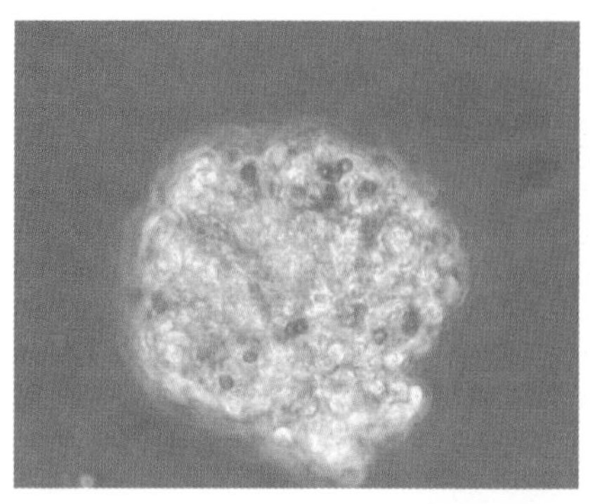

Figure 2.3 A glomerulus is shown after the purification procedure using a magnetic catcher. The brown-colored ferromagnetic beads are visible within the glomerular capillaries (100× magnification).

glomerular protein amount of 10–20 μg can be extracted from remnant kidneys 10 weeks after surgical procedure.

2.2.1.2 Isolation of Human Glomeruli

The glomeruli can be isolated from human kidney using laser microdissection technique. The biopsy tissue is mounted in Tissue-Tek OCT and snap frozen (−80 °C). Tissue blocks are serially sectioned (6-μm sections) and stored on dry ice. Sections are fixed, stained, and dehydrated (e.g., HistoGene LCM Frozen Section Staining Kit, Arcturus, USA). The glomeruli can then be isolated using an Automated LCM System (e.g., AutoPix Arcturus, USA) (Figure 2.4). The following laser beam settings are useful for human kidney material: laserpower 70 mW and lasertime 2500 μs. The isolated glomeruli are lysed, sonicated, centrifuged, and stored, as explained above (Section 2.2.1.1).

2.2.2 Protein Labeling

2.2.2.1 Labeling of Murine Proteins

For the reduction of cysteine residues, cell lysates (containing 3 μg protein each) are incubated with 1 nmol tris(2-carboxyethyl) phosphine hydrochloride (TCEP; Sigma) at 37 °C in the dark for 1 hour. Saturation CyDyes (GE Healthcare, Munich, Germany) are diluted with anhydrous DMF p.a. (2 nmol μl^{-1}; Sigma) and 2 nmol CyDye is added to the reduced TCEP sample [6]. The samples need to be vortexed, centrifuged briefly, and left at 37 °C in the dark for 30 minutes. The glomeruli and cortex lysates (3 μg each) are labeled with Cy5. For internal standardization, a pool of all analyzed samples (cortex and glomeruli) should be created and labeled with Cy3. The labeling reaction can be stopped by adding 2-μl DTT (1.08 g ml^{-1}; BioRad) and finally 2-μl Ampholine 2-4 (GE Healthcare) is added. Before isoelectric focusing (IEF), 3 μg of labeled internal standard is mixed with labeled glomeruli or cortex samples, respectively [6].

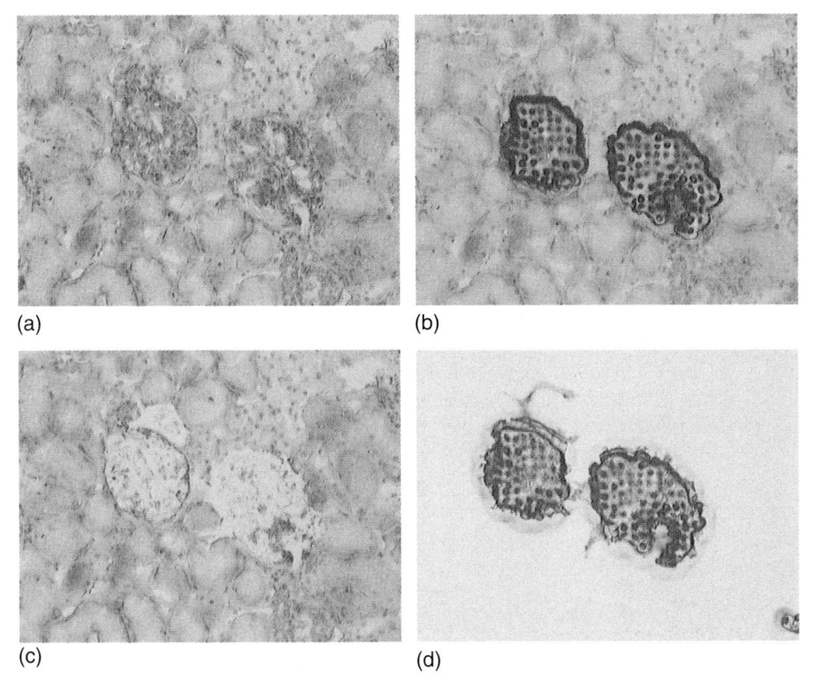

Figure 2.4 Human kidney cortex preparation focused on an area with two glomeruli before (a), during (b), and after (c) laser microdissection. After successful isolation of the glomeruli, the anatomical structure can be found on the extraction cap (d).

For preparative gels, 400 μg of reference proteome lysates (70% kidney cortex tissue and 30% glomerular extract) is reduced with 105 nmol TCEP and labeled with 210 nmol of Cy3.

2.2.2.2 Labeling of Human Proteins

For the DIGE saturation labeling of microdissected human glomeruli, protein extracts (0.5–4.5 μg) are reduced with 1-nmol TCEP, subsequently incubated with 2-nmol Cy5, and then processed as explained earlier (Figure 2.5).

2.2.3 2-D Gel Electrophoresis

Carrier ampholyte-based IEF is performed in an IEF chamber using tube gels (20 cm × 1.5 mm) as described elsewhere [10]. After running a 21.25-hour voltage gradient, the ejected tube gels are incubated in equilibration buffer (125 mM Tris, 40% (w/v) glycerol, 3% (w/v) SDS, 65 mM DTT, pH 6.8) for 10 minutes. The second dimension is performed in a Desaphor VA 300 system using polyacrylamide gels (15.2% total acrylamide and 1.3% bisacrylamide) as described elsewhere [10] (Figure 2.6). The IEF tube gels are placed onto the polyacrylamide gels (20 cm × 30 cm × 1.5 mm)

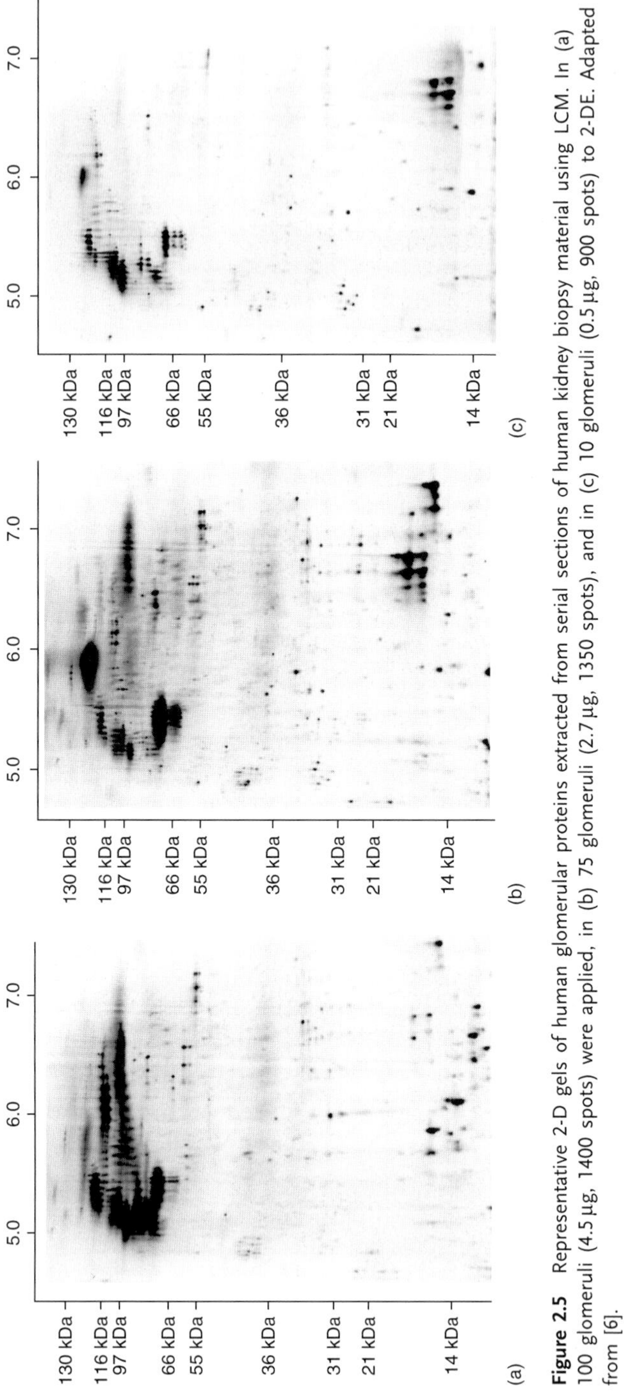

Figure 2.5 Representative 2-D gels of human glomerular proteins extracted from serial sections of human kidney biopsy material using LCM. In (a) 100 glomeruli (4.5 µg, 1400 spots) were applied, in (b) 75 glomeruli (2.7 µg, 1350 spots), and in (c) 10 glomeruli (0.5 µg, 900 spots) to 2-DE. Adapted from [6].

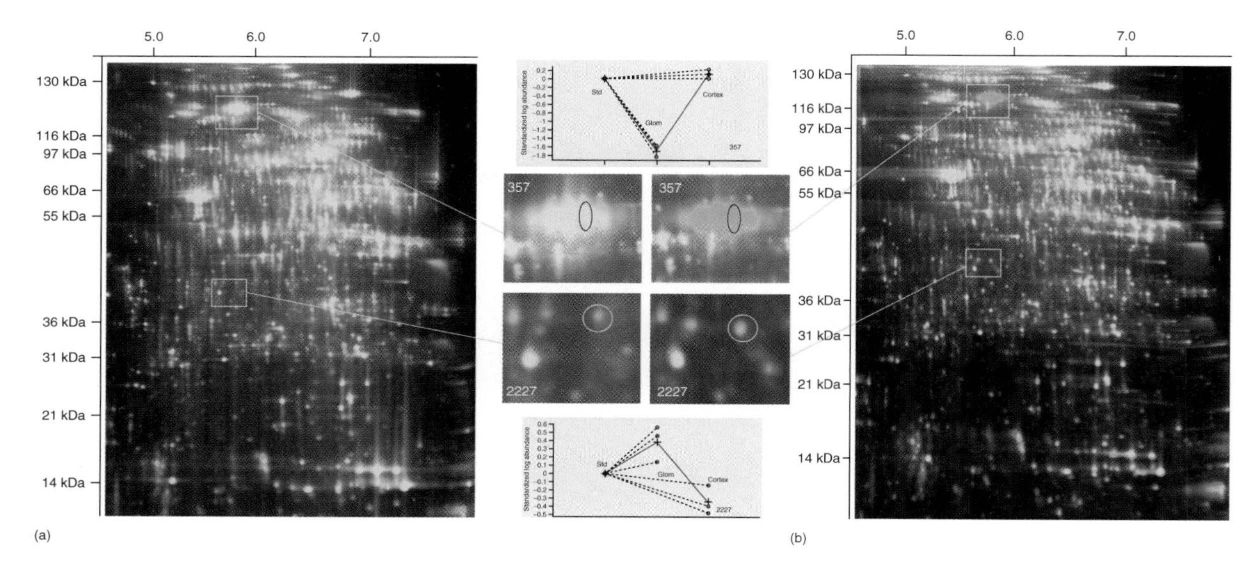

Figure 2.6 (a) A scanned gel loaded with 3.0-μg renal kidney cortical proteome (Cy5, red labeled) and 3.0-μg internal standard (Cy3, green labeled). (b) A scanned gel loaded with 3.0-μg glomerular proteome (Cy5, red labeled) and 3.0-μg internal standard (Cy3, green labeled). The figures in between show the glomerular (Glom) and cortical (Cortex) expression level of spot 357 and spot 2227 in comparison to the internal standard (Std) in a representative gel and statistically evaluated by DeCyder in three independent samples (three different kidney preparations). The upper evaluated spot 357 is almost equally abundant in cortex (red labeled) and reference (green labeled) proteome leading to a yellow overlay. Expression of spot 357 in glomerular (red labeled) proteome is barely detectable resulting in a green (reference) spot only. The lower evaluated spot 2227 shows an opposite expression pattern. Spot 2227 is low abundant in cortex (red labeled) compared to reference (green labeled) proteome leading to a green spot. Expression of spot 2227 in glomerular (red labeled) proteome is higher compared to reference (green labeled) proteome leading to a red spot. Adapted from [6].

and fixed with 1.0% (w/v) agarose containing 0.01% (w/v) bromophenol blue dye (Riedel deHaen, Seelze, Germany). For protein identification, the preparative-sized gel system (IEF: 20 cm × 1.5 mm, SDS-PAGE: 20 cm × 30 cm × 1.5 mm) needs to be applied under identical conditions. Silver poststaining is performed after gel scanning using a MS-compatible protocol as described elsewhere [11].

2.2.4 Scanning and Image Analysis

After 2-DE, the gels are left between the glass plates and images can be acquired using a scanner (e.g., Typhoon 9400 scanner, GE Healthcare). Excitation wavelengths and emission filters have to be chosen specific for each of the CyDyes according to the scanner user guide. Before image analysis with, for example, DeCyder software (GE Healthcare), the images are cropped with, for example, ImageQuant software (GE Healthcare). The intragel spot detection and quantification is performed using the differential in-gel analysis (DIA) mode. The estimated number of spots can be set to 3000. An exclusion filter should be applied to remove spots with a slope greater than 1.6. Images from different gels are matched using the BVA mode of the image analysis software. Protein spots represented in all analyzed gels with expression level greater then 1.5-fold change and $p < 0.05$ can be defined as being differentially expressed.

2.2.5 Mass Spectrometry: Nano-LC-ESI-MS/MS

Differentially expressed proteins are excised from the gel, tryptically digested and afterwards preconcentrated and separated in an LC packings HPLC system (e.g., Dionex LC Packings, Idstein, Germany). Peptides should be loaded on-line and preconcentrated with 0.1% TFA at a flow rate of 30 μl min^{-1} for 6 minutes on a microprecolumn (0.3 mm i.d. × 5 mm, 5 μm) using an isocratic loading pump. For separation by RP nano-HPLC (75 μm i.d. × 250 mm, 5 μm) a solvent system consisting of solvent A (0.1% v/v formic acid) and solvent B (0.1% v/v formic acid/84% v/v acetonitrile) is used. A gradient of 5–50% solution B in 35 minutes is carried out. Flow rate is generally adjusted to 250 nl min^{-1}. During separation, the second precolumn is washed with 50% v/v ACN/0.1% v/v TFA for 20 minutes and then with 84% v/v ACN/0.1% v/v TFA for 10 minutes using the isocratic loading pump as described elsewhere [12].

ESI-MS/MS spectra can be recorded using a hybrid triple quadrupole/linear ion trap LC/MS/MS mass spectrometer equipped with NanoSpray source and noncoated SilicaTips (e.g., 4000 Q Trap Applied Biosystems, Foster City, CA, USA). The needle voltage is set to 2800 V and the interface is heated to 150 °C. Nitrogen is used as curtain and collision gas. Each scan cycle consists of an MS scan (EMS), an enhanced resolution scan (ER), and up to three MS/MS scans (EPI) with an overall duration of 3.5 seconds. The mass range of the EMS is m/z 400–1400 and m/z

100–1750 in EPI mode. To determine more precisely the m/z-value and the charge state (+1 to +3), the ER scan can be performed using the three most intense ions.

For protein identification, uninterpreted ESI-MS/MS spectra are correlated with the NCBI protein database (http://www.ncbi.nlm.nih.gov) using the SEQUEST algorithm [13–15] with the following search parameters: variable modification due to methionine oxidation, two maximal missed cleavage sites in case of incomplete trypsin hydrolysis, and mass tolerances of ±1.5 Da for parent ions and ±0.5 Da for fragment ions. Proteins can be considered as identified if two peptides are explained by the spectra and have a SequestMetaScore (Proteinscape) greater than 3.

2.3
Conclusions

Murine glomeruli can be extracted using ferromagnetic Dynabeads. Embolization of 4.5-μm diameter ferromagnetic particles within the glomerular capillaries allows the isolation of glomeruli with a magnetic catcher. Compared to the small organ size, a reasonable amount of glomerular protein can be provided even in subtotally (5/6) nephrectomized mice, a model for chronic renal failure [5b].

Extraction of human glomeruli using ferromagnetic beads is not applicable since renal biopsy is the only material that can be used for proteomic analysis. LCM has been shown to be a useful tool to perform segmental analysis of renal tissue [6, 16]. The major limiting factor in the past was the low amount of protein obtained by LCM. Using the above-presented protocols, it now becomes possible that the glomeruli picked from serial 6-μm-thick sections of human renal biopsy material are suitable for further proteomic analyzes [6].

To analyze the low amount of human and murine glomerular protein in a range of 0.5–3.0 μg, the demonstrated fluorescence dye saturation labeling technique has to be used. With these techniques, sensitivity of 2-DE analysis is increased to a dimension that enables glomerular proteome analysis of one single section with 10 human glomeruli. This gives us the opportunity to plan functional proteomics study of glomerular diseases in humans and mice to find out more about the cellular mechanisms leading to end-stage renal failure and the molecular background of new pharmacological approaches.

References

1. Rump, L.C., Amann, K., Orth, S., and Ritz, E. (2000) Sympathetic overactivity in renal disease: a window to understand progression and cardiovascular complications of uraemia? *Nephrol Dial Transplant*, **15**, 1735–1738.
2. Thongboonkerd, V., Chutipongtanate, S., Kanlaya, R., Songtawee, N. *et al.* (2006) Proteomic identification of alterations in metabolic enzymes and signaling proteins in hypokalemic nephropathy. *Proteomics*, **6**, 2273–2285.
3. Fogo, A.B. (2003) Renal fibrosis: not just PAI-1 in the sky. *J Clin Invest*, **112**, 326–328.
4. Zhao, H.J., Wang, S., Cheng, H., Zhang, M.Z. *et al.* (2006) Endothelial

nitric oxide synthase deficiency produces accelerated nephropathy in diabetic mice. *J Am Soc Nephrol*, **17**, 2664–2669.

5. (a)Yoshida, Y., Miyazaki, K., Kamiie, J., Sato, M. *et al.* (2005) Two-dimensional electrophoretic profiling of normal human kidney glomerulus proteome and construction of an extensible markup language (XML)-based database. *Proteomics*, **5**, 1083–1096; (b)Potthof, S., Sitek, B., Stegbauer, J., Schulenborg, T., Marcus, K., Ouack, J., Rump, L.C., Meyer, H.E., Stuehler, K., and Vonend, O. (2008) The glomerular proteome in a model of chronic disease disease. *Proteomics-Clinical Applications*, **2** (7-8), 1127–1139.
6. Sitek, B., Potthoff, S., Schulenborg, T., Stegbauer, J. *et al.* (2006) Novel approaches to analyse glomerular proteins from smallest scale murine and human samples using DIGE saturation labelling. *Proteomics*, **6**, 4337–4345.
7. Sitek, B., Luttges, J., Marcus, K., Kloppel, G. *et al.* (2005) Application of fluorescence difference gel electrophoresis saturation labelling for the analysis of microdissected precursor lesions of pancreatic ductal adenocarcinoma. *Proteomics*, **5**, 2665–2679.
8. Vonend, O., Stegbauer, J., Sojka, J., Habbel, S. *et al.* (2005) Noradrenaline and extracellular nucleotide cotransmission involves activation of vasoconstrictive P2X(1,3)- and P2Y6-like receptors in mouse perfused kidney. *Br J Pharmacol*, **145**, 66–74.
9. Vonend, O., Apel, T., Amann, K., Sellin, L. *et al.* (2004) Modulation of gene expression by moxonidine in rats with chronic renal failure. *Nephrol Dial Transplant*, **19**, 2217–2222.
10. Klose, J. and Kobalz, U. (1995) Two-dimensional electrophoresis of proteins: an updated protocol and implications for a functional analysis of the genome. *Electrophoresis*, **16**, 1034–1059.
11. Nesterenko, M.V., Tilley, M., and Upton, S.J. (1994) A simple modification of Blum's silver stain method allows for 30 minute detection of proteins in polyacrylamide gels. *J Biochem Biophys Methods*, **28**, 239–242.
12. Schaefer, H., Chervet, J.P., Bunse, C., Joppich, C. *et al.* (2004) A peptide preconcentration approach for nano-high-performance liquid chromatography to diminish memory effects. *Proteomics*, **4**, 2541–2544.
13. Eng, J.K., McCormack, A.L., and Yates, J.R. III (1994) An approach to correlate tandem mass spectral data of peptides with amino acid sequences in a protein database. *Am Soc Mass Spectrom*, **5**, 976–989.
14. Yates, J.R. III, Eng, J.K., and McCormack, A.L. (1995) Mining genomes: correlating tandem mass spectra of modified and unmodified peptides to sequences in nucleotide databases. *Anal Chem*, **67**, 3202–3210.
15. Yates, J.R. III, Eng, J.K., McCormack, A.L., and Schieltz, D. (1995) Method to correlate tandem mass spectra of modified peptides to amino acid sequences in the protein database. *Anal Chem*, **67**, 1426–1436.
16. Kohda, Y., Murakami, H., Moe, O.W., and Star, R.A. (2000) Analysis of segmental renal gene expression by laser capture microdissection. *Kidney Int*, **57**, 321–331.

3
Isolation and Analysis of Thick Ascending Limb of Henle's Loop (TALH) Cells

Hassan Dihazi and Gerhard A. Müller

3.1
Introduction

The kidney is a heterogeneous organ. The functional unit is the nephron, which is composed of a multitude of cell types with widely differing structures and functions. To understand the function of each cell type at the cellular and subcellular levels, the isolation and separation of homogeneous cell population in preparative amounts are necessary. Only the isolation of the different parts of nephron allowed the characterization of the different cells of each part [1–3]. For example, the isolation of the medulla showed that this part of the nephron contains different cell types, including cells of the pars recta of the proximal tubules, cell of the thick ascending limb of Henle's loop (TALH), cells of the collecting ducts, as well as medullary nephron segments. The TALH represents an interesting and important tubular segment for the analysis of the molecular mechanisms of sodium chloride (NaCl) transport and mode of the action of diuretics. The TALH segment generates concentrated urine in antidiuresis and dilute urine in water diuresis [1]. Because of its ability the reabsorb NaCl and its impermeability to the water reabsorption, the TALH segment generates a hypertonic interstitial environment. The higher osmolarity in tubular lumen can be used in the case of antidiuresis to remove water from the collecting ducts and to generate a highly concentrated urine [1].

In outer renal medulla, high extracellular osmolarity has been described [4] due to the elevated NaCl and urea concentrations during antidiuresis. In particular, the TALH cells participate to a considerable degree in the regulation of the osmotic gradient due to their strong NaCl transport activities. Therefore, it is conceivable that they possess specific osmoregulatory mechanisms to tolerate and counteract changes of extracellular osmolarity. The TALH cells are protected from the strong osmotic pressure in the interstitium by accumulating organic osmolytes, such as sorbitol, betaine, inositol, glycerophosphocholine (GPC), and taurine [5]. Indeed, investigations of different kidney segments have shown that there is a gradient of total tissue sorbitol content from a maximum level at the papillary tip to very low contents in the renal cortex [6]. These organic osmolytes are involved in

Renal and Urinary Proteomics: Methods and Protocols. Edited by Visith Thongboonkerd

ISBN: 978-3-527-31974-9

maintaining cell volume and electrolyte contents without perturbing the protein structure and function over a wide concentration range [7].

3.2 Methods for Isolation of TALH Cells

Despite the large number of protocols established to isolate TALH cells, all these previous efforts to culture TALH cells have relied on two different methods. The first method uses microdissection of single tubules fragment for the isolation of specific types of renal epithelial cells. Burg *et al.* [8] used explants of rabbit kidneys to isolate specific TALH cells. The isolated primary cells grew in cell culture conditions. They showed a furosemide-sensitive transepithelial voltage and maintained transepithelial voltages characteristic of the differentiated function of cells from the tick ascending limbs [9].

The second method is based on density gradient centrifugation in a Ficoll gradient. Scott *et al.* [10] established culture of rabbit TALH cells from those isolated by density gradient centrifugation in a Ficoll gradient. In these latter cultures, high levels of Na^+-K^+-ATPase activity were detected, and respiration was inhibited with both ouabain and furosemide. Drugge *et al.* [11] established TALH culture from cells isolated by centrifugation elutriation. Electron microscopic analysis of the isolated cells showed typical TALH-cell characteristics. The presence of Tamm–Horsfall protein, a surface membrane protein of TALH cells, in 90–95% of the cells confirmed the homogeneity of the culture.

3.3 The Role of Organic Osmolytes in the Osmoregulation of TALH Cells

Changes in extracellular medium osmolarity require an adequate reaction of the TALH cells to survive and achieve their normal functions. The reaction on changes in medium osmolarity involves initially fast water movements accompanied by changes in intracellular ion concentration especially sodium and chloride. Long-term adaptation in TALH cells is achieved by the adjustment of the intracellular concentration of "organic osmolytes," such as sorbitol, myo-inositol, glycerophosphorylcholine, and betaine, through changes in the rate of efflux of these metabolites from the cell (Table 3.1). All these organic osmolytes were found in significant amounts in TALH cells: myo-inositol = betaine > GPC > sorbitol [12]. The taurine concentration was found to be 10 times lower in TALH cells (below 1 mM) [13] compared to the thin descending limb of Henle (about 11 mM) [14].

3.3.1 Sorbitol and Aldose Reductase in TALH Cells

Sorbitol is one of the principal organic osmolytes in the renal medulla (see Table 3.2). Its intracellular content correlates well with the extracellular osmolarity.

Table 3.1 Major mechanisms in osmotic regulation of intracellular organic osmolytes in TALH cells.

	Sorbitol	Myo-inositol	GPC	Betaine	Taurine
TALH-hypertonicity	Increased synthesis via induction of aldose reductase	Increased rate of apical uptake via synthesis of sodium-dependent transport system (SMIT)	Increased synthesis	Increased synthesis	No role as osmolyte
TALH-hypotonicity	Increased apical release. Decreased synthesis via reduced amount of aldose reductase	Increased release	Increased release	Increased release	No role as osmolyte

Table 3.2 Content of organic osmolytes in TALH cells. The amount of organic osmolytes is given in (millimole per gram protein) for both 300 and 600 $mOsmol\,l^{-1}$.

	Osmolarity ($mOsmol\,l^{-1}$)	Sorbitol ($mmol\,g^{-1}$ protein)	Myo-inositol ($mmol\,g^{-1}$ protein)	GPC ($mmol\,g^{-1}$ protein)	Betaine ($mmol\,g^{-1}$ protein)	Taurine ($mmol\,g^{-1}$ protein)
TALH	300	20	30	<10	280	<10
	600	630	410	40	430	–

It is alreadyestablished that a high extracellular concentration of NaCl or glucose induces sorbitol accumulation in the cells of the inner renal medulla and TALH cells. This increase in intracellular sorbitol content is accompanied and augmented by an induction of aldose reductase (AR), which reduces D-glucose in the polyol pathway [15]. In addition to the AR expression regulation, intracellular sorbitol content is also regulated by the modification of membrane permeability [16]. Changes in membrane permeability require just a few minutes, whereas adaptation of the intracellular synthesis needs several days. Sorbitol efflux occurs mainly across the luminal membrane in TALH cells of the rabbit kidney [17]. The sorbitol uptake into TALH cells is markedly stimulated by the presence of Ca^{2+} and Mg^{2+}-ATP. Inhibitor studies suggest the involvement of a Ca^{2+}-calmodulin protein kinase [17].

3.3.2
Myo-Inositol in TALH Cells

Myo-inositol is the second important organic substance; its uptake seems to be very important for osmoregulation of TALH cells [18]. Renal myo-inositol depletion *in vivo* causes acute renal failure with severe injury of the thick ascending limb, resulting from a disturbed osmoregulation [18]. Expression of the sodium-dependent myo-inositol transporter (SMIT) was confirmed in TALH of rat kidney by *in situ* hybridization [19]. Its expression level correlates well with the osmotic gradient in the rat kidney [19]. On the other hand, the osmotic regulation is dependent on the increased transcription of the SMIT gene [20] and on the subsequent synthesis of more transport proteins.

3.3.3
Betaine in TALH Cells

TALH cells react on increased extracellular osmolarity with increase in betaine content. The cells possess a betaine transport system that is mainly located in the luminal membrane. For the osmotic regulation, a change in betaine uptake and a cooperation between proximal tubule cells and TALH have been proposed [12]. However, the betaine synthesis in TALH cells is under osmotic control and betaine degradation could not be found [21, 22]. Betaine content can also be altered by changes in membrane permeability. Hypoosmotic exposure leads to a rapid release of more than 30% of the intracellular betaine content [21].

3.3.4
Glycerophosphorylcholine in TALH Cells

Glycerophosphorylcholine (GPC) concentration in TALH cells is also dependent on extracellular osmolarity. TALH cells show significant choline kinase activity [23], which is the mean enzyme responsible for GPC synthesis using extracellular choline [12]. The presence of choline kinase in TALH cells is an indicator for the presence of the complete GPC synthetic pathway. Decreasing extracellular osmolarity elevates the membrane permeability for GPC [12].

3.3.5
Taurine in TALH Cells

In contrast to the proximal tubule, taurine uptake in TALH cells occurs via a luminal sodium- and partly chloride-dependent, furosemide-sensitive transport system [24]. Luminal transport is regulated by transcription (at the mRNA level) in rat [25].

3.4
Proteomic Mapping of Adaptation Mechanisms of TALH Cells to Hyperosmotic Stress

Medullary cells of the TALH participate to a considerable degree in the regulation of the osmotic gradient due to their strong NaCl transport activity and low water conductivity. Thus, they have to possess specific osmoregulatory mechanisms in order to tolerate and counteract changes of extracellular osmolality [26]. Exposition of TALH cells to an increased extracellular tonicity of 600 mosmol kg^{-1} resulted in shrinkage of the cell as a consequence of osmotically induced water loss (Figure 3.1a–c). Apparent morphological changes in TALH-STD cells were observed after a few hours of exposition and significant distinctive changes were observed after 24 hours. TALH cells exhibiting resistance to 600 mosmol kg^{-1} were more resistant to higher osmolarity (900 and 1200 mosmol kg^{-1}); under these conditions, morphological changes occurred later (first after 12 hours) than in the case of TALH cells shifted to 600 mosmol kg^{-1} and furthermore, they were less intense (Figure 3.1b,c). Despite the cell shrinkage as a fast reaction on changes in environment osmolarity, TALH cells are protected from osmotic effect of concentrated sodium ions, chloride ions, and urea in the interstitium by accumulating organic osmolytes [5]. These organic osmolytes are involved in long-term cell reaction for maintaining cell volume and electrolyte contents without perturbing the protein structure and function over a wide concentration range [7].

The high extracellular concentration of NaCl or glucose can induce sorbitol accumulation in the cells of the inner renal medulla and TALH cells, and the increased intracellular sorbitol content is accompanied and augmented by an induction of AR, which reduces D-glucose in the polyol pathway [27, 28]. In addition, recent investigations on renal cells have disclosed that osmotic stress is also a trigger for enhanced expression of heat shock proteins. So far, five target

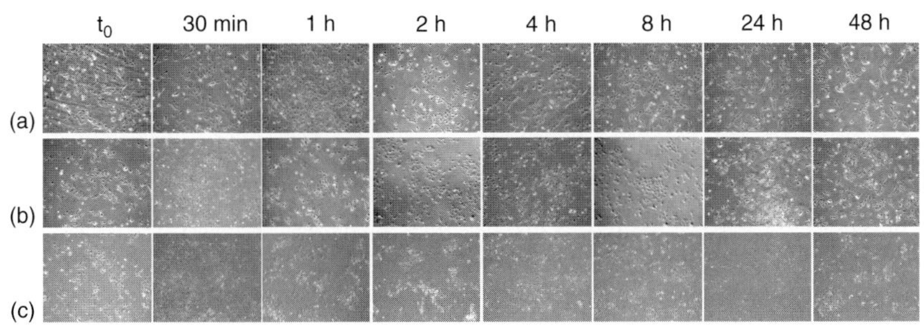

Figure 3.1 Solute effect on the morphology and volume of TALH cells exposed to hyperosmotic stress. (a) TALH cells were cultivated at 300 mosmol kg^{-1} H_2O and were exposed to hyperosmotic stress with 600 mosmol kg^{-1} H_2O for different time periods (30 minutes, 1, 2, 4, 8, 24, and 48 hours). (b) TALH cells were cultivated at 600 mosmol kg^{-1} H_2O and were exposed to hyperosmotic stress with 900 mosmol kg^{-1} H_2O for different time periods (30 minutes, 1, 2, 4, 8, 24, and 48 hours). (c) TALH cells were cultivated at 900 mosmol kg^{-1} H_2O and were exposed to hyperosmotic stress with 1200 mosmol kg^{-1} H_2O for different time periods (30 minutes, 1, 2, 4, 8, 24, and 48 hours) (Dihazi *et al.* unpublished).

HSPs for this kind of stress have been identified, including two members of the HSP70 family, two small HSPs, HSP25/27, α-crystallin B chain, and a new member of the HSP110 subfamily, the so-called osmotic stress protein OSP94, which is highly inducible by hyperosmolality [29]. Modulation of intracellular contents of organic osmolytes, elevated activity of AR, and the expression of specific HSPs are aspects of the complex response of medullary renal cells, including TALH cells to widely fluctuating extracellular solute concentration [29].

Proteomic analysis of renal epithelial cells exposed long term to hyperosmotic stress was performed to investigate the impact of the osmolarity change on the proteome pattern in TALH cells. Several new proteins from different functional groups were found to be differentially expressed under osmotic stress, revealing a complex response mechanism. Fifteen were down-regulated, whereas 25 were up-regulated. Only some of these proteins have been previously linked to renal osmotic stress resistance [20–23]. This highlights a complex response mechanism and the importance of proteomics approaches in understanding the mechanism of osmoregulation in TALH cells.

The morphological changes of the TALH cells under osmotic stress were very intense (Figure 3.1) and were accompanied by changes in the cell proteome (Figures 3.2 and 3.3). Analysis of TALH-cell proteome derived from conventional 2-D gel electrophoresis in the pH range of 3–10 identified more than 2000 (2552 for TALH-NaCl and 2654 for TALH-STD) protein spots in each stressed and nonstressed cells. Proteomic analysis also revealed significant differences between the protein compositions from stressed and nonstressed TALH cells. The 25 up-regulated proteins were from three different protein classes, namely, HSPs, metabolism proteins, and structure proteins (Table 3.3). The 15 down-regulated

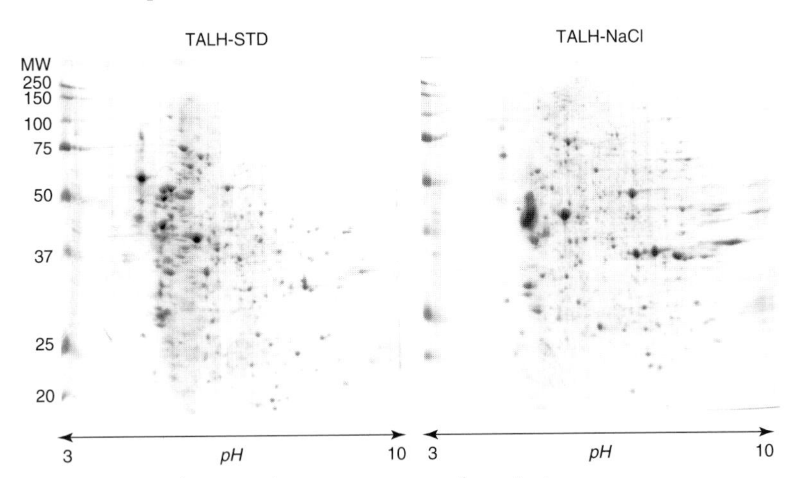

Figure 3.2 Two-dimensional proteome pattern of a variant of the TALH-STD-cell line and TALH-NaCl-cell line exhibiting higher NaCl resistance. TALH-cell extracts from both cell lines were prepared and the proteins were separated by 2-D PAGE (pH range of 3.0–10.0). The protein spots were visualized by colloidal Coomassie Brilliant Blue G-250.

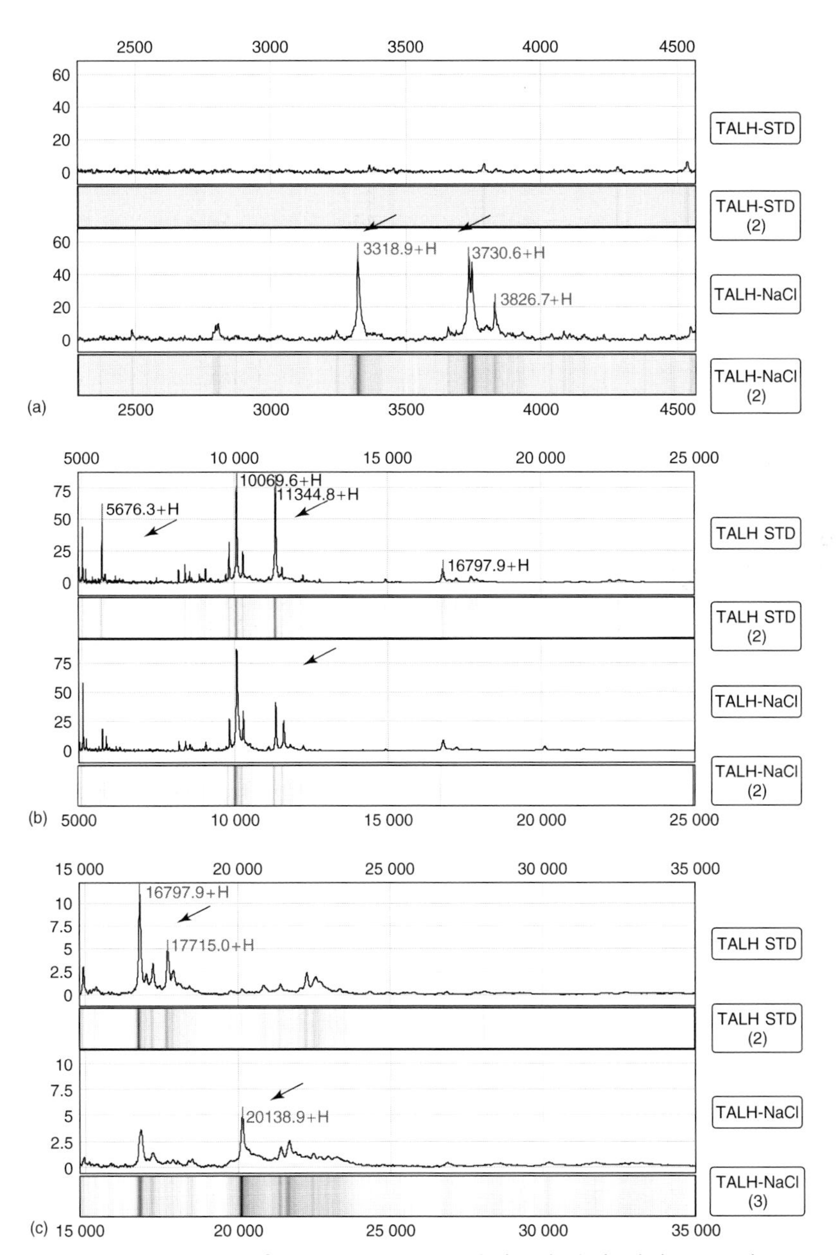

Figure 3.3 Comparison of SELDI-TOF MS spectra from a reverse-phase ProteinChip array demonstrating differences in protein expression between stressed and nonstressed TALH cells. TALH-cell extracts from stressed and nonstressed cells were prepared and applied to the hydrophobic H50 Chip array. (a)–(c) Spectra were obtained from different molecular mass ranges. The arrow indicates differentially expressed peaks. The peptide expression profiles are also represented as "gel view" (Dihazi *et al.* unpublished).

Table 3.3 Significantly up-regulated proteins (pH 3–10) in TALH cells exposed to hyperosmotic stress with 600 mOsmol kg^{-1} H_2O (TALH-NaCl) compared to control cells cultivated at 300 mOsmol kg^{-1} H_2O. Protein name, molecular weight, pI-value, MS/MS scores, and fingerprint scores are given [30].

Identified protein	Molecular weight (Da)	pI-value	Score MS/MS	Score fingerprint
3-Hydroxy CoA dehydrogenase	27 140	8.45	197	–
Adenylate kinase 3	25 610	8.69	228	–
Aldose reductase	35 763	6.48	296	204
α-Enolase	47 172	6.16	616	122
α-Crystallin chain B	20 024	6.76	205	110
Annexin I	38 711	6.28	336	111
Annexin II	38 448	8.36	360	100
Annexin V	35 914	4.94	89	120
ATP synthase β-chain	56 318	5.18	729	–
Creatine kinase	42 648	5.34	193	119
Cytokeratin 19	44 609	5.21	668	101
Dnak-type molecular chaperone hsc73	73 808	6.04	699	132
EndoA cytokeratin	53 210	5.42	173	–
Glutathione-S-transferase	26 617	6.82	175	152
Glyceraldehyd-3-phosphate-dehydrogenase	35 827	8.51	512	120
G-protein β subunit like protein	34 504	7.72	141	–
GTP-binding protein Ran	22 339	9.16	111	–
Heat shock protein 8	70 828	5.28	129	120
HSP90 α	84 664	4.96	143	95
Lactate dehydrogenase	36 615	5.71	303	132
Long-chain-acyl-CoA dehydrogenase	47 842	7.63	335	89
Malate dehydrogenase	35 589	8.93	467	82
S-Adenosylhomocysteine hydrolase	47 663	5.88	–	163
Tropomyosin 4	33 149	4.73	1247	–
Vimentin	53 569	5.06	1057	214

proteins belonged to endoplasmic reticulum (ER) stress proteins, structure proteins (tubulin), and others (Table 3.4).

3.4.1
Involvement of Carbohydrate Metabolism Enzymes in Osmotic Adaptation

Cells adapt to hyperosmotic stress by a variety of mechanisms that restore cell volume by regulating intracellular salt and osmolyte concentrations [31]. The most

Table 3.4 Significantly down-regulated proteins (pH 3–10) in TALH cells exposed to hyperosmotic stress with 600 mOsmol kg^{-1} H_2O (TALH-NaCl) compared to control cells cultivated at 300 mOsmol kg^{-1} H_2O. Protein name, molecular weight, pI-value, MS/MS scores, and fingerprint scores are given [30].

Identified protein	Molecular weight (Da)	pI-value	Score MS/MS	Score fingerprint
Calreticulin	67 277	4.30	321	102
Dnak-type molecular chaperone grp75	73 744	5.87	562	227
Glucose-regulated protein GRP78	72 334	5.07	–	113
Glucose-regulated protein GRP94	82 557	4.90	246	120
Lamin C	65 096	6.54	210	152
Nucleophosmin	32 540	4.62		161
Protein disulfide isomerase	56 761	4.77	838	218
Protein disulfide isomerase A4 (ERp72)	72 932	4.96	692	139
Triose-phosphate isomerase	26 609	7.1	242	100
Tropomyosin 1	32 708	4.71	1065	–
Tropomyosin 2	32 836	4.66	1048	–
Tropomyosin 3	28 467	4.60	1233	–
Tubulin α-chain	49 877	4.96	696	145
Tubulin β-2 chain	49 639	4.79	–	160
Tumor rejection antigen GRP96	92 432	4.74	729	121

evident change observed in TALH cells was overexpression of AR because of its role in sorbitol synthesis. However, the extent of the increase in protein expression (more than 15-fold) (Figure 3.4) [30] was found to be much higher when compared with the increase in mRNA expression in inner medulla cells (more than fourfold) [32]. High extracellular osmolarity induces renal sorbitol synthesis and accumulation in TALH cells (Figure 3.5c) by increasing the AR activity (Figure 3.5a,b) and decreasing membrane permeability [6, 15]. The expression of the *AR* gene in mammalian renal medulla has been shown to be osmotically regulated. In this tissue, *AR* mRNA and sorbitol content increase during dehydration or antidiuresis and decrease during diuresis [33]. Moreover, experiments on PAP-H25 cells derived from rabbit renal papillae have demonstrated that external hyperosmolarity enhances transcription of the *AR* gene [34], resulting in a rapid increase in *AR* mRNA [35], followed by an increase in AR activity and sorbitol content [36].

Another interesting enzyme from the carbohydrate metabolism, which was regulated by osmotic stress is α-enolase, is also described as a heat shock protein in yeast since it shares some degree of homology with dnaK [37]. Interestingly, it has been

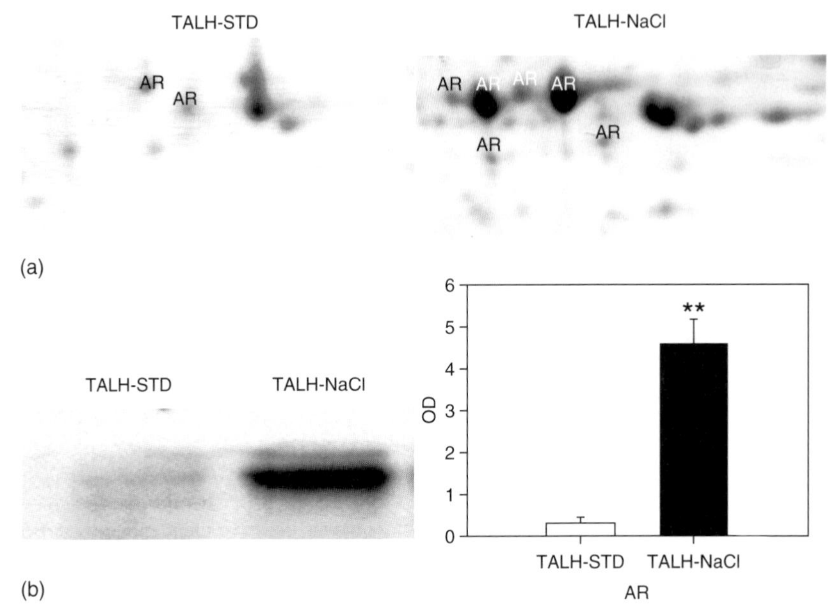

Figure 3.4 Differential expression of aldose reductase (AR) in nonstressed TALH-STD and their stressed counterparts TALH-NaCl. (a) Close-up of the differentially expressed AR spots. (b) Western blot analysis of soluble proteins from crude homogenates of TALH-STD and TALH-NaCl probed with the antiserum to AR. The degree of differential expression in the TALH-cell lines is shown in the histogram. Each bar represents mean ± SD of band intensity level obtained from three independent experiments.

demonstrated to be an early target of oxidative damage by carbonylation in different cell systems, ranging from yeast [38] to humans [39]. An involvement of enolase in oxidative stress in lens has previously been suggested by Paron *et al.* [40]. However, the role of the overexpression of α-enolase under osmotic stress (Figure 3.6) remained unclear. Glyceraldehyde-3-phosphate dehydrogenase (GAPDH) is a central glycolytic protein with a pivotal role in energy production. However, studies have demonstrated that GAPDH or some of its isoforms display a number of activities that are unrelated to its glycolytic function [41–45]. Its importance in oxidative stress and apoptosis has already been demonstrated [46]. In the case of osmotic stress, the pivotal role in energy production may be the main function of GAPDH. Lactate dehydrogenase (LDH) and malate dehydrogenase (MDH) are enzymes involved in carbohydrate metabolism. One of the reasons for the overexpression of these proteins (Figure 3.6) is the activation of the gluconeogenesis pathway to provide enough glucose for the synthesis of sorbitol in the polyol pathway.

Among the overexpressed proteins, adenylate kinase and creatine kinase are known to maintain ATP level in tissues with high energy demands, such as brain and muscle [47]. The overexpression of these proteins in TALH-NaCl highlights the higher energy needed in stressed TALH cells and the role of the energy balance under hyperosmotic stress conditions.

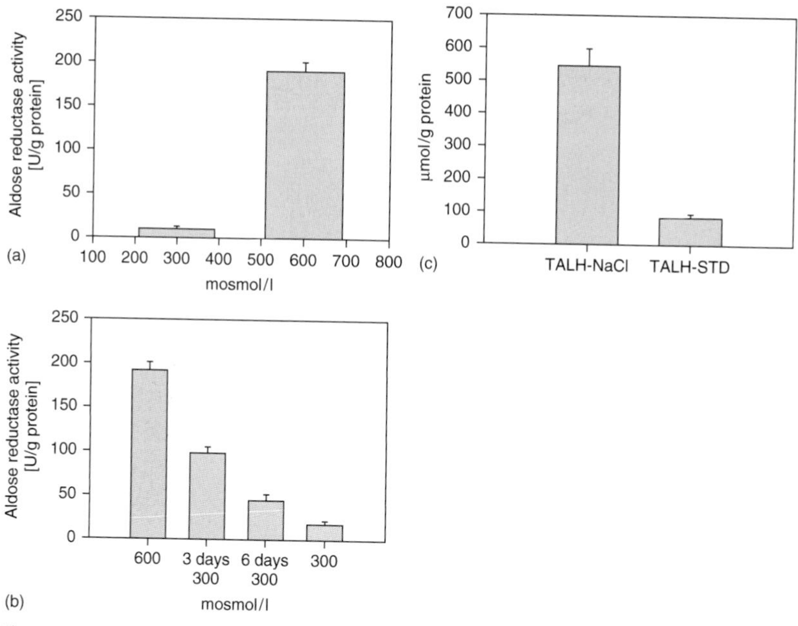

Figure 3.5 Effect of extracellular osmolarity on aldose reductase activity in TALH cells. (a) Aldose reductase activities in U/g protein of TALH cells, which were grown in 300 (control) or 600 mosmol l^{-1} NaCl medium are shown. (b) Effect of decreasing medium osmolarity on aldose reductase activity after three and six days from 600 (from cells adapted to 600 mosmol l^{-1} NaCl medium) to 300 mosmol l^{-1}. (c) Effect of extracellular osmolarity on sorbitol in TALH cells. TALH cells grown in 300 mOsm l^{-1} (control) and 600 mOsm l^{-1} NaCl medium. Each bar represents mean ± SD of 6–10 observations.

3.4.2
The Role of Cytoskeleton and Cytoskeleton-Associated Proteins in Maintenance of Cell Integrity under Hyperosmotic Stress

On the basis of its properties and subcellular organization, vimentin is considered to be the major contributor to the mechanical integrity of cells and tissues. The unique viscoelastic properties of vimentin render it more resistant than either microtubules or microfilaments to deformation and other external physical stress [48]. The increase in vimentin expression in stressed TALH cells (Figure 3.7) seems to play an important role for the stabilization and reinforcement of cells exposed to a high salt concentration [30].

Tropomyosins encompass a large family of actin regulatory proteins that are expressed by muscle and nonmuscle cells. The transition of lens epithelial cells from the undifferentiated to the differentiated state is characterized by a shift in tropomyosin isoform expression from high molecular weight (TM1, TM2, TM3, and TM4) to low molecular weight (TM5) and by a resulting reorganization of actin [49–51]. The down-regulation of TM1-3 and up-regulation of TM4 (Figure 3.7)

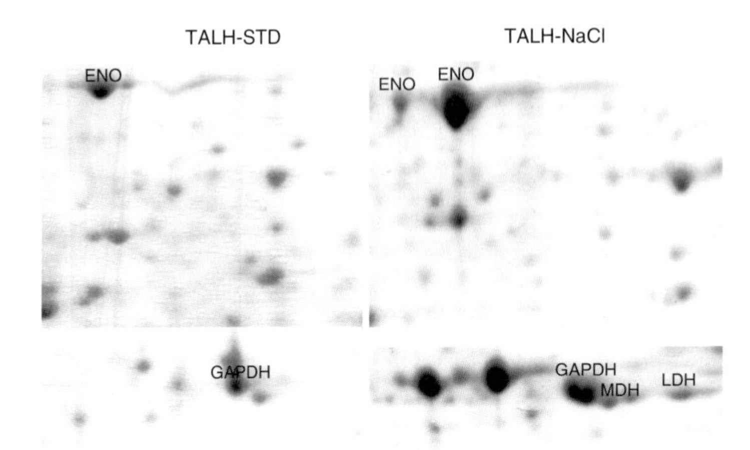

Figure 3.6 Close-up of the region of the gels showing differential expression of enolase (ENO), glyceraldehyde-3-phosphate dehydrogenase (GAPDH), malate dehydrogenase (MDH), and lactate dehydrogenase (LDH) in nonstressed TALH-STD and their stressed counterparts TALH-NaCl.

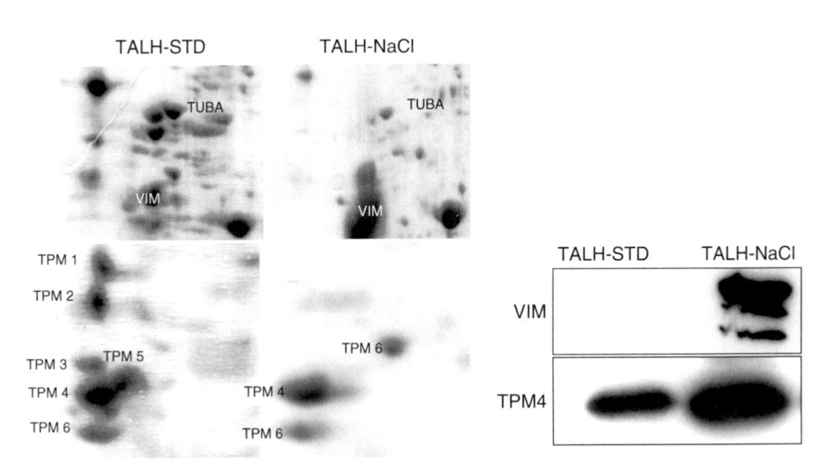

Figure 3.7 Close-up of the region of the gels showing differential expression of cytoskeleton proteins in nonstressed TALH-STD and their stressed counterparts TALH-NaCl. VIM: vimentin, TUBA: tubulin α-chain, and TPM1-6: tropomyosin 1-6. Right panel shows Western blot analysis of TPM4 and VIM showing the up-regulation of both proteins.

seem to be an indication of the strong reorganization necessity of actin in TALH cells under osmotic stress condition.

3.4.3 Expression of Molecular Chaperones is Affected by Hyperosmotic Stress

One of the major protein group found to be influenced by osmotic stress in TALH cells is the chaperone group, including HSPs. HSPs are a group of highly conserved proteins, some of which are expressed constitutively and/or induced by stress. Constitutively expressed HSPs participate in protein folding and assembly, elimination of misfolded proteins, and stabilization of newly synthesized proteins in various intracellular compartments [52]. Exposure of cells to diverse physical or chemical stressors such as heat, heavy metals, hypoxia, arsenite, and amino acid analogs, evokes strong induction of HSP synthesis [29]. High osmotic stress exhibits differential expression in several major HSPs in TALH-STD and its counterpart TALH-NaCl (Figure 3.8a,b). In TALH-NaCl cells, there was an increased expression of the small stress protein α-crystallin chain B and an up-regulation of members of the HSP70

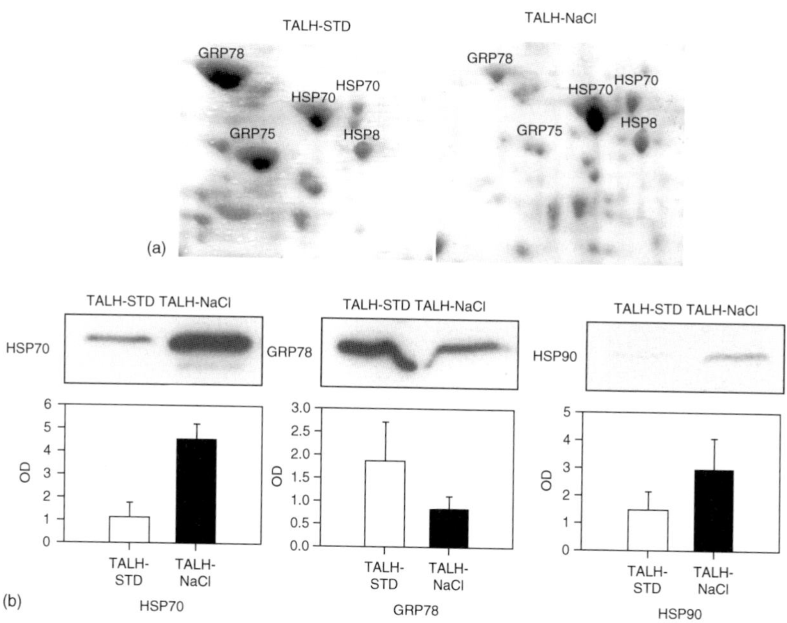

Figure 3.8 Close-up of the region of the gels showing differential expression of molecular chaperons in nonstressed TALH-STD and their stressed counterparts TALH-NaCl (a). HSP70: heat shock protein 70, HSP90: heat shock protein 90, HSP8: heat shock protein 8, and GRP75: glucose-regulated protein 75, glucose-regulated protein 78. (b) Western blot analysis of selected molecular chaperons found to be up- or down-regulated. The degree of differential protein expression in the TALH-cell lines is shown in the histogram. Each bar represents mean ± SD of band intensity level obtained from three independent experiments.

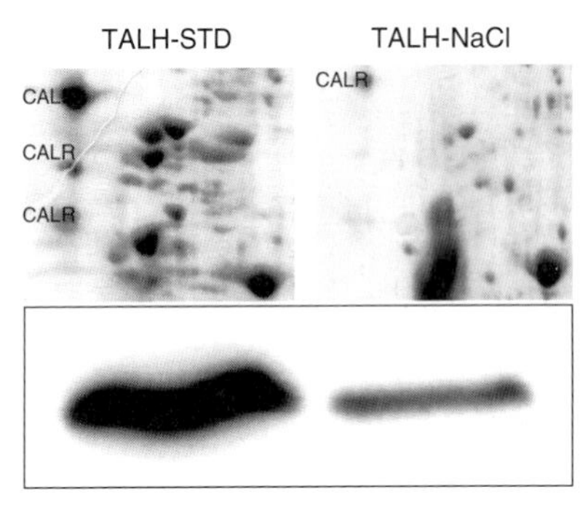

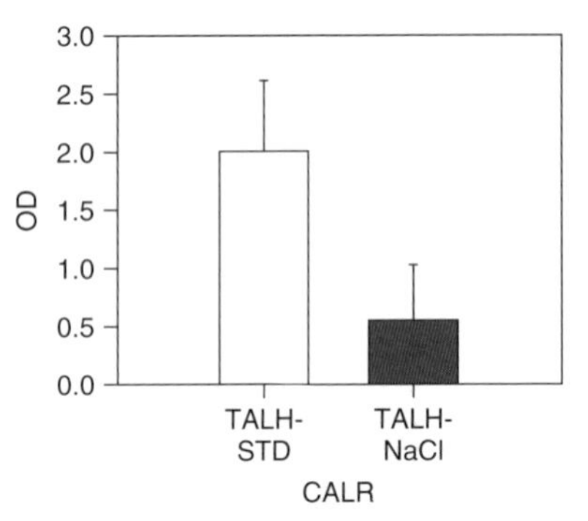

Figure 3.9 Close-up of the region of the gels showing differential expression of calreticulin (CALR) in nonstressed TALH-STD and their stressed counterparts TALH-NaCl. Lower panel shows Western blot analysis of calreticulin, whereas the degree of differential protein expression in the TALH cells is shown in the histogram. Each bar represents mean ± SD of band intensity level obtained from three independent experiments.

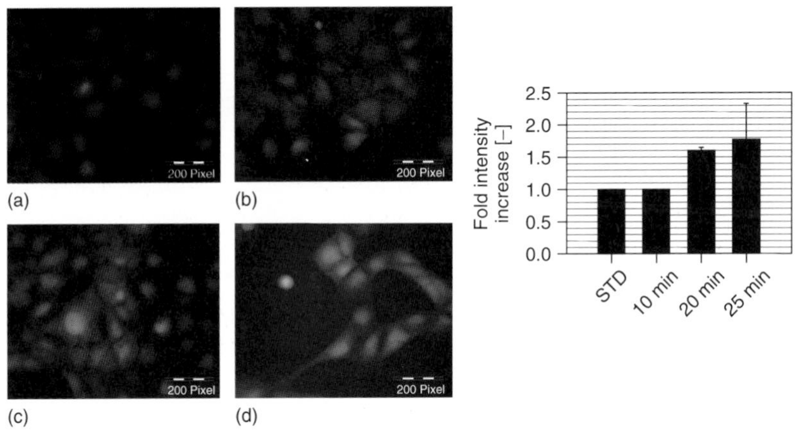

Figure 3.10 Time-dependent increase in free calcium in TALH cells exposed to hyperosmotic stress. Left panel: (a) control cells in isoosmotic medium (STD), (b) cells exposed for 10 minutes to NaCl stress, (c) Cells exposed for 20 minutes to NaCl stress (d), and cells exposed for 25 minutes to NaCl stress. Right panel: quantitative analysis of fluorescence intensity in fura-2-stained TALH cells after osmotic stress treatment.

family: hsc73 and HSP8a. Furthermore, proteome analysis revealed up-regulation of HSP90 (Figure 3.8b) in TALH-NaCl cells [30]. HSPs enhance survival of the stressed cells by acting as molecular chaperones when normal cellular protein synthesis is inhibited through denaturation. Owing to recent investigations on both renal and nonrenal cells, similar effects of osmotic stress concerning the expression of α-crystallin B chain and HSP70 under stress situation have been reported [53–55].

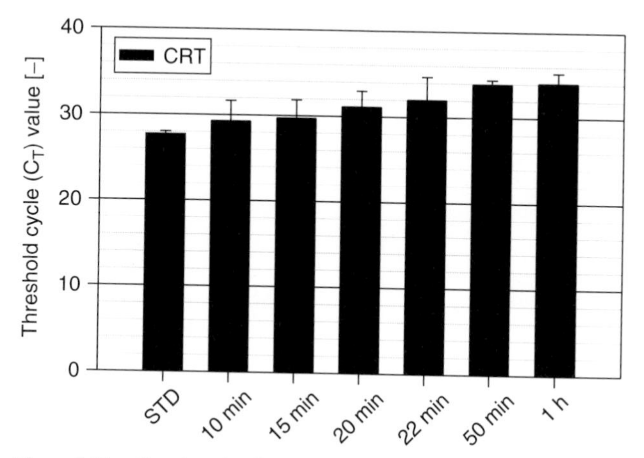

Figure 3.11 C_T value developing in real-time PCR for cDNA samples with calreticulin primer.

3.4.4
Down-Regulation of Endoplasmic Reticulum Stress Proteins and Their Roles in the Calcium Homeostasis under Hyperosmotic Stress

Besides up-regulated proteins, osmotic stress resulted in down-regulation of 15 proteins. An interesting finding is the involvement of the endoplasmatic stress proteins in osmotic stress resistance. The glucose-regulated proteins (GRPs) are a family of ER molecular chaperones and Ca^{2+}-binding stress proteins. These proteins are also induced under ER stress [56, 57]. Induction of GRPs by ER stress protects cells against a variety of toxic insults including Ca^{2+} ionophores, oxidative stress, topoisomerase inhibitors, and cytotoxic T-cells [58–61]. In our case, GRP78, GRP94, and GRP96, protein disulfide isomerase, and calreticulin (CALR) were found to be down-regulated [30]. The role of the CALR down-regulation (Figure 3.9) in Ca^{2+} homeostasis was investigated using real-time PCR, Western blot analysis, and calcium imaging. The results could confirm that the down-regulation of CALR is important for the increase in cytosolic-free calcium and in the prevention of the disturbance of intracellular Ca^{2+} homeostasis under osmotic stress (Figures 3.10 and 3.11).

References

1. Burg, M., Stoner, L., Cardinal, J., and Green, N. (1973) Furosemide effect on isolated perfused tubules. *Am J Physiol*, **225**, 119–124.
2. Burg, M.B. and Green, N. (1973) Function of the thick ascending limb of Henle's loop. *Am J Physiol*, **224**, 659–668.
3. Rocha, A.S. and Kokko, J.P. (1973) Sodium chloride and water transport in the medullary thick ascending limb of Henle. Evidence for active chloride transport. *J Clin Invest*, **52**, 612–623.
4. Bagnasco, S.M., Uchida, S., Balaban, R.S., Kador, P.F., and Burg, M.B.

(1987) Induction of aldose reductase and sorbitol in renal inner medullary cells by elevated extracellular NaCl. *Proc Natl Acad Sci U S A*, **84**, 1718–1720.
5. Garcia-Perez, A. and Burg, M.B. (1991) Renal medullary organic osmolytes. *Physiol Rev*, **71**, 1081–1115.
6. Grunewald, R.W. and Kinne, R.K. (1989) Sorbitol metabolism in inner medullary collecting duct cells of diabetic rats. *Pflugers Arch*, **414**, 346–350.
7. Yancey, P.H., Clark, M.E., Hand, S.C., Bowlus, R.D., and Somero, G.N. (1982) Living with water stress: evolution of osmolyte systems. *Science*, **217**, 1214–1222.
8. Burg, M., Grantham, J., Abramow, M., and Orloff, J. (1966) Preparation and study of fragments of single rabbit nephrons. *Am J Physiol*, **210**, 1293–1298.
9. Burg, M., Green, N., Sohraby, S., Steele, R., and Handler, J. (1982) Differentiated function in cultured epithelia derived from thick ascending limbs. *Am J Physiol*, **242**, C229–C233.
10. Scott, D.M., MacDonald, C., Brzeski, H., and Kinne, R. (1986) Maintenance of expression of differentiated function of kidney cells following transformation by SV40 early region DNA. *Exp Cell Res*, **166**, 391–398.
11. Drugge, E.D., Carroll, M.A., and McGiff, J.C. (1989) Cells in culture from rabbit medullary thick ascending limb of Henle's loop. *Am J Physiol*, **256**, C1070–C1081.
12. Grunewald, R.W. and Kinne, R.K. (1999) Osmoregulation in the mammalian kidney: the role of organic osmolytes. *J Exp Zool*, **283**, 708–724.
13. Trachtman, H., Lu, P., and Sturman, J.A. (1993) Immunohistochemical localization of taurine in rat renal tissue: studies in experimental disease states. *J Histochem Cytochem*, **41**, 1209–1216.
14. Amiry-Moghaddam, M., Nagelhus, E., and Ottersen, O.P. (1994) Light- and electronmicroscopic distribution of taurine, an organic osmolyte, in rat renal tubule cells. *Kidney Int*, **45**, 10–22.
15. Bagnasco, S.M., Murphy, H.R., Bedford, J.J., and Burg, M.B. (1988) Osmoregulation by slow changes in aldose reductase and rapid changes in sorbitol flux. *Am J Physiol*, **254**, C788–C792.
16. Eckstein, A. and Grunewald, R.W. (1996) Osmotic regulation of sorbitol in the thick ascending limb of Henle's loop. *Am J Physiol*, **270**, F275–F282.
17. Kinne-Saffran, E. and Kinne, R.K. (1997) Sorbitol uptake in plasma membrane vesicles isolated from immortalized rabbit TALH cells: activation by a Ca2+/calmodulin-dependent protein kinase. *J Membr Biol*, **159**, 231–238.
18. Kitamura, H., Yamauchi, A., Sugiura, T., Matsuoka, Y., Horio, M., Tohyama, M. *et al.* (1998) Inhibition of myo-inositol transport causes acute renal failure with selective medullary injury in the rat. *Kidney Int*, **53**, 146–153.
19. Wiese, T.J., Matsushita, K., Lowe, W.L. Jr, Stokes, J.B., and Yorek, M.A. (1996) Localization and regulation of renal Na+/myo-inositol cotransporter in diabetic rats. *Kidney Int*, **50**, 1202–1211.
20. Yamauchi, A., Nakanishi, T., Takamitsu, Y., Sugita, M., Imai, E., Noguchi, T. *et al.* (1994) In vivo osmoregulation of Na/myo-inositol cotransporter mRNA in rat kidney medulla. *J Am Soc Nephrol*, **5**, 62–67.
21. Grunewald, R.W. and Eckstein, A. (1995) Osmotic regulation of the betaine metabolism in immortalized renal cells. *Kidney Int*, **48**, 1714–1720.
22. Moeckel, G.W. and Lien, Y.H. (1997) Distribution of de novo synthesized betaine in rat kidney: role of renal synthesis on medullary betaine accumulation. *Am J Physiol*, **272**, F94–F99.
23. Wirthensohn, G., Vandewalle, A., and Guder, W.G. (1982) Choline

kinase activity along the rabbit nephron. *Kidney Int*, **21**, 877–879.

24. Silbernagl, S., Volker, K., Lang, H.J., and Dantzler, W.H. (1997) Taurine reabsorption by a carrier interacting with furosemide in short and long Henle's loops of rat nephrons. *Am J Physiol*, **272**, F205–F213.
25. Matsell, D.G., Bennett, T., Han, X., Budreau, A.M., and Chesney, R.W. (1997) Regulation of the taurine transporter gene in the S3 segment of the proximal tubule. *Kidney Int*, **52**, 748–754.
26. Grunewald, R.W., Fahr, M., Fiedler, G.M., Jehle, P.M., and Muller, G.A. (2001) Volume regulation of thick ascending limb of Henle cells: significance of organic osmolytes. *Exp Nephrol*, **9**, 81–89.
27. Gullans, S.R., Blumenfeld, J.D., Balschi, J.A., Kaleta, M., Brenner, R.M., Heilig, C.W., and Hebert, S.C. (1988) Accumulation of major organic osmolytes in rat renal inner medulla in dehydration. *Am J Physiol*, **255**, F626–F634.
28. Bagnasco, S., Balaban, R., Fales, H.M., Yang, Y.M., and Burg, M. (1986) Predominant osmotically active organic solutes in rat and rabbit renal medullas. *J Biol Chem*, **261**, 5872–5877.
29. Beck, F.X., Neuhofer, W., and Muller, E. (2000) Molecular chaperones in the kidney: distribution, putative roles, and regulation. *Am J Physiol Renal Physiol*, **279**, F203–F215.
30. Dihazi, H., Asif, A.R., Agarwal, N.K., Doncheva, Y., and Muller, G.A. (2005) Proteomic analysis of cellular response to osmotic stress in thick ascending limb of Henle's loop (TALH) cells. *Mol Cell Proteomics*, **4**, 1445–1458.
31. Burg, M.B. (1995) Molecular basis of osmotic regulation. *Am J Physiol*, **268**, F983–F996.
32. Grunewald, R.W., Wagner, M., Schubert, I., Franz, H.E., Muller, G.A., and Steffgen, J. (1998) Rat renal expression of mRNA coding for aldose reductase and sorbitol dehydrogenase and its osmotic regulation in inner medullary collecting duct cells. *Cell Physiol Biochem*, **8**, 293–303.
33. Cowley, B.D. Jr, Ferraris, J.D., Carper, D., and Burg, M.B. (1990) In vivo osmoregulation of aldose reductase mRNA, protein, and sorbitol in renal medulla. *Am J Physiol*, **258**, F154–F161.
34. Smardo, F.L. Jr, Burg, M.B., and Garcia-Perez, A. (1992) Kidney aldose reductase gene transcription is osmotically regulated. *Am J Physiol*, **262**, C776–C782.
35. Garcia-Perez, A., Martin, B., Murphy, H.R., Uchida, S., Murer, H., Cowley, B.D. Jr *et al.* (1989) Molecular cloning of cDNA coding for kidney aldose reductase. Regulation of specific mRNA accumulation by NaCl-mediated osmotic stress. *J Biol Chem*, **264**, 16815–16821.
36. Uchida, S., Garcia-Perez, A., Murphy, H., and Burg, M. (1989) Signal for induction of aldose reductase in renal medullary cells by high external NaCl. *Am J Physiol*, **256**, C614–C620.
37. Tamarit, J., Cabiscol, E., and Ros, J. (1998) Identification of the major oxidatively damaged proteins in Escherichia coli cells exposed to oxidative stress. *J Biol Chem*, **273**, 3027–3032.
38. Cabiscol, E., Piulats, E., Echave, P., Herrero, E., and Ros, J. (2000) Oxidative stress promotes specific protein damage in Saccharomyces cerevisiae. *J Biol Chem*, **275**, 27393–27398.
39. Castegna, A., Aksenov, M., Thongboonkerd, V., Klein, J.B., Pierce, W.M., Booze, R. *et al.* (2002) Proteomic identification of oxidatively modified proteins in Alzheimer's disease brain. Part II: dihydropyrimidinase-related protein 2, alpha-enolase and heat shock cognate 71. *J Neurochem*, **82**, 1524–1532.
40. Paron, I., D'Elia, A., D'Ambrosio, C., Scaloni, A., D'Aurizio, F., Prescott, A. *et al.* (2004) A proteomic approach to identify early molecular targets of

oxidative stress in human epithelial lens cells. *Biochem J*, **378**, 929–937.

41. Kawamoto, R.M. and Caswell, A.H. (1986) Autophosphorylation of glyceraldehydephosphate dehydrogenase and phosphorylation of protein from skeletal muscle microsomes. *Biochemistry*, **25**, 657–661.
42. Durrieu, C., Bernier-Valentin, F., and Rousset, B. (1987) Microtubules bind glyceraldehyde 3-phosphate dehydrogenase and modulate its enzyme activity and quaternary structure. *Arch Biochem Biophys*, **252**, 32–40.
43. Glaser, P.E. and Gross, R.W. (1995) Rapid plasmenylethanolamine-selective fusion of membrane bilayers catalyzed by an isoform of glyceraldehyde-3-phosphate dehydrogenase: discrimination between glycolytic and fusogenic roles of individual isoforms. *Biochemistry*, **34**, 12193–12203.
44. Brune, B. and Lapetina, E.G. (1996) Nitric oxide-induced covalent modification of glycolytic enzyme glyceraldehyde-3-phosphate dehydrogenase. *Methods Enzymol*, **269**, 400–407.
45. Nagy, E. and Rigby, W.F. (1995) Glyceraldehyde-3-phosphate dehydrogenase selectively binds AU-rich RNA in the NAD(+)-binding region (Rossmann fold). *J Biol Chem*, **270**, 2755–2763.
46. Dastoor, Z. and Dreyer, J.L. (2001) Potential role of nuclear translocation of glyceraldehyde-3-phosphate dehydrogenase in apoptosis and oxidative stress. *J Cell Sci*, **114**, 1643–1653.
47. Janssen, E., Terzic, A., Wieringa, B., and Dzeja, P.P. (2003) Impaired intracellular energetic communication in muscles from creatine kinase and adenylate kinase (M-CK/AK1) double knock-out mice. *J Biol Chem*, **278**, 30441–30449.
48. Mountain, I., Waelkens, E., Missiaen, L., and van Driessche, W. (1998) Changes in actin cytoskeleton during volume regulation in C6 glial cells. *Eur J Cell Biol*, **77**, 196–204.
49. Tilly, B.C., Edixhoven, M.J., Tertoolen, L.G., Morii, N., Saitoh, Y., Narumiya, S., and de Jonge, H.R. (1996) Activation of the osmo-sensitive chloride conductance involves P21rho and is accompanied by a transient reorganization of the F-actin cytoskeleton. *Mol Biol Cell*, **7**, 1419–1427.
50. Chou, Y.H. and Goldman, R.D. (2000) Intermediate filaments on the move. *J Cell Biol*, **150**, F101–F106.
51. Fischer, R.S., Lee, A., and Fowler, V.M. (2000) Tropomodulin and tropomyosin mediate lens cell actin cytoskeleton reorganization in vitro. *Invest Ophthalmol Vis Sci*, **41**, 166–174.
52. Ellis, R.J. and Hemmingsen, S.M. (1989) Molecular chaperones: proteins essential for the biogenesis of some macromolecular structures. *Trends Biochem Sci*, **14**, 339–342.
53. Cohen, D.M., Wasserman, J.C., and Gullans, S.R. (1991) Immediate early gene and HSP70 expression in hyperosmotic stress in MDCK cells. *Am J Physiol*, **261**, C594–C601.
54. Head, M.W., Corbin, E., and Goldman, J.E. (1994) Coordinate and independent regulation of alpha B-crystallin and hsp27 expression in response to physiological stress. *J Cell Physiol*, **159**, 41–50.
55. Kegel, K.B., Iwaki, A., Iwaki, T., and Goldman, J.E. (1996) AlphaB-crystallin protects glial cells from hypertonic stress. *Am J Physiol*, **270**, C903–C909.
56. Lee, A.S. (1992) Mammalian stress response: induction of the glucose-regulated protein family. *Curr Opin Cell Biol*, **4**, 267–273.
57. Brostrom, M.A., Cade, C., Prostko, C.R., Gmitter-Yellen, D., and Brostrom, C.O. (1990) Accommodation of protein synthesis to chronic deprivation of intracellular sequestered calcium. A putative role for GRP78. *J Biol Chem*, **265**, 20539–20546.
58. Little, E. and Lee, A.S. (1995) Generation of a mammalian cell line deficient in glucose-regulated

protein stress induction through targeted ribozyme driven by a stress-inducible promoter. *J Biol Chem*, **270**, 9526–9534.

59. Gomer, C.J., Ferrario, A., Rucker, N., Wong, S., and Lee, A.S. (1991) Glucose regulated protein induction and cellular resistance to oxidative stress mediated by porphyrin photosensitization. *Cancer Res*, **51**, 6574–6579.

60. Hughes, C.S., Shen, J.W., and Subjeck, J.R. (1989) Resistance to etoposide induced by three glucose-regulated stresses in Chinese hamster ovary cells. *Cancer Res*, **49**, 4452–4454.

61. Shen, J., Hughes, C., Chao, C., Cai, J., Bartels, C., Gessner, T., and Subjeck, J. (1987) Coinduction of glucose-regulated proteins and doxorubicin resistance in Chinese hamster cells. *Proc Natl Acad Sci U S A*, **84**, 3278–3282.

4
Proteomic Analysis of the Renal Inner Medulla and Collecting Ducts

Dietmar Kültz and Beverly J. Gabert

4.1
Introduction

The renal inner medulla (IM) is composed mainly of epithelial cells constituting collecting ducts and loops of Henle, interstitial cells, and endothelial cells of the vasa recta. Mammalian renal IM cells are routinely exposed to osmotic fluctuations resulting from the renal concentrating mechanism. This mechanism develops only after birth and its development coincides with the formation of the IM [1]. Urine osmotic concentration and volume depend on the degree of systemic hydration and salt load. Urine osmolality and volume are adjusted by changing the NaCl and urea concentrations (and consequently the interstitial osmolality) in the renal IM, regulating the water permeability and water reabsorption of collecting ducts, and regulating NaCl reabsorption. The urinary concentrating mechanism represents a fundamental function of all mammalian kidneys. It serves to sequester and excrete excessive NaCl and urea and thereby maintains osmotic balance and eliminates nitrogenous waste. Thus, a major goal of proteomic studies focusing on the IM is to better understand the molecular physiology of the urinary concentrating mechanism and the pathology of disorders in which this mechanism is compromised.

A second equally strong rationale for proteomic studies of the IM is the extreme stress tolerance of cells in that part of the kidney. In humans, the osmolality in the IM ranges from 600 (diuresis) to 1200–1400 mosmol kg^{-1} (antidiuresis); and in rodents, the maximal osmotic concentration can reach several thousand milliosmols per kilogram. Such hyperosmotic stress is far beyond the tolerance range of most other mammalian cell types [2]. In addition to the extremely high osmotic stress tolerance, IM cells are also unusually tolerant to hypoxic stress (ischemia) [3], and to nephrotoxins, which are highly concentrated in the IM along with other urinary solutes [4]. This high stress tolerance of the renal IM is reflected in the lesser frequency of diseases associated with the IM compared to the renal cortex or outer medulla (e.g., renal cancers). Because IM cells share the same genome with other cells in any given organism, the increased tolerance toward environmental stress is based on their cell-type-specific proteome. This

Renal and Urinary Proteomics: Methods and Protocols. Edited by Visith Thongboonkerd

ISBN: 978-3-527-31974-9

poses the question as to what proteins and intracellular pathways/networks confer unusually high environmental stress tolerance to IM cells. Obviously, identification of such proteins and protein networks promises to have outstanding therapeutic potential for applications in diseases and clinical interventions (e.g., surgery stress) in which it is desirable to increase the stress tolerance of particular cells, tissues, or organs.

4.2
Comparative Proteomics of Renal Inner Medulla versus Cortex

Proteomic studies of the IM have been performed using rodent models. Witzmann and coworkers [5] compared proteomes of male Sprague–Dawley rat kidney cortex and medulla by two-dimensional (2-D) gel electrophoresis using colloidal Coomassie Blue G-250 (CCB-G250) staining and image analysis with Kepler software. Proteins that differed significantly between cortex and medulla were identified by Western immunoblotting or by in-gel tryptic digestion followed by either matrix-assisted laser desorption/ionization time-of-flight mass spectrometry (MALDI-TOF MS) or electrospray ionization coupled to tandem mass spectrometry (ESI-MS/MS) [5]. This study identified four proteins that were more than twofold enriched in renal medulla compared to cortex. These proteins are glutathione *S*-transferase, heat shock cognate 70 (HSC70), heat shock protein 70 (HSP70), and HSP90. In another study, four additional proteins that were more than twofold enriched in the renal medulla relative to cortex of Sprague–Dawley rats were detected by silver staining of 2-D gels and bioinformatic analysis using BioImage software [6]. After in-gel tryptic digestion, MALDI-TOF MS and peptide mass fingerprinting, these proteins were identified as aflatoxin B1 aldehyde reductase, alpha-B crystallin, BH3 interacting domain death agonist, and glucose-regulated protein precursor 78 kDa (GRP78). Enrichment of alpha-B-crystallin and GRP78 in renal medulla was confirmed by Western blot analysis in this study.

In addition to these 8 proteins, we recently identified 18 proteins that were at least twofold enriched in IM compared to cortex of C57/BL6 mice (Gabert and Kültz; manuscript in preparation). These proteins were identified by a 2-D gel-based approach, CCB G250 staining, Delta 2-D quantitation of protein expression using 10 internal standards, and MALDI-TOF/TOF tandem mass spectrometry of in-gel tryptic digested protein spots (see Section 4.5 below). We subjected the combined set of 26 IM proteins identified in these three studies to a bioinformatic pathway modeling approach using the protein analysis through evolutionary relationships (PANTHERs) database. This analysis was performed as previously described [7] and revealed four biological processes that were significantly ($P < 0.05$, Bonferroni correction) overrepresented in renal IM. These processes are protein folding, stress response, other protein targeting/localization, and immunity/defense. In addition, two molecular functions (chaperone and transfer/carrier protein functions) were significantly overrepresented in the renal IM proteome. Oxidoreductases were also abundant but not overrepresented (Figure 4.1). These results clearly support the

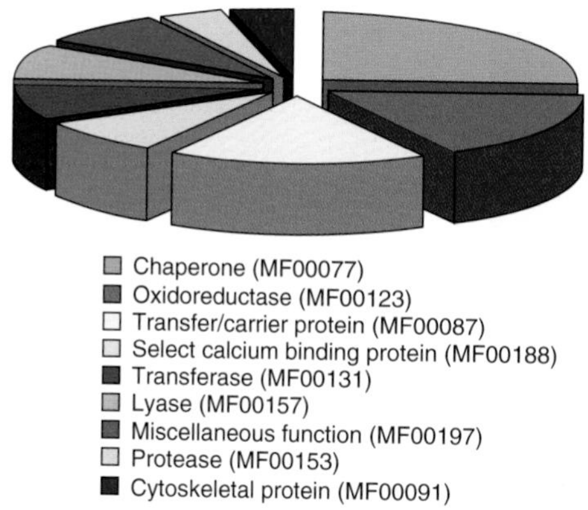

Figure 4.1 Molecular functions of 26 proteins that were identified by comparative proteomics as being at least twofold enriched in rat and mouse renal inner medulla relative to cortex. Molecular functions were determined with the protein analysis through evolutionary relationships (PANTHER) program. Please refer to the text for details.

notion that the unique features of the proteome of renal IM cells are responsible for their extreme stress tolerance (see above).

We have recently investigated region-specific differences in the renal proteome by MALDI imaging with emphasis on detecting IM-specific proteins. This technique is capable of analyzing native tissue sections while maintaining high-resolution spatial information about the distribution of hundreds of proteins and peptides [8]. In contrast to high-resolution fluorescence imaging, no prior labeling of proteins is required. MALDI imaging is most often performed on frozen tissue sections to minimize interference of fixatives but recent work shows that this technique is also suitable for traditional formalin-fixed, paraffin-embedded tissue sections when a suitable matrix is applied [9]. MALDI imaging of mouse kidney sections using a Bruker Ultraflex III TOF/TOF mass spectrometer clearly demonstrates that the proteomes of renal IM and cortex differ substantially. An example of such differences is shown in Figure 4.2. Thus, it is possible to rapidly survey not only changes in abundance or posttranslational modifications but also detailed spatial distribution of proteins in the kidney in response to environmental and physiological challenges or renal diseases. Because of the high resolution, very high information content, and speed of MALDI imaging, it is increasingly being used for investigating proteome changes associated with diseases. We anticipate that this trend will extend to the investigation of proteome changes associated with kidney diseases in the near future. One current limitation of MALDI proteome imaging is the difficulty of obtaining protein sequence information on target because intact proteins are being analyzed. Therefore, in most cases, the mass limit for successful MS/MS fragmentation is surpassed. However, the rapid evolution of ever more capable mass spectrometers utilizing innovative ionization and fragmentation approaches and on-target trypsin digestion promises to overcome

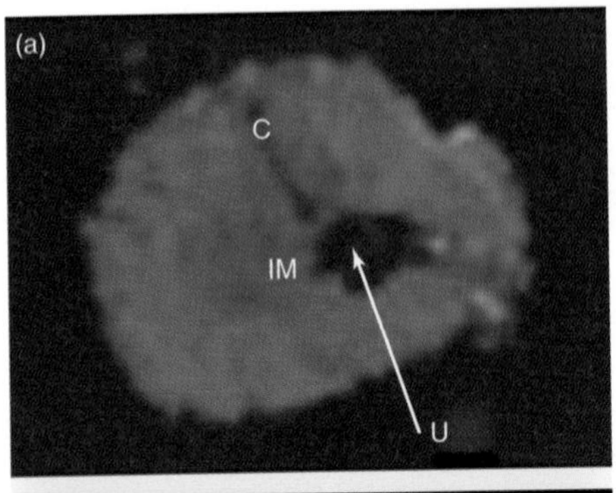

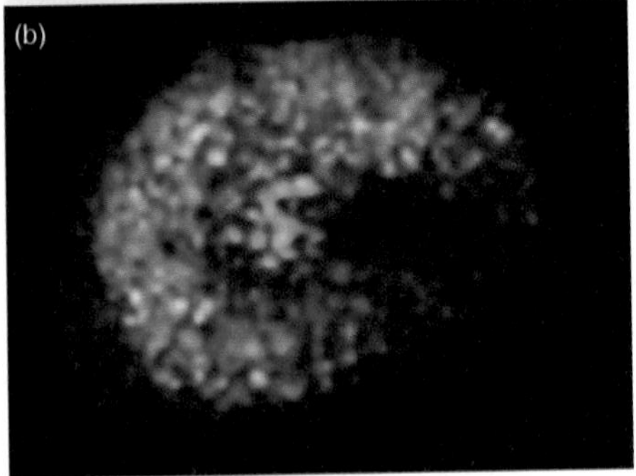

Figure 4.2 MALDI imaging of mouse kidney. (a) Micrograph of a mouse kidney section obtained by densitometry. C, renal cortex; IM, renal inner medulla; U, urinary space. (b) MALDI image generated by a Bruker Ultraflex III MALDI-TOF/TOF mass spectrometer. The peak intensity for two peptides is color-coded (m/z 6221 ± 6 Da = purple, m/z 9043 ± 10 Da = green). Note that the peptide with m/z of 6221 is highly expressed in renal cortex, whereas the peptide with m/z of 9043 is highly expressed in renal inner medulla.

this current drawback in the future and further increase the utility of MALDI proteome imaging for renal proteomics.

4.3 Proteomics of Inner Medullary Collecting Ducts

Although the renal IM contains many different cell types including epithelial cells of the loop of Henle, endothelial vasa recta cells, and interstitial cells, it is particularly densely packed with collecting ducts. Inner medullary collecting duct (IMCD) cells are critical for multiple key functions of the kidney including maintenance of body water/salt homeostasis, and acid–base regulation. Therefore, the proteome of IMCD cells and its regulation during diuresis and antidiuresis has been a recent focus of interest to renal physiologists. Purified IMCD preparations can be obtained by microdissection or biochemical purification strategies as described previously in detail [10–12]. The proteome of such IMCD preparations has been compared to that of inner medullary non-IMCD cells. Such comparison was done in male Sprague–Dawley rats using 2-D difference gel electrophoresis (DIGE) and MALDI-TOF mass spectrometry [11]. This study showed that HSP27 and HSP70 were primarily expressed in IMCD cells, while annexins were highly expressed in non-IMCD cells. Another study by the same group expanded this type of analysis by performing bioinformatic pathway analysis to identify additional, low-abundance proteins associated with biochemical pathways in IMCD cells that could not be detected directly on 2-D gels [13]. IMCD preparations were used to examine the changes in protein expression following escape from ddAVP-induced antidiuresis (ddAVP is a vasopressin receptor agonist). The protein changes observed were indicative of nitric oxide, calcium and cAMP signalings, receptor internalization, and endoplasmic reticulum processes. Additional bioinformatic pathway analysis of this protein set revealed transcription factors and low-abundance proteins induced by ddAVP; and some of these predicted changes were then experimentally confirmed by Western blot analysis [13]. Another proteomics approach that was used for analysis of the IMCD proteome is the in vacuo isotope-coded alkylation technique (IVICAT), which represents a variation of the isotope-coded affinity tagging (ICAT) technique. For IVICAT, a methylation reaction is promoted at low pressure yielding stable quaternary trimethylammonium groups at the amino-terminus of peptides, adding a net positive charge to the peptides. IVICAT analysis identified 10 proteins that were more than twofold enriched in IMCD over other kidney cells [14].

In addition to the analysis of the overall IMCD proteome, several recent studies have focused on the proteome analysis of subcellular compartments of IMCD cells. Because kidney function relies heavily on membrane transport proteins but membrane proteins are generally underrepresented in proteomics studies, special emphasis was placed on the enrichment of IMCD membrane proteins. The IMCD membrane compartments were isolated from rat kidneys by immunopurification with aquaporin 2 (AQP2) antibodies. AQP2 is a transmembrane protein forming water channels and a well-characterized marker of IMCD. It is located in intracellular membrane vesicles and the plasma membrane. Barile and coworkers [15] employed such an immunoprecipitation strategy to isolate AQP2 vesicles from rat IMCD. They found that AQP2 was mostly present in endosomes, membranes of the trans-Golgi network, and the rough endoplasmic reticulum. Proteins associated with the immunoprecipitated AQP2 complex were then analyzed by MALDI-TOF MS. The proteins identified as part of the AQP2 complex (SPA-I, actin, calcium-binding adapter molecule 2, myosin regulatory light chain smooth muscle isoforms 2-A and 2-B, alpha-tropomyosin 5b, annexin A2 and A6, scinderin, gelsolin, alpha-actinin 4, alpha-II spectrin, and myosin heavy chain nonmuscle type A) are thought to form a "force generator" complex that is responsible for movement of AQP2 vesicles between the cell interior and the plasma membrane [16]. Such subcellular relocalization of AQP2 is critical for physiological regulation of water reabsorption in the IMCD and urine concentration depending on systemic hydration state.

Protein differences between the apical and basolateral membranes of IMCD cells, which are highly polarized, were identified after labeling and enrichment of membrane proteins by biotinylation followed by streptavidin affinity chromatography [17]. Biotin labeling of apical membranes was achieved via perfusion of IMCD segments. Basolateral membranes were labeled via incubation of isolated cell suspensions. Membrane proteins were then extracted from those membrane vesicles and identified with liquid chromatography coupled to tandem mass spectrometry (LC-MS/MS). Most of the identified proteins were peripheral membrane proteins that were easier to extract but 17 integral apical membrane proteins and 23 integral basolateral membrane proteins were also identified in this study [17].

4.4 Regulation of the IMCD Proteome in Response to Physiological and Pathological Signals

Because IMCD cells are frequently exposed to severe stress and many regulatory signals under physiological conditions, proteome changes in IMCD cells have been investigated in response to physiologically significant stresses and stimuli. Proteome changes in response to the vasopressin analog, ddAVP, which promotes water reabsorption in IMCD and urine concentration, were observed in Brattleboro rats using an ICAT-based LC-MS/MS approach [18]. In addition to the well-known effects of vasopressin on the expression of AQP2 and gamma-epithelial Na channel (gamma-ENaC), five other proteins were substantially affected by ddAVP treatment (syntaxin-7, Rap1, GAPDH, HSP70, and cathepsin D). In a similar study using an alternative approach (DIGE followed by MALDI-TOF MS), long-term effects of the vasopressin analog ddAVP on the IMCD proteome were investigated [19]. Nitric oxide synthase type 2, cathepsin D, PDI A6 precursor, acetyl/L-xylulose reductase, and HSP27 were increased by at least twofold in response to ddAVP, while ATP synthase beta-subunit, triosephosphate isomerase 1, and NADPH oxidase beta-subunit were decreased by at least 50%. The authors of this study concluded that nitric oxide signaling might play a significant role in vasopressin escape [19]. Of interest, the ICAT approach yielded 5.5% integral membrane proteins compared to only 1.6% for DIGE relative to the total number of proteins identified in these two studies.

Changes in IMCD proteome in response to osmotic stress have been analyzed using the mouse IMCD3 cell line because hypertonicity is one of the most peculiar conditions unique to the IM and essential for the urinary concentrating mechanism. A proteome map for this IMCD cell line was generated through fusion of 2-D gels produced from different sample preparations in order to obtain the fullest representation of abundant mIMCD3 proteins and to minimize losses resulting from different types of sample preparation [20]. The identified mIMCD3 proteins function in cellular stress response (i.e., HSPs and antioxidant proteins), cytoskeleton dynamics, energy metabolism, protein synthesis, and RNA processing. When mIMCD3 cells were exposed to hyperosmolality, the abundance of molecular chaperones was increased consistent with the need for counteractive measures against the deleterious effects of hyperosmolality on 3-D protein structure [20].

Several proteins that are increased in abundance during progression of polycystic kidney disease (PKD) have been identified by comparative proteomics in the juvenile cystic kidney (jck) mouse model. In PKD, renal collecting duct cells are characterized by abnormal proliferation, differentiation, and polarity. Using a proteomics approach based on 2-D gels and MALDI-TOF/TOF mass spectrometry, we have shown that such cellular changes were accompanied by increases in galectin 1, sorcin, vimentin, and major urinary proteins (MUPs) [21]. The increases in these proteins were confirmed by quantitative immunohistochemistry and laser scanning cytometry, implicating their roles in the pathogenesis of PKD. Another proteomic study based on DIGE followed by LC-MS/MS of mIMCD3 cells adapted to increasing levels of hypertonicity revealed several proteins that were up-regulated in response to hyperosmotic stress. S100A4 protein was one of the most highly up-regulated proteins identified in this study. This protein was subsequently shown to be highly enriched also in renal IM, where hyperosmotic stress is prevalent and its abundance was further increased in renal IM after mice were thirsted for 36 hours, a treatment that increases IM osmolality [22].

Most proteomic studies, to date, have focused on changes in protein abundance as these are often evident when comparing samples. However, protein levels take time to change and thus are less powerful indicators for examining the mechanisms underlying short-term responses. The analysis of posttranslational modifications provides a powerful and very rapid alternative to protein regulation independent of changes in abundance. Therefore, proteomic studies increasingly focus on posttranslational modifications of proteins in addition to abundance changes. In particular, protein phosphorylation is a common event in response to many stresses and physiological signals that can be very rapid and greatly alter the function of a protein. Using a proteomics approach based on 2-D gel analysis and phosphoprotein staining (Pro-Q Diamond) coupled with MALDI-TOF/TOF tandem mass spectrometry, we were able to detect hundreds of phosphoproteins and quantify changes in phosphorylation of HSP27 in response to hyperosmotic stress in mIMCD3 cells [20]. An alternative approach that has been used for identification of hundreds of phosphoproteins in rat IMCD cells is immobilized metal affinity chromatography coupled with LC-MS neutral loss scanning [23]. Among the phosphorylation sites identified in this study were eight sites in key transporter/channel proteins of the IMCD, namely AQP2, AQP4, and urea transporters A1 and A3. In addition, this study identified vasopressin-induced changes in phosphorylation of AQP2, Bclaf1, LRRC47, Rgl3, and SAFB2 [23].

The application of a variety of different high-throughput proteomics approaches as briefly reviewed above has demonstrated that we can obtain valuable insights into physiological adaptations, pathological alterations, and molecular mechanisms underlying the unique ability of renal IM cells to survive in a hostile extracellular environment and perform urinary concentration as well as other functions that are vital for the organism. The proteomic studies performed on the renal IM, to date, have begun to shed light on biological processes and molecular pathways responsible for IM cell function. We anticipate that, with the continued advancement of mass spectrometry and other proteomic tools, proteomics approaches will increasingly be adopted and become more influential for the study of renal IM function.

4.5 Methods and Protocols

The following is a selection of proteomics protocols used in our laboratory. Additional detailed protocols for the analysis of the renal IM proteome are included in many of the references cited in this chapter.

4.5.1 Sample Preparation

Renal tissue pieces are shock-frozen in liquid nitrogen immediately after collection. Proteins are then extracted from tissues in either 2DRIPA buffer or UT buffer by homogenizing tissues in four buffer volumes with tight-fitting glass homogenizers (Wheaton) on ice. We use these two different buffers complementary because there are differences in the representation of various proteins between these two extraction buffers. 2DRIPA buffer is composed of 50 mM Tris-HCl (pH7.6), 150 mM NaCl, 1%(w/v) CHAPS, 1% (v/v) NP-40, 1× Complete Protease inhibitor cocktail (Roche), 5 mM NaF, and 2 mM activated Na_3VO_4. UT buffer is composed of 7 M urea, 2 M thiourea, 2% CHAPS, 1 mM dithioerythrit (DTE), 5 mM NaF, 2 mM activated Na_3VO_4, 0.002% bromophenol blue, and 0.5% ampholyte solution (appropriate to the pH range of the isoelectric focusing strip). Cells cultured as monolayers (e.g., mIMCD3 cells) are rinsed three times with phosphate-buffered saline (PBS) and lysed in 2DRIPA buffer or UT buffer by freezing at −80 °C for 30 minutes and then scraping off the dish with a cell lifter. Tissue and cell homogenates are cleared from debris and unsoluble material by centrifugation for 5 minutes at 19 000 g and 4 °C. Supernatants are then aliquoted into siliconized low-retention microcentrifuge tubes and protein concentration is determined by BCA assay (Pierce) or 2DQuant assay (GE Healthcare). When using 2DRIPA buffer, proteins are precipitated to remove salt and other contaminants in precooled (−20 °C) acetone + 10% trichloro-acetic acid for 30 minutes at −20 °C. After centrifugation for 5 minutes at 19 000 g and 4 °C protein pellets are air-dried for 5 minutes and resuspended at a concentration of 2 mg ml^{-1} in UT buffer [20].

4.5.2 Two-Dimensional Gel Electrophoresis

Samples (450 μl, 0.9 mg protein) are rehydration-loaded to 24-cm immobilized pH gradient (IPG) strips by overnight incubation at 20 °C. IPG strips are then processed by isoelectric focusing in an IEFCell (Biorad) for 60 000 Vh. IPG strips can be stored frozen at $-80\,^{\circ}C$ for up to one month after isoelectric focusing. Second dimensional separation of proteins is preceded by protein reduction (using dithiothreitol, DTT) followed by alkylation (using iodoacetamide, IAA) both of which are carried out in equilibration buffer. Equilibrated IPG strips are loaded on top of uniform 11% SDS/ polyacrylamide/ Bis gels and overlaid with 0.5% low melting point agarose (FMC) containing 0.002% bromophenol blue. Six second dimensional gels ($25 \times 20 \times 0.15$ cm) are electrophoresed simultaneously by SDS-PAGE at 100 W and $10\,^{\circ}C$ with EttanDalt-Six (GE Healthcare). After the bromophenol blue dye front has reached the bottom of the gels, staining is performed with CCB. CCB is prepared as follows to maximize sensitivity. First, Coomassie Blue G-250 is mixed with ultrapure water (3.23 g per 1.61 l) while stirring. Then, 89.5 ml of concentrated H_2SO_4 is added. This solution is stirred overnight (seal Erlenmeyer cylinder with parafilm to prevent evaporation) and then filtered through Whatman 3-mm filter paper. Next, 355 ml of 10 N NaOH (= 142 g/ 355 ml ultrapure water) is added while stirring and finally 500 ml of 100% trichloroacetic acid (500 g made up to 500 ml by dissolving in ultrapure water) is added. This solution is green and can be stored at room temperature protected from light for up to one week. During storage, some CCB will fall out of the colloidal phase and thus it is best to use the solution immediately. The staining solution can be reused but sensitivity of protein spot detection will decline. Stained gel images are recorded as 16-bit TIFF files with an Epson 1680 or an equivalent densitometer [20].

4.5.3 2-D Gel Image Analysis

Protein spots on 2-D gels are detected and matched using Delta 2-D gel analysis software (Decodon GmbH). In our experience, this software is more accurate and reliable than other 2-D gel analysis software packages. First, all gels in each experimental group are aligned to a single reference gel chosen for each group. This is done using Delta 2D automatic warping guided by manual inspection to assure stringent quality control during warping. When all gels within an experimental group have been warped to the reference gel of that group, a fusion gel is created for each group that represents each spot present on any of the gels in that group at its maximal abundance. Using this approach, it can be guaranteed that all spots on any gel can be detected and quantified. The fusion gels are then warped as described above to enable accurate alignment of gels from different experimental groups. A master gel is created by merging all group-specific fusion gels. This master gel contains all spots present on any gel and each spot at its maximal intensity. Spot detection is performed on the master gel and spot boundaries are exported from the master gel to all individual gels to eliminate variability in spot quantification due to spot shape. Ten internal reference spots that are evenly distributed throughout the gel and are equally abundant on all gels (as determined by Delta2D) are selected as internal standards and spot volume of all spots is subsequently normalized to these 10 internal standards. Every spot on each gel is quantified, group-specific averages for each spot calculated, and statistics performed to identify proteins that are common and different between experimental groups. Fold difference and statistical significance (*t*-test) is automatically calculated for each spot by Delta2D [21].

4.5.4
In-Gel Tryptic Digestion

During 2-D gel electrophoresis, staining, and preparation of samples for mass spectrometry, caution must be exercised to avoid contamination of samples with keratin present in dust, skin, or hair. Therefore, it is important to wear clean nitrile gloves and a clean laboratory coat. Spots of interest are extracted from 2-D gels using a 1.5-mm diameter tissue microarray puncher (Beecher Instruments) in a dust-free Hepa-filtered benchtop enclosure. They should be picked from gels as soon as possible after staining to avoid contamination with dust or microorganisms. Picked gel spots can be stored in 0.6-ml siliconized, dust-free, low-retention tubes that have been rinsed with acetonitrile. All solutions used for the trypsin digestion should be stored in small aliquots and used only once. All solutions containing acetonitrile and ammonium bicarbonate must be prepared fresh. All chemicals (including water) should be HPLC or sequencing grade.

Gel material is washed with 100% of 200-μl acetonitrile while shaking the tube for 10 minutes three times at 37 °C, 200 rpm in a shaking incubator. CCB destaining and gel dehydration are then performed by shrinking the gel material with 70 μl of 200 mM ammonium bicarbonate (NH_4HCO_3), pH 7.8, in acetonitrile (4 : 6) while shaking for 30 minutes at 37 °C, 200 rpm in a shaking incubator. The gel material is then rehydrated with 70 μl of 200 mM NH_4HCO_3, pH 7.8 while shaking 30 minutes at 37 °C, 200 rpm in a shaking incubator. Gel material is again shrunk in 70-μl acetonitrile for 2 minutes at room temperature and dried completely in a SpeedVac at low setting afterward. Before using the SpeedVac, the lid of the sample tube is perforated with a clean needle to create small openings (<0.5 mm) and closed to avoid loss of gel material. In-gel tryptic digestion is performed with Promega Trypsin Gold (MS grade). A stock solution of 20-μg trypsin reconstituted in 100 μl of 50 mM acetic acid is prepared and stored in 5-μl aliquots at −30 °C. For protein digestion, one aliquot of trypsin stock solution (5 μl = 1 μg) is mixed with 245 μl of 50 mM NH_4HCO_3 and 5 μl (= 0.02 μg trypsin) of the resulting solution is then added to each dehydrated sample (just enough to fully cover the gel piece when completely hydrated). After replacing the perforated lids with unperforated lids to prevent evaporation, samples are incubated at 37 °C for 16 hours at 250 rpm in a shaking incubator. The supernatant is collected in a clean siliconized, low-retention tube and peptides are further eluted from the gel during two rounds of 30 minutes incubation with 10 μl of 0.3% TFA/60% acetonitrile solution at room temperature and 250 rpm in a shaking incubator. The supernatants are pooled with the trypsin supernatant (25 μl total). The peptides are concentrated in a SpeedVac at low speed down to a volume of 1 μl. Nine microliters of 1% formic acid are added and the peptide mixture is either used immediately for zip-tipping and MS analysis or stored at −30 °C.

4.5.5
MALDI-TOF/TOF Tandem Mass Spectrometry

Before spotting the peptide mixture on a MALDI target, a clean-up step is performed to remove salts and other contaminants. A ZipTip (Millipore) is washed twice with 10 μl of 100% acetonitrile followed by equilibration twice with 10 μl of 1% formic acid. A volume of 10 μl of peptide mixture is then loaded onto the ZipTip by pipetting up and down seven times. The ZipTip is washed three times with 10 μl HPLC-grade water. Then the peptides are eluted into 1-μl elution solution (1% formic acid/50% acetonitrile) by pipetting up and down three times into a clean 0.6-ml siliconized low-retention tube. The entire eluate is then immediately spotted in two 0.5-μl aliquots on a MALDI plate and each sample spot is overlaid with 0.5 μl of 2 mg ml^{-1} MALDI matrix. The MALDI matrix, α-cyano-4-hydroxycinnamic acid, is recrystallized from 70 : 30 acetonitrile : H_2O before use. Samples are analyzed with a 4700 Proteomics Analyzer from Applied Biosystems (Foster City, CA) using both MS and MS/MS operating modes [7]. Peptide fragmentation in MS/MS mode is achieved either by postsource decay (PSD) or collision-induced dissociation (CID). After collection of mass spectra, protein identification is carried out with GPS Explorer software (Applied Biosystems) using the Mascot search algorithm [24] and with 4700 Explorer software (Applied Biosystems) using the DeNovo Explorer module and MSBLAST-P2 search algorithm [25]. Six parameters are considered for successful identification of proteins: (i) matched molecular weight; (ii) matched pI on 2-D gels; (iii) Mascot score for peptide mass fingerprinting taking into account, during matching, sequence of MS/MS peptides generated by CID; (iv) Mascot score for peptide mass fingerprinting taking into account, during matching, sequence of MS/MS peptides generated by PSD; (v) MSBLASTP2 score based on de novo sequencing of multiple MS/MS peptides generated by CID; and (vi) MSBLASTP2 score based on de novo sequencing of multiple MS/MS peptides generated by PSD.

4.5.6
MALDI Imaging

MALDI imaging of mouse kidney sections can be efficiently performed on a Bruker Ultraflex III MALDI-TOF/TOF mass spectrometer. Sample preparation is best performed from frozen sections to minimize interference of formaldehyde or other fixatives with peptide and matrix ionization. Mouse kidneys are shock-frozen in liquid nitrogen and then cut frozen into 10-μm sections using a cryostat-microtome. Sections are adhered to conductive microscope slides and stored frozen until coating with matrix. Matrix application is performed after briefly rinsing the slides three times with 70% ethanol and three times with 90% ethanol. The MALDI matrix (sinapinic acid in 50% acetonitrile/0.1% TFA) is applied by repeated fine spraying onto the slide in about 1 minute increments for 30–40 repeated cycles until homogeneous and even coating of the entire slide is achieved. The slide is marked for orientation and then scanned using a densitometer to create a micrograph image that is used to orient the sample in the mass spectrometer. It is then mounted on a Bruker Ultraflex III MALDI target plate and introduced to the mass spectrometer. MALDI imaging is performed with the Ultraflex III in 100-μm scan increments at 200-Hz laser firing rate within a mass range of 2000–50 000 m/z. Finally, peptide peaks are detected and images generated with FLEX-Analysis and FLEX-Image software (Bruker Daltonics).

Acknowledgments

We would like to thank Drs Gongyi Shi and Gary Kruppa for their assistance with MALDI imaging and for providing access to a Bruker Ultraflex III MALDI-TOF/TOF imaging mass spectrometer. This work was supported by grants from NIDDK (R01DK59470) and NIEHS (2P42ES04699).

References

1. Schlondorff, D., Weber, H., Trizna, W., and Fine, L.G. (1978) Vasopressin responsiveness of renal adenylate cyclase in newborn rats and rabbits. *Am J Physiol*, **234**, F16–F21.
2. Kultz, D. (2005) Molecular and evolutionary basis of the cellular stress response. *Annu Rev Physiol*, **67**, 225–257.
3. Zou, A.P. and Cowley, A.W. Jr (2003) Reactive oxygen species and molecular regulation of renal oxygenation. *Acta Physiol Scand*, **179**, 233–241.
4. Dahlmann, A., Dantzler, W.H., Silbernagl, S., and Gekle, M. (1998) Detailed mapping of ochratoxin A reabsorption along the rat nephron in vivo: the nephrotoxin can be reabsorbed in all nephron segments by different mechanisms. *J Pharmacol Exp Ther*, **286**, 157–162.
5. Witzmann, F.A., Fultz, C.D., Grant, R.A., Wright, L.S., Kornguth, S.E., and Siegel, F.L. (1998) Differential expression of cytosolic proteins in the rat kidney cortex and medulla: preliminary proteomics. *Electrophoresis*, **19**, 2491–2497.
6. Arthur, J.M., Thongboonkerd, V., Scherzer, J.A., Cai, J., Pierce, W.M., and Klein, J.B. (2002) Differential expression of proteins in renal cortex and medulla: a proteomic approach. *Kidney Int*, **62**, 1314–1321.
7. Perroud, B., Lee, J., Valkova, N., Dhirapong, A., Lin, P.Y., Fiehn, O., Kultz, D., and Weiss, R.H. (2006) Pathway analysis of kidney cancer using proteomics and metabolic profiling. *Mol Cancer*, **5**, 64.
8. McDonnell, L.A. and Heeren, R.M. (2007) Imaging mass spectrometry. *Mass Spectrom Rev*, **26**, 606–643.
9. Lemaire, R., Desmons, A., Tabet, J.C., Day, R., Salzet, M., and Fournier, I. (2007) Direct analysis and MALDI imaging of formalin-fixed, paraffin-embedded tissue sections. *J Proteome Res*, **6**, 1295–1305.
10. Miller, R.L., Zhang, P., Chen, T., Rohrwasser, A., and Nelson, R.D. (2006) Automated method for the isolation of collecting ducts. *Am J Physiol Renal Physiol*, **291**, F236–F245.
11. Hoffert, J.D., van Balkom, B.W., Chou, C.L., and Knepper, M.A. (2004) Application of difference gel electrophoresis to the identification of inner medullary collecting duct proteins. *Am J Physiol Renal Physiol*, **286**, F170–F179.
12. Chou, C.L., DiGiovanni, S.R., Luther, A., Lolait, S.J., and Knepper, M.A. (1995) Oxytocin as an antidiuretic hormone. II. Role of V2 vasopressin receptor. *Am J Physiol*, **269**, F78–F85.

13. Hoorn, E.J., Hoffert, J.D., and Knepper, M.A. (2005) Combined proteomics and pathways analysis of collecting duct reveals a protein regulatory network activated in vasopressin escape. *J Am Soc Nephrol*, **16**, 2852–2863.
14. Simons, B.L., Wang, G., Shen, R.F., and Knepper, M.A. (2006) In vacuo isotope coded alkylation technique (IVICAT); an N-terminal stable isotopic label for quantitative liquid chromatography/mass spectrometry proteomics. *Rapid Commun Mass Spectrom*, **20**, 2463–2477.
15. Barile, M., Pisitkun, T., Yu, M.J., Chou, C.L., Verbalis, M.J., Shen, R.F., and Knepper, M.A. (2005) Large scale protein identification in intracellular aquaporin-2 vesicles from renal inner medullary collecting duct. *Mol Cell Proteomics*, **4**, 1095–1106.
16. Noda, Y., Horikawa, S., Katayama, Y., and Sasaki, S. (2005) Identification of a multiprotein "motor" complex binding to water channel aquaporin-2. *Biochem Biophys Res Commun*, **330**, 1041–1047.
17. Yu, M.J., Pisitkun, T., Wang, G., Shen, R.F., and Knepper, M.A. (2006) LC-MS/MS analysis of apical and basolateral plasma membranes of rat renal collecting duct cells. *Mol Cell Proteomics*, **5**, 2131–2145.
18. Pisitkun, T., Bieniek, J., Tchapyjnikov, D., Wang, G., Wu, W.W., Shen, R.F., and Knepper, M.A. (2006) High-throughput identification of IMCD proteins using LC-MS/MS. *Physiol Genomics*, **25**, 263–276.
19. van Balkom, B.W., Hoffert, J.D., Chou, C.L., and Knepper, M.A. (2004) Proteomic analysis of long-term vasopressin action in the inner medullary collecting duct of the Brattleboro rat. *Am J Physiol Renal Physiol*, **286**, F216–F224.
20. Valkova, N. and Kultz, D. (2006) Constitutive and inducible stress proteins dominate the proteome of the murine inner medullary collecting duct-3 (mIMCD3) cell line. *Biochim Biophys Acta*, **1764**, 1007–1020.
21. Valkova, N., Yunis, R., Mak, S.K., Kang, K., and Kultz, D. (2005) Nek8 mutation causes overexpression of galectin-1, sorcin, and vimentin and accumulation of the major urinary protein in renal cysts of jck mice. *Mol Cell Proteomics*, **4**, 1009–1018.
22. Rivard, C.J., Brown, L.M., Almeida, N.E., Maunsbach, A.B., Pihakaski-Maunsbach, K., Andres-Hernando, A., Capasso, J.M., and Berl, T. (2007) Expression of the calcium-binding protein S100A4 is markedly up-regulated by osmotic stress and is involved in the renal osmoadaptive response. *J Biol Chem*, **282**, 6644–6652.
23. Hoffert, J.D., Pisitkun, T., Wang, G., Shen, R.F., and Knepper, M.A. (2006) Quantitative phosphoproteomics of vasopressin-sensitive renal cells: regulation of aquaporin-2 phosphorylation at two sites. *Proc Natl Acad Sci U S A*, **103**, 7159–7164.
24. Perkins, D.N., Pappin, D.J., Creasy, D.M., and Cottrell, J.S. (1999) Probability-based protein identification by searching sequence databases using mass spectrometry data. *Electrophoresis*, **20**, 3551–3567.
25. Shevchenko, A., Sunyaev, S., Loboda, A., Shevchenko, A., Bork, P., Ens, W., and Standing, K.G. (2001) Charting the proteomes of organisms with unsequenced genomes by MALDI-quadrupole time-of-flight mass spectrometry and BLAST homology searching. *Anal Chem*, **73**, 1917–1926.

5
Proteomic Analysis of Mesangial Cells

Karl-Friedrich Beck and Josef Pfeilschifter

5.1
The Biology of Glomerular Mesangial Cells

Renal mesangial cells play a key role in the pathogenesis of glomerular renal diseases. These smooth muscle-like pericytes are important to maintain the structure and function of the glomerular capillary tuft [1, 2]. Quiescent mesangial cells are embedded in the glomerular extracellular matrix (ECM). They regulate the glomerular ultrafiltration apparatus and produce and degrade as much ECM as necessary to maintain a functional glomerular structure. In the healthy glomerulus, mesangial cells do not proliferate and their renewal rate is below 1% in a life span [3]. Although cultured mesangial cells respond to serum and growth factors such as platelet-derived growth factor (PDGF) or basic fibroblast growth factor (bFGF) with massive proliferation *in vitro*, the situation *in vivo* is different. In the intact glomerulus, mesangial cell growth is suppressed by a growth-inhibitory action of the glomerular environment, but in an inflammatory setting, the behavior of mesangial cells changes significantly.

A progressive glomerular renal disease has roughly four phases (excellently reviewed by Anders *et al.* [4]). The initiation phase is characterized by an injury of glomerular cells, followed by an influx of neutrophils and macrophages, which produce large amounts of various inflammatory mediators. In the amplification phase, resident glomerular cells – mainly mesangial cells – react with an immense production of inflammatory mediators such as interleukin-1β (IL-1β) and tumor necrosis factor a (TNF-α), subsequently inducing an autocrine inflammatory loop that is followed by the secretion of cytotoxic amounts of reactive oxygen species (ROS) and nitric oxide (NO). In this phase, rapid mesangial cell proliferation is accompanied by an extensive production of ECM. This results in a complex autocrine and paracrine signaling network between different resident cells and monocytes/macrophages, which in turn decides whether the inflamed glomerulus enters the progression phase, followed by an irreversible damage of the glomerular structure and loss of renal function, or whether inflammatory signaling unbridles mesangial cell proliferation and ECM production is stopped, resulting in a resolution phase where the removal of dispensable cells by apoptosis and degradation of

Renal and Urinary Proteomics: Methods and Protocols. Edited by Visith Thongboonkerd

ISBN: 978-3-527-31974-9

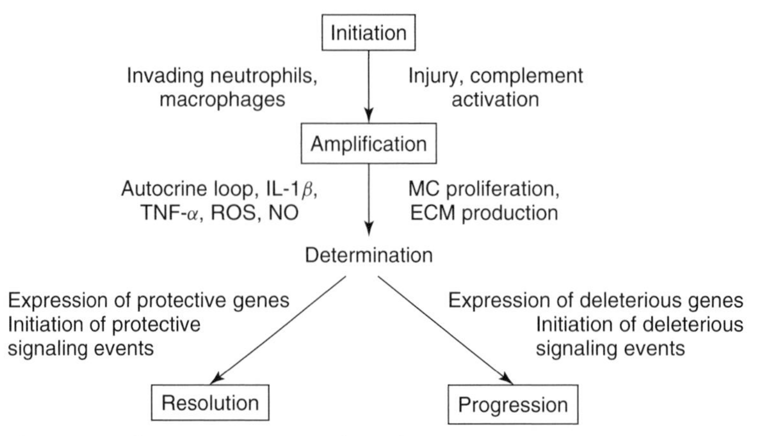

Figure 5.1 Schematic presentation of signaling events occurring in glomerular injury. For further details, see text.

ECM may result in a complete restoration of renal function (Figure 5.1). Naturally, this decision point between organ damage and healing is the most prominent target for therapeutic interventions and, therefore, elucidating the molecular mechanisms involved in these decisive signaling processes is of high scientific interest.

5.2 Nonhypothesis-Driven Approaches to Analyze Gene and Protein Expression in Mesangial Cells

As mentioned above, in particular mesangial cells are regarded as one of the key players in the course of an inflammatory process in the glomerulus [5]. Activated mesangial cells produce inflammatory mediators such as cytokines, chemokines, NO, ROS, and products of the arachidonic acid metabolism [6–10]. *In vivo*, these mediators can act in a complex cross-talk, resulting in a positive inflammatory feedback loop that – if not terminated spontaneously or by therapeutic intervention – results in loss of glomerular function. In this context, it seems worth to make an attempt to identify genes that are expressionally controlled in the course of an inflammatory setting or to characterize posttranslational modifications that trigger deleterious or beneficial signaling events. To elucidate the molecular mechanisms triggered by these complex signaling processes, cultured mesangial cells represent a valuable model. The knowledge of differentially expressed genes or proteins in mesangial cells can be used: (i) to define new biomarkers for glomerular diseases; (ii) to elucidate common molecular mechanisms (e.g., transcription factors, posttranscriptional mechanisms, or posttranslational modifications) that may provide potential therapeutic targets for the treatment of glomerular diseases; and (iii) to develop strategies for a pharmacological or genetic inhibition of a specific deleterious mediator.

During the last two decades, several so-called nonhypothesis-driven approaches have been developed and improved to analyze cellular signaling events and their impact on gene or protein expression in cultured cells or tissues. These techniques include the following: (i) complementary cDNA subtraction also referred to as representational difference analysis or suppression subtractive hybridization (SSH) [11, 12]; (ii) mRNA fingerprinting methods such as differential display RT-PCR (DDRT-PCR) [13], RNA arbitrarily primed PCR (RAP-PCR) [14], or serial analysis of gene expression (SAGE) [15]; and (iii) several microarray techniques [16, 17] and different variants of 2-D protein gel electrophoresis (2-D gel electrophoresis) [18, 19]. The first report that describes the identification of genes by such a nonhypothesis-driven approach in mesangial cells was published in 1997 [20]. Using the DDRT-PCR technique, Holmes *et al.* [20] compared the mRNA patterns of human mesangial cells that were kept under normal or high glucose conditions and demonstrated an up-regulation of mRNAs encoding the α-subunit of prolyl 4-hydroxylase and thrombospondin-1 by high glucose. Subsequent 33 articles further describe the identification of differentially expressed genes by mRNA-based approaches in glomerular mesangial cells (Table 5.1). Most of these studies were performed to detect genes that are related to inflammation, proliferation, or diabetes. Interestingly, the first studies (published during 1997–1999) were performed using DDRT-PCR technique, which was later replaced by the related RAP-PCR method or by representational difference analysis. Since 2004, differences in the mRNA expression pattern in mesangial cells have been extensively analyzed using microarrays, indicating that this valuable technique will be the dominant approach at least for the next couple of years.

5.3 Analysis of the Mesangial Cell Proteome

Among 39 articles dealing with nonhypothesis-driven approaches to analyze gene expression or function in mesangial cells published within the last decade (see Table 5.1), only six manuscripts report a successful application of 2-D gel electrophoresis. In 2003, Kuncewicz *et al.* published a functional proteomics approach in murine mesangial cells [21]. To this end, cytosolic and membrane-associated proteins derived from a murine mesangial cell line were treated with the NO-releasing compound *S*-nitrosoglutathione and the resulting patterns of nitrosated proteins were compared to respective controls. To detect nitrosation of proteins, the authors modified an elegant method, the so-called biotin-switch method [22]. Nitrosated proteins were then analyzed by matrix-assisted laser desorption/ionization time-of-flight mass spectrometry (MALDI-TOF-MS). Using this protocol, a total of 34 nitrosated proteins were identified. The S-nitrosation of three of them, namely, uroguanylin, peroxisome proliferator activated receptor γ (PPAR-γ), and NADPH cytochrome-P450 oxidoreductase could also be demonstrated in intact cells forced with the cytokine IL-1β to endogenously produce high amount of NO. S-nitrosation of cystein-containing proteins by

Table 5.1 Recent reports regarding the analysis of differential gene and protein expression in mesangial cells.

Year	References	Origin	Products identified	Method	Conditions/stimulus
1997	[28]	hMC	β-defensin-1	DDRT-PCR	High glucose
1997	[20]	hMC	Thrombospondin, prolyl 4-hydroxylase α	DDRT-PCR	High glucose (long term)
1998	[29]	rMC	Osteopontin	DDRT-PCR	2-D/3-D culture
1998	[30]	hMC	HGRG-14	DDRT-PCR	High glucose (long term)
1998	[31]	hMC	Hmunc13	DDRT-PCR	High glucose
1999	[32]	hMC	ZNF236	DDRT-PCR	High glucose
1999	[33]	hMC	CTGF	SSH	High glucose
2000	[34]	rMC	HSP90beta	SSH	Thy1.1 GN
2000	[35]	rMC	SPARC	RAP-PCR	IL-1β, NO
2001	[36]	rMC	Growth factors	Microarray	IGA Nephropathie
2001	[37]	rMC	ILK	RAP-PCR	IL-1β, NO
2001	[38]	rMC	MIP-2	RAP-PCR	IL-1β, NO
2002	[39]	hMC	EGF family members	Microarray	Serum
2002	[40]	rMC	CTGF	SSH	NO
2002	[41]	rMC	LIMK-1	RAP-PCR	C6-ceramide
2002	[42]	rMC	P8	Microarray	ET-1
2002	[43]	rMC	MDR1 and others	Microarray	COX-2 overexpression
2002	[44]	hMC	About 200 different genes	SSH	High glucose
2003	**[21]**	**mMC**	**34 S-nitrosated proteins**	**Proteomics**	**NO**
2003	**[23]**	**rMC**	**Mn-SOD**	**Proteomics**	**NO**
2003	[45]	hMC	VDUP-1	Microarray	High glucose
2003	[46]	hMC	AngRem104	SSH	AngII
2003	[47]	hMC	Growth factors	Microarray	ET-1
2003	[48]	hMC	AngRem104	SSH	AngII
2003	[49]	rMC	Biglycan	RAP-PCR	NO
2003	[50]	hMC	33 genes	Microarray	AngRem104 overexpr.
2003	[51]	rMC	Several growth fact	Microarray	S1P vs PDGF
2004	[52]	mMC	RANTES VCAM-1	Microarray	Age
2005	**[24]**	**rMC**		**Proteomics**	**None**
2005	[53]	rMC	Several	Microarray	Anti-Thy glom. PDGF-MC
2005	[54]	hMC	Several profibrotic	Microarray	PDGF, LXA(4)
2005	[55]	mMC	IRF and others	Microarray	LPS
2006	**[25]**	**rMC**	**HSP70, HSP90 Grp94, and others**	**Proteomics**	**Akt activation**
2006	[56]	rMC	VEGF, IL-6	Microarray	Glut1 overexpressing cells

Table 5.1 *(Continued)*

Year	References	Origin	Products identified	Method	Conditions/stimulus
2006	[57]	mMC	389 genes	Microarray	High glucose
2006	[58]	rMC	NGFI-B, Gadd45γ	Microarray	Thy1.1 GN
2006	[59]	hMC	MMP-13, 14, Cytochrome B5, cathepsin L	Microarray	PDGF-BB PDGF-DD
2007	**[26]**	**hMC**	**Enolase 1, Calmodulin, and others**	**Proteomics**	**High glucose (long term)**
2007	[27]	**rMC**	**14-3-3-zeta, Akt substrates**	**Proteomics**	**IGF-1**

Proteomic approaches are displayed in bold.

changing their activity may represent one important mechanism by which NO triggers signaling cascades. Here, the authors provide valuable data regarding new targets of NO in mesangial cells. However, the functional consequence of the observed S-nitrosation has to be analyzed by further experiments.

The analysis of NO-triggered signaling events was also the basis of the first report that described 2-D gel electrophoresis for the analysis of the protein expression pattern in mesangial cells. Here, Keller *et al.* [23] compared the proteome of rat mesangial cells treated with or without (Z)-1-[2-(2-aminoethyl)-*N*-(2-ammonioethyl)amino]-diazen-1-ium-1,2-diolate (DETA-NO), a nitric oxide donor, and they combined 2-D gel electrophoresis and MALDI-TOF-MS to detect a nitric-oxide-dependent up-regulation of the protein for manganese superoxide dismutase (Mn-SOD, also referred to as SOD-2). Further examinations revealed that the up-regulation of Mn-SOD was accompanied by an increase in Mn-SOD activity and that enhanced protein expression and activity was due to an NO-dependent increase in Mn-SOD mRNA. Furthermore, in a rat model of endotoxemia NO-induced Mn-SOD mRNA expression could also be confirmed *in vivo*, indicating a protective role for NO-mediated gene expression.

In 2005, Jiang *et al.* [24] performed 2-D gel electrophoresis to obtain a map of the proteome as well as the phosphoproteome of rat mesangial cells. Proteins derived from primary mesangial cells (passage 4–7) were subjected to 2-D gel electrophoresis. Phastgel Blue R stained spots of interest were excised and analyzed by either MALDI-TOF-MS or liquid chromatography-electrospray ionization-ion trap mass spectrometry (LC-ESI-IT MS). For the analysis of the phosphoproteome, 2-D gels were treated with a fluorescent stain that recognizes solely phosphorylated proteins (Pro-Q Diamond). Using this technique, they identified 118 proteins and 28 phosphoproteins that represent a comprehensive reference map of the mesangial cell proteome. Furthermore, these proteins were annotated regarding their function, physicochemical properties, and subcellular localization. The resulting database will serve as a valuable tool for other groups dealing with 2-D gel electrophoresis in rat mesangial cells.

A further report describing the analysis of the mesangial cell proteome has been published recently [25]. This group was interested in the role of PKB/Akt serine/threonine kinase pathway for the control of mesangial cell proliferation, migration, and apoptosis. Barati *et al.* [25] used two different experimental settings. First, they incubated cell extracts obtained from an immortalized rat mesangial cell line with recombinant Akt in the presence of (γ-^{32}P-ATP). A critical step for these investigations was to determine appropriate urea concentrations that inhibited endogenous protein kinase activity and allowed exogenously administered recombinant Akt activity. Separation of these *in vitro* phosphorylated proteins and comparison of the autoradiograph with the respective Coomassie-stained proteins and peptide mass fingerprinting using MALDI-TOF-MS identified six substrates of Akt even in a 1-D gel electrophoresis or SDS-PAGE approach. For a better resolution, the same protein samples were also subjected to 2-D gel electrophoresis. In this setting, additionally five protein spots representing Akt substrates were identified. Second, mesangial cells were treated with recombinant PDGF, a potent inducer of PKB/Akt signaling pathway, and protein lysates were compared with that of untreated controls by immunoblot analysis using an antibody that recognizes the Akt phospho-motif. Using both the above-mentioned approaches, a total of 20 proteins were characterized as Akt substrates. Remarkably, three of them were found by both, the radioactive *ex vivo* kinase assay as well as the immunoblot approach, indicating a high reliability of both techniques for the analysis of protein kinase substrates. Interestingly, eight candidates found in this study are chaperones. Therefore, the authors performed an *in vitro* assay by incubating recombinant chaperone proteins with Akt and by the analysis of changes in the isoelectric points of proteins in cells stimulated with PDGF. Using these approaches, four chaperones, namely, HSP70, HSP90, Grp94, and PDI, could be verified as the substrates for Akt.

In an attempt to analyze changes in the protein expression pattern by elevated glucose levels Ramachandrarao *et al.* [26] compared the 2-D proteome patterns of cultured human mesangial cells kept under normal (5 mM) and high (30 mM) glucose conditions. To obtain a better resolution of the proteins, this group separated membrane-bound and cytosolic proteins in a similar manner as described by Kuncewicz *et al.* [21] for the analysis of S-nitrosated proteins. To evaluate differentially expressed proteins, the DeCyder software (GE Healthcare) was used. Protein spots were analyzed by surface-enhanced laser desorption/ionization time-of-flight mass spectrometry (SELDI-TOF MS). In this set of experiments, a total of four proteins were identified to be down-regulated in the membrane subproteome, and three proteins were down-regulated, whereas two other proteins were up-regulated in the cytosolic fraction under the high glucose condition. In further experiments, the authors focused on the expression of calmodulin (CaM), one of the proteins that was found to be up-regulated under the high glucose condition in the cytosolic subproteome. Remarkably, they could block TGF-β- and insulin-induced glucose uptake by CaM inhibitors, indicating a role for CaM in diabetes.

A more recent paper describes the use of the 2-D gel electrophoresis approach to identify proteins that were under regulatory control by the insulin-like growth factor (IGF-1)–Akt/PKB signaling pathway [27]. Akt/PKB phosphorylated and, thereby, inactivated glycogen synthase kinase 3β (GSK-3β) resulting in hypertrophy and overexpression of matrix-associated proteins in mesangial cells. To analyze IGF-1/GSK-3β-mediated gene expression, immortalized rat mesangial cells were stimulated with IGF-1 and SB216763, an inhibitor of GSK-3β, and the 2-D protein expression patterns were compared with respective controls. Remarkably, to be able to do a statistical evaluation of their data, the authors used a very redundant experimental setting. First, the experiments were performed in triplicate and second, two different compounds were used to inhibit GSK-3β activity. Protein spots were visualized with Colloidal Coomassie and analyzed using the PDQuest (BioRad) software, and 25 spots were excised by an automatic gel cutter. Identification of the spots was achieved by MALDI-TOF-ms or ms/ms mass spectrometry. Nearly all of the proteins were regulated by IGF-1 and SB216763 – as expected – into the same direction. In a further attempt, the authors focused on expression of the signal adaptor protein 14-3-3ζ that was found to be up-regulated by the inhibition of GSK-3β and that conveyed signals by binding to phosphorylated serine or threonine residues. It is worth noting that 14-3-3ε another protein of this family was found by a functional proteomics approach as a substrate for Akt in mesangial cells [25]. In another study, Singh *et al.* [27] have demonstrated that not only the treatment of mesangial cells with IGF-1 resulted in an up-regulation of the adaptor protein 14-3-3ζ but also 14-3-3ζ was associated with phospho-GSK-3β and overexpression of 14-3-3ζ augmented cell viability, indicating an important role for 14-3-3ζ in IGF-1 signaling.

5.3.1
2-D Gel Electrophoresis for the Investigation of Mesangial Cell Proteome

Overall, these valuable reports clearly document the successful application of 2-D gel electrophoresis for the investigation of mesangial cell biology. These approaches are complementary to the analysis of gene expression, functional proteomics, and the use of 2-D gel electrophoresis to establish the mesangial cell proteome and phosphoproteome maps. Although most of the groups used a similar set of 2-D gel electrophoresis procedures, the preparation of protein extracts was quite variable. Two groups performed the so-called subcellular proteome analysis, in which cytosolic and membrane proteins were prefractionated before 2-D gel electrophoresis. This approach yielded a higher resolution of the proteins and reduced the number of overlapping protein spots. Analysis of the 2-D proteome pattern was achieved by different staining methods. Differentially expressed or posttranslationally modified proteins were detected by using an appropriate computer software, such as PDQuest or Decyder, simply by visually considering the intensity of the spots, or – as demonstrated for the analysis of substrates for Akt – by immunoblotting. The identity of the spots was determined using different mass spectrometric

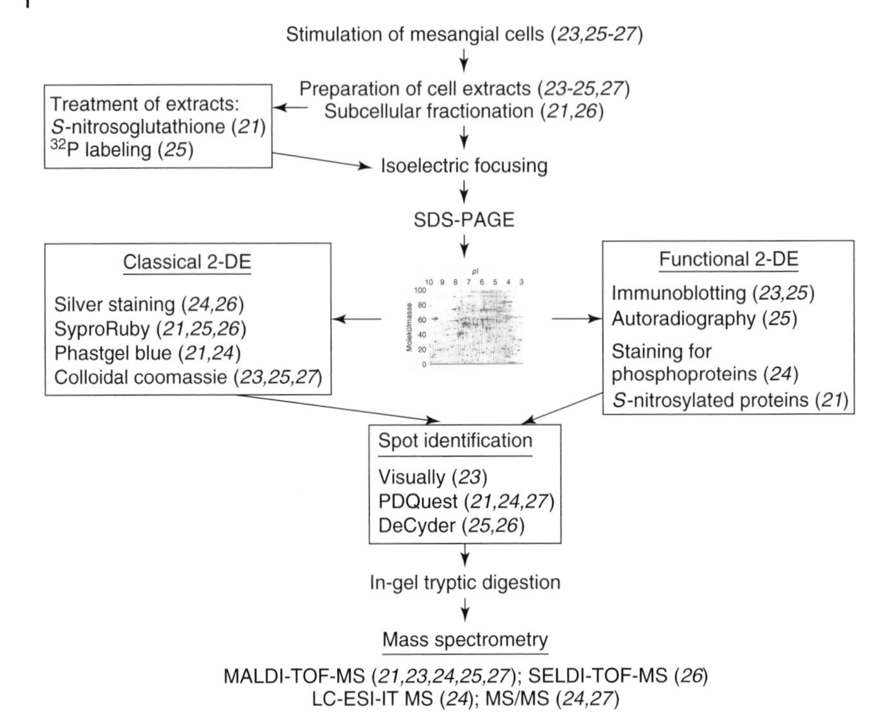

Figure 5.2 Recapitulation of the strategies and methods for 2-D gel electrophoresis used by the itemized working groups for the investigation of mesangial cells. For further details, see text.

approaches such as MALDI-TOF-MS, SELDI-TOF MS, or LC-ESI-IT MS. A flow diagram, which represents the methodological procedures itemized for the different groups, is displayed in Figure 5.2.

5.4 Future Role of 2-D Gel Electrophoresis in the Analysis of Protein Expression and Function in Mesangial Cells

As summarized here, 2-D gel electrophoresis has been successfully applied in order to analyze the mesangial cell proteome. It is worth noting that most reports have described a functional proteomics approach to identify signaling processes that include protein modifications such as S-nitrosation and phosphorylation that subsequently result in changes of basic cellular processes such as proliferation or apoptosis. Considering the development of the 2-D gel electrophoresis approach in the recent years, it turns out that the power of 2-D gel electrophoresis is based more on the ability to analyze protein modifications than simply in the analysis of gene expression. The reason for this evolution may be based on the competition with the analysis of the mRNA expression pattern that had been developed immensely in the last decade with the adoption of the new chip technologies. Most of these modern approaches are less time consuming and less expensive than 2-D gel electrophoresis, and many investigators accept the disadvantage that regulation at the mRNA level may not necessarily imply the regulation of the respective gene products. Furthermore, competition also occurs for the direct analysis of protein expression. The so-called chip-ELISA technique evolves rapidly and allows the simultaneous analysis of hundreds of proteins. However, in comparison to 2-D gel electrophoresis, this method is dependent on the availability and the quality of the antibodies used to cover the chips. During the last two decades, mRNA fingerprinting methods and 2-D gel electrophoresis have been the useful tools not only to identify differentially expressed genes or proteins but also to detect so far unknown genes or gene products. However, the knowledge of the complete genome of humans and most of the species commonly used as laboratory animals will soon be supplemented by the knowledge of the whole proteome and, therefore, hunting for yet unknown genes or gene products will be put aside within the next couple of years. Although mRNA- and protein-based chip technologies are based on known mRNAs/cDNAs and proteins, their high performance will also yield in a series of unexpected or surprising results. Nevertheless, functional 2-D gel electrophoresis as a potent method that enables investigators to directly analyze protein modifications and to detect disease markers without the need of high-quality antibodies will reasonably complement other methods for the analysis of gene and protein expression. In mesangial cells, 2-D gel electrophoresis will help to decipher the complex cross-talk of signaling cascades, which determine whether an inflamed glomerulus will undergo fibrosis with loss of function or resolution of disease and complete healing. Therefore, 2-D gel electrophoresis will eventually help to develop strategies to treat inflammatory glomerular diseases.

Acknowledgments

This work was supported by the Deutsche Forschungsgemeinschaft (SFB 553, GRK 757, FOG 784) and by the EU (grant no. LSHM-CT-2004-005033, EICOSANOX).

References

1. Pfeilschifter, J. (1989) Cross-talk between transmembrane signalling systems: a prerequisite for the delicate regulation of glomerular haemodynamics by mesangial cells. *Eur J Clin Invest*, **19**, 347–361.
2. Schlöndorff, D. (1996) Roles of the mesangium in glomerular function. *Kidney Int*, **49**, 1583–1585.
3. Pabst, R. and Sterzel, R.B. (1983) Cell renewal of glomerular cell types in normal rats. An autoradiographic analysis. *Kidney Int*, **24**, 626–631.
4. Anders, H.J., Vielhauer, V., and Schlöndorff, D. (2003) Chemokines and chemokine receptors are involved in the resolution or progression of renal disease. *Kidney Int*, **63**, 401–415.
5. Pfeilschifter, J. (1994) Mesangial cells orchestrate inflammation in the renal glomerulus. *News Physiol Sci*, **9**, 271–276.

6. Lovett, D.H., Szamel, M., Ryan, J.L., Sterzel, R.B., Gemsa, D., and Resch, K. (1986) Interleukin 1 and the glomerular mesangium. I. Purification and characterization of a mesangial cell-derived autogrowth factor. *J Immunol*, **136**, 3700–3705.
7. Zoja, C., Wang, J.M., Bettoni, S., Sironi, M., Renzi, D., Chiaffarino, F., Abboud, H.E., Van Damme, J., Mantovani, A., and Remuzzi, G. (1991) Interleukin-1 beta and tumor necrosis factor-alpha induce gene expression and production of leukocyte chemotactic factors, colony-stimulating factors, and interleukin-6 in human mesangial cells. *Am J Pathol*, **138**, 991–1003.
8. Pfeilschifter, J. and Schwarzenbach, H. (1990) Interleukin 1 and tumor necrosis factor stimulate cGMP formation in rat renal mesangial cells. *FEBS Lett*, **273**, 185–187.
9. Radeke, H.H., Meier, B., Topley, N., Floege, J., Habermehl, G.G., and Resch, K. (1990) Interleukin 1-alpha and tumor necrosis factor-alpha induce oxygen radical production in mesangial cells. *Kidney Int*, **37**, 767–775.
10. Pfeilschifter, J., Pignat, W., Vosbeck, K., and Marki, F. (1989) Interleukin 1 and tumor necrosis factor synergistically stimulate prostaglandin synthesis and phospholipase A2 release from rat renal mesangial cells. *Biochem Biophys Res Commun*, **159**, 385–394.
11. Hedrick, S.M., Cohen, D.I., Nielsen, E.A., and Davis, M.M. (1984) Isolation of cDNA clones encoding T cell-specific membrane-associated proteins. *Nature*, **308**, 149–153.
12. Hubank, M. and Schatz, D.G. (1994) Identifying differences in mRNA expression by representational difference analysis of cDNA. *Nucleic Acids Res*, **22**, 5640–5648.
13. Liang, P. and Pardee, A.B. (1992) Differential display of eukaryotic messenger RNA by means of the polymerase chain reaction. *Science*, **257**, 967–971.
14. Welsh, J., Chada, K., Dalal, S.S., Cheng, R., Ralph, D., and McClelland, M. (1992) Arbitrarily primed PCR fingerprinting of RNA. *Nucleic Acids Res*, **20**, 4965–4970.
15. Velculescu, V.E., Zhang, L., Vogelstein, B., and Kinzler, K.W. (1995) Serial analysis of gene expression. *Science*, **270**, 484–487.
16. Chalifour, L.E., Fahmy, R., Holder, E.L., Hutchinson, E.W., Osterland, C.K., Schipper, H.M., and Wang, E. (1994) A method for analysis of gene expression patterns. *Anal Biochem*, **216**, 299–304.
17. Schena, M., Shalon, D., Davis, R.W., and Brown, P.O. (1995) Quantitative monitoring of gene expression patterns with a complementary DNA microarray. *Science*, **270**, 467–470.
18. Unlu, M., Morgan, M.E., and Minden, J.S. (1997) Difference gel electrophoresis: a single gel method for detecting changes in protein extracts. *Electrophoresis*, **18**, 2071–2077.
19. Wilkins, M.R., Pasquali, C., Appel, R.D., Ou, K., Golaz, O., Sanchez, J.C., Yan, J.X., Gooley, A.A., Hughes, G., Humphery-Smith, I., Williams, K.L., and Hochstrasser, D.F. (1996) From proteins to proteomes: large scale protein identification by two-dimensional electrophoresis and amino acid analysis. *Biotechnology (NY)*, **14**, 61–65.
20. Holmes, D.I., Abdel Wahab, N., and Mason, R.M. (1997) Identification of glucose-regulated genes in human mesangial cells by mRNA differential display. *Biochem Biophys Res Commun*, **238**, 179–184.
21. Kuncewicz, T., Sheta, E.A., Goldknopf, I.L., and Kone, B.C. (2003) Proteomic analysis of S-nitrosylated proteins in mesangial cells. *Mol Cell Proteomics*, **2**, 156–163.
22. Jaffrey, S.R., Erdjument-Bromage, H., Ferris, C.D., Tempst, P., and Snyder, S.H. (2001) Protein S-nitrosylation: a physiological signal for neuronal nitric oxide. *Nat Cell Biol*, **3**, 193–197.

23. Keller, T., Pleskova, M., McDonald, M.C., Thiemermann, C., Pfeilschifter, J., and Beck, K.F. (2003) Identification of manganese superoxide dismutase as a NO-regulated gene in rat glomerular mesangial cells by 2D gel electrophoresis. *Nitric Oxide*, **9**, 183–193.
24. Jiang, X.S., Tang, L.Y., Cao, X.J., Zhou, H., Xia, Q.C., Wu, J.R., and Zeng, R. (2005) Two-dimensional gel electrophoresis maps of the proteome and phosphoproteome of primitively cultured rat mesangial cells. *Electrophoresis*, **26**, 4540–4562.
25. Barati, M.T., Rane, M.J., Klein, J.B., and McLeish, K.R. (2006) A proteomic screen identified stress-induced chaperone proteins as targets of Akt phosphorylation in mesangial cells. *J Proteome Res*, **5**, 1636–1646.
26. Ramachandrarao, S.P., Wassell, R., Shaw, M.A., and Sharma, K. (2007) Profiling of human mesangial cell sub-proteomes reveals a role for calmodulin in glucose uptake. *Am J Physiol Renal Physiol*, **292**, F1182–F1189.
27. Singh, L.P., Jiang, Y., and Cheng, D.W. (2006) Proteomic identification of 14-3-3ζ as an adapter for IGF-1 and Akt/GSK-3β signaling and survival of renal mesangial cells. *Int J Biol Sci*, **3**, 27–39.
28. Page, R., Morris, C., von Williams, J., Ruhland, C., and Malik, A.N. (1997) Isolation of diabetes-associated kidney genes using differential display. *Biochem Biophys Res Commun*, **232**, 49–53.
29. Pröls, F., Loser, B., and Marx, M. (1998) Differential expression of osteopontin, PC4, and CEC5, a novel mRNA species, during in vitro angiogenesis. *Exp Cell Res*, **239**, 1–10.
30. Abdel Wahab, N., Gibbs, J., and Mason, R.M. (1998) Regulation of gene expression by alternative polyadenylation and mRNA instability in hyperglycaemic mesangial cells. *Biochem J*, **336**, 405–411.
31. Song, Y., Ailenberg, M., and Silverman, M. (1998) Cloning of a novel gene in the human kidney homologous to rat munc13s: its potential role in diabetic nephropathy. *Kidney Int*, **53**, 1689–1695.
32. Holmes, D.I., Wahab, N.A., and Mason, R.M. (1999) Cloning and characterization of ZNF236, a glucose-regulated Kruppel-like zinc-finger gene mapping to human chromosome 18q22-q23. *Genomics*, **60**, 105–109.
33. Murphy, M., Godson, C., Cannon, S., Kato, S., Mackenzie, H.S., Martin, F., and Brady, H.R. (1999) Suppression subtractive hybridization identifies high glucose levels as a stimulus for expression of connective tissue growth factor and other genes in human mesangial cells. *J Biol Chem*, **274**, 5830–5834.
34. Pieper, M., Rupprecht, H.D., Bruch, K.M., De Heer, E., and Schöcklmann, H.O. (2000) Requirement of heat shock protein 90 in mesangial cell mitogenesis. *Kidney Int*, **58**, 2377–2389.
35. Walpen, S., Beck, K.F., Eberhardt, W., Apel, M., Chatterjee, P.K., Wray, G.M., Thiemermann, C., and Pfeilschifter, J. (2000) Downregulation of SPARC expression is mediated by nitric oxide in rat mesangial cells and during endotoxemia in the rat. *J Am Soc Nephrol*, **11**, 468–476.
36. Katsuma, S., Nishi, K., Tanigawara, K., Ikawa, H., Shiojima, S., Takagaki, K., Kaminishi, Y., Suzuki, Y., Hirasawa, A., Ohgi, T., Yano, J., Murakami, Y., and Tsujimoto, G. (2001) Molecular monitoring of bleomycin-induced pulmonary fibrosis by cDNA microarray-based gene expression profiling. *Biochem Biophys Res Commun*, **288**, 747–751.
37. Beck, K.F., Walpen, S., Eberhardt, W., and Pfeilschifter, J. (2001) Downregulation of integrin-linked kinase mRNA expression by nitric oxide in rat glomerular mesangial cells. *Life Sci*, **69**, 2945–2955.
38. Walpen, S., Beck, K.F., Schaefer, L., Raslik, I., Eberhardt, W.,

Schaefer, R.M., and Pfeilschifter, J. (2001) Nitric oxide induces MIP-2 transcription in rat renal mesangial cells and in a rat model of glomerulonephritis. *FASEB J*, **15**, 571–573.

39. Mishra, R., Leahy, P., and Simonson, M.S. (2002) Gene expression profiling reveals role for EGF-family ligands in mesangial cell proliferation. *Am J Physiol Renal Physiol*, **283**, F1151–F1159.
40. Keil, A., Blom, I.E., Goldschmeding, R., and Rupprecht, H.D. (2002) Nitric oxide down-regulates connective tissue growth factor in rat mesangial cells. *Kidney Int*, **62**, 401–411.
41. Shabahang, S., Liu, Y.H., Huwiler, A., and Pfeilschifter, J. (2002) Identification of the LIM kinase-1 as a ceramide-regulated gene in renal mesangial cells. *Biochem Biophys Res Commun*, **298**, 408–413.
42. Goruppi, S., Bonventre, J.V., and Kyriakis, J.M. (2002) Signaling pathways and late-onset gene induction associated with renal mesangial cell hypertrophy. *EMBO J*, **21**, 5427–5436.
43. Patel, V.A., Dunn, M.J., and Sorokin, A. (2002) Regulation of MDR-1 (P-glycoprotein) by cyclooxygenase-2. *J Biol Chem*, **277**, 38915–38920.
44. Clarkson, M.R., Murphy, M., Gupta, S., Lambe, T., Mackenzie, H.S., Godson, C., Martin, F., and Brady, H.R. (2002) High glucose-altered gene expression in mesangial cells. Actin-regulatory protein gene expression is triggered by oxidative stress and cytoskeletal disassembly. *J Biol Chem*, **277**, 9707–9712.
45. Kobayashi, T., Uehara, S., Ikeda, T., Itadani, H., and Kotani, H. (2003) Vitamin D3 up-regulated protein-1 regulates collagen expression in mesangial cells. *Kidney Int*, **64**, 1632–1642.
46. Liang, X., Zhang, H., Zhou, A., Hou, P., and Wang, H. (2003) Screening and identification of the up-regulated genes in human mesangial cells exposed to angiotensin II. *Hypertens Res*, **26**, 225–235.
47. Mishra, R., Leahy, P., and Simonson, M.S. (2003) Gene expression profile of endothelin-1-induced growth in glomerular mesangial cells. *Am J Physiol Cell Physiol*, **285**, C1109–C1115.
48. Liang, X., Zhang, H., Zhou, A., and Wang, H. (2003) AngRem104, an angiotensin II-induced novel upregulated gene in human mesangial cells, is potentially involved in the regulation of fibronectin expression. *J Am Soc Nephrol*, **14**, 1443–1451.
49. Schaefer, L., Beck, K.F., Raslik, I., Walpen, S., Mihalik, D., Micegova, M., Macakova, K., Schonherr, E., Seidler, D.G., Varga, G., Schaefer, R.M., Kresse, H., and Pfeilschifter, J. (2003) Biglycan, a nitric oxide-regulated gene, affects adhesion, growth, and survival of mesangial cells. *J Biol Chem*, **278**, 26227–26237.
50. Liang, X.B., Zhang, H., Zhou, A.Y., and Wang, H.Y. (2003) Screening for functional genes related to a novel gene, AngRem104, in human mesangial cells by cDNA microarray. *Biotechnol Lett*, **25**, 139–142.
51. Katsuma, S., Hada, Y., Shiojima, S., Hirasawa, A., Tanoue, A., Takagaki, K., Ohgi, T., Yano, J., and Tsujimoto, G. (2003) Transcriptional profiling of gene expression patterns during sphingosine 1-phosphate-induced mesangial cell proliferation. *Biochem Biophys Res Commun*, **300**, 577–584.
52. Zheng, F., Cheng, Q.L., Plati, A.R., Ye, S.Q., Berho, M., Banerjee, A., Potier, M., Jaimes, E.A., Yu, H., Guan, Y.F., Hao, C.M., Striker, L.J., and Striker, G.E. (2004) The glomerulosclerosis of aging in females: contribution of the proinflammatory mesangial cell phenotype to macrophage infiltration. *Am J Pathol*, **165**, 1789–1798.
53. Sadlier, D.M., Ouyang, X., McMahon, B., Mu, W., Ohashi, R., Rodgers, K., Murray, D., Nakagawa, T., Godson, C., Doran, P., Brady, H.R., and Johnson, R.J. (2005) Microarray and bioinformatic detection of novel and established genes expressed in experimental anti-Thy1 nephritis. *Kidney Int*, **68**, 2542–2561.

54. Rodgers, K., McMahon, B., Mitchell, D., Sadlier, D., and Godson, C. (2005) Lipoxin A4 modifies platelet-derived growth factor-induced pro-fibrotic gene expression in human renal mesangial cells. *Am J Pathol*, **167**, 683–694.

55. Fu, Y., Xie, C., Yan, M., Li, Q., Joh, J.W., Lu, C., and Mohan, C. (2005) The lipopolysaccharide-triggered mesangial transcriptome: Evaluating the role of interferon regulatory factor-1. *Kidney Int*, **67**, 1350–1361.

56. Pfafflin, A., Brodbeck, K., Heilig, C.W., Haring, H.U., Schleicher, E.D., and Weigert, C. (2006) Increased glucose uptake and metabolism in mesangial cells overexpressing glucose transporter 1 increases interleukin-6 and vascular endothelial growth factor production: role of AP-1 and HIF-1alpha. *Cell Physiol Biochem*, **18**, 199–210.

57. Cheng, D.W., Jiang, Y., Shalev, A., Kowluru, R., Crook, E.D., and Singh, L.P. (2006) An analysis of high glucose and glucosamine-induced gene expression and oxidative stress in renal mesangial cells. *Arch Physiol Biochem*, **112**, 189–218.

58. Xu, J.H., Qiu, W., Wang, Y.W., Xu, J., Tong, J.X., Gao, L.J., Xu, W.H., and Wu, Y.Q. (2006) Gene expression profile and overexpression of apoptosis-related genes (NGFI-B and Gadd 45 gamma) in early phase of Thy-1 nephritis model. *Cell Tissue Res*, **326**, 159–168.

59. van Roeyen, C.R., Ostendorf, T., Denecke, B., Bokemeyer, D., Behrmann, I., Strutz, F., Lichenstein, H.S., LaRochelle, W.J., Pena, C.E., Chaudhuri, A., and Floege, J. (2006) Biological responses to PDGF-BB versus PDGF-DD in human mesangial cells. *Kidney Int*, **69**, 1393–1402.

6
Proteomic Analysis of Podocytes

Maria Pia Rastaldi and Harry Holthofer

6.1
Introduction

The glomerular podocyte is a highly differentiated cell provided with extensive ramifications. From the cell body that bulges into the Bowman's urinary space of the glomerulus, primary processes extend and further ramify into secondary (foot) processes. Processes coming from a podocyte and its neighbors intertwine to completely envelop the glomerular capillary (Figure 6.1) and, in doing this, they adhere on the basal side to the glomerular basement membrane and laterally interconnect among themselves by the specialized adhesion known as the slit diaphragm [1].

Until 1998, namely until a milestone advance was made with the discovery of nephrin [2], our knowledge of podocyte molecular and biochemical properties was very limited and the podocyte was mainly considered as a passive bystander to the events undergoing in the glomerulus. In recent years, thanks to the enormous progresses of molecular biology, genetics, and bioinformatics, our view of podocytes has completely changed and these cells are now regarded as the most important players in the maintenance of the glomerular filtration barrier. The absence or dysfunction of single podocyte proteins is, in fact, sufficient to determine important protein loss into the urine in animal models and has been demonstrated as causative of familial and sporadic nephrotic syndrome in humans [3–5]. Furthermore, we know that podocytes intervene either primarily or secondarily in the pathogenesis of most if not all proteinuric conditions, their role extending beyond primary glomerulonephritis. It follows that a complete knowledge of podocyte biology is mandatory to understand the mechanisms underlying glomerular diseases.

Among the technological advances that nowadays are enriching our investigative approach, proteomics, the study of the complete set of proteins, is not only powerful but also indispensable and complementary to genomics (the complete set of genes), transcriptomics (the complete set of mRNA molecules), and metabolomics (the low molecular weight intermediates) in trying to reach a whole comprehension of cell and organ biology.

Renal and Urinary Proteomics: Methods and Protocols. Edited by Visith Thongboonkerd

ISBN: 978-3-527-31974-9

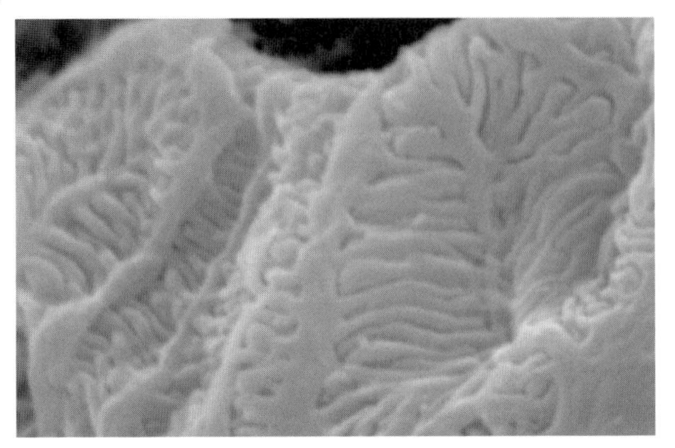

Figure 6.1 Scanning electron microscopy shows primary podocyte processes and their ramification in foot processes that interdigitate to cover the glomerular capillary.

As the field is expanding rapidly, it is important to exactly define proteomic applications. The term global proteomics is used when talking about research of all of the proteins encoded by a genome, targeted proteomics refers to the study of protein expression within a specific group of proteins or selected cell type, and cell-mapping proteomics concerns the use of proteomics to study protein–protein interactions.

6.2
Targeted Podocyte Proteomics

In the recent years, we have contributed toward the exponential development of efforts aimed at studying the kidney and the urinary proteome. Linked to the HUPO initiative, the database of the glomerular proteome is available online (http://www.hkupp.org) and is providing an enormous amount of information about glomerular protein contents available at this stage, while the database is continuously expanding. The database, which contains almost 7000 entries obtained from 1-D and 2-D gel analyses of human glomeruli, is very user friendly and queries can also be made by protein functional classification.

Indeed, only a very limited number of studies are strictly focused on podocytes. The complexity of the podocyte cell structure and its location in the renal glomerulus still limit the possibility of obtaining sufficient amounts of cells for proteomic analysis (Table 6.1). A PubMed search for "podocyte" and "proteomic" or "podocyte" and "mass spectrometry" returned less than 20 publications at the time of this study, among them only two specifically dealing with targeted podocyte proteomics [6, 7].

The authors of both the above-mentioned reports have utilized 2-D gel electrophoresis and matrix-assisted laser desorption/ionization time-of-flight (MALDI-TOF) analysis on conditionally immortalized cell cultures of mouse [6] and human [7] origin. The availability of conditionally immortalized cell lines

Table 6.1 Starting material in podocyte proteomics: advantages and disadvantages.

Starting material	Advantages	Disadvantages
Conditionally immortalized cell lines	Adequate quantities of starting material can be obtained any time.	These cells are different from the *in vivo* podocyte.
Primary podocyte cell cultures	Theoretically, they show more proximity to the *in vivo* phenotype.	It is not always possible to get adequate quantity of starting material. Cells have to be isolated from the animal each time and each preparation can be very different from the other, depending on the method used and the experience of the operator.
Glomerular extracts	Cells are studied in their physiological location.	The material necessarily comprises all the other glomerular components.
Podocytes obtained by laser capture microdissection or by micromanipulation of isolated glomeruli	Cells are taken from their physiological location.	Scarcity of the starting material. The cell body is present, but the processes are under-represented. The purity of the preparation is highly operator dependent.

has been a revolutionary tool in the study of podocytes: these cells are, in fact, temperature (and IFN-γ) sensitive and are able to proliferate in permissive conditions (33 °C + IFN-γ), whereas they enter a distinct differentiation pathway when cultivated in nonpermissive conditions, when the temperature of 37 °C and the absence of IFN-γ inactivate the SV40 large T-antigen [8]. The present availability of these cells that makes feasible the conduct of an enormous amount of experiments that could not be practically exploited on primary podocyte cell cultures, because of the limited capacity of the podocyte to proliferate *in vitro* and on the differences resulting from different methods of preparation [8, 9].

Ransom *et al.* [6] have used proteomic analysis to study molecules induced by glucocorticoids and identified by mass spectrometry 92 protein spots in cultured murine podocytes. Among them, the authors focused on six molecules whose expression was changed by dexamethasone and, especially, they were able to detect and confirm by Western blotting the differential expression of ciliary neurotrophic factor (CNTF), αB-crystallin, and heat shock protein 27 (HSP27), three molecules with a known role in protecting cells from injury.

More recently, Viney *et al.* have provided for the first time an example of a human podocyte proteome map, where 75 proteins were identified by MALDI-TOF, of which 43% were associated with the cytoskeleton, whereas the remaining belonged to protein folding and general metabolism, and a minority involved in secretory pathways, protein synthesis, redox regulation, RNA regulation, ion transport, and cell proliferation [7]. The authors have also examined differences between wild-type human podocytes and WT1-mutant podocytes derived from a patient affected by Denys-Drash syndrome. To this purpose, they applied the differential gel electrophoresis (DIGE) technique, in which proteins to be compared are labeled by different *N*-hydroxy succinimyl ester derivatives of the cyanine (Cy) fluorescent dyes, namely, Cy2, Cy3, and Cy5. With this approach, they identified several proteins involved in the regulation of the cytoskeletal architecture, which were differentially expressed in mutant podocytes.

Although these first results are of fundamental importance and pave the way for successive podocyte analysis, it has become increasingly clear that the cultured podocytes suffer from limited resemblance with their *in vivo* counterparts. It is also worth noting that in both cases no specific podocyte proteins, particularly nephrin, podocin, or synaptopodin presently considered as key players in the maintenance of the glomerular filtration barrier, have emerged from the analysis.

6.3
Podocyte-Mapping Proteomics

Providing the possibility of rapidly identifying proteins by peptide mass fingerprinting, proteomic techniques have also revolutionized the research aimed at discovering protein–protein interactions. In this area, the number of studies focusing on podocyte proteins has been increasing from the beginning (date back to 1999), when the group of Salant used mass spectrometry to identify nephrin as the target of the nephritogenic antibody mAb 5-1-6 [10].

Protein interaction experiments typically start with protein precipitation, by immunoprecipitation or pull-down assays. Thereafter the resulting protein complex, which in the past was mainly characterized by Western blotting, can now be rapidly identified by mass spectrometry and peptide mass fingerprinting. The advantage of this technique in the study of the podocyte is demonstrated by several important results obtained during the last years.

Ahola *et al.* [11], after immuno-precipitating nephrin from rat and human glomerular protein lysates, separated the samples by 1-D gel electrophoresis and analyzed an excised 200-kD band by MALDI-TOF MS. Subsequent peptide mass fingerprinting led to the identification of densin as a protein that colocalizes with nephrin, as confirmed by immunogold electron microscopy.

Pull-down assays with GST-nephrin, followed by SDS-PAGE and mass spectrometry, were used by Liu *et al.* [12] and Lehtonen *et al.* [13, 14] to identify several other proteins interacting with nephrin, in the attempt to better understand the precise function of nephrin at the slit diaphragm. From these studies, it emerged that nephrin intracellular domain interacts with membrane-associated guanylate kinase inverted2/synaptic scaffolding molecule (MAGI-2/S-SCAM), IQ motif-containing

GTPase-activating protein1 (IQGAP1), calcium/calmodulin-dependent serine protein kinase (CASK), α-actinin, αII spectrin, βII spectrin, p120 catenin, ZO-1, and CD2AP, all of which are scaffolding proteins associated with cell junctions.

Following a line of research that pursues similarities between podocytes and neurons, our group initially detected the presence in podocytes of the small GTPase Rab3A, a molecule whose expression is restricted to cells capable of processes of highly regulated exocytosis, thereby most abundantly found in neurons and neuroendocrine cells [15]. Its presence in podocytes raised a number of questions while suggesting fundamental similarities between these two morphologically similar, complex cells, warranting a new research line by exploiting proteomic techniques to reveal specific Rab3A interacting proteins in the podocytes. Therefore, we immunoprecipitated Rab3A from mouse glomeruli and primary podocytes and also from mouse brain lysates for comparison. From this analysis, it appeared that in podocytes Rab3A is associated with other synaptic proteins and this led to the discovery that podocytes contain synaptic-like vesicles able to undergo processes of spontaneous and regulated exocytosis with glutamate release [16]. Furthermore, peptide mass fingerprinting identified the mitochondrial molecule ATP5b (Figure 6.2) (unpublished data). This could be very important, especially considering the need of energy required by exocytosis processes and this main point remains to be studied in podocytes. A preliminary screening aimed at verifying the possible relationship of Rab3A with mitochondria has been then conducted in our laboratory by using indirect immunofluorescence on primary podocytes. The double staining with Rab3A and a mitochondrial marker revealed several spots of colocalization of the two antibodies, mainly in the cell body around the nucleus, but especially interesting because the two markers colocalize in small dots also along podocyte processes, where the images suggest the proximity of mitochondria to exocytotic vesicles (Figure 6.3). Further studies are ongoing to provide the kinetic data of the possible interaction of Rab3A and specific elements of the mitochondrion in podocytes as well as in the neurons.

Mass spectrometry has also become an essential tool in identifying posttranslational modifications and their location in proteins. Several mass-spectrometry-based approaches in combination with enzymatic pretreatment of proteins have been developed, and very recently they have been applied to identify nephrin N-linked glycosylation sites [17].

6.4 Methods and Protocols

6.4.1 Primary Podocyte Cell Culture

Kidneys are taken from mice aged between 7 and 10 days and are immediately immersed in a culture medium, which is DME: F12 medium supplemented with 10% FCS, 5 $\mu g\,ml^{-1}$ transferrin, 10^{-7} M hydrocortisone, 5 $ng\,ml^{-1}$ sodium selenite,

Version 4.10.5

ProFound - Search Result Summary — The Rockefeller University Edition

Protein Candidates for search B8B3460E-04E0-A94FD6CE [79362 sequences searched]							
Rank	Probability	Est'd Z	Protein Information and Sequence Analyse Tools (T)	%	pI	kDa	®
+1	1.0e+000	2.26	T gi|28302366|gb|AAH46616.1| ATP synthase, H+ transporting mitochondrial F1 complex, beta subnit [Mus musculus]	24	5.2	56.28	®

Details for rank 1 candidate in search B8BF46DE-04D4A95BD6DA

gi|28302366|gb|AAH46616.1| ATP synthase, H+ transporting mitochondrial F1 complex, beta subunit [Mus musculus]

gi|20455479|sp|P56480|ATPB_MOUSE ATP synthase beta chain, mitochondrial precursor

gi|26350917|dbj|BAC39095.1| unnamed protein product [Mus musculus]

gi|31980648|ref|NP_058054.2| ATP synthase, H+ transporting mitochondrial F1 complex, beta subunit; ATP synthase, H+ transporting mitochondrial F1 complex, alpha subunit [Mus musculus]

gi|12845667|dbj|BAB26846.1|unnamed protein product [Mus musculus]

Sample ID : [Pass : 0]
Measured peptides : 25
Matched peptides : 10
Min. sequence coverage : 28%

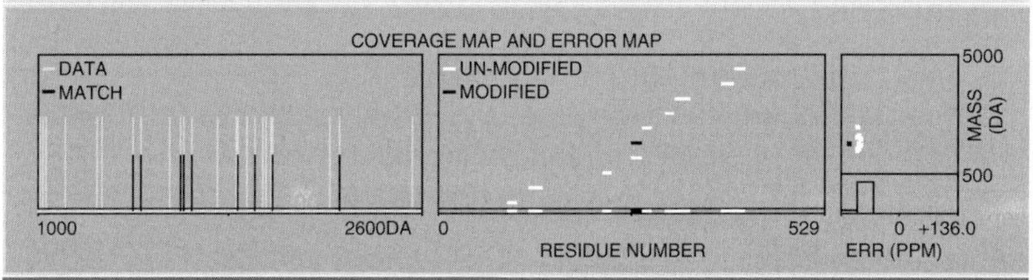

COVERAGE MAP AND ERROR MAP

DATA
MATCH
1000
2600DA

UN-MODIFIED
MODIFIED
0
529
RESIDUE NUMBER

5000
MASS (DA)
500
0 +136.0
ERR (PPM)

Measured Mass (M)	Avg/ Mono	Computed Mass	Error (ppm)	Residues Start	Residues To	Missed Cut	Peptide sequence
1405.539	M	1405.673	−96	226	239	0	AHGGYSVFAGVGER
1434.614	M	1434.746	−92	311	324	0	FTQAGSEVSALLGR
1438.642	M	1438.781	−97	282	294	0	VALTGLTVAEYFR
1600.658	M	1600.802	−90	265	279	0	VALVYGQMNEPPGAR
1616.614	M	1616.797	−114	265	279	0	VALVYGQMNEPPGAR (1)+ 0@M;
1649.757	M	1649.909	−92	95	109	0	LVLEVAQHLGESTVR
1841.710	M	1841.872	−88	407	422	0	IMDPNIVGNEHYDVAR
1917.905	M	1918.088	−95	125	143	1	VLDSGAPIKIPVGPETLGR
1986.844	M	1987.025	−91	388	406	0	AIAELGIYPAVDPLDSTSR
2264.856	M	2265.075	−97	325	345	0	IPSAVGYQPTLATDMGTMQER

Figure 6.2 Results obtained by the ProFound analysis of peptide masses.

0.12 U ml^{-1} insulin, 100 U ml^{-1} penicillin, 100 μg ml^{-1} streptomycin, and 2 mM L-glutamine. They have to be decapsulated before passing through a series of sieves of decreasing mesh size, that is, 100, 75, 50, and 36 μm. Glomeruli are collected from the top of the 50- and 36-μm sieves, and centrifuged at 280g for 10 minutes at 4 °C. The pellet is resuspended in the medium and further purified manually under a stereomicroscope by removing debris by hand-pipetting. Then, the glomeruli are seeded at 37 °C with 5% CO_2 humidified atmosphere in culture flasks precoated with collagen type IV. After 3–5 days, primary podocytes begin to grow and between 7 and 10 days the glomeruli are ready to be detached by treatment with trypsin-EDTA, followed by filtering through the 36-μm mesh sieve to separate the glomeruli from cells. Part of these second passage podocytes, which should result >90% purity by observation under an inverted microscope, can be seeded on thermanox-plastic-collagen-coated coverslips to be characterized by

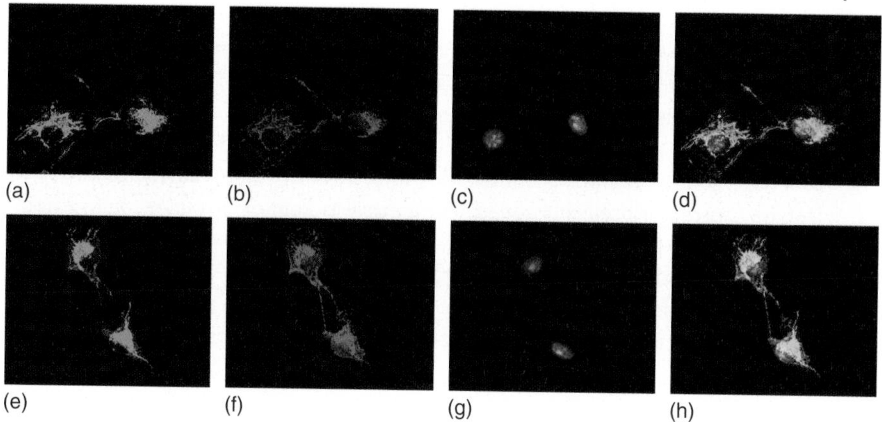

Figure 6.3 Primary cultured podocytes double stained for the mitochondrial marker using mouse monoclonal antibody to mitochondrial MTCO2 (Abcam, Cambridge, UK) (a and e) and the polyclonal rabbit anti-Rab3A (Synaptic System, Goettingen, Germany) (b and f). Nuclear counterstaining by DAPI (Sigma-Aldrich, Milan, Italy) is shown (c and g). (d and h) are obtained by merging and show colocalization in yellow. Interestingly, yellow dots are present not only in the cell body, but also along the cell processes (indirect immunofluorescence, 400×).

immunocytochemistry (we use nephrin, podocin, cytokeratin, CD31, α-SMA, and CD45).

6.4.2 Protein Extraction

Tissue samples are added with modified Ripa buffer (10 mM Tris-HCl pH 7.5, 10 mM EDTA, 0.5% Nonidet P-40, 0.5% DOC, 0.1% SDS) and a protease inhibitor cocktail. Brain tissue is homogenized manually with a pestle. The glomeruli are sonicated with multiple short bursts maintaining the sample immersed in an ice bath to prevent excessive heating. Primary podocytes are washed thrice with PBS, then added with 1× lysis buffer and scrubbed from the flask. Lysis buffer stock solution (2×) contains 2% Triton ×100, 150 mM NaCl, 100 mM Hepes pH 7.4, 2 mM EGTA, 2 mM Na_3VO_4, and protease inhibitor cocktail. All samples are transferred to 1.5-ml tubes, where they are repeatedly mixed by pipetting and incubated on ice for 10–20 minutes. Samples are then centrifuged at 13 000 rpm at 4 °C for 10 minutes. The supernatant is transferred to a new 1.5-ml tube and can be frozen at −20 °C until use.

6.4.3 Immunoprecipitation

For each sample, three tubes are prepared: one for the specific antibody, one for the isotype control, and one without primary antibody. Protein G beads immobilized on agarose are dissolved in ethanol. The aliquot needed for the experiment is washed thrice with 1× lysis buffer, and each time the samples are centrifuged at 5000 rpm for 3–5 minutes to discard the supernatant. Totally 400 μg protein lysates are precleared with 30 μl protein G beads immobilized on agarose in 500 μl lysis buffer for 1 hour at 4 °C and then centrifuged at 13 000 rpm for 10 minutes at 4 °C. The supernatant is transferred into a 1.5-ml tube and the antibody (or isotype control, or buffer) is added. The sample is kept on a rotator for 1–2 hours at 4 °C, then 30 μl beads are added, and the mixture is incubated on a rotator for 1 hour at 4 °C and centrifuged at 13 000 for 10 minutes at 4 °C. The supernatant is discarded and the pellet washed thrice with 1 ml lysis buffer (1×). Samples can be kept at −20 °C until use. Western blot analysis will confirm the specificity of the procedure.

6.4.4 Sample Preparation before MALDI-TOF MS Analysis

Protein concentration of the immunoprecipitate is measured by the Bradford method. Totally 20 μg protein samples are loaded on a 10% SDS-PAGE gel and silver-stained, according to the method published by O'Connell and Stults [18]. Selected protein bands are excised from the gel and cut into small pieces. The stain is removed with 15 mM potassium ferricyanide in 50 mM sodium thiosulfate, followed by washing with water, shrinking with acetonitrile (ACN), and drying in a vacuum centrifuge. Proteins are then reduced with 20 mM dithiothreitol in 0.1 mM NH4HCO3 for 15 minutes at room temperature, washed thrice with water with ACN shrinking in between, followed by final ACN shrinking and drying in vacuum. In-gel trypsin digestion is performed with 0.05 μg μl^{-1} sequencing-grade modified trypsin in 10 mM NH4CO3/10%ACN on an ice bath for 10 minutes, followed by overnight incubation at 37 °C. The digested peptides are extracted with 0.1%TFA/60%ACN. At this point, given the limited amount of material recovered, we find very useful to preconcentrate the sample to attain the highest possible sensitivity [19]. We use the ZipTip tool (Millipore) that consists of a pipette tip filled with a reversed-phase resin (bed volume of about 1 μl).

6.4.5 MALDI-TOF MS Analysis

We conduct peptide mass fingerprinting for the extracted peptides with a Biflex MALDI-TOF mass spectrometry (2-GHz digitizer; Bruker, Rheinstetten, Germany) at the Protein Chemistry Unit of the Biomedicum in Helsinki. Positive ion reflector mode is used with an accelerating voltage of 19 000 V and delayed extraction of 2 nanoseconds. Internal peptide calibration standards (Bruker Daltons, Bremen, Germany) are applied to obtain higher peptide mass accuracy. Results are then analyzed by the ProFound search engine: http://bioinformatics.genomicsolutions.com/prowl/profound/profound_E_adv.html. Very important, along all procedures, is the issue of contamination, especially from the operator (skin and hair). To avoid it, precautions have to be taken, such as the frequent change of gloves and wearing surgical mask and bonnet.

6.5 Perspectives in Podocyte Proteomics

The usefulness of proteomic methods in studying podocyte biology is evident, and the research on this field has just started. Understanding the podocyte proteome will bridge the gap still existing in expression and genetic studies and will provide the necessary information to understand podocyte biology and pathology. It is expected that developing better cultured podocytes, more strictly reflecting the functions of *in vivo* terminally differentiated cells, will rapidly advance our understanding of this key cell of the glomerular filtration barrier. Technological advances, such as the already applicable mass-spectrometry-based analysis directly on tissue sections [20] and the use of much less quantities of starting material for proteomic analysis, will soon overcome the difficulties that have slowed down proteomic application in podocyte research and will lead the transfer of proteomics from podocyte research to podocyte disease diagnosis.

References

1. Kerjaschki, D. (2001) Caught flat-footed: podocyte damage and the molecular bases of focal glomerulosclerosis. *J Clin Invest*, **108**, 1583–1587.
2. Kestila, M., Lenkkeri, U., Mannikko, M., Lamerdin, J., McCready, P., Putaala, H., Ruotsalainen, V., Morita, T. *et al.* (1998) Positionally cloned gene for a novel glomerular protein – nephrin – is mutated in congenital nephrotic syndrome. *Mol Cell*, **1**, 575–582.
3. Patari-Sampo, A., Ihalmo, P., and Holthofer, H. (2006) Molecular basir of the glomerular filtration: nephrin and the emerging protein complex at the podocyte slit diaphragm. *Ann Med*, **38**, 483–492.
4. Michaud, J.L. and Kennedy, C.R. (2007) The podocyte in health and disease: insights from the mouse. *Clin Sci*, **112**, 325–335.
5. Tryggvason, K., Patrakka, J., and Wartiovaara, J. (2006) Hereditary proteinuria syndromes and mechanisms of proteinuria. *N Engl J Med*, **354**, 1387–1401.
6. Ransom, R.F., Vega-Warner, V., Smoyer, W.E., and Klein, J. (2005) Differential proteomic analysis of proteins induced by glucocorticoids in cultured murine podocytes. *Kidney Int*, **67**, 1275–1285.
7. Viney, R.L., Morrison, A.A., van den Heuvel, L.P., Ni, L., Mathieson, P.W., Saleem, M.A., and Ladomery, M.R. (2007) A proteomic investigation of glomerular podocytes from a Denys-Drash syndrome patient with a mutation in the Wilms tumor suppressor gene WT1. *Proteomics*, **7**, 804–815.
8. Shankland, S.J., Pippin, J.W., Reiser, J., and Mundel, P. (2007) Podocytes in culture: past, present, and future. *Kidney Int*, **72** (1), 26–36.
9. Krtil, J., Platenik, J., Kazderova, M., Tesar, V., and Zima, T. (2007) Culture methods of Glomerular Podocytes. *Kidney Blood Press Res*, **30**, 162–174.
10. Topham, P.S., Kawachi, H., Haydar, S.A., Chugh, S., Addona, T.A., Charron, K.B., Holzman, L.B., Shia, M.

et al. (1999) Nephritogenic mAb 5-1-6 is directed at the extracellular domain of rat nephrin. *J Clin Invest*, **104**, 1559–1566.

11. Ahola, H., Heikkila, E., Astrom, E., Inagaki, M., Izawa, I., Pavenstadt, H., Kerjaschki, D., and Holthofer, H. (2003) A novel protein, densin, expressed by glomerular podocytes. *J Am Soc Nephrol*, **14**, 1731–1737.
12. Liu, X.L., Kilpelainen, P., Hellman, U., Sun, Y., Wartiovaara, J., Morgunova, E., Pikkarainen, T., Yan, K. *et al.* (2005) Characterization of the interactions of the nephrin intracellular domain. *FEBS J*, **272**, 228–243.
13. Lehtonen, S., Lehtonen, E., Kudlicka, K., Holthofer, H., and Farquhar, M.G. (2004) Nephrin forms a complex with adherens junction proteins and CASK in podocytes and in Madin-Darby canine kidney cells expressing nephrin. *Am J Pathol*, **165**, 923–936.
14. Lehtonen, S., Ryan, J.J., Kudlicka, K., Iino, N., Zhou, H., and Farquhar, M.G. (2005) Cell junction-associated proteins IQGAP1, MAGI-2, CASK, spectrins, and alpha-actinin are components of the nephrin multiprotein complex. *Proc Natl Acad Sci*, **102**, 9814–9819.
15. Rastaldi, M.P., Armelloni, S., Berra, S., Li, M., Pesaresi., M., Poczewski, H., Langer, B., Kerjaschki, D. *et al.* (2003) Glomerular podocytes possess the synaptic vesicle molecule Rab3A and its specific effector rabphilin-3a. *Am J Pathol*, **163**, 889–899.
16. Rastaldi, M.P., Armelloni, S., Berra, S., Calvaresi, N., Corbelli, A., Giardino, L.A., Li, M., Wang, G.Q. *et al.* (2006) Glomerular podocytes contain neuron-like functional synaptic vesicles. *FASEB J*, **20**, 976–978.
17. Khoshnoodi, J., Hill, S., Tryggvason, K., Hudson, B., and Friedman, D.B. (2007) Identification of N-linked glycosylation sites in human nephrin using mass spectrometry. *J Mass Spectrom*, **42**, 370–379.
18. O'Connell, K.L. and Stults, J.T. (1997) Identification of mouse liver proteins on two-dimensional electrophoresis gels by matrix-assisted laser desorption/ionization mass spectrometry of in situ enzymatic digests. *Electrophoresis*, **18**, 349–359.
19. Pluskal, M.G. (2000) Microscale sample preparation. *Nat Biotechnol*, **18**, 104–105.
20. Chaurand, P., Sanders, M.E., Jensen, R.A., and Caprioli, R.M. (2004) Proteomics in diagnostic pathology. *Am J Pathol*, **165**, 1057–1068.

7
Proteomics of Renal Peroxisomes

Thomas Gronemeyer, Sebastian Wiese*, Christian Stephan, Rob Ofman, Ronald J.A. Wanders, Helmut E. Meyer, and Bettina Warscheid*
(Both authors contributed equally to this work)*

7.1
Introduction

7.1.1
Organellar Proteomics

Eukaryotic cells contain a membrane-surrounded nucleus and a well-defined intracellular membrane network that encloses distinct compartments, the so-called organelles. This compartmentalization is a key factor for the regulation of protein activities as contrary processes, such as anabolic and catabolic pathways, can be operated simultaneously inside the cell. Organelles thus represent separated microenvironments for the different proteins inside the cell in order to provide adequate conditions for their specific functions and activities. Accordingly, detailed knowledge about the subcellular localization of proteins represents an important key to understand both their distinct functions and individual contributions to the manifold metabolic pathways as well as signal transduction pathways inside the cell. To fulfill this task, various biochemical and cell biological methods are routinely applied. Apart from imaging approaches (e.g., fluorescence microscopy) that facilitate the localization of proteins inside the living cell, enrichment methods such as subcellular fractionation by density gradient centrifugation in combination with enzyme activity measurements and/or immunoblot analyses are widely used to obtain specific information about protein localization. However, all these approaches follow the principle "you have to know what you are looking for," that is, previous knowledge about the protein of interest such as its amino acid sequence and/or its 3-D structure is required. The potential of all such methods eventually depends on the availability of highly specific antibodies and/or highly specific activity assays. Moreover, these classical approaches in cell biology do usually neither allow the simultaneous identification of a large number of proteins within a single experiment nor facilitate the direct identification of so far unknown constituents of an organelle.

Renal and Urinary Proteomics: Methods and Protocols. Edited by Visith Thongboonkerd

ISBN: 978-3-527-31974-9

Subcellular fractionation combined with mass spectrometry (MS)-based analysis, referred to as organellar proteomics provides a most powerful tool for the establishment of comprehensive protein catalogs of isolated organelles (reviewed in [1–3]). When following this approach, antibodies or assays for distinct marker proteins for the organelles of interest should be at hand, but only for monitoring their successful purification. Organellar proteomics can be universally applied to yeast and mammalian cells, without previous knowledge about the content of the target compartment and without previous invasive manipulations such as the expression of fusion proteins. A major benefit of organellar proteomics is that the complexity of subcellular preparations is generally compatible with both sensitivity and dynamic range of current MS-based methods, enabling the identification of low abundant proteins. Nonetheless, proper purification of subcellular structures remains a key factor in obtaining significant data in such an approach. This is particularly true for low abundant organelles such as peroxisomes, which contribute to only 1–5% of the cell volume depending on tissue type and metabolic state. Moreover, mitochondrial and ER membranes, which account for the largest fraction of total intracellular membranes, can significantly impede the detection of low abundant peroxisomal membrane proteins (PMPs) even at their minor presence in preparations of peroxisomal membranes [4]. Taken together, challenges in the proteomic analysis of peroxisomes mainly rely on their low abundance combined with the limited ability to purify this organelle [5].

Since the inability to purify subcellular structures to homogeneity is a general, limiting problem in organellar proteomics, significant effort was made to allow for the reliable discrimination between true constituents of an organelle and copurifying contaminants in subcellular preparations. To this end, strategies were developed that enable to quantitatively monitor hundreds of proteins through various fractions of a density gradient by MS either in combination with stable isotope labeling [6] or without [7]. This quantitative data in combination with statistical analysis now facilitate the accurate prediction of the cellular location of proteins in a global manner, thereby providing an excellent means by which new insights into the proteomes and functions of subcellular structures can be obtained [7–11].

7.2
Mammalian Peroxisomes and Proteomics

7.2.1
General Considerations

Peroxisomes are ubiquitous single membrane-bound organelles present in virtually all eukaryotic cells. They harbor approximately 50 different enzymes that enable them to execute a fast array of metabolic functions such as α- and β-oxidation of fatty acids, ether-phospholipid biosynthesis, degradation of reactive oxygen species, or detoxification of glyoxylate in mammals (reviewed in [12]). For the biogenesis of peroxisomes complex processes such as membrane assembly, import of matrix proteins, and division of mature peroxisomes are required. These processes are enabled by the concerted action of a complex network composed of more than 20 different proteins, the so-called peroxins [13–15]. Proteins destined for peroxisomes are generally synthesized on free ribosomes in the cytoplasm; they usually contain a peroxisomal targeting signal (PTS) in order to be routed to peroxisomal membranes as well as to be posttranslationally imported into the organelle [16].

In response to the metabolic state as well as external factors, mammalian cells are able to modulate both the number and the enzymatic composition of peroxisomes. Moreover, peroxisome function, and thus protein composition, is organ specific and varies between species [17]. A significant increase in the number of peroxisomes is generally observed as a consequence of, for example, alcohol abuse, a lipid-rich diet, or the intake of lipid-lowering drugs, whereas peroxisome number is drastically reduced by catalase inhibitors and in fatty liver disease, for example. Failure in peroxisomal biogenesis or deficiencies in the function of single peroxisomal proteins ultimately lead to serious, often lethal diseases such as Zellweger syndrome or X-linked adenoleukodystrophy [12]. To date, more than 15 inherited peroxisomal diseases are known in human, underscoring the metabolic importance of this small organelle. Nonetheless, our picture of mammalian peroxisomes is still limited and deeper insight into the biochemistry and function of this organelle is certainly needed to improve our understanding of its role in human health and disease.

7.2.2
Proteomics of Peroxisomes

In silico proteomics approaches have widely been applied to identify putative peroxisomal proteins or even to predict the entire peroxisomal proteome [18–20]. Common *in silico* strategies of peroxisomal proteomics comprise comparative genomics tools (i.e., the identification of conserved domains and homologs via BLAST searches) and/or the prediction as well as the identification of peroxisomal targeting sequences, that is, PTS1 and PTS2 for peroxisomal matrix proteins and mPTS for PMPs [20]. Nonetheless, apart from the fact that the PTS1 is a comparatively well-defined consensus sequence, namely, serine-lysine-leucin (SKL), located at the extreme C-terminus of peroxisomal proteins [21], the latter approach is hampered for several reasons: (i) the less common PTS2 motif contained within the N-terminal region of selected peroxisomal matrix proteins is rather variable and not well characterized [22]; (ii) some proteins such as phosphomevalonate kinase exhibit a consensus PTS but do not reside in peroxisomes [23], whereas (iii) other proteins of known peroxisomal location lack a consensus PTS [24]; (iv) among PMPs, only limited similarities between mPTS exist and so far no reliable conclusion on consensus motif could be drawn [25]; and (v) a limited set of PMPs may be directly targeted to peroxisomes via the ER [26]. In light of these findings, it becomes evident that experimental proteomics approaches have to be sought when aiming at comprehensively characterizing the proteome of peroxisomes.

In recent years, various proteomic investigations of peroxisomes purified from rat or mouse were performed in order to obtain extensive information on their proteome composition [27–30]. Common to all these studies was the use of either one-dimensional gel electrophoresis (1-D GE) or two-dimensional gel electrophoresis (2-D GE) for the efficient separation of proteins in peroxisomal preparations prior to MS analysis. To isolate peroxisomes from rat liver with high purity, Kikuchi *et al.* [28] performed density gradient centrifugation using Nycodenz, followed by immunoaffinity chromatography employing an antibody against PMP70, a highly abundant protein in peroxisomal membranes. Proteomic analysis of such peroxisomal preparations using 1-D GE in combination with LC/tandem MS resulted in the identification of more than 50 genuine peroxisomal proteins including a new isoform of Lon protease. Recently, 2-D gel-based investigations led to the identification of microsomal gluthatione-*S*-transferase as genuine component of rat liver peroxisomes [27] as well as a nudix hydrolase designated RP2p as new component of mouse kidney peroxisomes [30]. Three microsomal proteins, namely, aldehyde dehydrogenase, cytochrome b5 (Cyb5A), and its corresponding reductase were also found to be significantly enriched in peroxisomal preparations from rat liver [27, 28], suggesting their specific localization in mammalian peroxisomes. A very recent comparative study of peroxisomes from mouse kidney and mouse liver resulted in the identification of 31 known as well as 10 putative peroxisomal proteins. Moreover, proteins involved in the α- and β-oxidation of fatty acids as well as in both amino acid and nucleotide metabolism were shown to be more abundant in liver peroxisomes, whereas enzymes of straight-chain fatty acid β-oxidation were found in higher concentrations in kidney peroxisomes [29].

7.3
Toward the Complete Characterization of Mouse Kidney Peroxisomes

7.3.1
Establishment of Proteome Catalog of Mammalian Peroxisomes

The current literature lists approximately 75 proteins as genuine components of renal peroxisomes from mouse. Of these, 48 proteins reside in the matrix and 27 proteins in the membrane of peroxisomes. To establish a most complete proteome catalog of renal peroxisomes, we designed a comprehensive MS-based proteomics approach. For the purification of peroxisomal samples, mouse kidneys were homogenized in buffer and the homogenate was then subjected to differential and Nycodenz equilibrium gradient centrifugation. Moreover, a fraction of purified kidney peroxisomes was subjected to alkaline treatment (pH 11.5) to effectively increase the probability to detect PMPs of very low abundance by proteomic methodology.

Figure 7.1 illustrates the general workflow of our MS-based proteomic analyses of renal peroxisomes as well as the respective peroxisomal membrane fractions.

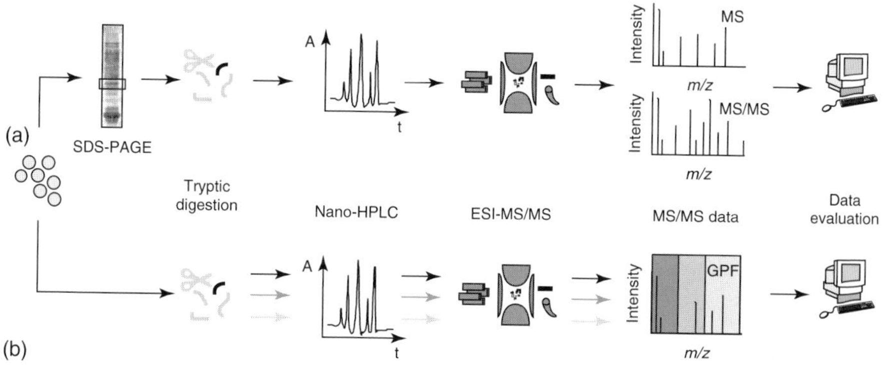

Figure 7.1 Schematic overview of the two proteomic tracks applied for the study of renal peroxisomes from mouse. After lysis of isolated peroxisomes, the proteome composition is analyzed by tandem mass spectrometry following (a) a gel-enhanced proteomics approach and (b) a shot-gun approach combined with gas-phase fractionation. In track A, proteins are separated by sodium dodecylsulfate polyacrylamide gel electrophoresis (SDS-PAGE), followed by Colloidal Coomassie staining. After cutting the gel in individual bands, proteins are subjected to in-gel tryptic digestion. In track B, proteins are digested in-solution using trypsin. The resulting peptide mixtures are then analyzed by nano-HPLC directly coupled to electrospray ionization tandem mass spectrometry (ESI-MS/MS). For gas-phase fractionation (GPF) experiments, the respective peptide samples are analyzed in triplicate using overlapping *m/z* ranges of 400–650, 600–850, and 800–1200 for the selection of precursor ions (track B).

Peroxisomal preparations were either separated by sodium dodecyl sulfate polyacrylamide gel electrophoresis (SDS-PAGE) followed by in-gel tryptic digestion of individual gel bands (Figure 7.1a) or directly subjected to in-solution tryptic digestion in an adequate buffer system (Figure 7.1b). The resulting tryptic peptide samples were then analyzed by nano-HPLC/ESI-MS/MS. For peptide identification and subsequent protein assembly, uniform MS/MS data sets were generated and then correlated with a composite version of the IPI database using MASCOT [31]. To increase the effective dynamic range of LC/tandem MS analyses of peptide samples obtained from in-solution digestion of peroxisomal preparations, we additionally applied gas-phase fractionation (GPF) (Figure 7.1b). This method was previously reported for MS-based analysis of yeast peroxisomes [32]. In GPF experiments, a sample is repeatedly analyzed by MS using several narrow *m/z* windows for the acquisition of MS survey scans from which peptides ions are then selected for MS/MS events. To effectively increase the rate of peptide sampling, we tested different conditions for GPF in the *m/z* dimension, resulting in the design of three slightly overlapping *m/z* windows of 400–650, 600–850, and 800–1200 [33].

On average, we could identify 110 unique proteins per analysis, of which up to 48% were annotated as peroxisomal proteins. As a result of all our MS-based proteomic analyses, we identified 252 nonredundant proteins, of which 64 proteins (25%) were reported in literature to be localized in peroxisomes. Accordingly, we successfully identified 85% of all proteins reported for renal peroxisomes in literature. Of these, 42 proteins are known peroxisomal matrix and 22 are known

PMPs. Most PMPs could already be identified through proteomic analysis of preparations of intact peroxisomes without further enrichment of the integral membrane components by alkaline treatment. However, four PMPs of very low abundance were only detected in the respective membrane preparations of renal peroxisomes, namely, adrenoleukodystrophy protein (ALDP) [34], Mpv17-like protein [35], as well as the two peroxins Pex13 [36] and Pex2 [37]. Pex13 represents not more than 0.1% of total membrane protein of mammalian peroxisomes [4, 38] and ALDP was estimated to contribute to approximately 2% of total PMP in rat liver [4].

In addition to ALDP, three other ABC transporters (namely adrenoleukodystrophy related (ALDR) protein, PMP70, and PMP69) are reported to reside in the membrane of mammalian peroxisomes [12, 39, 40]. Although there is strong evidence that these ABC transporters show some functional redundancy, their expression levels vary in different tissues [41]. For example, a high level of PMP70 but a low level of ALDR expression was found in mouse kidney [42]. In our proteomic study of renal peroxisomes, ALDP and PMP70 were successfully identified, whereas ALDR and PMP69 remained elusive [33]. In light of this finding, very low expression levels of the latter two proteins in mouse kidney is indicated. It is also important to note that the identification of low abundant peroxisomal receptor proteins, such as Pex7 and Pex19, is impeded due to their mainly cytosolic localization [43, 44]. If peripherally attached to the membrane, they are in all likelihood removed by carbonate treatment of the purified peroxisomes.

Failure in the detection of known peroxisomal matrix proteins may suggest their low expression in kidney peroxisomes. For example, rather low expression levels of the peroxisomal acyl-CoA thioesterase Ic (Pte1c) as well as the peroxisomal nudix hydrolase 7 (Nudt7) were found in kidney [45], thus providing us with a reasonable explanation for their nondetection by proteomics means. Kikuchi *et al.* [28] were able to detect bile acid-CoA : amino acid *N*-acyltransferase (BACAT) in rat liver peroxisomes. The latter enzyme was previously shown to be strongly expressed in the cytosolic, mitochondrial, and peroxisomal fraction of human liver preparations [46, 47]. However, BACAT was not detected in peroxisomal preparations from mouse kidney, suggesting its minor presence in kidney peroxisomes, if expressed at all. Two further peroxisomal matrix proteins, namely, *N*1-acetylated polyamine oxidase and malonyl-CoA decarboxylase, were elusive to our proteomic analyses and low expression levels of both proteins in kidney peroxisomes were reported before [48, 49].

In light of the above discussion, we argue that our proteomics approach enabled the detection of virtually all known proteins of renal peroxisomes from mouse. This established protein inventory is summarized in Table 7.1 and it also includes enzymes that were just recently designated as peroxisomal, such as the acyltransferase ACNAT1 [50], RP2p [30], and Lon protease [28]. In addition to the successful detection of 64 bona fide constituents of renal peroxisomes, proteomic investigation of the respective peroxisomal preparations resulted in the identification of a total of 123 proteins known to localize to other subcellular structures than peroxisomes as well as 65 proteins of unknown localization. One can assume that a very high percentage of these proteins represent copurified contaminants of rather low

Table 7.1 Catalog of peroxisomal proteins identified by proteomic analyses of mouse kidney peroxisomes.

Protein name	Gene symbol	SwissProt accession	Protein name	Gene symbol	SwissProt accession
Acyl-CoA oxidase 1, plamitoyl	Acox1	Q4KN64	Testosterone-regulated RP2 protein	Nudt19	P11930
Peroxisomal multifunctional enzyme type II (D-BP)	Hsd17b4	P51660	Serine hydrolase	Serhl	Q3U3G8
Peroxisomal-3-keto-acyl-CoA thiolase A	Acaa1a	Q921H8	Peroxisomal Lon Protease	1300002A08Rik	Q3TEG8
Peroxisomal-3-keto-acyl-CoA thiolase B	Acaa1b	Q8VCH0	Epoxide hydrolase	Ephx2	P34914
Steroid delta-isomerase	Hsd3b3	P26150	NADPH-dependent carbonyl reductase/NADP-retinol DH	Dhrs4	Q99LB2
Δ-3,5-Δ-2,4-Dienoyl-CoA isomerase	Ech1	O35459	Glutathion S-transferase Class Kappa	Gstk1	Q9DCM2
Putative peroxisomal 2,4-dienoyl-CoA reductase	Decr1	Q9WV68	Isocitrate dehydrogenase 2 (NADP+)	Idh2	Q8C2R9
Peroxisomal 3,2-*trans*-enoyl-CoA isomerase	Peci	Q9WUR2	Glycerol-3-phosphate dehydrogenase 1 (NAD+)	Gpd1	P13707
Peroxisomal trans-2-enoyl-CoA reductase	Pecr	Q99MZ7	Acyl-Coenzyme A dehydrogenase family, member 11[a]	Acad11	Q80XL6
Acyl-CoA oxidase 2	Acox2	Q9QXD1	Zinc-binding alcohol dehydrogenase domain containing protein 2[a]	Zadh2	Q8BGC4
Sterol carrier protein 2	Scp2	P32020	Acbd5 protein[a]	Acbd5	Q5CZX6
α-Methylacyl-CoA racemase	Amacr	Q9DCW6	Adrenoleukodystrophy protein (ALDP)	Abcd1	P48410
Acyl-CoA: *N*-acyltransferase 1	Acnat1	Q0QBA1	PMP70	Abcd3	P55096
Phytanoyl-CoA α-hydroxylase	Phyh	O35386	PMP22	Pxmp2	P42925

(continued overleaf)

Table 7.1 (Continued)

Protein name	Gene symbol	SwissProt accession	Protein name	Gene symbol	SwissProt accession
2-Hydroxyphytanoyl-CoA lyase	Hpcl	Q9QXE0	Mpv17-like protein	Mpv17l	Q8CC11
Pristanoyl-CoA oxidase	Acox3	Q7TPP6	Peroxisomal Ca-dependent solute carrierlike protein	Prdx5	P99029
Very long chain acyl-CoA synthetase	Slc27a2	O35488	Peroxisomal membrane protein 4	Pxmp4	Q9JJW0
Acyl-CoA synthetase long chain family member 1	Acsl1	Q6GTG6	PMP34	Slc25a17	O70579
Peroxisomal carnitine octanoyltransferase	Crot	Q9DC50	Tetratricopeptide repeat domain 11	Fis1	Q9CQ92
Carnitine *O*-acetyltransferase	Crat	P47934	2810439K08RIK PROTEIN (PMP52)[a]	Tmem135	Q8BSY5
Catalase	Cat	P24270	Pex1	Pex1	Q5BL07
Peroxisomal multifunctional enzyme type I (L-BP)	Ehhadh	Q91W49	Pex2	Pex2	Q3UJB7
Hydroxyacid oxidase 2	Hao3	Q9NYQ2	Pex3	Pex3	Q9QXY9
Fatty acyl-CoA reductase 1	Mlstd2	Q922J9	Pex5	Pex5	Q91YC7
Peroxisomal acyl-coenzyme A thioester hydrolase 1a	acot8	P58137	Pex6	Pex6	Q99LC9
Peroxisomal acyl-coenzyme A thioester hydrolase 1b	acot4	Q8BWN8	Pex10	Pex10	–
Peroxisomal acyl-coenzyme A thioester hydrolase 6	acot6	Q32Q92	Pex11α	Pex11a	Q9Z211
Peroxisomal acyl-CoA thioester hydrolase 2a	acot3	Q9QYR7-2	Pex11β	Pex11b	Q9D090
Dihydroxyacetone phosphate acyltransferase	Gnpat	P98192	Pex11γ	Pex11c	Q6P6M5-1
Alkyldihydroxyacetonephosphate synthase	Agps	Q8BKC4	Pex12	Pex12	Q8VC48
Peroxisomal sarcosine oxidase	Pipox	Q9D826	Pex13	Pex13	Q8CCW5
D-aminoacid oxidase 1	Dao1	Q91WH3	Pex14	Pex14	Q9R0A0
D-aspartate oxidase	Ddo	Q922Z0	Pex16	Pex16	Q3TRJ6
Peroxisomal NADH pyrophosphatase NUDT12	Nudt12	Q9DCN1-1	Pex26	Pex26	Q8BGI5-2

[a] New peroxisomal proteins identified by protein correlation profiling and validated by immunofluorescence microscopy.

abundance, since peroxisomes as other subcellular structures cannot be purified to homogeneity in general. However, to reliably distinguish between contaminants and genuine constituents of peroxisomes, a quantitative proteomics approach has to be followed. Moreover, such an approach provides a most powerful tool by which new peroxisomal candidates can be reliably identified.

7.3.2 Identification of New Peroxisomal Proteins by Protein Correlation Profiling

To improve our knowledge about organellar proteomes, advanced proteomics approaches have to be sought by which proteins can be classified to different subcellular compartments without the need of purifying organelles to homogeneity. This important issue was addressed by the development of a (semi-)quantitative MS-based strategy referred to as *protein correlation profiling* (*PCP*) [7].

A schematic overview of PCP for MS-based protein localization studies is shown in Figure 7.2. Intact organelles such as lysosomes, mitochondria, and peroxisomes present in a cell lysate are separated by differential and density gradient centrifugation. Following the fractionation of the resulting density gradient, the respective protein samples are subjected to in-solution tryptic digestion. The corresponding tryptic peptide samples are then analyzed by nano-HPLC/ESI-MS/MS to facilitate peptide identification as well as protein assembly by database searches using, for example, MASCOT [51]. For quantitative analysis of data, peptides detected in MS spectra are plotted against their elution time from the HPLC column. In the resulting 3-D plot, the signal intensity of each peptide is on the z-axis, while its corresponding elution time and m/z ratio are shown on the x- and y-axis, respectively. Individual peptide peaks in the 3-D diagram are then linked to MS/MS data

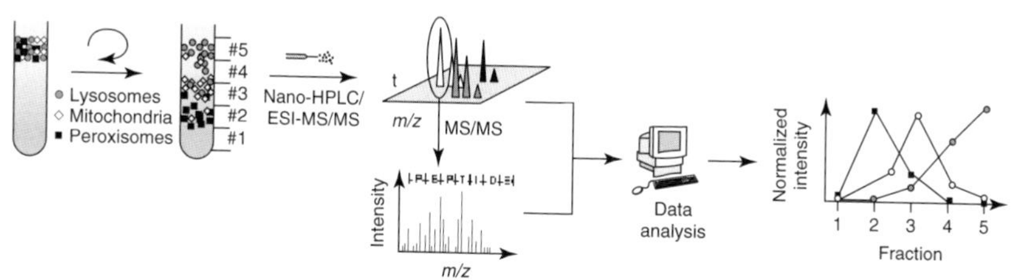

Figure 7.2 Workflow of the data evaluation via protein correlation profiling (PCP). Protein localization using (semi-)quantitative MS-based proteomic strategies. Intact organelles such as lysosomes, mitochondria, and peroxisomes present in a cell lysate are separated by density gradient centrifugation. Following the fractionation of the resulting gradient, enriched organelle fractions of interest are analyzed by ESI-MS/MS for protein identification by bioinformatics. To distinguish between genuine proteins of an organelle and copurified proteins from other cellular structures, a (semi-)quantitative MS approach is followed, referred to as protein correlation profiling (PCP). Protein profiles are established based on MS analysis of gradient fractions. Marker proteins for different organelles define the respective consensus profiles. Proteins that follow the characteristic consensus profile for an organelle are considered as resident proteins of this organelle, whereas deviations from the profile indicate contaminants. Further information on PCP is provided in the main text.

to confirm their sequence identity. Subsequently, abundance profiles of proteins are established by plotting the normalized peak intensities of the corresponding peptides against the respective gradient fraction numbers. Marker proteins for different organelles define the respective organellar consensus profiles. Proteins that follow the characteristic consensus profile for an organelle are considered as resident proteins of this organelle, whereas deviations from the profile indicate copurifying contaminants. Proteins may also feature characteristics of protein profiles of two or more distinct subcellular structures, indicating dual or even multiple localization sites [9].

In our experimental design, we focused on the ability to reliably discriminate between resident proteins of peroxisomes and other subcellular structures, in particular mitochondria, via PCP. The latter organelle is generally known to represent a major contaminant of peroxisomes purified by DGC. To this end, consecutive peptide MS/MS analyses of six gradient fractions were performed on a nano-HPLC/ESI-LTQ-FTICR system combined with both GPF and SIM scans [52]. A total of approximately 4000 peptides could be quantitatively followed across the six gradient fractions, enabling to establish the abundance profiles for more than 100 proteins identified in the peroxisomal peak fraction of the mouse kidney gradient [33]. Subsequently, protein profiles were statistically analyzed to eventually identify new candidates of mouse kidney peroxisomes. To this end, the correlation values (χ^2-values) between the abundance profiles of marker proteins for peroxisomes and mitochondria were calculated and then a "goodness of fit" was defined. We chose the peroxisomal bifunctional enzyme (PBE), the ATP-translocase, and the protein peroxiredoxin-5 as marker proteins, which are well known to localize to peroxisomes, to mitochondria, and to both organelles, respectively. Correlation values were then calculated for all the other proteins identified in the peroxisomal peak fraction. Since peroxisomal and mitochondrial proteins clearly have different χ^2-values, we could reliably discriminate between these two groups of proteins. Generally, proteins localized to peroxisomes exhibited great similarity to the profile of PBE and greatly differed from the profile of ATP-translocase. The established profiles of PBE and peroxiredoxin-5 showed great similarity in the first three fractions (2, 3, and 4), whereas major differences appeared in the gradient fractions of lower density (6, 8, and 10) due to the dual localization of the latter protein [33].

As a result of our PCP analysis, 15 new candidates of renal peroxisomes from mouse could be identified. Of these, six proteins, namely, the acyl-Coenzyme A dehydrogenase family member 11 (Acad11), the protein Acbd5, malate dehydrogenase 1 (Mdh1), Cyb5A, the corresponding reductase (Dia1), and the fatty aldehyde dehydrogenase (FALDH) variant form encoded by the gene *Aldh3a2*, were previously detected in peroxisome preparations from rat liver [27, 28]. PCP data thus point to the peroxisomal location of the three microsomal proteins Cyb5A, Dia1, and the FALDH variant form [27, 28, 53, 54]. The latter protein may be involved in peroxisomal lipid metabolism, such as the detoxification of fatty aldehydes. Moreover, mutations in the *Aldh3a2* gene cause Sjögren–Larsson syndrome, an inherited human neurocutaneous disorder characterized by ichthyosis, mental retardation, and spasticity. The pathogenesis of these symptoms is thought to result from abnormal lipid accumulation, defective metabolism of eicosanoids, or the increased formation of aldehyde adducts with lipids and/or proteins [55, 56]. Further proteins predicted to reside in the peroxisomal matrix were the zinc-binding alcohol dehydrogenase domain containing protein 2 (Zadh2), the abhydrolase containing protein 14b (Abhd14b), a protein similar to BACAT (RIKEN cDNA clone 1810022C23), the two acyl-CoA thioesterases Acot1 and Acot12 as well as the multifunctional protein ADE2. The latter protein may be involved in purine biosynthesis. Acot1 and Acot12 are members of a group of enzymes that hydrolyze CoA esters to the corresponding free acids and CoA [57, 58] and may therefore be involved in the regulation of peroxisomal lipid metabolism. As yet, however, Acot1 was reported to show high specificity for C_{12}–C_{20} acyl-CoA esters with high activity in the cytosol [59].

Apart from all these candidates located in the peroxisomal matrix, three new components of peroxisomal membranes, namely, PMP52, the MOCO sulfurase C-terminal domain containing 2 protein (Mosc2) and the ATPase family, AAA domain containing protein 1 (ATAD1), were identified by PCP. The latter protein is a member of the AAA-superfamily of ATPases, which function in the unfolding of proteins or disassembly of protein complexes and aggregates [60]. A potential role of ATAD1 in peroxisomal homeostasis is thus indicated. Since PMP52 is related to PMP24, these two proteins may exhibit similar, as yet, however, unknown functions. Mosc2, an iron–sulfur protein of unknown function, was so far reported to reside in mitochondrial membranes [61]. Nonetheless, PCP data clearly indicated its localization in both peroxisomes and mitochondria. At this point, it is important to note that the localization of four proteins (Acad11 protein, ZADH2 protein, ACBD5 protein, and PMP52) was additionally studied by immunofluorescence microscopy [33]. As a result, the peroxisomal localization of these proteins was confirmed, thus demonstrating the high potential of protein localization studies using PCP.

7.4 Methods and Protocols

7.4.1 Isolation of Peroxisomes from Mouse Kidney

We purify peroxisomes from kidneys of male Swiss mice, four to six months of age, as described previously [30]. Using this method, kidneys are homogenized in a buffer containing 10 mM morpholinopropanesulfonic acid (MOPS)-NaOH, 250 mM sucrose, 2 mM EDTA, and 0.1% ethanol (final pH 7.4). Then the homogenate is centrifuged at 600g for 10 minutes at 4 °C to obtain the postnuclear supernatant. Subsequently, peroxisomes are isolated using Nycodenz equilibrium gradient centrifugation as described in [62]. To assess the isolation of peroxisomes, fractions of 2 ml from the bottom of the gradient are taken and assayed for the marker enzymes catalase (peroxisomes), glutamate dehydrogenase (mitochondria), phosphoglucoisomerase (cytoplasm), β-hexosaminidase (lysosomes), and esterase (microsomes) as described (see [30] for more details and references therein).

Highest activity for catalase is measured in the high density part of the gradient. All the remaining enzymes showed highest activities at parts of lower density of the gradient [33]. The protein concentration of each gradient fraction is then measured according to Bradford [63] using bovine serum albumin (BSA) as standard. For subsequent proteomic analyses, a pellet from the peroxisomal peak fraction of the gradient is prepared by centrifugation at 16 000g for 10 minutes and stored at $-80\,^{\circ}$C. For PCP, pellets are prepared from 1 ml aliquots of several gradient fractions, which are then resuspended in sample buffer containing 30 mM Tris-HCl, 2 M thiourea, and 7 M urea (pH 8.5) to a final protein concentration of about $1\,\mu g\,\mu l^{-1}$ for proteomic analysis.

7.4.2 Preparation of Peroxisomal Membranes from Mouse Kidney

Purification of peroxisomal membranes is performed by alkaline extraction as described in [64]. For this, peroxisomes are resuspended in 0.12 M Na_2CO_3, pH 11.5, placed on ice for 30 minutes, and centrifuged at 100 000g (T1270, Sorvall) for 1 hour at $4\,^{\circ}$C. Membrane pellets can then be subjected to SDS-PAGE or in-solution digestion using trypsin as proteolytic agent.

7.4.3 Protein Separation by Gel Electrophoresis Followed by In-Gel Digestion

For SDS-PAGE separation, 25 µl of 4× SDS sample buffer is added to 75 µl of protein sample. Proteins are separated on a polyacrylamide gel (15.2% total acrylamide, 1.3% bisacrylamide) with a 4% stacking gel using a Desaphor VA 300 system (Invitrogen) according to the manufacturer's instructions and subsequently stained by colloidal Coomassie Brilliant Blue G-250. The gel is then equally cut in 2-mm slices and the resulting bands are destained by alternately incubating them with 20 µl of 10 mM ammonium hydrogencarbonate (NH_4HCO_3) and 20 µl of 5 mM NH_4HCO_3/50% ACN (v/v) for 10 minutes each. After the complete destaining of gel bands, in-gel digestion of proteins is performed overnight at $37\,^{\circ}$C under gentle shaking using trypsin (0.06 µg per gel slice) dissolved in 10 mM NH_4HCO_3 buffer at pH 7.8. The resulting peptides are extracted twice with 10 µl of ACN/5% FA (50 : 50, v/v). The extracts are then combined and ACN is removed in vacuo. For LC/ESI-MS/MS analyses, samples are acidified by addition of 5% FA to a final volume of 20 µl.

7.4.4 In-Solution Digestion of Proteins

Protein samples are dissolved in 50 mM NH_4HCO_3 to a final concentration of $0.1\,\mu g\,\mu l^{-1}$. Then, trypsin is added to result in a protein-to-trypsin ratio of 1 : 30

and enzymatic digestion is carried out for 6 hours at 37 °C, under gentle shaking. For LC/ESI-MS/MS analyses, the resulting peptide mixtures are diluted in 5% FA to a final concentration of 0.067 μg μl^{-1}.

7.4.5 Peptide Analysis by Nano-HPLC/ESI-MS/MS

We perform on-line reversed-phase capillary HPLC separations using the Dionex LC Packings HPLC systems (Dionex LC Packings, Idstein, Germany) as described previously [65]. ESI-MS/MS analyses are performed on a Bruker Daltonics HCT plus ion trap instrument (Bremen, Germany) equipped with a nanoelectrospray ion source (Bruker Daltonics) and distal-coated SilicaTips (FS360–20–10-D; New Objective, Woburn, MA, USA). Standard peptides are used for external mass calibration of the MS instrument. The general mass spectrometric parameters are as follows: CV, 1400 V; PO, 500 V; DG, 10.0 l min^{-1}; DT, 160 °C; aimed ICC 150 000; and maximal fill-time 500 milliseconds. Data-dependent software (HCT plus, Esquire Control, Bruker Daltonics) is then employed for the automatic acquisition of MS and MS/MS spectra of peptides. In general, MS spectra of tryptic peptides are acquired in the range from m/z 300 to 1500, whereas MS/MS spectra are recorded from m/z 100 to 2200. To generate sequence-informative MS/MS spectra, low-energy collision-induced dissociation (CID) is performed on isolated multiply charged peptide ions with a fragmentation amplitude of 0.6 V and helium as collision gas. In addition, exclusion limits are placed on previously selected mass-to-charge ratios for 1.2 minutes in order to obviate multiple MS/MS events on the same peptide species.

For peptide ESI-MS/MS experiments on a 7-T Finnigan LTQ-FT (Thermo Electron, Bremen, Germany), we apply the method described by Olsen *et al.* [52]. Using this method, survey MS spectra from m/z 300 to 1500 are acquired in the FTICR cell with $r = 25\,000$ at m/z 400 with a target accumulation value of 50 000 000. The three most intense ions are sequentially isolated for accurate mass measurements by an FTICR "SIM scan" applying a mass window ±5 Da and a target accumulation value of 100 000 at a mass resolution of 50 000. Subsequent to accurate mass measurements, peptide fragmentation is carried out in the linear ion trap using low-energy CID experiments at a target accumulation value of 10 000. The total cycle time is approximately 3 seconds. Former target ions selected for MS/MS are dynamically excluded for 45 seconds. The general parameters for LTQ-FT mass measurements of peptides are as follows: spray voltage, 1.8 kV; no sheath and auxiliary gas flow; ion transfer tube temperature, 200 °C; and normalized collision energy of 35% for MS/MS with activation $q = 0.25$ and activation time 30 milliseconds. Ion selection thresholds are usually 1000 counts for MS and 500 counts for MS/MS. For LC/ESI-MS/MS analyses combined with GPF, peptide samples are analyzed in triplicate using overlapping m/z ranges of 400–650, 600–850, and 800–1200 for selection of precursor ions (for more details on GPF, see [21]).

7.4.6
Mass Spectrometric Data Analysis

Peaklists of MS/MS spectra acquired on the ion trap (Bruker Daltonic) and LTQ-FT (Thermo Electron) instrument are generated using the software tools DataAnalysis 3.3 and Bioworks 3.1 SR 1, respectively. For peptide and protein identification, peaklists can be correlated with the mouse International Protein Index (Mouse IPI V3.15.1) (www.ebi.ac.uk) database using MASCOT (release version 2.0.04) [51]. Searches are performed with tryptic specificity, allowing two missed cleavages. In addition, methionine oxidation is considered in database searches as variable modification. MS/MS spectra acquired on the HCT ion trap instrument (Bruker Daltonics) are generally accepted with a MASCOT cut-off score of 22.5. The mass tolerances for MS and MS/MS experiments are 1.2 and 0.4 Da, respectively. LTQ-FT mass spectra are searched with a mass tolerance of 2 ppm for precursor ions and 0.4 Da for fragment ions. The respective MS/MS spectra are generally accepted with a minimum MASCOT score of 15. Proteins are assembled on the basis of two identified peptides at least using the ProteinExtractor Tool (version 1.0) in ProteinScape (version 1.3, Bruker Daltonics). This software automatically removes redundancies in protein entries, that is, only the protein of the lowest molecular weight is reported. Protein isoforms are listed only when unique peptides are identified. Subsequent to the assembly of proteins, the false-positive rate is calculated as the quotient of the number of all proteins identified in a specifically designed shuffled database and the sum of all protein identifications in both the mouse IPI database and its shuffled version (for details on the calculation of false-positive rates, see [31]). Protein hits up to an accumulated false-positive rate of 5% were considered as true-positive protein identifications. For the annotation of identified proteins, UniProt (http://www.pir.uniprot.org) and Marvester (http://harvester.embl.de) databases are used.

For the assembly of correlation profiles of proteins, LC/ESI-MS/MS runs are correlated using the software package DeCyder MS (version 1.0; GE Healthcare, Uppsala, Sweden). Generally, peptide peaks are detected with an average peak width of 0.5 minutes using the Pepdetect Module. Peak matching is then performed applying a mass accuracy of at least 0.01 Da for LTQ-FTMS measurements and a maximum time window of 4 minutes using the PepMatch software module. Abundances of peptides detected in individual fractions of the density gradient are calculated from the respective peak areas. In the case of overlapping peaks, these data are discarded. MS/MS spectra within a range of 20% to a peak are linked to those peaks and then exported for a secondary search using the SEQUEST algorithm (SEQUEST version 3.0) [66]. To assemble correlation profiles, the normalized abundances of all peptides assigned to a given protein across the gradient are averaged and blotted against the respective gradient fraction numbers. To perform a goodness-of-fit test, "χ^2-values" are calculated by $\chi^2 = \sum_i (x_i - x_p)^{2/x_p}$, in which i is the fraction number, X_i is the normalized value in fraction i, and X_P is the value of the reference protein in fraction i. Organelle marker proteins are used as reference proteins, such as the PBE for peroxisomes, ATP-translocase for mitochondria, and peroxiredoxin-5 for dual localization in peroxisomes and mitochondria.

7.5 Concluding Remarks

Organellar proteomics strives at dissecting proteins on a (sub)proteome-wide scale in order to draw molecular maps of proteins present in the various subcellular structures of eukaryotic cells. Such approaches, when combined with quantitative proteomics methodology, provide the capability to obtain dependable and most accurate information not only about the identity but also about the location, abundance as well as the dynamics of proteins in a distinct subcellular compartment in dependency on the species, tissue, and/or internal as well as external stimuli. Hence, organellar proteomics is a most powerful tool to add new dimensions to the characterization of subcellular compartments in living cells.

Using a (semi-)quantitative MS-based organellar proteomics screen, we could establish the so far most comprehensive catalog of mouse renal peroxisomes, which includes 15 new peroxisomal candidates, of which four proteins were additionally validated by immunocytochemistry. We expect that this set of new proteins, when functionally characterized in follow-up studies, will contribute to expand our knowledge about the biochemistry as well as the functions of mammalian peroxisomes in health and disease.

References

1. Yates, J.R. III *et al.* (2005) Proteomics of organelles and large cellular structures. *Nat Rev Mol Cell Biol*, **6** (9), 702–714.
2. Andersen, J.S. and Mann, M. (2006) Organellar proteomics: turning inventories into insights. *EMBO Rep*, 7 (9), 874–879.
3. Dreger, M. (2003) Subcellular proteomics. *Mass Spectrom Rev*, **22** (1), 27–56.
4. Gouveia, A.M. *et al.* (1999) Alkaline density gradient floatation of membranes: polypeptide composition of the mammalian peroxisomal membrane. *Anal Biochem*, **274** (2), 270–277.
5. Saleem, R.A., Smith, J.J., and Aitchison, J.D. (2006) Proteomics of the peroxisome. *Biochim Biophys Acta*, **1763** (12), 1541–1551.
6. Dunkley, T.P. *et al.* (2004) The use of isotope-coded affinity tags (ICAT) to study organelle proteomes in Arabidopsis thaliana. *Biochem Soc Trans*, **32** (Pt3), 520–523.
7. Andersen, J.S. *et al.* (2003) Proteomic characterization of the human centrosome by protein correlation profiling. *Nature*, **426** (6966), 570–574.
8. Gilchrist, A. *et al.* (2006) Quantitative proteomics analysis of the secretory pathway. *Cell*, **127** (6), 1265–1281.
9. Foster, L.J. *et al.* (2006) A mammalian organelle map by protein correlation profiling. *Cell*, **125** (1), 187–199.
10. Dunkley, T.P. *et al.* (2006) Mapping the Arabidopsis organelle proteome. *Proc Natl Acad Sci U S A*, **103** (17), 6518–6523.
11. Lilley, K.S. and Dupree, P. (2006) Methods of quantitative

proteomics and their application to plant organelle characterization. *J Exp Bot*, **57** (7), 1493–1499.

12. Wanders, R.J. and Waterham, H.R. (2006) Biochemistry of Mammalian peroxisomes revisited. *Annu Rev Biochem*, **75**, 295–332.
13. Purdue, P.E. and Lazarow, P.B. (2001) Peroxisome biogenesis. *Annu Rev Cell Dev Biol*, **17**, 701–752.
14. Brown, L.A. and Baker, A. (2003) Peroxisome biogenesis and the role of protein import. *J Cell Mol Med*, **7** (4), 388–400.
15. Heiland, I. and Erdmann, R. (2005) Biogenesis of peroxisomes. Topogenesis of the peroxisomal membrane and matrix proteins. *FEBS J*, **272** (10), 2362–2372.
16. Brocard, C. and Hartig, A. (2006) Peroxisome targeting signal 1: is it really a simple tripeptide? *Biochim Biophys Acta*, **1763** (12), 1565–1573.
17. Birdsey, G.M. *et al.* (2004) Differential enzyme targeting as an evolutionary adaptation to herbivory in carnivora. *Mol Biol Evol*, **21** (4), 632–646.
18. Emanuelsson, O. *et al.* (2003) In silico prediction of the peroxisomal proteome in fungi, plants and animals. *J Mol Biol*, **330** (2), 443–456.
19. Kurochkin, I.V. *et al.* (2005) Sequence-based discovery of the human and rodent peroxisomal proteome. *Appl Bioinformatics*, **4** (2), 93–104.
20. Schluter, A. *et al.* (2007) PeroxisomeDB: a database for the peroxisomal proteome, functional genomics and disease. *Nucleic Acids Res*, **35** (Database issue), D815–D822.
21. van der Klei, I.J., and Veenhuis, M. (2006) PTS1-independent sorting of peroxisomal matrix proteins by Pex5p. *Biochim Biophys Acta*, **1763** (12), 1794–1800.
22. Lazarow, P.B. (2006) The import receptor Pex7p and the PTS2 targeting sequence. *Biochim Biophys Acta*, **1763** (12), 1599–1604.
23. Hogenboom, S. *et al.* (2004) Phosphomevalonate kinase is a cytosolic protein in humans. *J Lipid Res*, **45** (4), 697–705.
24. Neuberger, G. *et al.* (2004) Hidden localization motifs: naturally occurring peroxisomal targeting signals in non-peroxisomal proteins. *Genome Biol*, **5** (12), R97.
25. Jones, J.M., Morrell, J.C., and Gould, S.J. (2001) Multiple distinct targeting signals in integral peroxisomal membrane proteins. *J Cell Biol*, **153** (6), 1141–1150.
26. Van Ael, E. and Fransen, M. (2006) Targeting signals in peroxisomal membrane proteins. *Biochim Biophys Acta*, **1763** (12), 1629–1638.
27. Islinger, M. *et al.* (2006) Insights into the membrane proteome of rat liver peroxisomes: microsomal glutathione-S-transferase is shared by both subcellular compartments. *Proteomics*, **6** (3), 804–816.
28. Kikuchi, M. *et al.* (2004) Proteomic analysis of rat liver peroxisome: presence of peroxisome-specific isozyme of Lon protease. *J Biol Chem*, **279** (1), 421–428.
29. Mi, J., Kirchner, E., and Cristobal, S. (2007) Quantitative proteomic comparison of mouse peroxisomes from liver and kidney. *Proteomics*, **7** (11), 1916–1928.
30. Ofman, R. *et al.* (2006) Proteomic analysis of mouse kidney

peroxisomes: identification of RP2p as a peroxisomal nudix hydrolase with acyl-CoA diphosphatase activity. *Biochem J*, **393** (Pt 2), 537–543.

31. Stephan, C. *et al.* (2006) Automated reprocessing pipeline for searching heterogeneous mass spectrometric data of the HUPO Brain Proteome Project pilot phase. *Proteomics*, **6** (18), 5015–5029.
32. Yi, E.C. *et al.* (2002) Approaching complete peroxisome characterization by gas-phase fractionation. *Electrophoresis*, **23** (18), 3205–3216.
33. Wiese, S., Gronemeyer, T. *et al.* (2007) Proteomic characterization of mouse kidney peroxisomes by tandem mass spectrometry and protein correlation profiling. *Mol Cell Proteomics*, **6** (12), 2045–2057.
34. Mosser, J. *et al.* (1994) The gene responsible for adrenoleukodystrophy encodes a peroxisomal membrane protein. *Hum Mol Genet*, **3** (2), 265–271.
35. Iida, R. *et al.* (2001) Cloning, mapping, genomic organization, and expression of mouse M-LP, a new member of the peroxisomal membrane protein Mpv17 domain family. *Biochem Biophys Res Commun*, **283** (2), 292–296.
36. Bjorkman, J. *et al.* (1998) Genomic structure of PEX13, a candidate peroxisome biogenesis disorder gene. *Genomics*, **54** (3), 521–528.
37. Shimozawa, N. *et al.* (1992) A human gene responsible for Zellweger syndrome that affects peroxisome assembly. *Science*, **255** (5048), 1132–1134.
38. Reguenga, C. *et al.* (2001) Characterization of the mammalian peroxisomal import machinery: Pex2p, Pex5p, Pex12p, and Pex14p are subunits of the same protein assembly. *J Biol Chem*, **276** (32), 29935–29942.
39. Theodoulou, F.L., Holdsworth, M., and Baker, A. (2006) Peroxisomal ABC transporters. *FEBS Lett*, **580** (4), 1139–1155.
40. Wanders, R.J. *et al.* (2007) The peroxisomal ABC transporter family. *Pflugers Arch*, **453** (5), 719–734.
41. Smith, K.D. *et al.* (1999) X-linked adrenoleukodystrophy: genes, mutations, and phenotypes. *Neurochem Res*, **24** (4), 521–535.
42. Berger, J. *et al.* (1999) The four murine peroxisomal ABC-transporter genes differ in constitutive, inducible and developmental expression. *Eur J Biochem*, **265** (2), 719–727.
43. Sacksteder, K.A. *et al.* (2000) PEX19 binds multiple peroxisomal membrane proteins, is predominantly cytoplasmic, and is required for peroxisome membrane synthesis. *J Cell Biol*, **148** (5), 931–944.
44. Rehling, P. *et al.* (1996) The import receptor for the peroxisomal targeting signal 2 (PTS2) in Saccharomyces cerevisiae is encoded by the PAS7 gene. *EMBO J*, **15** (12), 2901–2913.
45. Gasmi, L. and McLennan, A.G. (2001) The mouse Nudt7 gene encodes a peroxisomal nudix hydrolase specific for coenzyme A and its derivatives. *Biochem J*, **357** (Pt 1), 33–38.
46. O'Byrne, J. *et al.* (2003) The human bile acid-CoA:amino acid N-acyltransferase functions in the conjugation of fatty acids to glycine. *J Biol Chem*, **278** (36), 34237–34244.
47. Solaas, K. *et al.* (2000) Subcellular organization of bile acid amidation in human liver: a key issue

in regulating the biosynthesis of bile salts. *J Lipid Res*, **41** (7), 1154–1162.

48. Wu, T., Yankovskaya, V., and McIntire, W.S. (2003) Cloning, sequencing, and heterologous expression of the murine peroxisomal flavoprotein, N1-acetylated polyamine oxidase. *J Biol Chem*, **278** (23), 20514–20525.
49. Sacksteder, K.A. *et al.* (1999) MCD encodes peroxisomal and cytoplasmic forms of malonyl-CoA decarboxylase and is mutated in malonyl-CoA decarboxylase deficiency. *J Biol Chem*, **274** (35), 24461–24468.
50. Reilly, S.J. *et al.* (2007) A peroxisomal acyltransferase in mouse identifies a novel pathway for taurine conjugation of fatty acids. *FASEB J*, **21** (1), 99–107.
51. Perkins, D.N. *et al.* (1999) Probability-based protein identification by searching sequence databases using mass spectrometry data. *Electrophoresis*, **20** (18), 3551–3567.
52. Olsen, J.V., Ong, S.E., and Mann, M. (2004) Trypsin cleaves exclusively C-terminal to arginine and lysine residues. *Mol Cell Proteomics*, **3** (6), 608–614.
53. Miyauchi, K. *et al.* (1993) Microsomal aldehyde dehydrogenase or its cross-reacting protein exists in outer mitochondrial membranes and peroxisomal membranes in rat liver. *Cell Struct Funct*, **18** (6), 427–436.
54. Fowler, S. *et al.* (1976) Analytical study of microsomes and isolated subcellular membranes from rat liver. V. Immunological localization of cytochrome b5 by electron microscopy: methodology and application to various subcellular fractions. *J Cell Biol*, **71** (2), 535–550.
55. Rizzo, W.B. and Carney, G. (2005) Sjogren-Larsson syndrome: diversity of mutations and polymorphisms in the fatty aldehyde dehydrogenase gene (ALDH3A2). *Hum Mutat*, **26** (1), 1–10.
56. Gordon, N. (2007) Sjogren-Larsson syndrome. *Dev Med Child Neurol*, **49** (2), 152–154.
57. Hunt, M.C. and Alexson, S.E. (2002) The role Acyl-CoA thioesterases play in mediating intracellular lipid metabolism. *Prog Lipid Res*, **41** (2), 99–130.
58. Hunt, M.C. *et al.* (2005) A revised nomenclature for mammalian acyl-CoA thioesterases/hydrolases. *J Lipid Res*, **46** (9), 2029–2032.
59. Huhtinen, K. *et al.* (2002) The peroxisome proliferator-induced cytosolic type I acyl-CoA thioesterase (CTE-I) is a serine-histidine-aspartic acid alpha /beta hydrolase. *J Biol Chem*, **277** (5), 3424–3432.
60. Hanson, P.I. and Whiteheart, S.W. (2005) AAA+ proteins: have engine, will work. *Nat Rev Mol Cell Biol*, **6** (7), 519–529.
61. Da Cruz, S. *et al.* (2003) Proteomic analysis of the mouse liver mitochondrial inner membrane. *J Biol Chem*, **278** (42), 41566–41571.
62. Wanders, R.J. *et al.* (1994) 2-Hydroxyphytanic acid oxidase activity in rat and human liver and its deficiency in the Zellweger syndrome. *Biochim Biophys Acta*, **1227** (3), 177–182.
63. Bradford, M.M. (1976) A rapid and sensitive method for the quantitation of microgram quantities of protein utilizing the principle of protein-dye binding. *Anal Biochem*, **72**, 248–254.

64. Fujiki, Y. *et al.* (1982) Polypeptide and phospholipid composition of the membrane of rat liver peroxisomes: comparison with endoplasmic reticulum and mitochondrial membranes. *J Cell Biol*, **93** (1), 103–110.
65. Schaefer, H. *et al.* (2004) A peptide preconcentration approach for nano-high-performance liquid chromatography to diminish memory effects. *Proteomics*, **4** (9), 2541–2544.
66. Yates, J.R. III *et al.* (1995) Method to correlate tandem mass spectra of modified peptides to amino acid sequences in the protein database. *Anal Chem*, **67** (8), 1426–1436.

8
Applications of Tissue Microarrays in Renal Physiology and Pathology

Ali Mobasheri, Helen P. Cathro, Alex German, David Marples, Pablo Martín-Vasallo, and Cecilia M. Canessa

8.1
Introduction

The recent completion and annotation of the Human Genome Project (http://www.ornl.gov/sci/techresources/Human_Genome/home.shtml) has provided the basic structural information on all human genes. This information combined with the ongoing efforts to sequence and annotate the genomes of common experimental animals, companion animals, domesticated farm animals, fish, and other veterinary species (National Animal Genome Research Program) (http://www.animalgenome.org) will provide us with a much better understanding of the function of individual genes and the role of gene families in normal physiology and in pathology. The development and application of high-throughput postgenomic screening techniques has fundamentally transformed biomedical research. Large-scale academic and industrial efforts are presently under way to apply genomics and various related postgenomic applications, including proteomics for the identification of targets for new diagnostics and therapeutics (Figure 8.1). Identification, selection, validation, and prioritization of the best targets from tens of thousands of candidate genes and corresponding proteins are daunting tasks that require the application of multiple high-throughput techniques. Proteomics and associated high-throughput technologies are being developed to offer the scientific community a better understanding of the normal biology and physiology of cells, tissues, and organs, and to gain a better appreciation of the pathophysiology of common diseases in humans and animals [1]. Detailed proteomic and immunohistochemical analyses of molecular targets *in situ* at the cellular and tissue levels, quantitative assessment of their relative expression levels across all tissues and diseases, and evaluation of their clinical significance would therefore provide significant additional information and help prioritize target selection. Identification of novel biomarkers for early disease detection, prediction, and prognosis will reduce the morbidity and mortality associated with diseases and help define new therapeutic targets, drugs, and vaccines [1].

Renal and Urinary Proteomics: Methods and Protocols. Edited by Visith Thongboonkerd

ISBN: 978-3-527-31974-9

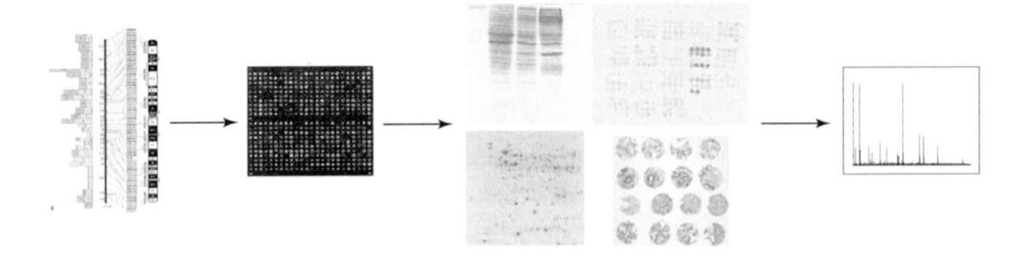

Figure 8.1 Hierarchical display of genomic and postgenomic technologies in biomedical research. Functional and integrative studies of the genome and the phenotype of a model organism are achieved through transcriptomics, proteomics (including application of TMAs), and metabolomics.

8.2
The Proliferation of Microarray Technologies

The development of high-throughput screening technologies has significantly proliferated in the last decade [2–6]. It is now possible to analyze the expression of thousands of genes at once from small quantities of tissue or cell samples [7]. Microarray technologies have pervaded all the medical, veterinary, and biological disciplines. These technologies allow investigators to determine the effects of hormones, cytokines, drugs, and environmental conditions on gene expression levels. The up- or down-regulation of subsets of genes in response to the stimulation of interest is then compared with unstimulated controls. Microarray technology can also be used to help understand cancer development, to improve patient treatment and management, and to identify those predisposed to develop cancer [8, 9]. Microarray expression patterns are currently being integrated with clinical data to identify new biomarkers to predict the potential aggressive behavior of tumors. Protein arrays are also becoming increasingly popular as the equivalent tool in the hands of the proteomics expert. However, proteomic approaches have received less interest from histopathologists and pathologists who are primarily interested in equivalent microscopic and histologic technologies, which will allow them to screen large numbers of tissue specimens for protein expression information and for discovering diagnostic and prognostic correlations. Basic scientists and clinical investigators have been using multiple tissue Northern and Western blots for many years for studies of gene and protein expression. However, these methods are expensive, laborious, and time consuming to establish. More importantly, these conventional approaches do not permit high-throughput analysis. Proteomics and tissue microarrays (TMAs) are receiving considerable attention from renal physiologists and pathologists, who are increasingly embracing these new technologies. The aim of this chapter is to bring nephrologists, urologists, and scientists from related disciplines up to date with the capabilities of TMAs as novel postgenomic innovations.

8.3
Structure and Function of the Mammalian Kidney

The kidney (Figure 8.2) is intimately involved in systemic homeostasis. Among its primary homeostatic functions are maintenance of the acid–base balance, regulation of electrolyte concentrations, control of blood volume and osmolarity, regulation of blood pressure, and secretion of several hormones (i.e., erythropoietin and renin). The kidney accomplishes these important homeostatic functions independently and through coordination with other organs, particularly those of the endocrine system. The kidney communicates with other organs through hormones secreted into the bloodstream. This organ is also involved in the excretion of a variety of waste products produced by metabolism; these include nitrogenous wastes: urea (from protein catabolism), uric acid (from nucleic acid metabolism), and metabolic water. The kidney plays an indispensable role in maintaining electrolyte and water balance, and its pathologies are a major focus of research in humans and veterinary species.

8.4
The Expanding Field of Renal Genomics and Proteomics

Proteomics is increasingly finding application in the field of nephrology [10]. Proteomic techniques are being used with increasing efficiency in the renal community [11]. Recent studies have focused on the urinary proteome, the renal proteome as a whole, and proteomes of individual renal structures (i.e., glomerulus, renal vasculature, individual nephron segments, isolated brush-border membranes, mesangial cells, and podocytes) [10]. Proteomics is also being used to study renal pathophysiology and help define novel biomarkers for various renal diseases (i.e., glomerular diseases, tubulointerstitial diseases, renal vascular disorders, and renal cancers) [1]. A number of techniques including two-dimensional gel electrophoresis, liquid chromatography/mass spectrometry, surface-enhanced laser desorption/ionization, capillary electrophoresis/mass spectrometry, antibody microarrays, and TMAs have been applied in experimental and clinical studies of the renal and urinary systems [11]. We have used a number of these postgenomic approaches (i.e., TMAs and protein microarrays) to identify and localize selected channels and transport proteins in different regions of the human kidney, equine kidney, and other human and animal organs [12–15]. Other investigators have used TMAs to associate the changes in the expression of certain proteins to the development and progression of renal cancer [16–23]. TMAs have not been used extensively for studies of inflammatory renal diseases and nephrotoxicity, and this represents opportunities for future research and development. High-throughput proteomic methodologies have also been used to study obstructive nephropathy [24], nephrotoxicity [2, 25, 26], and other kidney disease processes [27], including IgA mediated and diabetic nephropathy [28, 29]. There is also an extensive literature

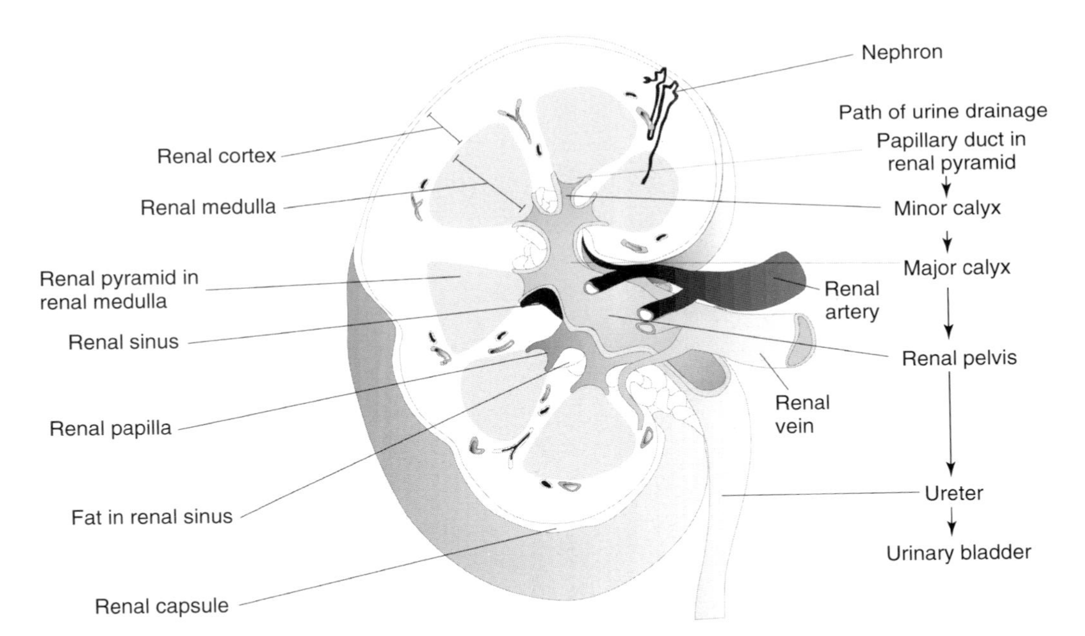

Figure 8.2 The anatomical structure of the mammalian kidney. In a normal human adult, each kidney is about 11 cm long, 5.5 cm in width, and about 3 cm thick, weighing 150 g. The outer portion of the kidney is called the renal cortex, and sits directly beneath the kidney's loose connective tissue/fibrous capsule. Deep to the cortex is the renal medulla, which is divided into 10–20 renal pyramids in humans. Each pyramid together with the associated overlying cortex forms a renal lobe. The tip of each pyramid (called a papilla) empties into a calyx, and the calices empty into the renal pelvis. The pelvis transmits urine to the urinary bladder via the ureter. On the medial aspect of each kidney is an opening, called the hilum, which admits the renal artery, the renal vein, nerves, and the ureter.

on applications of proteomic techniques to the discovery of diagnostic biomarkers in body fluids such as plasma and urine (for a recent review, see [30]).

8.5
Emergence of TMAs for Physiological and Pathophysiological Research

TMA technology is a relatively new technique, which allows rapid and simultaneous visualization of molecular targets in hundreds or thousands of tissue specimens [31] (see Figure 8.3). The molecular targets can be DNA, RNA, or protein, depending on the type of probe used for the analysis [32]. This methodology was established in response to an acute need for a reliable high-throughput technique for comprehensive screening of large numbers of human tumors [9, 31, 33]. Figure 8.4 illustrates the steps involved in producing TMAs. High-throughput TMA technology enables the clinical relevance of molecular markers to be assessed simultaneously in multiple tissue specimens [32]. The widespread adoption of TMAs in research laboratories worldwide has replaced the conventional and frustratingly tedious one-slide-one-section approach, in which individual archival clinical specimens were placed on separate microscope slides, with the ability to assess RNA, DNA, or protein biomarker expression in numerous individual tissue specimens in a single carefully controlled experiment [32]. This technology has found applications in almost all areas of human oncology and clinical medicine, and is significantly accelerating the pace of advances in translational research through more efficient assessment of new biomarkers. This technique has not only allowed pathologists to re-evaluate the use of existing tumor markers, but has also facilitated clinical practice by linking outcome and response to chemotherapy.

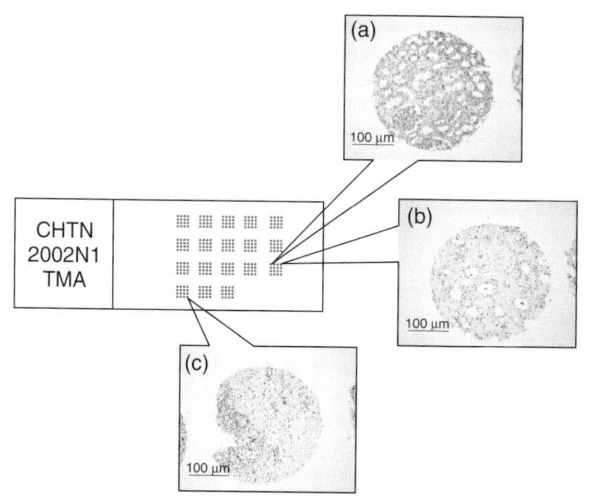

Figure 8.3 Example of high-density human TMAs showing hematoxylin and eosin stained cores from normal human kidney cortex (a), medulla (b), and urinary bladder (c).

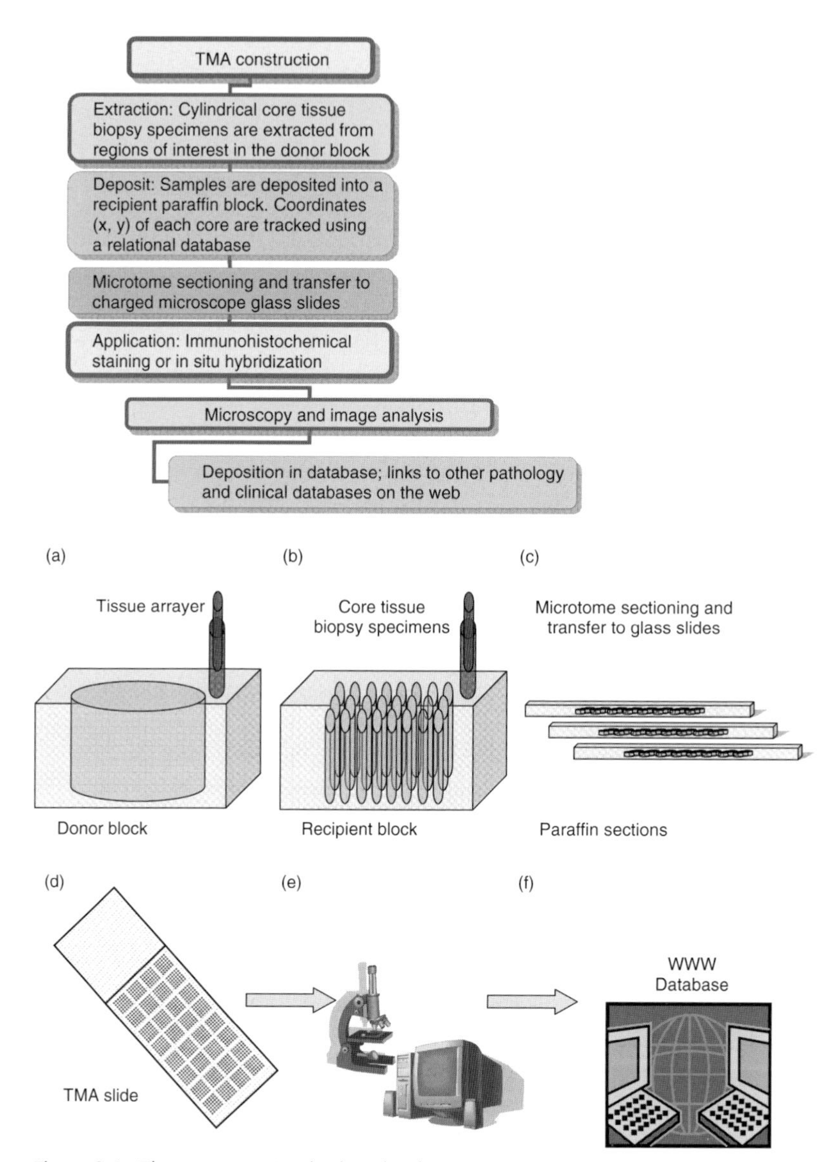

Figure 8.4 The processes involved in the design, construction, and manufacture of TMAs. The use of high-density TMAs allows validation of novel biomarker discoveries across human and animal tissues. Associating TMA data with data from clinical chemistry and pathology laboratories may permit validation of novel markers of disease pathogenesis.

8.6 TMAs for Comparative Studies of Renal Function

The potential of TMAs as tools for investigating the expression of renal markers in health and disease is only just being realized. Miyaji and coworkers [34] were the first investigators to develop frozen renal protein arrays using blocks of frozen histologic embedding compound containing an array of wells. The wells were filled with renal specimens, which were frozen and bonded to the block. Immunohistochemical staining of rat kidney cryosections with antibodies to the $\alpha 1$ subunit of Na, K-ATPase confirmed that renal ischemia decreases the abundance of this enzyme by 40% (an observation which was verified by Western blotting). This study confirmed that frozen protein arrays are a realistic platform for arraying tissue samples and that production of identical frozen protein arrays is easy, inexpensive, and requires only small quantities of tissues [34].

8.7 A Basic Method for Design of a Multispecies Renal TMA

Fresh kidneys were obtained from large animals immediately after slaughter. The kidneys were sourced from several local abattoirs, the Philip Leverhulme Large Animal Hospital (Leahurst, University of Liverpool, Neston, Cheshire) and several zoos, safari, and wildlife parks in the United Kingdom. None of the animals used had a history of renal disease and all were euthanased for unrelated clinical reasons. Fixed human kidneys were obtained from several sources including the Department of Pathology, University of Liverpool Medical School. Equine, canine, feline, and elephant kidneys were transported to the laboratory on ice within 1 hour of slaughter and were carefully dissected into cortex, medulla, and papilla before fixing in 10% neutral buffered formalin for 24 hours for histological and immunohistochemical studies. Samples were also frozen in liquid nitrogen for future SDS-PAGE, immunoblotting, and proteomic applications. Formalin fixation was used as standard but fixation was not allowed to continue beyond 24 hours to prevent drastic reduction of antigen recognition, which may occur for certain proteins exposed to formalin. Rats (Sprague–Dawley) and mice (BALB/cJ) were used in strict accordance with local ethical guidelines. They were fed standard chow and water ad libitum and were maintained under pathogen-free conditions. The animals were killed by raising CO_2 levels and death was confirmed by cervical dislocation. The kidneys were dissected and either frozen immediately in liquid nitrogen or fixed for 24 hours in 10% neutral buffered formalin and stored in 70% ethanol for subsequent immunohistochemical studies.

Renal TMAs containing samples of mouse, rat, feline, canine, human, equine, and elephant kidneys were prepared using an AbCam low density TMA builder (ab1802; Cambridge, UK). Using this basic TMA builder, samples of cortex, medulla, and papilla were carefully taken from wax embedded blocks of formalin-fixed samples from each species and placed alongside samples of renal cortex, medulla, and papilla from the above species (Figure 8.5). A total of three identical TMAs were designed for this study and were codenamed LUFVS TMA 24A/B/C. Since these arrays were intended to support comparative studies of normal kidney function, human and veterinary pathologists were recruited to confirm that the tissues represented on these TMAs were normal and free from any microscopical signs of renal pathology. Sections from these TMAs were then cut (7-μm thickness) using a microtome, and mounted on positively charged microscope glass slides and stained with Papanicolaou's hematoxylin and eosin or immunostained with polyclonal antibodies using a basic immunohistochemical protocol that has been described in several recent papers from our laboratory [12–15, 35].

Key:
Mouse Cortex and Medulla (A and B), Rat Cortex and Medulla (C and D)
Cat (Feline) Kidney (Cortex, E; Outer Medulla, F; Inner Medulla, G; Papilla, H)
Dog (Canine) Kidney (Cortex, I; Outer Medulla, J; Inner Medulla, K; Papilla, L)
Human Kidney (Cortex, M; Outer Medulla, N; Inner Medulla, O; Papilla, P)
Horse (*Equine caballus*) Kidney (Cortex, Q; Outer Medulla, R; Inner Medulla, S; Papilla, T) Elephant (*Loxodonta africana*) Kidney (Cortex, U; Outer Medulla, V; Inner Medulla, W; Papilla, X)

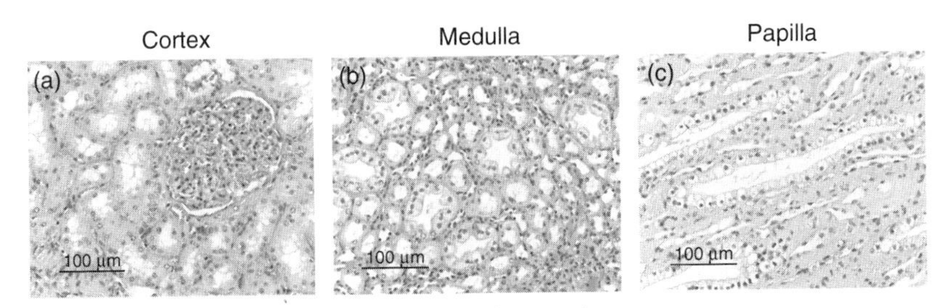

Figure 8.5 Design of a representative mammalian renal TMA consisting of 2-mm diameter spots of renal tissues from the cortex, medulla, and papilla of seven mammals in ascending size order: mouse, rat, cat, dog, human, horse, and elephant. Microscopic morphology of the equine kidney is illustrated using annotated hematoxylin- and eosin-stained images of normal equine renal cortex, medulla, and papilla.

8.8 Applications of TMAs to the Study of Renal Physiology: Case Studies

Over the last three decades, nephrologists have carried out extensive physiological characterization of water and electrolyte transport mechanisms along the renal tubule [36]. During the last 20 years, the majority of complementary DNAs encoding renal ATPase pumps, transporters, and channels involved in renal tubular electrolyte reabsorption have been cloned. The channels responsible for water and small solute transport across the nephron have been extensively studied. A large number of renal organic anion transporters have been cloned and characterized in the last 10 years. Molecular biology techniques and transcriptomics have been used to investigate physiological mechanisms of renal function at the molecular level [37]. The application of "physiological genomics" and "physiological proteomics" to the kidney has permitted a comprehensive analysis of renal water and electrolyte transporters.

Some of the current research in our laboratories is focused on basic, clinical, and comparative studies of renal transporters including Na, K-ATPase, aquaporin (AQP) water channels, and epithelial sodium channels (ENaCs) in kidneys of various domestic and exotic mammalian species. We have recently exploited a unique opportunity to develop and test new polyclonal antibodies raised against the major renal aquaporins (AQP1-4), Na, K-ATPase, and ENaC subunits and compare the distribution pattern of these channels and transporters in laboratory rodents with several other species including human, dog, cat, sheep, and horse using custom-designed renal TMAs. The information presented in the following three case studies illustrates how TMAs can be used to study the expression and distribution of renal transporters in humans and other species.

8.8.1 Renal TMA Case Study 1: Na, K-ATPase

The Na, K-ATPase belongs to the family of P-type cation transport ATPases, and to the subfamily of Na, K-ATPases. Na, K-ATPase is an integral membrane protein responsible for establishing and maintaining the electrochemical gradients of Na^+ and K^+ ions across the plasma membrane of almost all living cells [38]. These ion gradients are essential for osmoregulation, sodium-coupled transport of a variety of organic and inorganic molecules, and for electrical excitability of nerves and both cardiac and skeletal muscle. The enzyme is composed of two subunits, a large catalytic α subunit and a smaller regulatory β subunit glycoprotein [38]. Each subunit is encoded by four distinct genes and recent studies have shown that the four α subunit isoforms of Na, K-ATPase possess considerably different kinetic properties and modes of regulation, whereas the four β subunit isoforms modulate the activity, expression, and plasma membrane targeting of Na, K-ATPase isozymes. The $\alpha1$ subunit of Na, K-ATPase is thought to be the principal isoform expressed in the mammalian kidney and the distribution of this subunit along

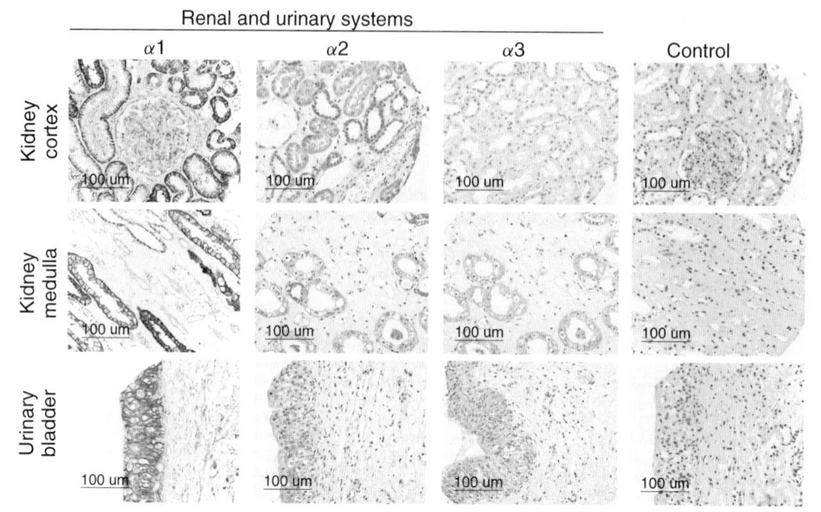

Figure 8.6 Renal TMA case study 1: immunohistochemical localization of the α subunits of Na, K-ATPase in human kidney and urinary bladder. The $\alpha 1$ subunit of Na, K-ATPase is abundantly expressed in the cortex and medulla. Lower levels of the $\alpha 2$ subunit of Na, K-ATPase were detected in the renal cortex and in the medulla. The $\alpha 3$ subunit of Na, K-ATPase was found to be expressed in the collecting duct cells. Low expression levels of both $\alpha 2$ and $\alpha 3$ subunits were detected in the urinary bladder. Solid bars represent 100 μm.

the nephron is highly segment-specific (see Figure 8.6). The presence of two isoforms of Na, K-ATPase, with different pharmacological and immunological properties, has been previously demonstrated in the rat kidney [39]. However, the authors of this rat study and subsequent studies in other species have not demonstrated the presence of any other α subunits in the kidney, and simply speculated that the two subpopulations might correspond to different isoforms of the α subunit of Na, K-ATPase ($\alpha 1$ and "$\alpha 3$-like;" [39]). Our TMA data confirms the existence of several α subunits of Na, K-ATPase in human kidney. We have used subunit-specific antibodies to demonstrate expression of the $\alpha 2$ and $\alpha 3$ subunits of Na, K-ATPase in the renal collecting duct and also in the urinary bladder (Figure 8.6). Proteomic and TMA studies are presently underway to compare the distribution of isoforms of this enzyme along the nephron in several mammalian species and examine potential changes in its expression in nephrotoxicity and in ischemic conditions.

8.8.2
Renal TMA Case Study 2: The Epithelial Sodium Channel (ENaC)

The apical membranes of many tight epithelia contain ENaCs that are pharmacologically characterized by their high affinity to the diuretic amiloride [40]. The amiloride-sensitive ENaC is a multimeric protein system consisting of three main subunits (α, β, and γ), and is involved in apical Na^+ uptake in a variety of epithelia

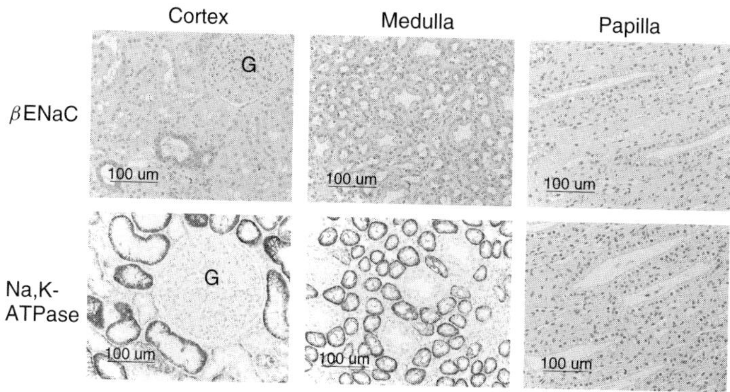

Figure 8.7 Renal TMA case study 2: immunohistochemical staining of the β subunit of ENaC and the $\alpha 1$ subunit of Na, K-ATPase in equine kidney. ENaC was predominantly expressed in the distal nephron (cortical region). Solid bars represent 100 µm. G, glomerulus.

[41, 42]. ENaC belongs to the MEC/DEG/ENaC superfamily whose members are involved in many diverse functions including acid sensing, maintenance of sodium homeostasis and transduction of mechanical stimuli and nociceptive pain [43]. ENaC represents the rate-limiting step for sodium reabsorption in the epithelia of the distal renal tubule [44]. This sodium reabsorption process is regulated by aldosterone and is essential for the maintenance of body salt and water homeostasis. ENaC also controls the reabsorption of sodium in the colon [45], lung [46], and sweat glands [47] and plays a key role in taste perception [44, 48].

The tissue distribution of ENaC has not been extensively studied. There is evidence of expression of ENaC subunits in neurons of the human and monkey telencephalon [49] and in the cardiovascular regulatory centers of the rat brain [50]. ENaC is also expressed in the bladder [51], eye [52, 53], inner ear [54], and in a few other tissues such as human articular cartilage [55] and epidermal (laminar) hoof tissues of the horse [56]. However, aside from studies in renal tissue from experimental rodents, the expression and tissue distribution of ENaC has not been extensively studied in the kidneys of humans and other large mammals. We have used the multispecies renal TMAs (shown in Figure 8.5) to study the distribution of ENaC in horse (Figure 8.7) and human (Figure 8.8) kidneys. Although the expression of ENaC in the kidney is restricted to the cortical and medullary collecting duct principal cell, a comprehensive comparative immunohistochemical study has not been performed until now. TMAs therefore provide a convenient and practical tool for studying expression of renal channels across multiple species.

8.8.3
Renal TMA Case Study 3: Aquaporins (AQPs)

AQPs are a family of water-channel proteins, which confer high intrinsic water permeability to epithelial cells of water-transporting tissues [57]. AQPs are involved

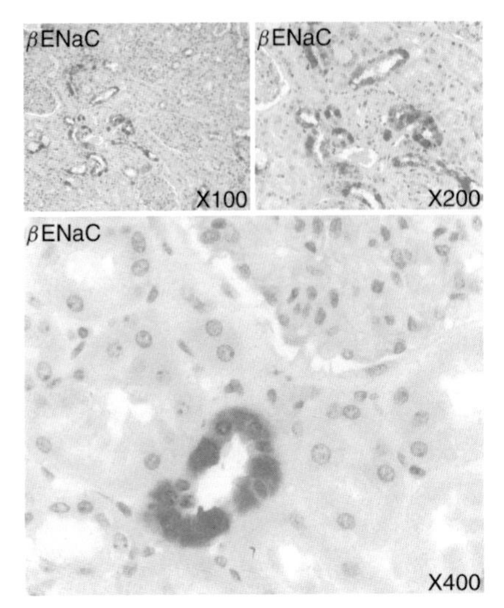

Figure 8.8 Renal TMA case study 2: immunohistochemical staining of the β subunit of ENaC in human kidney. ENaC was predominantly expressed in principal cells of the cortical distal nephron.

in transepithelial fluid absorption and secretion and rapid vectorial water movement driven by osmotic gradients [58]. At least seven AQPs are expressed in the kidney [58]. These renal AQPs, which include AQP1, AQP2, AQP3, AQP4, AQP6, AQP7, and AQP8, are differentially and strategically distributed along the nephron [59–61] to facilitate water reabsorption and concentration of urine. The cellular locations of four major AQPs (AQP1, AQP2, AQP3, and AQP4) have been well established in rodent and human kidneys. AQP1 is strongly expressed in apical and basolateral membranes of renal proximal tubules and descending thin limbs of the loop of Henle [62]. AQP1 is also present in podocytes within the glomerulus (although there is species variation in this) and in the visceral epithelium of the Bowman's capsule and the *vasa recta* [63]. In the collecting duct principal cell, AQP2 is the vasopressin-sensitive water channel, found in apical membranes and in subapical intracellular vesicles, which are inserted in apical plasma membranes when stimulated by vasopressin [59 64–68]; genetic defects in AQP2 or vasopressin type 2 receptors both result in severe nephrogenic diabetes insipidus [66 68–71]. AQP3 and AQP4 both reside in the basolateral membranes of collecting duct principal cells, providing an exit pathway for the water entering the cells apically via AQP2 channels [60, 64, 72]. AQP4 is also found in the proximal tubules (S3) of mouse, but not of rat, kidney. Significantly less is known about AQP6 and AQP7 in the kidney [73]. AQP6 is present in intracellular vesicles in intercalated cells, and AQP7 has been localized to the brush border of cells in the third segment of the proximal convoluted tubule [60, 61]. AQP8 has been demonstrated in vesicles within proximal tubules and the collecting duct [74].

We have recently utilized TMAs to study the distribution of renal AQPs in rats and horses. Immunohistochemical staining of custom-designed, multiple species, renal TMAs have allowed us to localize AQP1, AQP2, AQP3, and AQP4 in distinct regions of the equine kidney along with rodent kidneys (see Figure 8.9 and Figure 8.10). Current comparative studies in our laboratory are aimed at studying the distribution of AQPs in the renal and gastrointestinal systems of several desert-dwelling species, including camels and North American marmots.

8.9
TMAs for the Study of Renal Diseases: the Impact of TMAs on Renal Cancer Research

TMAs are being increasingly used for the study of human kidney diseases using immunohistochemistry and more sophisticated methods, because of the improved efficiency, preservation of tissue, and cost effectiveness, when compared to analysis on one slide per case basis. TMA technology holds great promise for facilitating collaborative studies, including those translating results from animal models to human diseases [75].

TMAs are constructed by sampling 0.6–2.00-mm cores of tissue from formalin-fixed paraffin-embedded blocks and inserting them into a fresh paraffin block in an orderly and carefully documented fashion [76]. Thus 50–600 specimens can be evaluated in a single immunohistochemical run, often with positive control tissue included in the same block. Staining conditions are identical across all the samples and usually two to three cores of each tumor are sampled to provide validity and allow for tissue loss. In addition to immunohistochemistry for protein expression, *in situ* hybridization for gene expression or chromosome abnormalities can be performed on this platform. Immunofluorescent studies can also be performed on TMAs, if the array has been constructed from frozen tissue. Fejzo and colleagues [77] demonstrated that frozen TMA technology can be used to analyze tumor proteins, as well as RNA and DNA. Not only surgical resection material, but also autopsy tissue, cell blocks prepared from cytologic specimens and cultured cell lines can be plugged into TMAs.

Since the initial description of TMAs in 1998 (reviewed in [31, 78]), this technique has been enthusiastically embraced by those performing both basic science and translational research [33]. For example, various types of renal cell carcinoma, chronic kidney disease (CKD), and autoimmune diseases have all been studied in this fashion. Moch and colleagues [79] evaluated the genetic profiles of the three most common types of renal cell carcinoma and oncocytomas using cDNA screening. Others have combined gene expression profiling with immunohistochemistry to subclassify better papillary renal cell carcinoma [80]. Some investigators have used immunohistochemical studies of human renal tumor TMAs to improve prognostication including treatment response [81].

However, TMAs are limited by a number of factors, not least the level of skill of those constructing the TMAs. The main problem is incorporation of tissue cores of differing lengths, usually the result of using partially exhausted donor

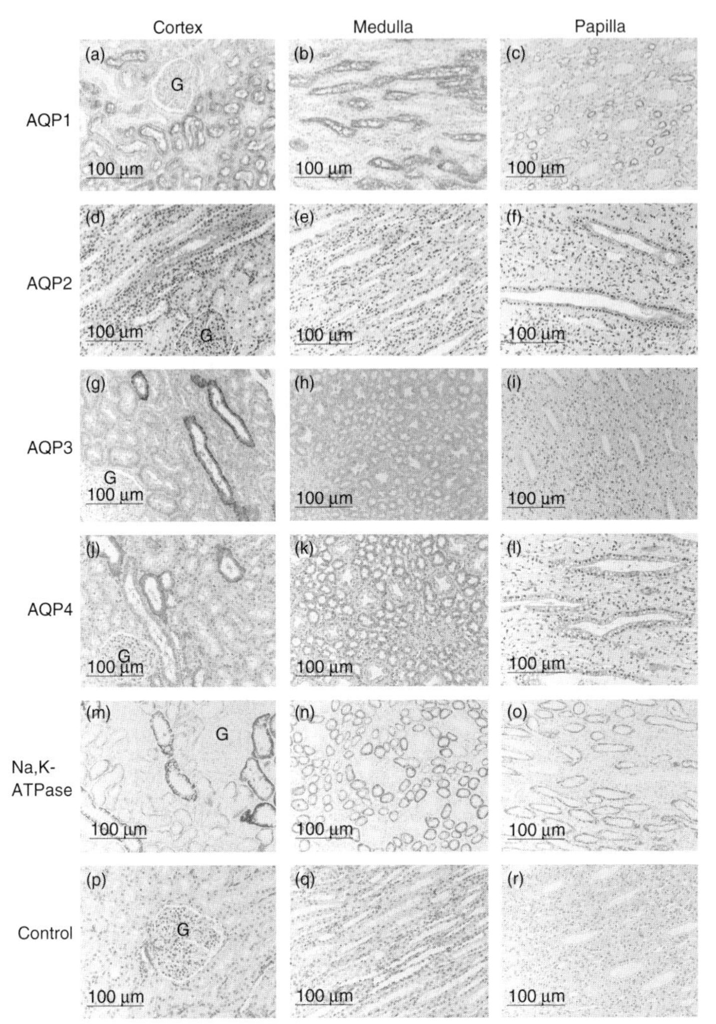

Figure 8.9 Renal TMA case study 3: immunohistochemical localization of AQP1, AQP2, AQP3, and AQP4 in the equine renal cortex, medulla, and papilla. AQP1 staining was observed in apical (brush border) and basolateral membranes of the cuboidal epithelial cell lining the proximal convoluted tubule (a) and squamous epithelial cells lining the descending loop of Henle (b and c). Weak but highly specific AQP2 immunostaining was restricted to principal cells of cortical, medullary, and papillary collecting ducts (d–f, respectively). AQP3 was strongly expressed in basolateral membranes of cells in cortical distal convoluted tubules and collecting ducts in the medulla and papilla (g–i, respectively). AQP4 was expressed in the basolateral membranes of distal convoluted tubule cells in the equine renal cortex (j) but weakly detected in medullary and papillary collecting ducts (k and l). Na, K-ATPase was highly expressed in distal convoluted tubules, thick ascending limbs of the loop of Henle, and collecting ducts of the equine kidney (m–o). Negative controls are shown in panels p, q, and r. Glomeruli are labeled G; bars represent 10 μm. Reproduced from [12] with copyright permission of the American Physiological Society.

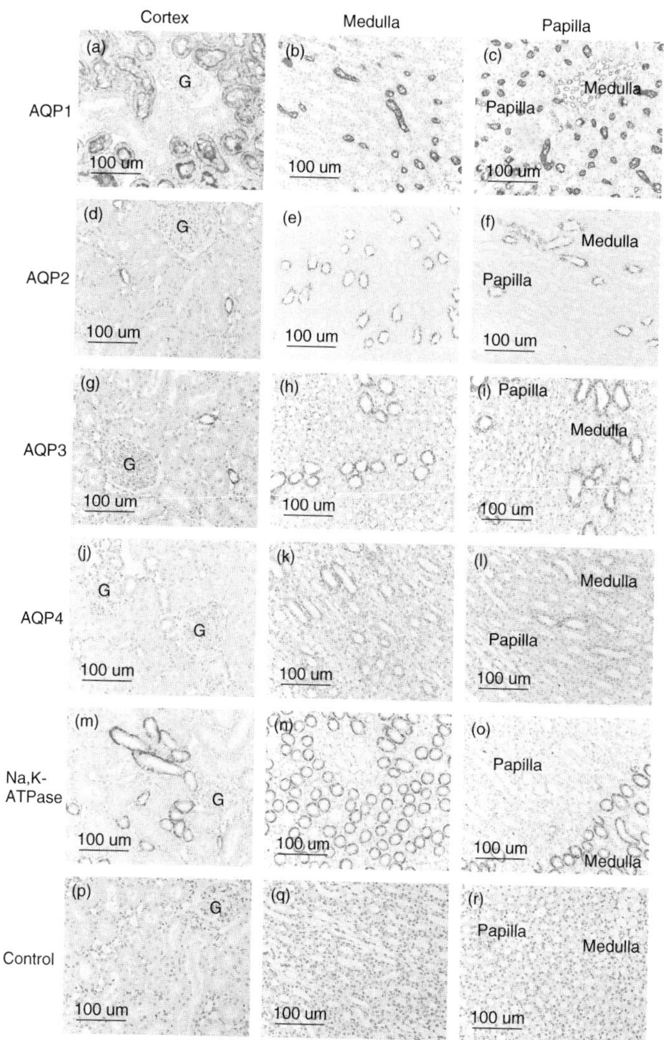

Figure 8.10 Renal TMA case study 3: immunohistochemical localization of AQP1, AQP2, AQP3, and AQP4 in the rat renal cortex, medulla, and medulla/papilla junction. AQP1 was present in apical and basolateral membranes of proximal convoluted tubule cells and in squamous epithelial cells of the descending loop of Henle (a–c). AQP2 expression was restricted to cells of the cortical and medullary collecting ducts, and distal convoluted tubules (d–f). AQP3 immunostaining was observed in distal convoluted tubules and collecting ducts (g–i). AQP4 was expressed in the basolateral membrane of distal convoluted tubules and collecting ducts (j–l). Na, K-ATPase was abundantly expressed in distal convoluted tubules, thick ascending limbs of the loop of Henle, and collecting ducts of the rat kidney (m–o). Na, K-ATPase expression was significantly lower in rat proximal tubules (m) and its immunostaining in the papilla (o) was significantly diminished compared to the cortex (m) and medulla (n). Negative controls are shown in panels p–r. Glomeruli are labeled G; bars represent 10 μm. Reproduced from [12] with copyright permission of the American Physiological Society.

tissue blocks. This leads to the loss of individual tumors at various levels on the recipient blocks. Those experienced in TMA construction can circumvent this problem by stacking multiple cores in the slot, but this takes true skill. Another problem is the difficulty of sampling some tumors because of subtle morphology or the variable presence of tumor throughout the donor block. Protein expression may be heterogeneous in a subset of neoplasms, limiting the utility of TMAs in their study. *In situ* hybridization may be troublesome on limited tissue samples, necessitating the use of full sections or frozen tissue TMAs. Frozen section TMAs can also circumvent the issue of variable duration of formalin fixation in donor tissue, hindering reproducibility of results. And, of course, an investigator using a TMA manufactured by a second party is dependent upon them for the integrity of the tissue, (i.e., primary untreated tumor material), clinical data, and follow-up.

Antigen and antibody microarrays are also finding application in basic and translational research. Antigen arrays are useful for the study of autoimmune diseases in particular, and also infectious diseases [82]. Antigens of various types including intact particles, proteins, lipids, carbohydrates, and peptides can all be placed on the same array [83, 84]. Autoantibodies occur in large concentrations in human autoimmune diseases; miniaturized fluorescence-based multiplex assays can reduce the need for large human samples to ensure sufficient antigen as well as large reagent volumes. This relatively low-cost method may allow rapid screening for early disease and treatment, characterization of specificity, diversity, and epitope spreading, determination of isotype subclass of antibodies, and may guide antigen-specific treatment. Minimal or complete lack of cross-reactivity to nonspecific proteins has already been achieved.

Antibody microarrays have demonstrated increased excretion of urinary cytokines in patients with CKD, providing insight into the initiation and progression of CKD. This technology allows simultaneous analysis of multiple cytokines – an absolute necessity when studying the complex interplay of multiple pathways. Capture-antibody microarrays use immobilized antibodies to trap specific antigens in sample solutions of various types. Detection can be performed by means of direct fluorophores attached to the antigens or by a sandwich immunoassay. Limitations of this technique include technical complexity, high cost, and the difficulty of quantifying target proteins [85].

8.10
Conclusions and Future Perspectives

The application of proteomics, TMAs, and related "omics" to nephrology will help us to improve our understanding of normal renal physiology and to explore the complexity of renal disease mechanisms, as well as enabling identification of novel biomarkers and new therapeutic targets [86]. In this chapter, we have provided our own examples and unique perspectives on how TMAs may be applied in experimental renal research, and cited the works of other investigators in clinical nephrology. We have also discussed how TMA data can be used to explore renal function in the context of normal water and electrolyte transport physiology. It is clear that proteomics and TMA technology have not been utilized extensively in veterinary medicine. The veterinary community has a unique opportunity to exploit these relatively straightforward technologies in research and diagnostics. The establishment of proteomic reference maps of plasma proteins, renal tissue proteins and renal tissue TMAs, from veterinary and domesticated species, may serve as a new and powerful tool in future studies. Changes in the protein repertoire in common renal diseases of companion animals can then be evaluated, and immunological responses to infectious diseases of the urinary system investigated. Utilizing proteomics and TMAs to increase our basic knowledge of renal processes will contribute to a more refined understanding of renal function. Proteomic approaches will also shed light on the significance of urinary biomarkers in inherited and acquired renal disorders, renal transplantation, renal replacement therapy (for acute renal failure and end-stage renal disease), and systemic diseases that affect the renal system. The rapid growth of development of these technologies along with the application of conventional physiological and biochemical techniques will facilitate new discoveries in nephrology and will help us identify and characterize better molecular markers for novel diagnostic kits and clinical assays.

Acknowledgments

Our renal research has received financial support from the Wellcome Trust, the BBSRC, and Novartis. The authors wish to thank Mr A.F. Brandwood and Mr S. Williams (Department of Veterinary Pathology, University of Liverpool) for excellent histology support.

References

1. Thongboonkerd, V. (2005) Proteomic analysis of renal diseases: unraveling the pathophysiology and biomarker discovery. *Expert Rev Proteomics*, **2** (3), 349–366.
2. Kennedy, S. (2002) The role of proteomics in toxicology: identification of biomarkers of toxicity by protein expression analysis. *Biomarkers*, **7** (4), 269–290.
3. Heller, M.J. (2002) DNA microarray technology: devices, systems, and applications. *Annu Rev Biomed Eng*, **4**, 129–153.
4. Kumble, K.D. (2003) Protein microarrays: new tools for pharmaceutical development. *Anal Bioanal Chem*, **377** (5), 812–819.
5. Sanchez-Carbayo, M. (2006) Antibody arrays: technical considerations and clinical applications in cancer. *Clin Chem*, **52** (9), 1651–1659.
6. Uttamchandani, M., Wang, J., and Yao, S.Q. (2006) Protein and small molecule microarrays: powerful tools for high-throughput proteomics. *Mol Biosyst*, **2** (1), 58–68.
7. Schena, M., Shalon, D., Davis, R.W., and Brown, P.O. (1995) Quantitative monitoring of gene expression patterns with a complementary DNA

microarray. *Science*, **270** (5235), 467–470.
8. Bubendorf, L., Kolmer, M., Kononen, J., Koivisto, P., Mousses, S., Chen, Y. *et al.* (1999) Hormone therapy failure in human prostate cancer: analysis by complementary DNA and tissue microarrays. *J Natl Cancer Inst*, **91** (20), 1758–1764.
9. Kallioniemi, O.P. (2001) Biochip technologies in cancer research. *Ann Med*, **33** (2), 142–147.
10. Thongboonkerd, V. (2004) Proteomics in nephrology: current status and future directions. *Am J Nephrol*, **24** (3), 360–378.
11. Janech, M.G., Raymond, J.R., and Arthur, J.M. (2007) Proteomics in renal research. *Am J Physiol Renal Physiol*, **292** (2), F501–F512.
12. Floyd, R.V., Mason, S.L., Proudman, C.J., German, A.J., Marples, D., and Mobasheri, A. (2007) Expression and nephron segment specific distribution of major renal aquaporins (AQP1-4) in Equus caballus, the domestic horse. *Am J Physiol Regul Integr Comp Physiol*, **293** (1), R492–R503.
13. Mobasheri, A., Airley, R., Hewitt, S.M., and Marples, D. (2005) Heterogeneous expression of the aquaporin 1 (AQP1) water channel in tumors of the prostate, breast, ovary, colon and lung: a study using high density multiple human tumor tissue microarrays. *Int J Oncol*, **26** (5), 1149–1158.
14. Mobasheri, A. and Marples, D. (2004) Expression of the AQP-1 water channel in normal human tissues: a semiquantitative study using tissue microarray technology. *Am J Physiol Cell Physiol*, **286** (3), C529–C537.
15. Mobasheri, A., Wray, S., and Marples, D. (2005) Distribution of AQP2 and AQP3 water channels in human tissue microarrays. *J Mol Histol*, **36** (1-2), 1–14.
16. Tretiakova, M.S., Sahoo, S., Takahashi, M., Turkyilmaz, M., Vogelzang, N.J., Lin, F. *et al.* (2004) Expression of alpha-methylacyl-CoA racemase in papillary renal cell carcinoma. *Am J Surg Pathol*, **28** (1), 69–76.
17. Ouban, A., Muraca, P., Yeatman, T., and Coppola, D. (2003) Expression and distribution of insulin-like growth factor-1 receptor in human carcinomas. *Hum Pathol*, **34** (8), 803–808.
18. Lam, J.S., Pantuck, A.J., Belldegrun, A.S., and Figlin, R.A. (2007) Protein expression profiles in renal cell carcinoma: staging, prognosis, and patient selection for clinical trials. *Clin Cancer Res*, **13** (2 Pt 2), 703s–708s.
19. Robb, V.A., Karbowniczek, M., Klein-Szanto, A.J., and Henske, E.P. (2007) Activation of the mTOR signaling pathway in renal clear cell carcinoma. *J Urol*, **177** (1), 346–352.
20. Warford, A., Flack, G., Conquer, J.S., Zola, H., and McCafferty, J. (2007) Assessing the potential of immunohistochemistry for systematic gene expression profiling. *J Immunol Methods*, **318** (1-2), 125–137.
21. Jin, J.S., Chen, A., Hsieh, D.S., Yao, C.W., Cheng, M.F., and Lin, Y.F. (2006) Expression of serine protease matriptase in renal cell carcinoma: correlation of tissue microarray immunohistochemical expression analysis results with clinicopathological parameters. *Int J Surg Pathol*, **14** (1), 65–72.
22. Paner, G.P., Srigley, J.R., Radhakrishnan, A., Cohen, C., Skinnider, B.F., Tickoo, S.K. *et al.* (2006) Immunohistochemical analysis of mucinous tubular and spindle cell carcinoma and papillary renal cell carcinoma of the kidney: significant immunophenotypic overlap warrants diagnostic caution. *Am J Surg Pathol*, **30** (1), 13–19.
23. Jubb, A.M., Pham, T.Q., Hanby, A.M., Frantz, G.D., Peale, F.V., Wu, T.D. *et al.* (2004) Expression of vascular endothelial growth factor, hypoxia inducible factor 1alpha, and carbonic anhydrase IX in human tumours. *J Clin Pathol*, **57** (5), 504–512.

24. Chevalier, R.L. (2004) Biomarkers of congenital obstructive nephropathy: past, present and future. *J Urol*, **172** (3), 852–857.
25. Betton, G.R., Kenne, K., Somers, R., and Marr, A. (2005) Protein biomarkers of nephrotoxicity; a review and findings with cyclosporin A, a signal transduction kinase inhibitor and N-phenylanthranilic acid. *Cancer Biomarkers*, **1** (1), 59–67.
26. Witzmann, F.A. and Li, J. (2004) Proteomics and nephrotoxicity. *Contrib Nephrol*, **141**, 104–123.
27. Dihazi, H. and Muller, G.A. (2007) Urinary proteomics: a tool to discover biomarkers of kidney diseases. *Expert Rev Proteomics*, **4** (1), 39–50.
28. Merchant, M.L. and Klein, J.B. (2005) Proteomics and diabetic nephropathy. *Curr Diab Rep*, **5** (6), 464–469.
29. Yasuda, Y., Horie, A., Odani, H., Iwase, H., and Hiki, Y. (2004) Application of mass spectrometry to IgA nephropathy: structural and biological analyses of underglycosylated IgA1 molecules. *Contrib Nephrol*, **141**, 170–188.
30. Pisitkun, T., Johnstone, R., and Knepper, M.A. (2006) Discovery of urinary biomarkers. *Mol Cell Proteomics*, **5** (10), 1760–1771.
31. Kallioniemi, O.P., Wagner, U., Kononen, J., and Sauter, G. (2001) Tissue microarray technology for high-throughput molecular profiling of cancer. *Hum Mol Genet*, **10** (7), 657–662.
32. Henshall, S. (2003) Tissue microarrays. *J Mammary Gland Biol Neoplasia*, **8** (3), 347–358.
33. Kononen, J., Bubendorf, L., Kallioniemi, A., Barlund, M., Schraml, P., Leighton, S. *et al.* (1998) Tissue microarrays for high-throughput molecular profiling of tumor specimens. *Nat Med*, **4** (7), 844–847.
34. Miyaji, T., Hewitt, S.M., Liotta, L.A., and Star, R.A. (2002) Frozen protein arrays: a new method for arraying and detecting recombinant and native tissue proteins. *Proteomics*, **2** (11), 1489–1493.
35. Simpson, D.M., Mobasheri, A., Haywood, S., and Beynon, R.J. (2006) A proteomics study of the response of North Ronaldsay sheep to copper challenge. *BMC Vet Res*, **2**, 36.
36. Knepper, M.A. and Masilamani, S. (2001) Targeted proteomics in the kidney using ensembles of antibodies. *Acta Physiol Scand*, **173** (1), 11–21.
37. Cheval, L., Virlon, B., Billon, E., Aude, J.C., Elalouf, J.M., and Doucet, A. (2002) Large-scale analysis of gene expression: methods and application to the kidney. *J Nephrol*, **15** (Suppl 5), S170–S183.
38. Mobasheri, A., Avila, J., Cozar-Castellano, I., Brownleader, M.D., Trevan, M., Francis, M.J. *et al.* (2000) Na+, K+-ATPase isozyme diversity; comparative biochemistry and physiological implications of novel functional interactions. *Biosci Rep*, **20** (2), 51–91.
39. Feraille, E., Barlet-Bas, C., Cheval, L., Rousselot, M., Carranza, M.L., Dreher, D. *et al.* (1995) Presence of two isoforms of Na, K-ATPase with different pharmacological and immunological properties in the rat kidney. *Pflugers Arch*, **430** (2), 205–212.
40. Horisberger, J.D., Puoti, A., Canessa, C., and Rossier, B.C. (1994) The amiloride receptor. *Clin Invest*, **72** (9), 695–697.
41. Canessa, C.M., Horisberger, J.D., and Rossier, B.C. (1993) Epithelial sodium channel related to proteins involved in neurodegeneration. *Nature*, **361** (6411), 467–470.
42. Canessa, C.M., Schild, L., Buell, G., Thorens, B., Gautschi, I., Horisberger, J.D. *et al.* (1994) Amiloride-sensitive epithelial Na+ channel is made of three homologous subunits. *Nature*, **367** (6462), 463–467.
43. Fyfe, G.K., Quinn, A., and Canessa, C.M. (1998) Structure and function of the Mec-ENaC family of ion channels. *Semin Nephrol*, **18** (2), 138–151.

44. Rossier, B.C., Canessa, C.M., Schild, L., and Horisberger, J.D. (1994) Epithelial sodium channels. *Curr Opin Nephrol Hypertens*, **3** (5), 487–496.
45. Coric, T., Hernandez, N., Alvarez de la Rosa, D., Shao, D., Wang, T., and Canessa, C.M. (2004) Expression of ENaC and serum- and glucocorticoid-induced kinase 1 in the rat intestinal epithelium. *Am J Physiol Gastrointest Liver Physiol*, **286** (4), G663–G670.
46. Matsushita, K., McCray, P.B. Jr, Sigmund, R.D., Welsh, M.J., and Stokes, J.B. (1996) Localization of epithelial sodium channel subunit mRNAs in adult rat lung by in situ hybridization. *Am J Physiol*, **271** (2 Pt 1), L332–L339.
47. Garty, H. and Palmer, L.G. (1997) Epithelial sodium channels: function, structure, and regulation. *Physiol Rev*, **77** (2), 359–396.
48. Lin, W., Finger, T.E., Rossier, B.C., and Kinnamon, S.C. (1999) Epithelial Na+ channel subunits in rat taste cells: localization and regulation by aldosterone. *J Comp Neurol*, **405** (3), 406–420.
49. Giraldez, T., Afonso-Oramas, D., Cruz-Muros, I., Garcia-Marin, V., Pagel, P., Gonzalez-Hernandez, T. *et al.* (2007) Cloning and functional expression of a new epithelial sodium channel delta subunit isoform differentially expressed in neurons of the human and monkey telencephalon. *J Neurochem*, **102** (4), 1304–1315.
50. Amin, M.S., Wang, H.W., Reza, E., Whitman, S.C., Tuana, B.S., and Leenen, F.H. (2005) Distribution of epithelial sodium channels and mineralocorticoid receptors in cardiovascular regulatory centers in rat brain. *Am J Physiol Regul Integr Comp Physiol*, **289** (6), R1787–R1797.
51. Watanabe, S., Matsushita, K., McCray, P.B. Jr, and Stokes, J.B. (1999) Developmental expression of the epithelial Na+ channel in kidney and uroepithelia. *Am J Physiol*, **276** (2 Pt 2), F304–F314.
52. Rauz, S., Walker, E.A., Murray, P.I., and Stewart, P.M. (2003) Expression and distribution of the serum and glucocorticoid regulated kinase and the epithelial sodium channel subunits in the human cornea. *Exp Eye Res*, **77** (1), 101–108.
53. Mirshahi, M., Nicolas, C., Mirshahi, S., Golestaneh, N., d'Hermies, F., and Agarwal, M.K. (1999) Immunochemical analysis of the sodium channel in rodent and human eye. *Exp Eye Res*, **69** (1), 21–32.
54. Zhong, S.X. and Liu, Z.H. (2004) Immunohistochemical localization of the epithelial sodium channel in the rat inner ear. *Hear Res*, **193** (1-2), 1–8.
55. Trujillo, E., Alvarez de la Rosa, D., Mobasheri, A., Gonzalez, T., Canessa, C.M., and Martin-Vasallo, P. (1999) Sodium transport systems in human chondrocytes. II. Expression of ENaC, Na+/K+/2Cl- cotransporter and Na+/H+ exchangers in healthy and arthritic chondrocytes. *Histol Histopathol*, **14** (4), 1023–1031.
56. Mobasheri, A., Critchlow, K., Clegg, P.D., Carter, S.D., and Canessa, C.M. (2004) Chronic equine laminitis is characterised by loss of GLUT1, GLUT4 and ENaC positive laminar keratinocytes. *Equine Vet J*, **36** (3), 248–254.
57. Agre, P., King, L.S., Yasui, M., Guggino, W.B., Ottersen, O.P., Fujiyoshi, Y. *et al.* (2002) Aquaporin water channels–from atomic structure to clinical medicine. *J Physiol*, **542** (Pt 1), 3–16.
58. Verkman, A.S. (2000) Physiological importance of aquaporins: lessons from knockout mice. *Curr Opin Nephrol Hypertens*, **9** (5), 517–522.
59. Nielsen, S. and Agre, P. (1995) The aquaporin family of water channels in kidney. *Kidney Int*, **48** (4), 1057–1068.
60. Nielsen, S., Frokiaer, J., Marples, D., Kwon, T.H., Agre, P., and Knepper, M.A. (2002) Aquaporins in the kidney: from molecules to medicine. *Physiol Rev*, **82** (1), 205–244.

61. Nielsen, S., Kwon, T.H., Christensen, B.M., Promeneur, D., Frokiaer, J., and Marples, D. (1999) Physiology and pathophysiology of renal aquaporins. *J Am Soc Nephrol*, **10** (3), 647–663.
62. Agre, P. and Nielsen, S. (1996) The aquaporin family of water channels in kidney. *Nephrologie*, **17** (7), 409–415.
63. Trujillo, E., Gonzalez, T., Marin, R., Martin-Vasallo, P., Marples, D., and Mobasheri, A. (2004) Human articular chondrocytes, synoviocytes and synovial microvessels express aquaporin water channels; upregulation of AQP1 in rheumatoid arthritis. *Histol Histopathol*, **19** (2), 435–444.
64. Knepper, M.A., Wade, J.B., Terris, J., Ecelbarger, C.A., Marples, D., Mandon, B. *et al.* (1996) Renal aquaporins. *Kidney Int*, **49** (6), 1712–1717.
65. Deen, P.M., van Balkom, B.W., and Kamsteeg, E.J. (2000) Routing of the aquaporin-2 water channel in health and disease. *Eur J Cell Biol*, **79** (8), 523–530.
66. Deen, P.M. and Knoers, N.V. (1998) Vasopressin type-2 receptor and aquaporin-2 water channel mutants in nephrogenic diabetes insipidus. *Am J Med Sci*, **316** (5), 300–309.
67. Fushimi, K., Sasaki, S., and Marumo, F. (1997) Phosphorylation of serine 256 is required for cAMP-dependent regulatory exocytosis of the aquaporin-2 water channel. *J Biol Chem*, **272** (23), 14800–14804.
68. Deen, P.M., Verdijk, M.A., Knoers, N.V., Wieringa, B., Monnens, L.A., van Os, C.H. *et al.* (1994) Requirement of human renal water channel aquaporin-2 for vasopressin-dependent concentration of urine. *Science*, **264** (5155), 92–95.
69. Deen, P.M., Croes, H., van Aubel, R.A., Ginsel, L.A., and van Os, C.H. (1995) Water channels encoded by mutant aquaporin-2 genes in nephrogenic diabetes insipidus are impaired in their cellular routing. *J Clin Invest*, **95** (5), 2291–2296.
70. van Lieburg, A.F., Verdijk, M.A., Knoers, V.V., van Essen, A.J., Proesmans, W., Mallmann, R. *et al.* (1994) Patients with autosomal nephrogenic diabetes insipidus homozygous for mutations in the aquaporin 2 water-channel gene. *Am J Hum Genet*, **55** (4), 648–652.
71. Mulders, S.M., Knoers, N.V., Van Lieburg, A.F., Monnens, L.A., Leumann, E., Wuhl, E. *et al.* (1997) New mutations in the AQP2 gene in nephrogenic diabetes insipidus resulting in functional but misrouted water channels. *J Am Soc Nephrol*, **8** (2), 242–248.
72. Ishibashi, K., Sasaki, S., Fushimi, K., Uchida, S., Kuwahara, M., Saito, H. *et al.* (1994) Molecular cloning and expression of a member of the aquaporin family with permeability to glycerol and urea in addition to water expressed at the basolateral membrane of kidney collecting duct cells. *Proc Natl Acad Sci U S A*, **91** (14), 6269–6273.
73. Verkman, A.S. (2002) Physiological importance of aquaporin water channels. *Ann Med*, **34** (3), 192–200.
74. Elkjaer, M.L., Nejsum, L.N., Gresz, V., Kwon, T.H., Jensen, U.B., Frokiaer, J. *et al.* (2001) Immunolocalization of aquaporin-8 in rat kidney, gastrointestinal tract, testis, and airways. *Am J Physiol Renal Physiol*, **281** (6), F1047–F1057.
75. Moch, H., Kononen, T., Kallioniemi, O.P., and Sauter, G. (2001) Tissue microarrays: what will they bring to molecular and anatomic pathology? *Adv Anat Pathol*, **8** (1), 14–20.
76. Mobasheri, A., Airley, R., Foster, C.S., Schulze-Tanzil, G., and Shakibaei, M. (2004) Post-genomic applications of tissue microarrays: basic research, prognostic oncology, clinical genomics and drug discovery. *Histol Histopathol*, **19** (1), 325–335.
77. Schoenberg Fejzo, M., and Slamon, D.J. (2001) Frozen tumor tissue microarray technology for analysis of tumor RNA, DNA, and proteins. *Am J Pathol*, **159** (5), 1645–1650.

78. Bubendorf, L. (2001) High-throughput microarray technologies: from genomics to clinics. *Eur Urol*, **40** (2), 231–238.
79. Moch, H., Schraml, P., Bubendorf, L., Mirlacher, M., Kononen, J., Gasser, T. (1999) High-throughput tissue microarray analysis to evaluate genes uncovered by cDNA microarray screening in renal cell carcinoma. *Am J Pathol*, **154** (4), 981–986.
80. Yang, X.J., Tan, M.H., Kim, H.L., Ditlev, J.A., Betten, M.W., Png, C.E. *et al.* (2005) A molecular classification of papillary renal cell carcinoma. *Cancer Res*, **65** (13), 5628–5637.
81. Atkins, M., Regan, M., McDermott, D., Mier, J., Stanbridge, E., Youmans, A. *et al.* (2005) Carbonic anhydrase IX expression predicts outcome of interleukin 2 therapy for renal cancer. *Clin Cancer Res*, **11** (10), 3714–3721.
82. Fathman, C.G., Soares, L., Chan, S.M., and Utz, P.J. (2005) An array of possibilities for the study of autoimmunity. *Nature*, **435** (7042), 605–611.
83. Robinson, W.H., DiGennaro, C., Hueber, W., Haab, B.B., Kamachi, M., Dean, E.J. *et al.* (2002) Autoantigen microarrays for multiplex characterization of autoantibody responses. *Nat Med*, **8** (3), 295–301.
84. Joos, T.O., Schrenk, M., Hopfl, P., Kroger, K., Chowdhury, U., Stoll, D. *et al.* (2000) A microarray enzyme-linked immunosorbent assay for autoimmune diagnostics. *Electrophoresis*, **21** (13), 2641–2650.
85. Liu, B.C., Zhang, L., Lv, L.L., Wang, Y.L., Liu, D.G., and Zhang, X.L. (2006) Application of antibody array technology in the analysis of urinary cytokine profiles in patients with chronic kidney disease. *Am J Nephrol*, **26** (5), 483–490.
86. Thongboonkerd, V. and Malasit, P. (2005) Renal and urinary proteomics: current applications and challenges. *Proteomics*, **5** (4), 1033–1042.

9
Proteomic and Mass Spectrometric Analyses of Formalin-Fixed Paraffin-Embedded Tissue

Brian M. Balgley, Weijie Wang, and Cheng S. Lee

9.1
Introduction and Overview of the Antigen Retrieval Method

Formalin-fixed and paraffin-embedded (FFPE) tissues represent the vast majority of archived tissues throughout the world. This is due to the ease, effectiveness, low cost, and historical use associated with this practice in contrast to other tissue preservation techniques. This practice has encouraged the development of a plethora of techniques using these tissues to conduct pathological evaluations. The most common of these is histopathological examination of FFPE sections, which are stained typically with hematoxylin and eosin, to reveal the morphological features of the tissue. Immunohistochemistry has come into widespread use over the past 20 years owing to its ability to specifically stain for a protein of interest based on recognition of an antigenic epitope on the protein by an antibody. However, formaldehyde-induced cross-linking can destroy antigenicity by modifying the reactive epitope. Early attempts were made to retrieve antigens from the fixed tissue section by proteolysis with trypsin [1, 2]. However, the method was not easily reproducible and the relatively short fragments retrieved were not necessarily the antigenic epitopes desired. In 1991, Shi *et al.* [3] demonstrated that heat could be used to greatly enhance antigenicity and it was proposed that the heating process might reverse the formaldehyde-induced cross-links. In 1998, Ikeda *et al.* [4] demonstrated that addition of SDS to the heating process further enhanced antigen recovery. More recently, in-depth studies have been conducted on standard proteins in an effort to characterize the effects of formalin fixation. Rait *et al.* [5] have proposed a mechanism by which formaldehyde-induced methylene bridges may be reversed, thus restoring the native protein sequence. Metz *et al.* [6] have characterized cross-linkages by mass spectrometry (MS) and have shown that only primary amino and thiol groups react with formaldehyde to form methylene bridges with arginine, asparagine, glutamine, histidine, tryptophan, and tyrosine. They note that other formaldehyde-induced modifications to methylol groups and Schiff bases are reversible and therefore generally not observed.

The ability to extract proteins with minimal modifications has opened up FFPE archived tissues to investigation by many standard protein analytical techniques.

Renal and Urinary Proteomics: Methods and Protocols. Edited by Visith Thongboonkerd

ISBN: 978-3-527-31974-9

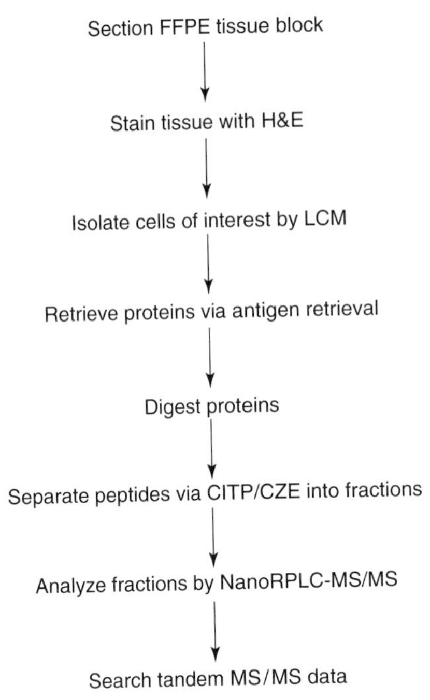

Figure 9.1 Protocol overview.

One such method is shotgun peptide sequencing by liquid chromatography-mass spectrometry (LC-MS). Prieto *et al.* [7] were the first to demonstrate this method applied to FFPE tissues using a commercially available extraction kit of unknown composition, but using a protocol similar to that of Ikeda *et al.* [4]. In collaboration with Shi, our group has applied a highly sensitive multidimensional separation technology to further evaluate the capabilities and limitations of the heat- and detergent-based antigen retrieval technique [8]. We have furthered this study by examining the technique's capability with regard to microdissected tissues [9]. An overview of the protocol is shown in Figure 9.1.

9.2 Methods and Protocols

9.2.1 Clinical Samples

Tissues and clinical (pathological) information must be obtained as part of a study approved by the Institutional Review Board. Tissues to be preserved should immediately be placed in neutral 10% buffered formalin, pH 7.0, and sent for routine processing to the hospital's pathology department. Fixation should proceed for no more than 24 hours followed by embedding in paraffin wax and storage at room temperature.

9.2.2 Materials and Reagents

Fused-silica capillaries (50-μm i.d./375-μm o.d. and 100-μm i.d./375-μm o.d.) may be acquired from Polymicro Technologies (Phoenix, AZ). Acetic acid, ammonium acetate, ammonium hydroxide, dithiothreitol (DTT), eosin, formic acid, hematoxylin, iodoacetamide (IAM), Scotts tap water substitute, and xylene can be obtained from Sigma (St. Louis, MO). Acetonitrile, hydroxypropyl cellulose (HPC, average MW 100 000), SDS, tris(hydroxymethyl)aminomethane (Tris), and urea can be purchased from Fisher Scientific (Pittsburgh, PA). Sequencing grade trypsin is obtained from Promega (Madison, WI). Pharmalyte 8–10 is obtained from GE Healthcare. All solutions are prepared using reverse-osmosis purified water filtered to 50 nm.

9.2.3 Tissue Staining and Laser-Capture Microdissection

Using a microtome, cut 5–10-μm sections from the FFPE tissue blocks. The number of sections to be cut will depend on the analysis to be performed and the number of cells available on each section representing the region of interest. Reports vary widely concerning the amount of protein available from given quantities of formalin-fixed tissues [10–14]. For the CE-based shotgun proteomics platform described in this protocol, typically 10 000 cells are utilized per analysis. However, it is often desirable to procure larger numbers of cells, if possible, to permit repeated or follow-on analyses and to allow for handling error. For a 5-μm section, approximately 10 000 cells can be obtained from an area of 6 mm^2, assuming an average cell diameter of 15 μm. In practice, it is preferable to dissect constant areas sample-to-sample rather than attempting to count cells, as area calculation is performed instantaneously by the laser-capture microscope. Mount the sections on polyethylene naphthalate membrane glass slides (Molecular Devices, Sunnyvale, CA). Then proceed to staining by placing the slides into a coplin jar containing the following solutions in sequence:

1. xylene, fresh, 5 minutes
2. xylene, fresh, 5 minutes
3. 100% ethanol, 15 seconds
4. 95% ethanol, 15 seconds
5. 70% ethanol, 15 seconds
6. ultrapure water, 15 seconds
7. Mayer's hematoxylin, 15 seconds
8. Ultrapure water, 15 seconds
9. Scott's tap water substitute, 10 seconds

10. 70% ethanol, 15 seconds
11. Eosin Y, 5 seconds
12. 95% ethanol, 15 seconds
13. 95% ethanol, 15 seconds
14. 100% ethanol, 15 seconds
15. 100% ethanol, 15 seconds
16. xylene, fresh, 60 seconds
17. xylene, fresh, 60 seconds
18. completely dry using a low-flow stream of ultrapure N_2.

The slide may be stored at room temperature prior to microdissection. Note that hemtatoxylin and eosin may be used at a fraction of their standard concentrations to improve both the brightness and the contrast of the noncoverslipped section images. However, in our experience, no differences are observed in either protein recovery or overall proteomic results between stained and unstained tissue sections.

Laser-capture microdissection has been described in great detail elsewhere [15] and is covered only briefly here. Outline areas that are relatively tumor-rich and avoid cells and/or areas characteristic of inflammation, necrosis, stroma, or endothelial or vascular proliferation. The area(s) to be captured may be circumscribed using the UV cutting laser to ensure that adjacent cells are not coincidentally captured due to cellular adhesion. Substantially irregular areas that infiltrate the area to be captured may be ablated using the UV laser to remove any chance of contamination with unwanted cells. Positioning the CapSure cap just above the tissue section, use the infrared laser to "tack" the CapSure polymer to the selected areas. The CapSure polymer will melt into contact at the tack points and the area circumscribed by the UV laser will be captured to the cap. The CapSure cap has a diameter of approximately 5.5 mm and can be used to capture up to approximately 30 000 cells from a 5-μm section depending on cellular heterogeneity. Following capture, place the CapSure cap firmly into a low-protein binding 0.6-ml microcentrifuge tube containing 100 μl of antigen retrieval buffer and set aside until all areas to be processed have been microdissected. The antigen retrieval buffer is 1% SDS in 20 mM Tris-HCl, pH 8.0.

9.2.4
Protein Extraction and Digestion for Shotgun Proteomics Using the Antigen Retrieval Method

Remove the CapSure cap from the tube and subsequently remove the CapSure polymer membrane from the cap using fine-point tweezers and place the membrane into the antigen retrieval buffer. Incubate the tube at 100 °C for 20 minutes on a heating block or in a water bath. Then incubate the tube at 60 °C for 2 hours. Following heating, briefly spin the tube. Reduce the protein solution by adding fresh DTT to 1 mM and incubating at RT for 1 hour under N_2 followed by carbamidomethylation with 5 mM IAM at RT for 1 hour in the dark. Remove the solution and place in a Slide-a-lyzer MINI dialysis unit with a 3.5-kDa cutoff membrane and cover. Dialyze against 1 l of 1 M urea, 50 mM Tris-HCl (pH 8.0) at 4 °C, overnight, and gently stir. Following dialysis, remove the solution to a new low-protein binding microcentrifuge tube and measure the protein concentration using the Bradford assay (Pierce). Add sequencing grade trypsin at a 1 : 50 enzyme to substrate ratio. This corresponds to approximately 200-ng trypsin per 10 000 cells. Incubate at 37 °C, overnight, with gentle agitation. Following digestion, desalt the peptide mixture on a polystyrene divinylbenzene reverse-phase column, such as a Macro- (200 µg capacity, 50 µl bed volume) or Cap-trap (20 µg capacity, 5 µl bed volume) (Michrom):

1. Condition column by flushing with 10 column volumes of 80% B, then 10 column volumes of 100% A.
2. Acidify sample to pH 3.0 with trifluoroacetic acid and dilute 1 : 1 with buffer A.
3. Load sample onto column and flush with 10 column volumes of 100% A.
4. Reverse the column orientation to back elute the sample with two column volumes each of 40% B, 60% B, and 80% B in sequence.
5. Collect the eluate in a low-protein binding microcentrifuge tube and place at −80 °C until frozen (approximately 30 minutes).
6. Lyophilize (SPD121P, Thermo Savant) the sample to dryness.
7. Clean and regenerate the column prior to the next sample loading by flushing with 10 column volumes of 70/30 concentrated formic acid/isopropanol.

9.2.5 Multidimensional Peptide Separation

The transient capillary isotachophoresis/capillary-zone electrophoresis apparatus operates using a coaxial sheath flow deposition system. It was constructed in-house using a CZE100R high voltage power supply (Spellman, Hauppauge, NY), a fiber optic UV detector (Ocean Optics, Dunedin, FL), a PHD2000 syringe pump (Harvard Apparatus, Holliston, MA), a robotic XYZ stage, and a coaxial sheath constructed from a microtee (Upchurch Scientific, Oak Harbor, WA) and a 5-cm 460-µm i.d. × 785-mm o.d. stainless steel tube (Upchurch). All electronic systems are under control of a custom LabView (National Instruments, Austin, TX) program. Fused-silica capillaries must be coated to suppress electroosmotic flow and analyte adsorption. To prepare a 100-µm o.d. capillary, dissolve HPC (average MW 100 000 Da) to 5% in ultrapure water and fill a 100-cm length of capillary. Purge the capillary with a stream of N_2 gas flowing at 20 psi. When all of the solution has visibly exited the capillary, as confirmed by inspection under a stereo microscope, place the capillary into a GC (HP5890A, Hewlett-Packard) oven with a temperature program of 60–140 °C at 5 °C min^{-1} followed by 20 minutes at 140 °C, while continuing to flow N_2 through the capillary. Cut the capillary to 80 cm, burn a 2-mm window for UV detection (MicroSolveCE, ThermoScientific) and mount the capillary in the capillary electrophoresis apparatus as shown in Figure 9.2. Dissolve the sample in a volume of 1% Pharmalyte 8–10 equal to 5 µl × the number of analyses to be

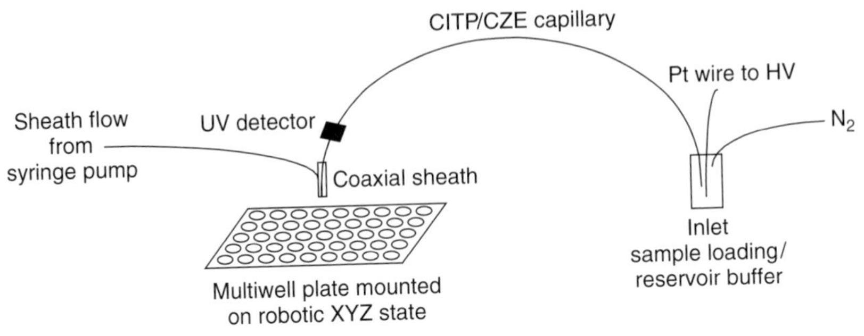

Figure 9.2 CITP/CZE setup: the setup is described in detail in the text. Abbreviations: Pt, platinum; HV, high voltage power supply; N_2, nitrogen gas line.

run. Freshly prepare 0.1 M acetic acid by diluting glacial acetic acid 175 × with ultrapure water and filling a 200-μl syringe and standard glass HPLC vial (Alltech, Deerfield, IL) and mounting as shown. Flush the coated capillary with 10 ml of 0.1 M acetic acid via N_2 displacement. Place 5 μl of sample into a low-volume HPLC vial insert (Alltech), seal with a silicone septa-fitted cap, replace the reservoir vial, and place the inlet end of the capillary through the septa to the bottom of the insert. Measure the amount of sample displaced from the capillary with a calibrated glass capillary (Fisher) and stop the flow of N_2 when 4 μl has been displaced. This indicates that the sample has filled the capillary to 60 cm. Replace the sample vial with the reservoir buffer vial and start the coaxial sheath flow at a flow rate of 1 μl min^{-1}. Apply a positive electric voltage of 24 kV, with current limited to 25 μA, to the reservoir vial to begin the transient CITP separation. As the separation shifts to CZE, the peptides will begin to elute from the capillary. When a sharp rise is seen in the UV absorbance being monitored at 280 nm, as shown in Figure 9.3, begin fraction collection in a low-protein-binding 384-microwell plate (Bio-Rad, Hercules, CA). Collect fractions every 2.5 minutes until 30 fractions are collected. Cover the microwell plate with a heat-sealing aluminum film (Bio-Rad) to prevent evaporation and place at −80 °C until ready for LC-MS.

9.2.6
Nano-RPLC-ESI-Tandem Mass Spectrometry

LC columns may be constructed in-house by packing 50-μm i.d. × 360-μm o.d. PicoFrit emitters (New Objective) with a slurry of 3.5 μm, 300 Å Zorbax StableBond C_{18} particles (Agilent) using a pressure bomb (Proxeon) connected to a high purity N_2 cylinder at 1000 psi as follows:

1. Mix 3.5-μm Agilent Zorbax SB C_{18} particles with MeOH/0.5% HAC solution, using a glass vial.
2. Vortex particle slurry briefly and sonicate 10 minutes.
3. Place glass vial with packing particles in pressure bomb with magnetic stirrer atop a stir plate.

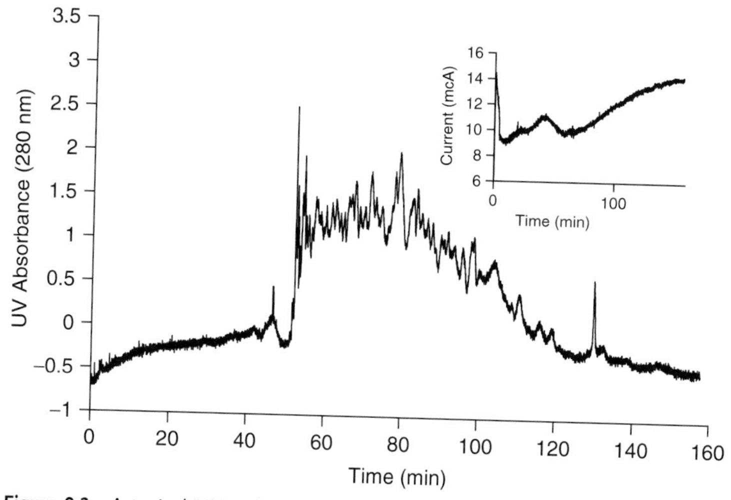

Figure 9.3 A typical UV and current (inset) trace from a CITP/CZE separation.

4. Attach capillary to bomb, and adjust pressure to ~1000 psi.
5. Pack column to ~11 cm, remove from bomb, and attach to LC.
6. Flush column at 300 bar, 80% acetonitrile, 0.1% TFA, 12 hours. Then decrease pressure to 0 bar over 6 hours.
7. Remove column from LC and cut to length.

Fractions are loaded using a FAMOS autosampler (Dionex, Sunnyvale, CA). Each fraction is loaded in total to a 10-μl loop using a custom sampling program. A Switchos (Dionex) loading pump delivers a 10 μl min^{-1} flow of buffer A (2% acetonitrile, 0.02% formic acid, 0.02% heptafluorobutyric acid) through the loop and to one of two Opti-Pak trap columns (Optimize Technologies, Oregon City, OR) packed with material identical to that used in the analytical columns and positioned on separate 10-port nano valves (Upchurch). LC separations are performed in parallel using staggered gradients delivered by separate pumps contained within an Ultimate Dual LC system (Dionex). The liquid flow setup is shown in Figure 9.4. Identical 120-minute gradients are run on each pump with a 60-minute offset. The gradient flows at 150 nl min^{-1} throughout, starting at 5% B (80% acetonitrile, 0.02% formic acid, 0.02% heptafluorobutyric acid) for 10 minutes, ramping to 45% B over 80 minutes, then to 100% B over 5 minutes, and back to 5% B in 5 minutes where it remains for the rest of the 120 minutes. During the gradient, the valve in-line with the column is in position A from 0–10 minutes and 110–120 minutes. The analytical columns are mounted side by side on a robotic stage approximately 1-cm apart. The columns are connected to the valve tubing using a microtee (Upchurch), which supplies the high voltage necessary to create an electrospray via a platinum wire (Fisher). Before the start of the run, each column is independently aligned, so each is optimally positioned to deliver the LC eluate by electrospray into the inlet of the mass spectrometer. During the run, the column at 30–90 minutes during the LC gradient is the column placed in-line with the MS inlet. The robotic

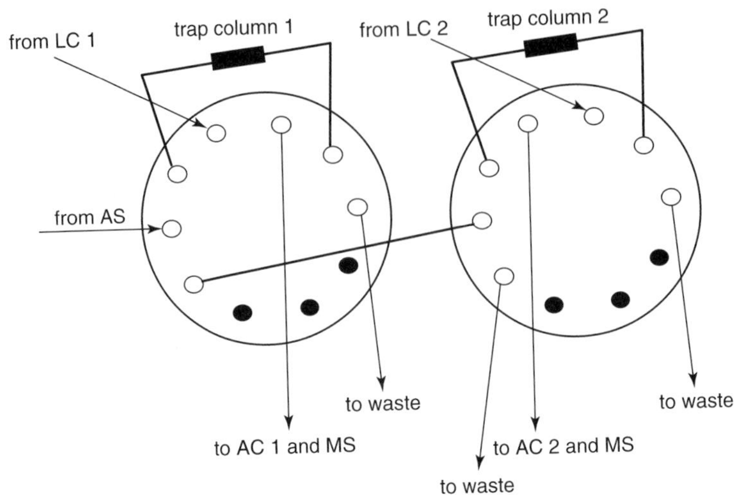

Figure 9.4 Liquid flow paths for parallel nano-RPLC separations. Black lines and arrows indicate fused-silica capillaries, all of which are 20-μm i.d. Open circles indicate ports through which liquid flows, whereas closed circles indicate unused ports. Abbreviations: LC1, HPLC pump number 1; LC2, HPLC pump number 2; AS, autosampler; AC1, analytical column number 1; AC2, analytical column number 2; MS, mass spectrometer.

stage moves the columns between the two positions when triggered by a contact closure event at 30 minutes in either LC gradient (corresponding to 90 minutes in the other gradient). This gradient has been optimized to elute the vast majority of peptides during the 60-minute window when the columns are in-line with the MS while simultaneously maximizing separation performance to deliver a high peak capacity. This strategy allows for an effective doubling of throughput per mass spectrometer. An LTQ (Thermo) linear ion trap mass spectrometer isused. The electrospray voltage is 1.8 kV. The acquisition program runs for 60 minutes when started by a contact closure event at 30 minutes in either LC gradient. Full scans are acquired from 400 to 1400 m/z in profile mode. Five data-dependent tandem MS scans are performed for the five most intense ions from the full scan. Dynamic exclusion is set to 30 seconds. MS and tandem MS scan times are set to a maximum of 100 ms. Automatic gain control settings are 30 000 for full MS scans and 10 000 for tandem MS scans.

9.2.7
Data Processing and Tandem Mass Spectrometry Search

Following data acquisition, RAW data files are converted to mzXML [16] format using readw [17]. Peaklists in the mgf format are generated from mzXML files using JRAP parser and an in-house Perl script. The open mass spectrometry search algorithm (OMSSA) [18] is used to search the peaklist files on a 14-node 28-CPU Linux cluster (Linux Networx, Bluffdale, UT) using the following parameters:

fully tryptic enzyme specificity, 1.5-Da precursor ion mass tolerance, monoisotopic precursor ion, 0.4-Da fragment ion mass tolerance, monoisotopic fragment ions, match b and y ions, require that one of the six most intense fragment ions match a predicted b or y ion, permit variable oxidation of methionine, and require cabamidomethylation of cysteine. Searches are performed against a target-decoy version of the UniProt protein sequence library. A target-decoy library is one in which the protein sequences within the original sequence library are reversed and concatenated to the original. This strategy creates an additional equally sized protein sequence library with the same protein lengths and amino acid compositions, which compete for matching to the tandem mass spectra. Additionally, this allows for an estimation of the false discovery rate at any given score threshold based on any given set of search inputs. Our laboratory typically sets a score cutoff equivalent to a 1% tandem mass spectra false discovery rate following the method of Elias *et al.* [19]. Search output XML files are parsed into an Oracle 10 g database using an in-house Java parser. Data extraction and analysis are performed using various scripts written in-house. It should be noted that the choice of search algorithm is not critical – most algorithms perform comparably so long as similar false discovery rates are utilized [20]. Most algorithms now include tools for creating and searching target-decoy protein sequence libraries. OMSSA and X!Tandem [21] are both open-source and freely available for use. While some of the data processing capabilities described may not be accessible to some laboratories, many of these functions are increasingly available through software from the mass spectrometer vendors such as BioSieve (Thermo), third-party developers such as ProteomeCenter (Proxeon, Odense, Denmark), and open-source developers such as the Trans-Proteomic Pipeline (Institute for Systems Biology, Seattle, WA).

9.3 Limitations and Future Directions

The antigen retrieval method, when first developed, opened up the vast archives of formalin-fixed tissues to more reproducible and sensitive immunohistochemistry. Antigen retrieval has recently performed a similar role for shotgun proteomics. In our laboratory, the technique permits the identification of thousands of proteins from a very small amount of starting tissue. If carefully performed, this analysis will allow the identification of approximately 85% of proteins identifiable by similar techniques using fresh-frozen tissue. Further refinements or alterations to the technique may be required to allow performance equivalent to fresh-frozen samples. Our studies have shown that extractions performed at different pH will retrieve somewhat different sets of proteins. The reasons for this remain under investigation. Another variable, typically beyond the control of the investigator, is the length of fixation time for any given sample. It is known that variable fixation times have an effect on antigen retrieval and thus on the results and reproducibility of immunohistochemistry [22]. Similar variability would be expected for shotgun proteomic results and is under investigation. Finally, one of the great advantages of the formalin-fixation method is that tissues remain preserved over great lengths of time. This allows for recursive investigation of samples collected during clinical trials, and so on, using techniques unavailable or unthought of at the time. It is known that immunohistochemistry can be performed on decades-old tissue with only moderate loss of sensitivity. Preliminary results in our laboratory using tissue blocks as old as 27 years indicate that this is also the case for shotgun proteomics.

The antigen retrieval approach applied to shotgun proteomics will directly lead to the creation of a database of proteins extractable by antigen retrieval under varying conditions and, perhaps more importantly, the epitopes exposed for immunodetection as defined by the peptide sequence coverage of any given protein. This will permit rational experimental design for immunohistochemistry based on this database. Additionally, intelligent antibody design may be undertaken on the basis of the sequences contained within the database. In the long term, the technique holds the promise of performing absolute quantitative assays for potentially hundreds of proteins from a small number of cells using rapid assays such as LC-SRM with synthetic stable isotope-labeled peptide standards [23, 24]. While this technique would not allow for the single cell resolution or sensitivity of IHC, it would permit absolute quantification of hundreds of proteins from a relatively small section area.

References

1. Huang, S.N., Minassian, H., and More, J.D. (1976) Application of immunofluorescent staining on paraffin sections improved by trypsin digestion. *Lab Invest*, **35**, 383–390.
2. Battifora, H. and Kopinski, M. (1986) The influence of protease digestion and duration of fixation on the immunostaining of keratins. A comparison of formalin and ethanol fixation. *J Histochem Cytochem*, **34**, 1095–1100.
3. Shi, S.R., Key, M.E., and Kalra, K.L. (1991) Antigen retrieval in formalin-fixed, paraffin-embedded tissues: an enhancement method for immunohistochemical staining based on microwave oven heating of tissue sections. *J Histochem Cytochem*, **39**, 741–748.
4. Ikeda, K., Monden, T., Kanoh, T., Tsujie, M., Izawa, H., Haba, A., Ohnishi, T., Sekimoto, M. *et al.* (1998) Extraction and analysis of diagnostically useful proteins from formalin-fixed, paraffin-embedded tissue sections. *J Histochem Cytochem*, **46**, 397–403.
5. Rait, V.K., Xu, L., O'Leary, T.J., and Mason, J.T. (2004) Modeling formalin fixation and antigen retrieval with bovine pancreatic RNase A II. Interrelationship of cross-linking, immunoreactivity, and heat treatment. *Lab Invest*, **84**, 300–306.
6. Metz, B., Kersten, G.F., Hoogerhout, P., Brugghe, H.F., Timmermans, H.A., de Jong, A., Meiring, H., ten Hove, J. *et al.* (2004) Identification of formaldehyde-induced modifications in proteins: reactions with model peptides. *J Biol Chem*, **279**, 6235–6243.
7. Prieto, D.A., Hood, B.L., Darfler, M.M., Guiel, T.G., Lucas, D.A., Conrads, T.P., Veenstra, T.D., and Krizman, D.B. (2005) Liquid Tissue: proteomic profiling of formalin-fixed tissues. *Biotechniques*, **38** (Suppl), 32–35.
8. Shi, S.R., Liu, C., Balgley, B.M., Lee, C., and Taylor, C.R. (2006) Protein extraction from formalin-fixed, paraffin-embedded tissue sections: quality evaluation by mass

spectrometry. *J Histochem Cytochem*, **54**, 739–743.

9. Guo, T., Wang, W., Rudnick, P.A., Song, T., Li, J., Zhuang, Z., Weil, R.J., DeVoe, D.L. *et al.* (2007) Proteome analysis of microdissected formalin-fixed and paraffin-embedded tissue specimens. *J Histochem Cytochem*, **55**, 763–772.
10. Palmer-Toy, D.E., Krastins, B., Sarracino, D.A., Nadol, J.B. Jr, and Merchant, S.N. (2005) Efficient method for the proteomic analysis of fixed and embedded tissues. *J Proteome Res*, **4**, 2404–2411.
11. Hood, B.L., Darfler, M.M., Guiel, T.G., Furusato, B., Lucas, D.A., Ringeisen, B.R., Sesterhenn, I.A., Conrads, T.P. *et al.* (2005) Proteomic analysis of formalin-fixed prostate cancer tissue. *Mol Cell Proteomics*, **4**, 1741–1753.
12. Rahimi, F., Shepherd, C.E., Halliday, G.M., Geczy, C.L., and Raftery, M.J. (2006) Antigen-epitope retrieval to facilitate proteomic analysis of formalin-fixed archival brain tissue. *Anal Chem*, **78**, 7216–7221.
13. Hwang, S.I., Thumar, J., Lundgren, D.H., Rezaul, K., Mayya, V., Wu, L., Eng, J., Wright, M.E. *et al.* (2007) Direct cancer tissue proteomics: a method to identify candidate cancer biomarkers from formalin-fixed paraffin-embedded archival tissues. *Oncogene*, **26**, 65–76.
14. Jiang, X., Jiang, X., Feng, S., Tian, R., Ye, M., and Zou, H. (2007) Development of efficient protein extraction methods for shotgun proteome analysis of formalin-fixed tissues. *J Proteome Res*, **6**, 1038–1047.
15. Espina, V., Wulfkuhle, J.D., Calvert, V.S., VanMeter, A., Zhou, W., Coukos, G., Geho, D.H., Petricoin, E.F. III, *et al.* (2006) Laser-capture microdissection. *Nat Protoc*, **1**, 586–603.
16. Pedrioli, P.G., Eng, J.K., Hubley, R., Vogelzang, M., Deutsch, E.W., Raught, B., Pratt, B., Nilsson, E. *et al.* (2004) A common open representation of mass spectrometry data and its application to proteomics research. *Nat Biotechnol*, **22**, 1459–1466.
17. Keller, A., Eng, J., Zhang, N., Li, X.J., and Aebersold, R. (2005) A uniform proteomics MS/MS analysis platform utilizing open XML file formats. *Mol Syst Biol*, **1**, 2005.
18. Geer, L.Y., Markey, S.P., Kowalak, J.A., Wagner, L., Xu, M., Maynard, D.M., Yang, X., Shi, W. *et al.* (2004) Open mass spectrometry search algorithm. *J Proteome Res*, **3**, 958–964.
19. Elias, J.E., Gibbons, F.D., King, O.D., Roth, F.P., and Gygi, S.P. (2004) Intensity-based protein identification by machine learning from a library of tandem mass spectra. *Nat Biotechnol*, **22**, 214–219.
20. Balgley, B.M., Laudeman, T., Yang, L., Song, T., and Lee, C.S. (2007) Comparative evaluation of tandem MS search algorithms using a target-decoy search strategy. *Mol Cell Proteomics*, **6**, 1599–1608.
21. Craig, R. and Beavis, R.C. (2004) TANDEM: matching proteins with tandem mass spectra. *Bioinformatics*, **20**, 1466–1467.
22. Shi, S.R., Liu, C., and Taylor, C.R. (2007) Standardization of immunohistochemistry for formalin-fixed, paraffin-embedded tissue sections based on the antigen-retrieval technique: from experiments to hypothesis. *J Histochem Cytochem*, **55**, 105–109.
23. Gerber, S.A., Rush, J., Stemman, O., Kirschner, M.W., and Gygi, S.P. (2003) Absolute quantification of

proteins and phosphoproteins from cell lysates by tandem MS. *Proc Natl Acad Sci U S A*, **100**, 6940–6945.

24. Kirkpatrick, D.S., Gerber, S.A., and Gygi, S.P. (2005) The absolute quantification strategy: a general procedure for the quantification of proteins and post-translational modifications. *Methods*, **35**, 265–273.

10
Selective Tissue Procurement for Renal Tumor Proteomics

Zhengping Zhuang, Jie Li, Harry Mushlin, and Alexander O. Vortmeyer

10.1
Introduction

Proteomic analysis is being used increasingly to identify potential therapeutic targets and biomarkers in body fluids, like serum or plasma [1–5], urine [6–8], CSF [9–13], and saliva [14–18]. Tumor tissues represent a rich source of distinct proteins that may closer characterize the biology of tumor development. These proteins can be correlated with various clinicopathological features, including diagnosis, prognosis, and drug response. In addition, they may represent novel therapeutic targets. To obtain more complete proteotypic spectra of tumors, however, tumor tissue itself needs to be subjected to proteomic analysis. Principally, any tumor tissue can be screened by various high-throughput proteomic techniques. However, tumor tissue removed at surgery not only consists of neoplastic cells but also a large number of diverse "normal" cells including vascular cells, fibrous cells, and inflammatory cells. Without careful histological evaluation, attempts to isolate "tumor" and "normal" tissue from surgical specimens are of limited value, as the exact nature of the cells is unknown at the time of gross tissue procurement. In addition to various immunological, vascular, and fibroproliferative responses, tumor itself may exhibit considerable intratumoral variability in regard to tumor cell biology. For example, when malignant tumor growth is not accompanied by sufficient neovascularization, subsets of tumor cells may undergo pathways of hypoxic stress response and finally apoptosis and/or necrosis [19–22]. Therefore, results may differ significantly even among tissue samples obtained from the same tumor.

Careful selection of tumor cells, preceded by light-microscopic visualization of stained tissue sections, is therefore essential to permit an accurate and reproducible comparison of the proteotypes among different specimens. In the past, the major disadvantage of tissue sections for histological control of tumor tissue analysis has been twofold: first, histological control is performed on 10-μm thick sections from which only small amounts of tissue can be recovered by microdissection. Second, histological tissue sections need to be stained with dyes, the chemical nature of which may substantially interfere with subsequent protein analysis [23–25].

Renal and Urinary Proteomics: Methods and Protocols. Edited by Visith Thongboonkerd

ISBN: 978-3-527-31974-9

We have developed selective dissection techniques, which allow procuring and analyzing specific areas of neoplastic cells [26–31]. In this chapter, we describe the selective dissection technique under full morphological control and subsequent proteomic analysis. Then, we review some implications of this approach of proteomic analysis for the detection and identification of proteins of interest. Finally, we review recent important progress in the field of renal tumor proteomics.

10.2 Methods and Protocols

10.2.1 Tissue Preparation

In agreement with the Institutional Review Board, surgically procured renal and/or tumor tissue has to be evaluated by a pathologist, and any tumor parts that are relevant for tissue diagnosis need to be processed separately for diagnostic evaluation. Additional tumor tissue can be procured for research, fresh, frozen, or in varied fixation for further studies. We recommend using paraffin-embedded tissue for pathological evaluation, and only tissues with an unequivocal histological diagnosis should be considered for further proteomic analysis. From all frozen tissue samples, a single 10-μm section is taken and stained with hematoxylin and eosin (H&E) for histological evaluation.

10.2.2 Selective Tissue Microdissection

From the tumors with unequivocal histopathological diagnosis, a single 10-μm frozen section is taken and stained with H&E for special histological evaluation. A semiquantitative cell count is performed on tumor-rich areas that are uncompromised by inflammation, necrosis, or reactive fibrosis; subsequently, these areas will be subjected to selective tumor dissection from serial sections from the same block. The primary goal is to obtain 100 000 viable tumor cells, which are derived from about 1–10 consecutive sections taken from the selected blocks. Procurement of normal kidney tissue, or areas of inflammation, necrosis, or hemorrhage, is strictly avoided. Tissue dissection is performed manually, as described previously, to avoid possible heating artifacts induced by laser-assisted technology. However, unstained serial sections are used to avoid chemical artifacts induced by tissue staining and the H&E staining is performed after every three consecutive cuts to guarantee the histological consistency [28, 32].

10.2.3 Proteomic Analysis

With recent improvements in proteomic techniques, proteome profiling has been a powerful complementary approach to nucleic acid-based molecular profiling (cDNA microarray) in large-scale gene analyses implicated in tumorigenesis [33–35].

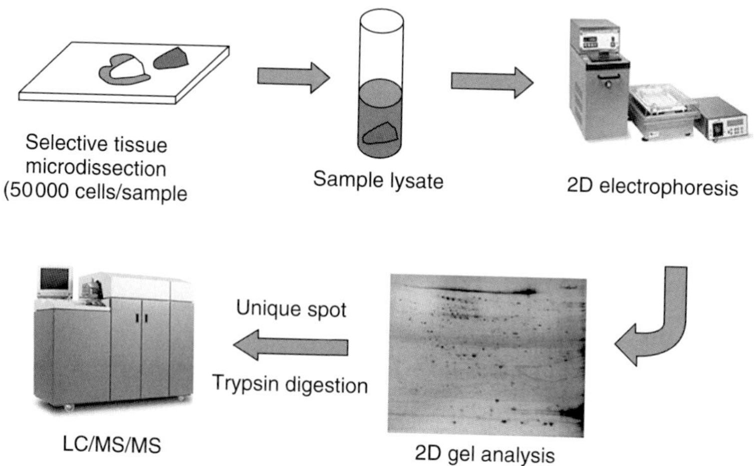

Figure 10.1 A summary of work flow in 2-D PAGE analysis of microdissected tissue.

Current proteomic profiling tools include two-dimensional polyacrylamide gel electrophoresis (2-D PAGE) or nongel/shotgun-based multidimensional liquid chromatography protein separation, followed by peptide sequencing [34, 35]. Generally, gel-based proteomic techniques provide a quantitative view of the proteome; while nongel-based tools permit a higher resolution and sensitivity in peptide separation with the same protein loading, they are difficult to create a quantitative profile of proteins.

Proteome profiling work is based on "comparison." Nongel-based proteomic comparisons identify the differences of proteome between two comparable groups of cell when some proteins are express in an "on/off" way, whereas gel-based comparisons can provide additional information on proteomic changes when some proteins are varied between groups in a quantitative way. Since, within the latter technique, much less instruments (in terms of the amount and the expense) and much simpler procedures are requested, we summarize a general protocol for 2-D PAGE proteomic comparison as follows (see Figure 10.1).

10.2.3.1 **Protein Preparation**

All samples are run in triplicate. The dissected cells are dissolved into Extraction Buffer II containing 8 M urea, 4% (w/v) Bio-Lyte 4/7, and 2 mM tributyl phosphine (Bio-Rad, Hercules, CA), vigorously vortexed and centrifuged. The supernatant is combined with a rehydration buffer mixture containing Rehydration Buffer (8 M urea, 2% CHAPS, 50 mM DTT, and 0.2% (w/v) Bio-Lyte 4/7 ampholytes; Bio-Rad), Immobilized pH gradient (IPG) buffer (Amersham Biosciences, Piscataway, NJ), and bromophenol blue, and subsequently rehydrated overnight with Immobiline Drystrips (pH 4–7, 11 cm; Amersham Biosciences) on a Reswelling Tray (Amersham Biosciences).

10.2.3.2 First-Dimensional Protein Separation

This step is to separate proteins according to one of protein's natural characters, which is the isoelectric point (pI). The isoelectric focusing for the first-dimensional electrophoresis is performed with a Multiphore II Electrophoresis System (Amersham Biosciences). The strips are subjected to voltages at 300–3500 V.

10.2.3.3 Second-Dimensional Protein Separation

This step is to separate proteins according to another of protein's natural characters, the molecular weight. IPG strips are equilibrated with Equilibration Buffer I containing 6 M urea, 2% SDS, 375 mM Tris-HCL (pH 8.8), 20% glycerol, and 2% (w/v) DTT; and Buffer II containing 6 M urea, 2% SDS, 375 mM Tris-HCL (pH 8.8), 20% glycerol, and 2.5% (w/v) iodoacetamide (Bio-Rad, Hercules, CA). Precast ExcelGel SDS gels (12–14% gradient gel; pH4–7, 245 × 180 × 0.5 mm, Amersham Biosciences) are used for the second dimension of protein separation by a Multiphor II Flated System (Amersham Biosciences) under a constant voltage of 700 V.

10.2.3.4 Protein Staining

Multiple staining methods can be used for highlighting proteins on the second-dimensional gel, including Coomassie blue staining, fluorescent staining, and silver staining. Considering the sensitivity, cost, and instruments requested, the silver staining is the easiest for the beginners. A silver staining kit (Amersham Biosciences) can be used to detect protein spots according to the manufacturer's instructions.

10.2.3.5 Computerized Imaging and Statistical Analysis

Digital images are acquired for each sample gel and the intensity of protein spots is taken to be proportional to the protein concentration. The protein spots are detected, quantified, and matched using Proteomweaver software (Definiens, Munich, Germany). The images are also manually analyzed to avoid improper image alignment from the computerized program. Individual spot volumes are normalized against total spot volumes for a given gel. Differences in apparent protein expression level are considered potentially significant when the matched spots exhibit at least a statistical difference ($P < 0.05$) in their averaged normalized intensity levels. Since the two groups are independent entities and their respective data are verified by the F-text, the results obtained for the groups are then compared using unpaired Student's *t*-test.

10.2.3.6 Preparation of Peptides for Sequencing

Protein spots of interest that are significantly different between the two groups ($P < 0.05$) are excised from the gel and subjected to in-gel digestion with trypsin based on a previously described procedure [35].

10.2.3.7 Mass Spectrometry-Based Peptide Sequencing

The identification of proteins by correlation with sequence databases relies on the availability of constraining parameters which distinguish specific matches from all the other sequences in the database. Among the parameters (protein amino acid composition, stretches of amino acid sequence, protein and peptide masses, and peptide fragment masses), the accurate mass of a molecule is particularly attractive because it is highly constraining and can be determined with great accuracy, rapidity, and sensitivity by mass spectrometry (MS). As an example of MS protein sequencing in our laboratory, peptides from in-gel digests are analyzed by LC-MS/MS on a ProteomeX LC/MS system (ThermoElectron, San Jose, CA) operated in the high-throughput mode. Reversed phase HPLC is carried out using a BioBasic-18 column ($0.18 \times 150\,\mathrm{mm}$, ThermoElectron) eluted at $1\text{–}2\,\mu\mathrm{l\,min}^{-1}$ with a gradient of 2–50% B over 30 minutes. Mobile phase A is H_2O (0.1% FA) and mobile phase B is CH_3CN (0.1% FA). Column effluent is analyzed on the LCQ Deca XP Plus (ThermoElectron) operating in the "Top Five" mode. Uninterpreted MS/MS spectra are searched against database of a certain species (match the species of your samples) utilizing the BioWorks and SEQUEST programs (ThermoElectron). A protein identification is accepted when MS/MS spectra of at least two peptides from the same protein exhibit at a minimum of the default Xcorr versus charge values set by the program (for $Z = 1$, 1.50; for $Z = 2$, 2.00; for $Z = 3$, 2.50).

10.3 An Example of Selective Tissue Procurement for Proteomic Analysis of Renal Tumors

The following example provides an overview of our work applying above principles to proteomics of human renal tissue samples [36].

- **Objective:** Correlation of disease phenotype with protein profile (proteotype) is a significant challenge for biomedical research. We present a proteomic approach that combines enhanced detection sensitivity with selective tissue dissection from frozen tumor tissue, followed by 2-D gel analysis and protein identification with MS.
- **Materials and methods:** A series of primary renal tumors was used including oncocytoma ($n = 3$), clear cell carcinoma ($n = 3$), Wilms' tumor ($n = 3$), and papillary carcinoma ($n = 3$). These renal tumors are histologically distinct types of renal cancer with well-defined, characteristic morphological features. Details of proteomic procedures are as aforementioned.
- **Results:** The data are shown and described in Figures 10.2–10.5 and Table 10.1.
- **Conclusions:** The proteomic technique allows not only for sensitive identification of specific protein patterns that correspond to a histological tumor phenotype but also for identification of specific disease-associated protein targets.

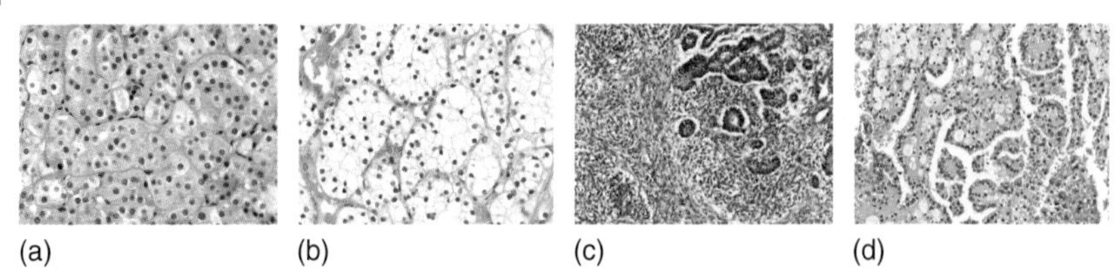

Figure 10.2 Four different types of renal tumors with markedly distinct histopathological phenotypes were selected for the study. (a) Renal oncocytoma consists of cells with abundant, granular, and eosinophilic cytoplasm that are tightly packed in solid nests; the nuclei are regular and round with no or small nucleoli. (b) Clear cell carcinoma is characterized by an alveolar architectural pattern with nests of cells separated by thin-walled vascular septae. (c) Wilms' tumor shows a triphasic pattern and contains blastemal, stromal, and epithelial cell types of differentiation. (d) Papillary carcinoma is composed of fibrovascular stalks filled with abundant lipid-laden macrophages and lined by neoplastic cells.

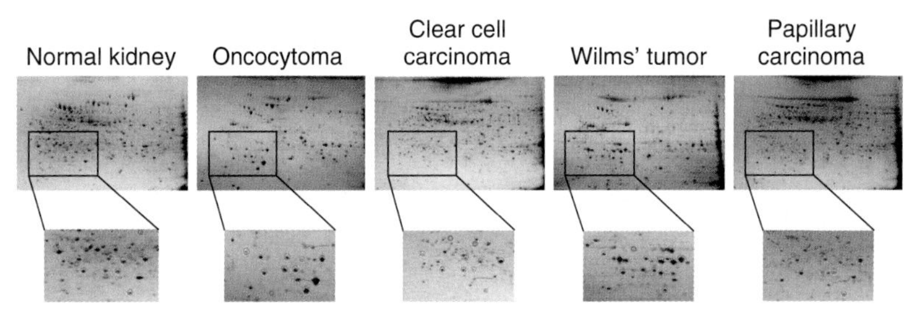

Figure 10.3 2-D PAGE analysis revealed individual proteotypic profiles for oncocytoma, clear cell carcinoma, Wilms' tumor, and papillary carcinoma. In the magnified areas, red circles indicate proteins that were consistently present in all types of tumors. Green circles indicate proteins that were differentially expressed in individual tumor types.

10.4 Important Developments in Renal Tumor Proteomics

Cancer cell proteome profiling allows for the identification of differentially expressed proteins in tumors in comparison to normal tissue. The use of proteomics in nephrology has made important advances over the last few years, particularly in research pertaining to renal cell carcinoma (RCC). RCC accounts for 95 000 deaths a year worldwide, 11 000 of those deaths being in the United States [37, 38]. Identification of RCC differentially expressed proteins give physician's potential diagnostic, prognostic, and therapeutic biomarkers for RCC. When RCC is diagnosed at the metastatic stage, which is 20–30% of the time, the five-year survival is only 9%. However, if RCC is diagnosed at its early stage, the five-year survival rate is increased to 89%, thus highlighting the importance of developing diagnostic tests for recognizing RCC early on [39].

2-D PAGE is the most commonly used proteomic technique and has frequently been used to compare RCC with normal tissue [40]. For example, a number of

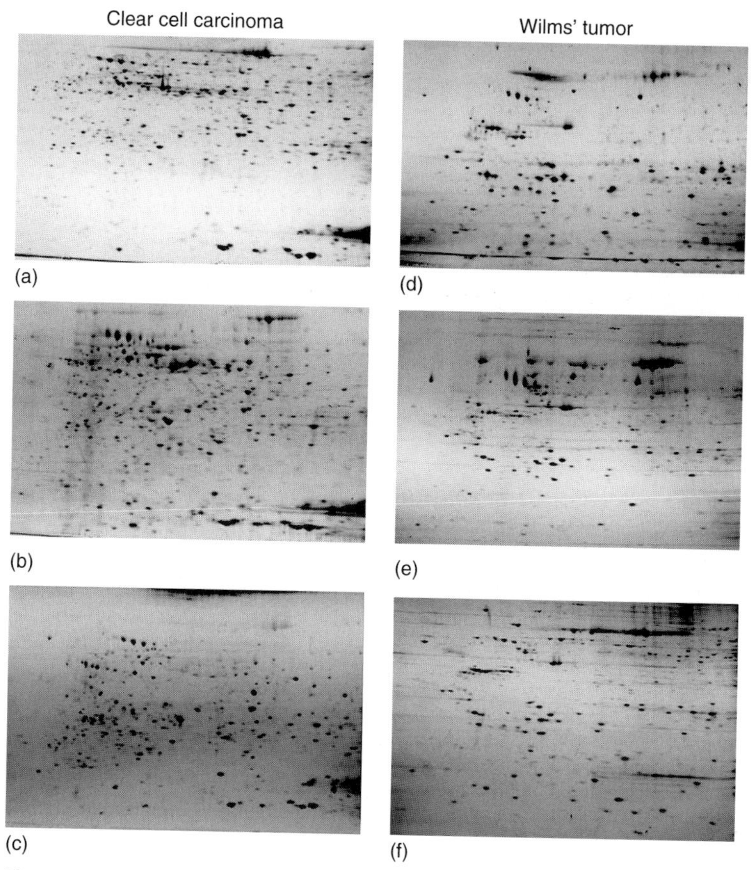

Figure 10.4 Different tumors of the same histopathological phenotype show highly consistent proteotypic profiles after 2-D PAGE analysis. (a–c) show 2-D gels of three different clear cell carcinomas, (d–f) show 2-D gels of three different Wilms' tumors.

up-regulated proteins in RCC have been found using 2-D PAGE and peptide mass fingerprinting with MALDI-TOF MS, including tumor rejection antigen-1 and alpha-1 antitrypsin precursor, along with down-regulated proteins such as ribosomal P0 protein, aminoacylase-1, aldehyde reductase, tropomyosin, and ketohexokinase [38, 41]. Overall, current findings of RCC-specific proteins that are differentially expressed when compared to normal renal tissue may be summarized into the following categories.

10.4.1
Metabolism-Related Proteins

Perroud *et al.* [37] used 2-D PAGE and MS to compare proteomes between RCC and normal renal tissue, and identified 31 proteins that differed significantly. Many of

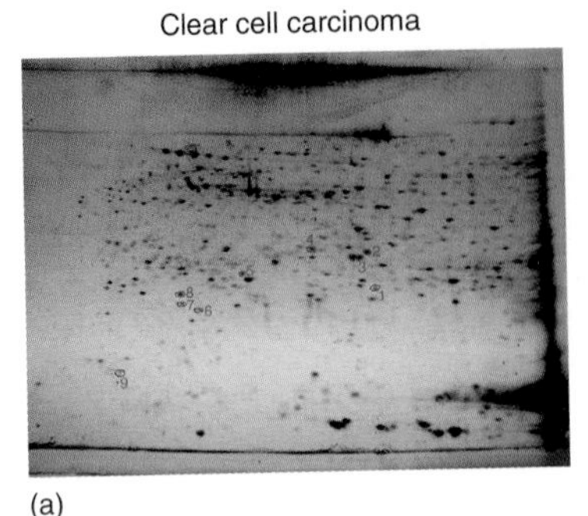

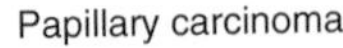

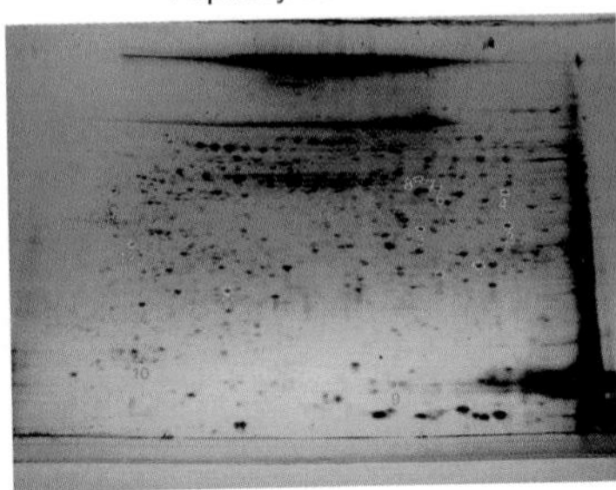

(a) (b)

Figure 10.5 Differentially expressed proteins in clear cell carcinoma (a) versus papillary carcinoma (b) were selected for protein identification. From clear cell carcinoma, spots 1–9 (corresponding to "C1"–"C9" in Table 10.1) were separately excised, destained, digested, and analyzed by capillary LC-MS/MS. Similarly, spots 1–12 were analyzed from papillary carcinoma.

these distinguishing proteins are related to metabolic pathways such as glycolysis, propanoate metabolism, pyruvate metabolism, urea cycle, and arginine/proline metabolism. Of note, metabolic pathway analysis added evidence to previous research revealing the role of TNF-alpha in RCC pathogenesis. Also, this study added to the previous proteomic evidence for the Warburg effect in RCC cell by demonstrating an increase of proteins associated with the glycolytic pathway, decrease of proteins associated with gluconeogenic pathway, and decrease in a number of mitochondrial enzymes [42].

10.4.2
Heat Shock Proteins (HSPs)

Heat shock proteins (HSPs) are another group of proteins that have been found by proteomic analysis to be differentially expressed in RCC compared to normal tissue. HSP expression can be induced by environmentally stressful conditions, such as heat shock and oxidative stress [43] and has been linked to a variety of cancers including breast and ovarian cancers. Sarto *et al.* [44] identified a number of HSP27 isoforms using 2-D PAGE analysis to be more highly expressed in RCC than in normal tissue. Posttranscription modulation of HSP27 determines its cellular interactions with proteins that modulate apoptosis and thus the fate of the cell, as well as translocation toward actin filaments. In addition to HSP27, other HSP proteins, such as HSP75 and HSP90, have been identified as being overexpressed in RCC [38].

10.4.3
Cytoskeleton Proteins

These proteins may correspond to changes in cytoskeletal organization during malignant transformation and progression. Intermediate filaments are one family

Table 10.1 Proteomics identifies proteins associated with various renal cancers. Each tumor sample obtained from the same type of renal tumor revealed "differentially expressed" protein spots that were absent in other types of renal tumor.

Type	Protein name	Functional characteristics of the protein (OMIM ID and chromosome location of the gene, if known)
Clear cell cancers		
C1.	Annexin A2	A member of the annexin family of calcium-dependent phospholipids and membrane-binding proteins. Substrate of the SRC tyrosine kinase. A possible autocrine factor. (151740; 15q21–q22)
C2.	Fibrinogen gamma-A chain precursor	A member of the fibrinogen family, which are thought to play a role in cell adhesion and platelet aggregation. (134850; 4q28)
C3.	Annexin A4	A member of the annexin or lipocortin family of proteins. Has been suggested to play a role in regulation of water and protein permeability of cell membranes in an aquaporin-independent manner. (106491; 2p13)
C4.	Cathepsin D preprotein	A lysosomal aspartic proteinase involved in degradation of proteins, antigen processing, and involved in mediation of apoptosis. Has been suggested to play a role in growth and metastatic potential of epithelial cancers, especially breast Ca. (116840; 11p15)
C5.	Alpha enolase	An isoform of the glycolytic enzyme enolase, which is involved in basic energy and metabolism, as well as plasminogene binding and activation. (607098)
C6.	Proteasome activator subunit 1	Activates multicatalytic protease, which degrades protein conjugated to ubiquitin. Up-regulated by inferno gamma. (600654; 14q11.2)
C7.	Glutathione transferase omega	Acts as a stress response protein involved in cellular redox homeostasis.
C8.	Tubulin, beta 2	Member of the tubulin family, critical component of microtubules, which are critical to cell division and proper segregation of chromosomal abnormal function leads to aneuploidy. (602660)
Papillary cancers		
P1.	NADH dehydrogenase FeS protein	Involved in the electron transport chain of mitochondrial oxidative phosphorylation.

(continued overleaf)

Table 10.1 *(Continued)*

Type	Protein name	Functional characteristics of the protein (OMIM ID and chromosome location of the gene, if known)
P2.	Dihydrolipoamide succinyltransferase	Member of a multienzyme complex within the mitochondrion involved in catalyzing oxidative decarboxylation of alpha-keto acids. (126063)
P3.	Phosphotriesterase-related gene	A zinc metalloenzyme catalyzing phosphodiester compounds; may be involved in hydrolyzing organophosphate compounds. Expressed in normal renal tubules. (604446; 10p12)
P5.	Annexin I	Member of lipocortin family. Anti-inflammatory action of glucocorticoids attributed to lipocortin's inhibition of phospholipase A2. Found in renal medullary cells. (151690; 9q11–q22)
P6.	Capping protein, actin filament, gelsolin-like	A member of gelsolin family of proteins that caps actin filament ends in nonmuscle cells to help alter shape during movement. Found in renal tubular cells. (153615; 2cen-q24)
P7.	Lamin A precursor	One of three known members of lamin family, which are located on inner membrane of mitochondrion. Associated with rare forms of muscular dystrophy, lipodystrophy, and familial dilated cardiomyopathy. (150330; 1q21.2)
P8.	Gamma enolase	A member of the enolase family of glycolytic pathway enzymes. Major form of enolase found in neurons. (131360; 12p13)
P9.	Beta actin	The major form of nonmuscle cytoskeletal actin. (102630; 7p220p12)
P10.	Alpha-tubulin	Alternatively spliced alpha-tubulin, preferentially expressed in testis. Interacts to form microtubules. (191110; 2q)
Wilms' tumor		
W1.	Ariadne-2 protein homolog	A nuclear RING finger protein with a conserved, cystein-rich domain, of uncertain function. (605615; chromosome 3)
W2.	Zinc finger protein 267	A nuclear transcription factor that binds DNA, RNA. (604752; chromosome 16)
W3.	Makorin 1	Encodes a novel class of zinc finger protein. Conserved from nematodes, to fruitflies, to mouse, to humans. (607754)

Table 10.1 *(Continued)*

Type	Protein name	Functional characteristics of the protein (OMIM ID and chromosome location of the gene, if known)
W4.	Krueppel-related zinc finger protein (zinc finger 184)	A nuclear transcription factor. (602277; 6p21.3)
W5.	Golgi-apparatus protein 1 precursor	Exact function not defined.
Renal oncocytoma		
O1.	Metallothionein-1A	A metal-binding protein, thought to detoxify metals (zinc, copper), to control metal levels during development, and to protect against oxidative stress. (156350; 16q13)
O2.	Copper transport protein	ATOX1 plays a critical role in copper homeostasis. In the kidney normally located in the cortex and the medullary loop of Henle. May play an antioxidant role as well. (602270; 5q32)
O3.	Metallothionein-1L	Member of metallothionein family of low molecular weight. Heavy metal-binding proteins. (156358; 16q13)
O4.	Pyruvate dehydrogenase E1 component, beta	A member of the multisubunit pyruvate dehydrogenase family involved in intermediate metabolism.
Clear RCC		
R1.	Unidentified protein	A protein of unknown function.
R2.	Cathepsin D	A lysosomal aspartic proteinase involved in degradation of proteins, antigen processing, and involved in mediation of apoptosis. May play a role in growth and metastatic potential of epithelial cancers, especially breast Ca. (116840; 11p15.5)
R3.	Fibrinogen gamma	A member of the fibrinogenfamily, which are thought to play a role in cell adhesion and platelet aggregation. (134850; 4q28)
R4.	Phosphoglycerate kinase 1	A major component of glycolytic pathway, to catalyze conversion of 1,3 diphosphoglycerate to 3-phosphoglycerate yield one molecule of ATP. Postulated to be secreted by tumor cells and to play a role in tumor angiogenesis. (311800; Xq13)

(continued overleaf)

Table 10.1 *(Continued)*

Type	Protein name	Functional characteristics of the protein (OMIM ID and chromosome location of the gene, if known)
Papillary RCC		
P-R1.	NADH-ubiquinone oxireductase 24 kDa subunit	Member of the mitochondrial respiratory chain. Deficiencies in allelic variants of this protein have been implicated in parkinson's disease and hypertrophic cardiomyopathy. (600532; 11p11.31–p11.2)
P-R2.	AF-6 protein	Contains GLGF motif shared by other proteins that are thought to play a role in signal transduction as specialized cell–cell junctions. AF-6 is associated with translocation sin leukemia and epithelial ovarian cancers. (159559; 6q27)
P-R3.	Tumor rejection antigen-1	A cell surface glycoprotein that elicits tumor-specific immunity. May play a role in chaperoning peptides to MHC class molecules on antigen-presenting immune cells and in inducing dendritic cells to express B7 antigen and to secrete interleukin-12 and tumor necrosis factor. (191175; 12q24.2–q24.3)

of cytoskeleton proteins that have been found to contain members such as vimentin and cytokeratins, whose simultaneous expression has been implicated in tumor migration, invasion, and differentiation [45, 46]. More specifically, a heterogeneous expression of cytokeratin 8, stathmin, and vimentin have been identified in RCC subtypes [38]. These proteins may serve as biomarkers for distinguishing RCC from normal renal cells.

10.4.4 Metabolic Enzymes

Recent findings indicate that some critical enzymes are differentially expressed in RCC versus normal tissue. For example, manganese superoxide dismutase (Mn-SOD), thioredoxin (THIO), and ubiquitin carboxyl-terminal hydrolase L1 (UCHL1) have been shown to be overexpressed in RCC [47, 48]. On the other hand, the expression of some enzymes declines in RCC, including aldehyde dehydrogenase-1, enoyl-CoA hydratase, α-glycerol-3-phosphate dehydrogenase, plasma glutathione peroxidase, glucose-6-phosphatase, NADH-ubiquinone oxidoreductase complex I, and ubiquinone cytochrome-C reductase [38].

10.4.5 Other Reports

Recent proteomic analysis by Adam *et al.* [49] revealed the up-regulation of a type II transmembrane receptor known as in RCC cells. CD70 was identified as an

RCC biomarker and found to be possibly useful for cytotoxic immunotherapy to treat RCC. Recent expressional differentiation analysis has also been carried out by Craven *et al.* [50] who found 43 up-regulated proteins and 29 down-regulated proteins in a majority of RCC cell lines. Overall, the data offered validating evidence for previously recognized up-regulated proteins such as HSP27 and vimentin in addition to the novel findings such as radixin, moesin, fascin, and actin bundling protein. Seliger *et al.* [51] found UCHL1 to be down-regulated in primary RCC, but its expression is also dependent on the RCC subtype, tumor grade, and the von Hippel–Lindau phenotype. Importantly, higher proliferation and migration rates of RCC were exhibited by cells with UCHL1 gain of function in comparison to UCHL1-negative RCC cells. GRIM-19 has been shown to induce cell death when overactivated, but promote cell growth when inactivated. Alchanati *et al.* [52] discovered via MS the down-regulation and sometimes complete loss of GRIM-19 within primary RCC, confirming the tumor suppressor role of GRIM-19. One important tumor suppressor function to note is that GRIM-19 behaves much like the VHL protein and inactivates STAT3; activation of STAT3 is conjunctive with the loss of GRIM 19 and promotes tumorigenesis.

10.5 Conclusions

The combination of selective tissue microdissection and proteomics to study renal tumors has revealed a multitude of exciting discoveries that will hopefully lead to a more promising outcome for renal tumor patients. Proteomics research offers physicians a chance to diagnose earlier and, more specifically, to define the type of renal tumor, while also having more effective therapeutic options at their disposal. With the continuous improvement of proteomic tools in terms of their sensitivity and accuracy, the future use of selective tissue dissection and proteomics to study renal tumors is promising and will continue to provide beneficial discoveries particularly as the technology becomes more advanced.

References

1. Maurya, P., Meleady, P., Dowling, P., and Clynes, M. (2007) Proteomic approaches for serum biomarker discovery in cancer. *Anticancer Res*, **27**, 1247–1255.
2. Petricoin, E.F., Ardekani, A.M., Hitt, B.A., Levine, P.J., Fusaro, V.A., Steinberg, S.M., Mills, G.B., Simone, C., Fishman, D.A., Kohn, E.C., and Liotta, L.A. (2002) Use of proteomic patterns in serum to identify ovarian cancer. *Lancet*, **359**, 572–577.
3. Petricoin, E.F. III, Ornstein, D.K., Paweletz, C.P., Ardekani, A., Hackett, P.S., Hitt, B.A., Velassco, A., Trucco, C., Wiegand, L., Wood, K., Simone, C.B., Levine, P.J., Linehan, W.M., Emmert-Buck, M.R., Steinberg, S.M., Kohn, E.C., and Liotta, L.A. (2002) Serum proteomic patterns for detection of prostate cancer. *J Natl Cancer Inst*, **94**, 1576–1578.
4. Adam, B.L., Qu, Y., Davis, J.W., Ward, M.D., Clements, M.A., Cazares, L.H., Semmes, O.J.,

Schellhammer, P.F., Yasui, Y., Feng, Z., and Wright, G.L. Jr (2002) Serum protein fingerprinting coupled with a pattern-matching algorithm distinguishes prostate cancer from benign prostate hyperplasia and healthy men. *Cancer Res*, **62**, 3609–3614.
5. Wetterhall, M., Palmblad, M., Hakansson, P., Markides, K.E., and Bergquist, J. (2002) Rapid analysis of tryptically digested cerebrospinal fluid using capillary electrophoresis-electrospray ionization-Fourier transform ion cyclotron resonance-mass spectrometry. *J Proteome Res*, **1**, 361–366.
6. Kreunin, P., Zhao, J., Rosser, C., Urquidi, V., Lubman, D.M., and Goodison, S. (2007) Bladder cancer associated glycoprotein signatures revealed by urinary proteomic profiling. *J Proteome Res*, **6**, 2631–2639.
7. Rao, P.V., Lu, X., Standley, M., Pattee, P., Neelima, G., Girisesh, G., Dakshinamurthy, K.V., Roberts, C.T. Jr, and Nagalla, S.R. (2007) Proteomic identification of urinary biomarkers of diabetic nephropathy. *Diabetes Care*, **30**, 629–637.
8. Valmu, L., Paju, A., Lempinen, M., Kemppainen, E., and Stenman, U.H. (2006) Application of proteomic technology in identifying pancreatic secretory trypsin inhibitor variants in urine of patients with pancreatitis. *Clin Chem*, **52**, 73–81.
9. Huang, J.T., McKenna, T., Hughes, C., Leweke, F.M., Schwarz, E., and Bahn, S. (2007) CSF biomarker discovery using label-free nano-LC-MS based proteomic profiling: technical aspects. *J Sep Sci*, **30**, 214–225.
10. Jin, T., Hu, L.S., Chang, M., Wu, J., Winblad, B., and Zhu, J. (2007) Proteomic identification of potential protein markers in cerebrospinal fluid of GBS patients. *Eur J Neurol*, **14**, 563–568.
11. Khwaja, F.W., Nolen, J.D., Mendrinos, S.E., Lewis, M.M., Olson, J.J., Pohl, J., Van Meir, E.G., Ritchie, J.C., and Brat, D.J. (2006) Proteomic analysis of cerebrospinal fluid discriminates malignant and nonmalignant disease of the central nervous system and identifies specific protein markers. *Proteomics*, **6**, 6277–6287.
12. Khwaja, F.W., Reed, M.S., Olson, J.J., Schmotzer, B.J., Gillespie, G.Y., Guha, A., Groves, M.D., Kesari, S., Pohl, J., and Meir, E.G. (2007) Proteomic identification of biomarkers in the cerebrospinal fluid (CSF) of astrocytoma patients. *J Proteome Res*, **6**, 559–570.
13. Romeo, M.J., Espina, V., Lowenthal, M., Espina, B.H., Petricoin, E.F. III, and Liotta, L.A. (2005) CSF proteome: a protein repository for potential biomarker identification. *Expert Rev Proteomics*, **2**, 57–70.
14. Brinkman, B.M. and Wong, D.T. (2006) Disease mechanism and biomarkers of oral squamous cell carcinoma. *Curr Opin Oncol*, **18**, 228–233.
15. Hu, S., Loo, J.A., and Wong, D.T. (2007) Human saliva proteome analysis. *Ann N Y Acad Sci*, **1098**, 323–329.
16. Huang, C.M. (2004) Comparative proteomic analysis of human whole saliva. *Arch Oral Biol*, **49**, 951–962.
17. Millea, K.M., Krull, I.S., Chakraborty, A.B., Gebler, J.C., and Berger, S.J. (2007) Comparative profiling of human saliva by intact protein LC/ESI-TOF mass spectrometry. *Biochim Biophys Acta*, **4**, 897–906.
18. Streckfus, C.F. and Dubinsky, W.P. (2007) Proteomic analysis of saliva for cancer diagnosis. *Expert Rev Proteomics*, **4**, 329–332.
19. Charalambous, C., Chen, T.C., and Hofman, F.M. (2006) Characteristics of tumor-associated endothelial cells derived from glioblastoma multiforme. *Neurosurg Focus*, **20**, E22.
20. Charalambous, C., Hofman, F.M., and Chen, T.C. (2005) Functional and phenotypic differences between glioblastoma multiforme-derived and normal human brain endothelial cells. *J Neurosurg*, **102**, 699–705.
21. Fischer, I., Gagner, J.P., Law, M., Newcomb, E.W., and Zagzag, D.

(2005) Angiogenesis in gliomas: biology and molecular pathophysiology. *Brain Pathol*, **15**, 297–310.

22. Kaur, B., Tan, C., Brat, D.J., Post, D.E., and Van Meir, E.G. (2004) Genetic and hypoxic regulation of angiogenesis in gliomas. *J Neurooncol*, **70**, 229–243.
23. Craven, R.A. and Banks, R.E. (2001) Laser capture microdissection and proteomics: possibilities and limitation. *Proteomics*, **1**, 1200–1204.
24. Craven, R.A., Totty, N., Harnden, P., Selby, P.J., and Banks, R.E. (2002) Laser capture microdissection and two-dimensional polyacrylamide gel electrophoresis: evaluation of tissue preparation and sample limitations. *Am J Pathol*, **160**, 815–822.
25. Bova, G.S., Eltoum, I.A., Kiernan, J.A., Siegal, G.P., Frost, A.R., Best, C.J., Gillespie, J.W., Su, G.H., and Emmert-Buck, M.R. (2005) Optimal molecular profiling of tissue and tissue components: defining the best processing and microdissection methods for biomedical applications. *Mol Biotechnol*, **29**, 119–152.
26. Glasker, S., Lonser, R.R., Okamoto, H., Li, J., Jaffee, H., Oldfield, E.H., Zhuang, Z., and Vortmeyer, A.O. (2007) Proteomic profiling: a novel method for differential diagnosis? *Cancer Biol Ther*, **6**, 343–345.
27. Okamoto, H., Li, J., Vortmeyer, A.O., Jaffe, H., Lee, Y.S., Glasker, S., Sohn, T.S., Zeng, W., Ikejiri, B., Proescholdt, M.A., Mayer, C., Weil, R.J., Oldfield, E.H., and Zhuang, Z. (2006) Comparative proteomic profiles of meningioma subtypes. *Cancer Res*, **66**, 10199–10204.
28. Vortmeyer, A.O., Devouassoux-Shisheboran, M., Li, G., Mohr, V., Tavassoli, F. and Zhuang, Z. (1999) Microdissection-based analysis of mature ovarian teratoma. *Am J Pathol*, **154**, 987–991.
29. Wang, Y., Rudnick, P.A., Evans, E.L., Li, J., Zhuang, Z., Devoe, D.L., Lee, C.S., and Balgley, B.M. (2005) Proteome analysis of microdissected tumor tissue using a capillary isoelectric focusing-based multidimensional separation platform coupled with ESI-tandem MS. *Anal Chem*, **77**, 6549–6556.
30. Li, J., Yin, C., Okamoto, H., Jaffe, H., Oldfield, E.H., Zhuang, Z., Vortmeyer, A.O., and Rushing, E.J. (2006) Proteomic analysis of inclusion body myositis. *J Neuropathol Exp Neurol*, **65**, 826–833.
31. Li, J., Zhuang, Z., Okamoto, H., Vortmeyer, A.O., Park, D.M., Furuta, M., Lee, Y.S., Oldfield, E.H., Zeng, W., and Weil, R.J. (2006) Proteomic profiling distinguishes astrocytomas and identifies differential tumor markers. *Neurology*, **66**, 733–736.
32. Weil, R.J., Wu, Y.Y., Vortmeyer, A.O., Moon, Y.W., Delgado, R.M., Fuller, B.G., Lonser, R.R., Remaley, A.T., and Zhuang, Z. (1999) Telomerase activity in microdissected human gliomas. *Mod Pathol*, **12**, 41–46.
33. DeRisi, J., Penland, L., Brown, P.O., Bittner, M.L., Meltzer, P.S., Ray, M., Chen, Y., Su, Y.A., and Trent, J.M. (1996) Use of a cDNA microarray to analyse gene expression patterns in human cancer. *Nat Genet*, **14**, 457–460.
34. Washburn, M.P., Ulaszek, R., Deciu, C., Schieltz, D.M., and Yates, J.R. III (2002) Analysis of quantitative proteomic data generated via multidimensional protein identification technology. *Anal Chem*, **74**, 1650–1657.
35. Furuta, M., Weil, R.J., Vortmeyer, A.O., Huang, S., Lei, J., Huang, T.N., Lee, Y.S., Bhowmick, D.A., Lubensky, I.A., Oldfield, E.H., and Zhuang, Z. (2004) Protein patterns and proteins that identify subtypes of glioblastoma multiforme. *Oncogene*, **23**, 6806–6814.
36. Zhuang, Z., Huang, S., Kowalak, J.A., Shi, Y., Lei, J., Furuta, M., Lee, Y.S., Lubensky, I.A., Rodgers, G.P., Cornelius, A.S., Weil, R.J., Teh, B.T., and Vortmeyer, A.O. (2006) From tissue phenotype to proteotype: sensitive protein identification in

microdissected tumor tissue. *Int J Oncol*, **28**, 103–110.

37. Perroud, B., Lee, J., Valkova, N., Dhirapong, A., Lin, P.Y., Fiehn, O., Kultz, D. and Weiss, R.H. (2006) Pathway analysis of kidney cancer using proteomics and metabolic profiling. *Mol Cancer*, **5**, 64.
38. Seliger, B., Lichtenfels, R., and Kellner, R. (2003) Detection of renal cell carcinoma-associated markers via proteome- and other 'ome'-based analyses. *Brief Funct Genomics Proteomics*, **2**, 194–212.
39. Weiss, R.H. and Lin, P.Y. (2006) Kidney cancer: identification of novel targets for therapy. *Kidney Int*, **69**, 224–232.
40. Janech, M.G., Raymond, J.R., and Arthur, J.M. (2007) Proteomics in renal research. *Am J Physiol*, **292**, F501–F512.
41. Hwa, J.S., Park, H.J., Jung, J.H., Kam, S.C., Park, H.C., Kim, C.W., Kang, K.R., Hyun, J.S., and Chung, K.H. (2005) Identification of proteins differentially expressed in the conventional renal cell carcinoma by proteomic analysis. *J Korean Med Sci*, **20**, 450–455.
42. Unwin, R.D., Craven, R.A., Harnden, P., Hanrahan, S., Totty, N., Knowles, M., Eardley, I., Selby, P.J., and Banks, R.E. (2003) Proteomic changes in renal cancer and co-ordinate demonstration of both the glycolytic and mitochondrial aspects of the Warburg effect. *Proteomics*, **3**, 1620–1632.
43. Sarto, C., Binz, P.A., and Mocarelli, P. (2000) Heat shock proteins in human cancer. *Electrophoresis*, **21**, 1218–1226.
44. Sarto, C., Valsecchi, C., Magni, F., Tremolada, L., Arizzi, C., Cordani, N., Casellato, S., Doro, G., Favini, P., Perego, R.A., Raimondo, F., Ferrero, S., Mocarelli, P., and Galli-Kienle, M. (2004) Expression of heat shock protein 27 in human renal cell carcinoma. *Proteomics*, **4**, 2252–2260.
45. Brattsand, G. (2000) Correlation of oncoprotein 18/stathmin expression in human breast cancer with established prognostic factors. *Br J Cancer*, **83**, 311–318.
46. Miettinen, M. and Fetsch, J.F. (2000) Distribution of keratins in normal endothelial cells and a spectrum of vascular tumors: implications in tumor diagnosis. *Hum Pathol*, **31**, 1062–1067.
47. Sarto, C., Frutiger, S., Cappellano, F., Sanchez, J.C., Doro, G., Catanzaro, F., Hughes, G.J., Hochstrasser, D.F. and Mocarelli, P. (1999) Modified expression of plasma glutathione peroxidase and manganese superoxide dismutase in human renal cell carcinoma. *Electrophoresis*, **20**, 3458–3466.
48. Lichtenfels, R., Kellner, R., Atkins, D., Bukur, J., Ackermann, A., Beck, J., Brenner, W., Melchior, S., and Seliger, B. (2003) Identification of metabolic enzymes in renal cell carcinoma utilizing PROTEOMEX analyses. *Biochim Biophys Acta*, **1646**, 21–31.
49. Adam, P.J., Terrett, J.A., Steers, G., Stockwin, L., Loader, J.A., Fletcher, G.C., Lu, L.S., Leach, B.I., Mason, S., Stamps, A.C., Boyd, R.S., Pezzella, F., Gatter, K.C., and Harris, A.L. (2006) CD70 (TNFSF7) is expressed at high prevalence in renal cell carcinomas and is rapidly internalised on antibody binding. *Br J Cancer*, **95**, 298–306.
50. Craven, R.A., Stanley, A.J., Hanrahan, S., Dods, J., Unwin, R., Totty, N., Harnden, P., Eardley, I., Selby, P.J., and Banks, R.E. (2006) Proteomic analysis of primary cell lines identifies protein changes present in renal cell carcinoma. *Proteomics*, **6**, 2853–2864.
51. Seliger, B., Fedorushchenko, A., Brenner, W., Ackermann, A., Atkins, D., Hanash, S., and Lichtenfels, R. (2007) Ubiquitin COOH-terminal hydrolase 1: a biomarker of renal cell carcinoma associated with enhanced tumor cell proliferation and migration. *Clin Cancer Res*, **13**, 27–37.
52. Alchanati, I., Nallar, S.C., Sun, P., Gao, L., Hu, J.,

Stein, A., Yakirevich, E., Konforty, D., Alroy, I., Zhao, X., Reddy, S.P., Resnick, M.B., and Kalvakolanu, D.V. (2006) A proteomic analysis reveals the loss of expression of the cell death regulatory gene GRIM-19 in human renal cell carcinomas. *Oncogene*, **25**, 7138–7147.

11
Proteomic Analysis of Primary and Established Cell Lines for the Investigation of Renal Cell Carcinoma

Rachel A. Craven and Rosamonde E. Banks

11.1
Introduction

Renal cell carcinomas (RCCs) [1, 2], which account for ~3% of adult malignancies, are epithelial tumors that arise from the renal parenchyma and include clear cell, papillary, and chromophobe histological subtypes, which account for 75, 15, and 5% of cases, respectively. Surgical resection of the primary tumor is the main treatment option for RCC and is associated with good survival for patients with organ-confined disease. However, the development of RCC is generally asymptomatic and a large number of patients have locally advanced or metastatic disease by the time of diagnosis. RCC is refractory to chemotherapy and radiotherapy regimes and routine treatment of advanced disease has therefore been limited to cytokine therapies, which give a response in only a subset (<20%) of patients. A number of novel therapies for RCC have been investigated, with the receptor tyrosine kinase inhibitors, sunitinib and sorafenib, whose targets include receptors for vascular endothelial growth factor (VEGF) and platelet-derived growth factor (PDGF), showing particularly promising results in clinical trials [3–6]. Despite such advances, there is still a clear need for new biomarkers and targets to improve management of patients with RCC.

The search for novel biomarkers and therapies for RCC has been facilitated by increased understanding of disease pathogenesis. This is illustrated by clear cell RCC, which is associated with loss of the *VHL* tumor suppressor gene. The well-studied function of VHL is as the substrate recognition subunit of an E3 ubiquitin ligase that targets hypoxia-inducible factor (HIF)-α subunits for degradation by the proteasome. Stabilization of HIF-α subunits under normoxic conditions via loss of VHL leads to induction of a hypoxic response including up-regulation of a number of HIF responsive genes such as glycolytic enzymes, the glucose transporter (GLUT)-1, carbonic anhydrase IX (CAIX), transforming growth factor (TGF)-α, PDGF, and VEGF [7]. This hypoxic response has been shown to be sufficient to promote tumorigenesis in mouse model systems. The HIF pathway has provided a number of molecules with potential clinical utility with CAIX being one of the most consistently up-regulated proteins currently identified

Renal and Urinary Proteomics: Methods and Protocols. Edited by Visith Thongboonkerd

ISBN: 978-3-527-31974-9

in RCC, and the signaling pathways downstream of VEGF and PDGF showing promise as targets for therapeutic intervention using the receptor tyrosine kinase inhibitors sunitinib and sorafenib [3–6]. Studies investigating the development of novel therapies for RCC have also exploited the immunogenic nature of the disease, concentrating on the development of a range of cell- and/or vaccine-based immunotherapies.

A large number of studies have investigated particular proteins or pathways in RCC, generally using antibody-based approaches such as immunohistochemistry or Western blotting to identify changes that accompany tumorigenesis, which may be exploited to generate biomarkers or targets. Biomarker discovery experiments can also take a global, untargeted approach, analyzing samples in different clinical groups, with the aim of identifying significant differences. Gene expression profiling at the mRNA level, generally using microarrays, which can simultaneously profile thousands of gene products, has been carried out extensively in RCC [8]; this analysis is high-throughput making global expression profiling of large number of samples a feasible strategy. Protein profiling offers a complementary approach for gene expression profiling, which overcomes the lack of correlation between mRNA and protein and allows the study of an added level of complexity in the form of posttranslational modifications that may alter the form of a protein rather than its level per se.

11.2 Primary and Established Cell Lines

Comparative analysis of the protein profiles of normal and tumor tissues has been used by a number of groups to identify changes that accompany the development of RCC. However, problems in interpretation of data can arise when whole tissue samples are used for analysis due to the complex mixture of cell types present in normal kidney cortex and RCC tissues. A number of strategies that are compatible with downstream analysis at the DNA, RNA, and protein levels have been employed to overcome the problem of tissue heterogeneity. Laser-assisted microdissection techniques, such as laser capture microdissection, allow areas of interest to be selected from tissue sections [9]; however, obtaining large amounts of material can be time consuming. The use of antibodies to selectively purify cells of interest from cell suspensions generated by mechanical disruption of tissue samples can also be employed, as illustrated by the positive selection of epithelial cells from normal kidney and RCC samples using antibodies such as BerEP4 [10] or depletion of B and T cells using antibodies to CD19 and CD2 [11]. An alternative strategy is the use of cell lines, which provide a model system for studying the tumorigenic process.

Enriched epithelial cell populations can be grown out of normal kidney cortex and RCC tissues, providing a surrogate for whole tissue in comparative analyses. Such primary cell lines have a finite life span and tend to be used at early passage. Established cell lines can also be generated, either by spontaneous transformation or immortalization, thereby allowing large amounts of material to be produced for analysis and facilitating the study of lower-abundance proteins. Cell lines also allow the effects of manipulations such as drug treatment or changes in the expression of particular genes to be investigated, with assays measuring phenotypes such as proliferation, apoptosis and invasion, or tumor formation in mouse models being used to assess the effect on tumorigenic potential and thereby unravel the mechanisms underlying tumor formation and behavior. Of particular significance, when considering the use of cell lines to study RCC, are cell-line pairs generated by transfection of VHL-defective RCC cell lines with vector control or *VHL*, which have been invaluable in defining VHL-dependent changes in gene expression and VHL functions such as its role as a subunit of an E3 ubiquitin ligase involved in regulation of HIF-α turnover. Although the advantages of cell lines are clear, they are obviously balanced by the potential introduction of *in vitro* artifacts resulting from adaptation to growth in culture, making downstream validation of findings in a more clinically relevant setting crucial to any experimental design.

11.3
Growth and Characterization of Primary Cell Lines

Primary cell lines have been grown successfully from renal tissue samples obtained from patients undergoing nephrectomy for RCC in a number of laboratories. Although cell sorting using antibody-based techniques can be used to purify cell populations prior to growth in culture, for example, to purify proximal or distal tubule cells from normal kidney cortex [12, 13], most primary and established cell lines that have been used in studies of RCC have been grown from explant cultures. A variety of different media have been used and some studies have opted to use collagenase to produce an initial cell suspension, but essentially very simple protocols have been adopted. In our studies, primary cell lines are grown from fresh blocks of matched samples of macroscopically viable tumor and distant normal cortical tissue excised immediately following nephrectomy.

11.4
Methods and Protocols for Growth and Characterization of Primary Cell Lines

11.4.1
Growth of Primary Cell Lines from Tissue Samples

1. Place excised tissue blocks in ice-cold RPMI medium for transport to the laboratory.
2. Cut tissues into small pieces using a scalpel and repeatedly agitate the resulting fragments by pipetting with medium (50 : 50 v/v RPMI 1640:DMEM containing 10% v/v FBS, 2 mM L-glutamine, 1.14 mM oxaloacetic acid, 0.45 mM sodium pyruvate, 0.2 IU ml^{-1} insulin, 50 IU ml^{-1} penicillin, and

50 g ml^{-1} streptomycin). Transfer to culture flasks and incubate at 37 °C in a humidified incubator with 5% CO_2.

3. After seven days, change the media, taking care not to disturb adherent tissue from the surface of the flask; make further routine media changes every three to four days. Passage adherent cells by incubating in 0.25% w/v trypsin with 0.2% w/v EDTA in HBSS for a few minutes at 37 °C.

The success rate for growth of primary cell lines from tissue samples in our studies are ~60% for tumor and >90% for normal kidney tissues. We generally use primary cell lines at early passage; however, whereas normal kidney-derived cell lines have a limited life span, the tumor-derived cell lines can grow on for many passages and become established.

Other groups have used very similar simple methods for growth of primary cell lines from normal and malignant renal tissues as well as metastatic tumors [14–20]. Many of these efforts have resulted in the generation of established cell lines derived from primary and metastatic tumors, as assessed by their continuous growth in culture over extended periods of time and >10 passages. Established cell lines generated from normal kidney cortex, although much less common, have been successfully produced as illustrated by the proximal tubule derived cell line HK2, which was immortalized using HPV16 E6/E7 [21] and the adenovirus 12-SV40-immortalized cell lines generated by Racusen and coworkers [22].

11.4.2 Characterization of Primary Cell Lines to Confirm Their Origin

The primary cell lines grown using our protocols do not undergo any cloning, and are thus likely to be heterogeneous cell populations. We routinely carry out initial characterization of primary renal cultures to confirm that the cells are epithelial in origin and this is achieved using immunostaining (Stanley and Banks, manuscript in preparation). Positive staining with antibodies specific for pan-cytokeratin (CK), CK8, CK18, and epithelial antigen (BerEP4) together with negative staining for CK20 and smooth muscle actin indicate successful growth of renal epithelial cells, with little or no contamination from fibroblasts. In addition, vimentin expression in normal-derived cells (as previously described in cultured tubular cells [23]), with higher levels in tumor–derived cells is consistent with the presumed normal and RCC origins of primary cell-line pairs. We use cytogenetic analyses carried out on metaphase spreads from colcemid-arrested cells using G-banding or FISH to confirm that normal primary cell lines have a normal karyotype. In contrast, tumor primary cell lines generally have an abnormal karyotype, which, as expected, often includes changes in 3p.

11.4.3 Assessing the Usefulness of Cell Lines as a Model System for Analyzing RCC

Although preliminary characterization can confirm the growth of epithelial cell populations with normal and abnormal karyotypes, this does not assess the extent of artifacts generated by growth in culture. This question must be addressed to ascertain whether cell lines are representative for the tissue compartment from which they were derived. This can be achieved by looking at the DNA level for conservation of mutations and at the mRNA or protein level for the maintenance of expression profiles. Only a very limited number of studies have done extensive analysis of the proteome to evaluate the extent of changes that occur in culture and these have not focused exclusively on RCC. In one study, the protein profiles of LCM-procured prostate cells and immortalized cell lines derived from the same tissue samples were reported to be very different [24], potentially reflecting both the relative levels of enrichment and the artifacts produced by the two sample-processing routes. In contrast, fresh bladder transitional cell carcinoma and short-term primary cultures gave much more similar profiles [25]; however, this study analyzed newly synthesized protein by ^{35}S labeling. Thus even the fresh sample was manipulated *in vitro* albeit for a much shorter time. Our results with primary cell lines derived from normal kidney cortex and RCC tissues fall between these two extremes with 50% of proteins being matched between the normal primary cell lines and the tissues from which they were derived and 65% for the corresponding RCC samples; as silver staining of total protein was used, loss of serum proteins in the 2-D gel profiles of primary cell lines was particularly significant. These studies are difficult to interpret, as differences will always reflect both enrichment and *in vitro* changes.

The usefulness of matched pairs of normal and tumor-derived cell lines as a model system can be evaluated more fully by looking at the maintenance of differences in gene expression patterns at the mRNA and protein levels. Studies with primary cell lines have only carried out small-scale analyses to look at proteins whose expression is known to be altered in RCC but the data is convincing, with a number of known changes in protein expression patterns being conserved. Manganese superoxide dismutase (Mn-SOD) and Hsp27 showed increased expression and isoform number in tumor-derived cell lines, analogous to results seen in tissues [18]. Similarly, differential expression of aldolase A, CAIX, and ubiquinol cytochrome c reductase complex core protein 1 were generally maintained in culture [26]. Taken together with the data from comparative analyses described below, these results clearly endorse the use of primary cell lines as a model system for studying tumor formation that is representative for RCC and similar conclusions can be drawn for established cell lines as illustrated by the study of VHL cell-line pairs. However, there will always be a need to validate findings using tissue samples and determine their clinical relevance in further studies.

11.5 Analysis of Cell Lines by 2-D Polyacrylamide Gel Electrophoresis (2-D PAGE)

The classical approach for global protein profiling, 2-D PAGE, which allows the simultaneous analysis of up to 2000 proteins in a standard format gel, remains a central separation tool in proteomic studies (for a review, see [27]).

In 2-D PAGE, proteins are separated on the basis of pI and molecular weight by isoelectric focusing and SDS-PAGE in the first and second dimensions, respectively. Detection of proteins in the resulting gel is achieved using stains such as silver, Coomassie, and Sypro Ruby generating a spot pattern that is characteristic for the sample under investigation. Many studies have elected to use whole-cell lysates for analysis and although this has the disadvantage that many of the proteins that are readily accessible for profiling are the more abundant often housekeeping proteins, the approach has been successful. To allow the study of less-abundant proteins, cells can be subjected to subcellular fractionation to focus on particular cellular compartments such as mitochondria, nuclei, or the plasma membrane, or fractionated to select for particular protein groups such as phosphoproteins or glycoproteins. Thus, initial sample handling can be very simple or involves significant prefractionation prior to protein analysis. Conditioned medium can also be used to profile proteins shed or secreted into the medium as a potential source of circulating markers shed or secreted by tumor cells. Here, we present protocols we use for preparation of whole-cell lysates and conditioned medium for 2-D PAGE.

11.6 Methods and Protocols for Analysis of Cell Lines by 2-D PAGE

11.6.1 Preparation of Whole-Cell Lysates for 2-D PAGE

We routinely harvest cell monolayers that are approaching confluence. Whole-cell lysates are prepared by washing cells to remove the medium and then lysing into urea/thiourea lysis buffer.

1. Wash cell monolayers three times with ice-cold PBS and then with ice-cold isotonic (250 mM) sucrose.
2. Add urea/thiourea lysis buffer (7 M urea, 2 M thiourea, 4% w/v CHAPS, 1% w/v DTT, 0.8% Pharmalyte pH3–10 (GE Healthcare), 1 mg ml^{-1} Pefabloc) containing 1 Complete mini protease inhibitor cocktail tablet (Roche) per 2.5 ml to the flask to cover the cells (~1 ml per T75), incubate for 2 minutes at room temperature, scrape the cells off the flask, and transfer to an ultracentrifuge tube.
3. Incubate lysates for 30 minutes at room temperature with intermittent vortexing, clear by centrifugation at 42 000*g* for 1 hour at 15 °C, and store in aliquots at −80 °C.

Protein concentration can be determined by a modified Bradford assay using the BioRad protein assay reagent with a standard curve constructed from BSA in urea/thiourea lysis buffer covering a range of 0–5 mg ml^{-1}.

11.6.2 Preparation of Conditioned Medium for 2-D PAGE

Media containing 10% FBS used in routine culture of cell lines contains high levels of proteins and so is unsuitable for use in preparation of conditioned medium for global protein profiling. Conditioned media is therefore collected from cells grown in serum-free medium with a limited number of supplements that are required to maintain good cell viability; growth conditions including cell confluence and media composition are optimized for each project. Analysis of cell proliferation and apoptosis is carried out to ensure that cells are growing well and therefore that the release of proteins into the medium as a result of cell death does not contribute significantly to the total protein content of conditioned medium.

The protein concentration of conditioned medium is relatively low and samples contain high levels of salts, which would interfere with downstream protein profiling. Conditioned media are therefore processed by concentration and buffer exchange into urea/thiourea lysis buffer.

1. Wash cell monolayers that have reached 70–80% confluence three times with serum-free medium and replace with serum-free medium supplemented with 5 μg ml^{-1} transferrin, 10 μg ml^{-1} insulin, and 5 ng ml^{-1} sodium selenite.
2. After 24 hours, harvest conditioned medium onto ice and add Complete mini protease inhibitor cocktail (1 tablet per 50 ml) prior to filtering through a 0.2-μm filter to remove cell debris and contaminating cells.
3. Concentrate conditioned medium 10-fold using a Gyrosep 300 stirred cell (Intersep) pressurized at 4 bar at 4 °C using a 10-kDa cut-off ultrafiltration membrane (Millipore) and a further 20-fold using a 15-ml 10-kDa cut-off Amicon Ultra centrifugal filter device (Millipore) at 1000g at 4 °C.
4. Buffer-exchange conditioned medium into 2D-PAGE lysis buffer by cycles of dilution and concentration using a 15-ml 10-kDa cut-off Amicon Ultra centrifugal filter device to achieve an approximate 150-fold dilution of starting buffer. Store in aliquots at −80 °C.

Typical yields of protein are 100–200 μg per T150 flask of which ~30% is transferrin (added as a supplement) and albumin (carried over from serum-containing medium). Transferrin contributes ~80% of this "contamination."

11.6.3 2-D PAGE Analysis

Methods for 2-D PAGE have been published by a number of groups and vary in their details. Here, we present protocols that we use routinely for 2-D PAGE using pH 3–10 NL gradients for isoelectric focusing, 10% resolving gels with a 4% stacking gel for SDS-PAGE, and silver staining for protein detection.

1. Dilute protein samples (generally 30 μg prepared in urea/thiourea lysis buffer) in reswelling buffer (7 M urea, 2 M thiourea, 4% w/v CHAPS, 0.46% w/v DTT,

0.2% Pharmalyte pH 3–10 with a trace of bromophenol blue) to give a final volume of 450 μl.

2. Apply to 18-cm pH 3–10 NL IPG strips (GE Healthcare) by in-gel rehydration (30 V, 13 hours) using an IPGphor (GE Healthcare) and carry out isoelectric focusing to give a total of 65 kVh (200 V (1 hour), 500 V (1 hour), 1000 V (1 hour), gradient incline to 8000 V (1 hour), then 8000 V to end).
3. Incubate IPG strips in equilibration buffer (6 M urea, 30% v/v glycerol, 2% w/v SDS, 50 mM Tris-HCl pH6.8) containing 1% w/v DTT for 15 minutes, then in equilibration buffer containing 4% w/v iodoacetamide for 10 minutes.
4. Rinse IPG strips in SDS-PAGE running buffer (24 mM Tris, 100 mM glycine, 0.1% w/v SDS) and seal in place on top of the second dimension gels with 1% w/v LMP agarose in SDS-PAGE running buffer. Electrophorese gels overnight (18 mA per gel) at 12.5 °C.
5. Fix gels in 50% methanol, 10% acetic acid, stain using OWL silver stain (Autogen Bioclear), and scan for downstream analysis using a Personal Densitometer (GE Healthcare).

Representative 2-D gel profiles of whole-cell lysates prepared from normal and tumor primary cell lines and the tissues from which they were derived are shown in Figure 11.1a and 2-D gels of conditioned medium and whole-cell lysate prepared from the RCC cell line HTB46 in Figure 11.1b.

Comparative analysis of 2-D gels can be carried out to identify protein species that are differentially expressed between sample groups. This is generally achieved using dedicated gel analysis software packages such as Melanie (GeneBio) or Progenesis (Nonlinear Dynamics). Proteins of interest can then be identified by mass spectrometry (MS). Preparative gels with an increased protein load (~1 mg of protein) are stained with PlusOne silver stain using a modified protocol that is compatible with MS [28]. Excised spots are then digested with trypsin or an alternative protease and the resulting peptides analyzed by MS. Protein identification can be achieved using the masses of the tryptic peptides (which are a peptide mass fingerprint of the protein under investigation) and searching against *in silico* digests of protein sequences in a database; alternatively, fragmentation of individual peptides in MS–MS can be used to generate spectra that are interpreted to obtain sequence information.

The 2-D PAGE protocols described here and other similar methods have been used extensively to study a number of diseases including RCC. However, the technology can be exploited further to improve and extend profiling studies. Firstly, 2-D difference gel electrophoresis (2-D DIGE), where samples are minimally labeled with one of three spectrally resolvable cyanine dyes prior to analysis, allows up to three samples (generally two samples of interest and a pooled internal standard that can be used to allow normalization between gels) to be run on a single gel, thereby simplifying gel analysis by improving the quality of gel matching and comparative analysis [29, 30]. 2-D DIGE is now routinely applied to profiling studies. Secondly, although most studies have elected to use broad-range pH gradients such as pH 3–10NL for isoelectric focusing, the use of narrow pH gradients, that is zoom gel

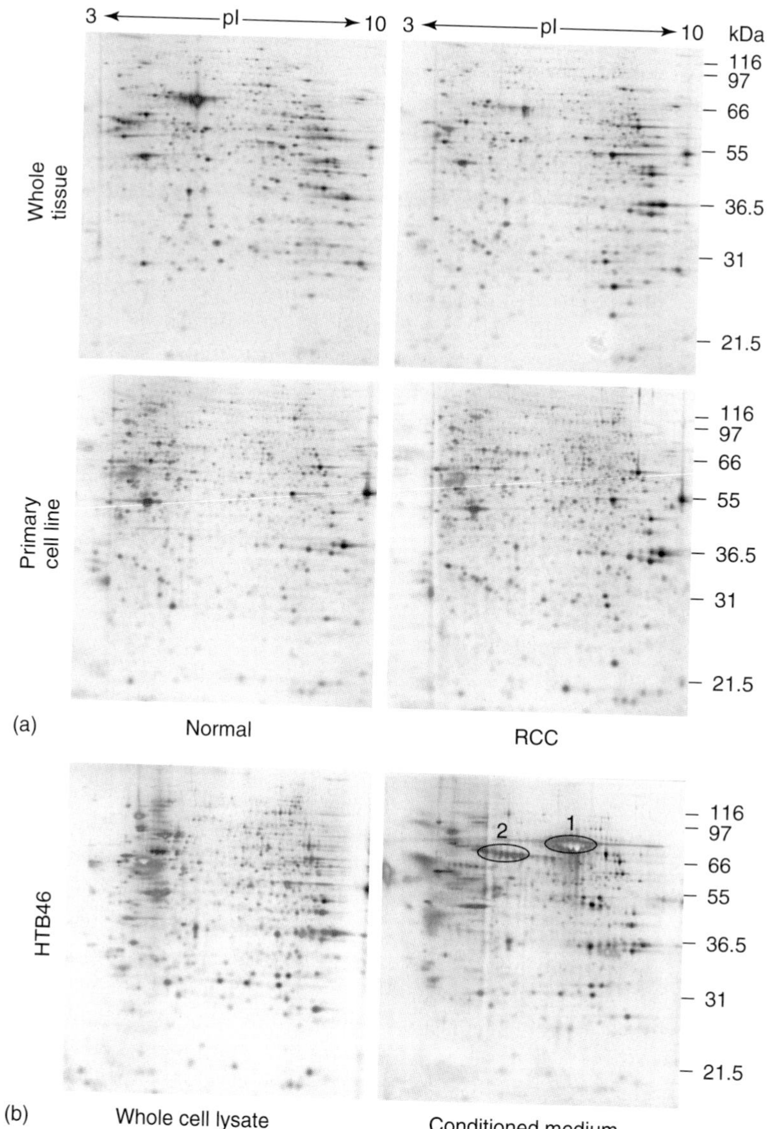

Figure 11.1 2-D gel profiles of tissues, cell lines, and conditioned medium. Representative silver stained 2-D gels of (a) whole-cell extracts of normal and RCC derived primary cell lines and the corresponding tissues from which they were derived and (b) whole-cell extracts and conditioned medium from the RCC cell line HTB46 are shown. Totally 30-μg protein (normalized in the case of conditioned medium to exclude protein contributed by transferrin and albumin, which are indicated on the gel as trains 1 and 2 respectively) was separated by 2-D PAGE using pH 3–10NL IPG strips for isoelectric focusing and 10% T gels for SDS-PAGE. Molecular weight markers (kDa) are indicated. Figure 11.1a is reproduced with permission from [26]. Copyright Wiley-VCH Verlag GmbH & Co. KGaA.

technology, increases the effective separation space, thereby allowing analysis of less-abundant molecules and improving proteome coverage. 2-D PAGE does have limitations such as the relatively large amounts of sample required especially for preparative gels, and under-representation of certain classes of proteins such as membrane proteins. However, ongoing improvements in sensitivity, for example, by the use of saturation labeling 2-D DIGE [31] and in sample preparation are likely to continue; thus, 2-D PAGE is likely to remain one of the most powerful strategies for proteome analysis.

Protein profiling and identification of molecules of interest is obviously only the first step in translational research programs; results must then be corroborated in initial validation studies using alternative techniques and findings with potential clinical utility then taken forward in larger studies to fully assess their usefulness. This generally involves the use of antibody-based techniques and must at some stage incorporate experiments using clinical samples to confirm that findings observed in cell line models are relevant *in vivo*. This topic is beyond the scope of this review but it must be considered given that few clinical research projects incorporate comprehensive validation.

11.7 Alternative Approaches for Protein Profiling

A number of alternative approaches are available for global, untargeted protein profiling, which incorporate different strategies for peptide/protein separation and quantitation with protein identification being achieved by MS. A few studies using cancer cell lines are referred to here to illustrate the application of these various techniques; studies on RCC are described in the next section.

Liquid chromatography (LC), for example, using chromatofocusing and reverse-phase chromatography to sequentially separate proteins on the basis of pI and hydrophobicity as achieved by the PF2-D platform available from Beckman Coulter, allows 2-D protein expression maps to be generated for comparative analysis. Like 2-D PAGE, 2-D LC is suited to the resolution of different protein forms and therefore to studying post-translational modifications. Profiling of serous ovarian cancer cell lines, immortalized ovarian surface epithelial cell lines or endometroid and breast epithelial cell lines by 2-D LC showed the potential of the approach for achieving protein-based sample classification [32]. In a study examining changes in protein expression that accompany modulation of the phenotype of J82 bladder cancer cells in response to growth on different extracellular matrix preparations, a number of protein identities were determined, highlighting a role for the myc pathway in the malignant phenotype [33]. 2-D LC has only been used in a relatively small number of studies and further downstream validation of findings is required, but it seems likely that it will complement 2-D PAGE in protein profiling studies.

Profiling intact proteins can also be achieved by matrix-assisted laser desorption/ionization time-of-flight (MALDI-TOF) MS as shown by surface-enhanced laser desorption/ionization (SELDI)-TOF MS, which uses ProteinChip arrays coated with chromatographic surfaces to selectively bind subsets of proteins from samples of interest that are then profiled by MALDI-TOF MS. SELDI-TOF MS is a sensitive and high-throughput technique that is ideally suited to profiling of proteins <20 kDa, which generally fall below the limit of 2-D PAGE. Expression profiling of proteins in a series of cancer cell lines by SELDI-TOF MS identified prothymosin-α as a marker for colon cancer cells [34] and by comparing the protein profiles with sensitivity of cells to a phosphatidylinositol 3-kinase inhibitor, alterations in the phosphorylation state of ribosomal P2 were found to correlate with chemosensitivity [35]. SELDI-TOF MS has also been used to analyze conditioned medium, as illustrated by a study analyzing the anti-apoptotic and mitogenic paracrine effect mediated by p21 expression in HT-1080 human fibrosarcoma cells using an inducible expression system. Cystatin C, pro-platelet basic protein (PPBP) and β_2-microglobulin were all found to be increased in conditioned medium, results that were corroborated by Western blotting [36].

Protein profiling can be also done by a bottom-up approach whereby protein extracts are digested, usually with trypsin, and the resulting peptides separated by multidimensional LC and analyzed by MS–MS, generating a catalog of proteins present in a sample. Although information on protein forms cannot always be elucidated in shotgun experiments as the link between peptide and protein is lost, hundreds of proteins can be identified in a single experiment using relatively small amounts of sample. Label-free strategies can be used to facilitate semiquantitative analysis of shotgun data. Using such an approach to analyze the effect of restoring expression of maspin in MDA-MB 435 mammary carcinoma cells, changes in ~27% of the detectable proteome were observed, including proteins involved in cytoskeletal organization and apoptosis. Western blotting confirmed altered expression of 20/22 proteins examined including up-regulation of paxillin and plectin-1 [37].

More formal quantitation can also be incorporated into shotgun proteomics via the use of stable isotopes and tags. For cell lines, one of the most readily applied methods is stable-isotope labeling by amino acids in cell culture (SILAC) [38]. Cells are grown in light and heavy versions of amino acids such as $^{12}C_6/^{13}C_6$ arginine and lysine and resulting peptide pairs are resolved as a doublet in MS, with the ratio of the peaks reporting on relative abundance. Membrane fractions of SILAC-labeled normal breast and malignant breast cancer cells were analyzed using this approach and many changes associated with tumorigenesis were identified [39]. Immunohistochemistry was used to confirm down-regulation of CD13 and up-regulation of osteoblast-specific factor (OSF)-2 [39]. Isobaric tags for relative and absolute quantitation (iTRAQ) [40] are an alternative approach for quantitation. Samples are labeled with one of four isobaric tagging reagents and the resulting labeled peptides are only distinguishable in MS–MS when each fragments to yield a characteristic reporter ion. Using iTRAQ, changes in the proteome that accompany increased migration and invasion seen in the TGF-β-induced epithelial–mesenchymal transition in lung cancer cells were analyzed. A number of changes, including increased expression of transglutaminase 2, β-1 integrin, and filamin A were confirmed by Western blotting [41].

11.8 Proteomic Studies of Cell Lines in RCC

Although a number of approaches are available for proteomic analysis of cell lines, only a small number of studies have been published describing protein profiling of cell lines in RCC. However, these studies are varied and have provided interesting results that endorse the use of cell lines as a model system for studying RCC.

Primary cell-line pairs grown from patient-matched normal kidney and RCC tissues have been studied in comparative analyses using 2-D PAGE to look for changes in the proteome that accompany tumorigenesis. The first study that adopted this approach analyzed 11 pairs of matched normal and RCC primary cell lines, 9 of which were from patients with clear cell RCC, and identified a number of changes, the most prevalent of which were up-regulation of Mn-SOD, α-B-crystallin and annexin IV [17]. Western blot analysis confirmed that these changes reflected those seen in the tissues from which the primary cell lines were derived. A similar study analyzing primary cell lines from patients with clear cell RCC identified 43 protein spots that were up-regulated and 29 that were down-regulated in at least 3/5 tumors [26]. Many of these changes were known to be associated with RCC, including up-regulation of glycolytic enzymes but others were novel, such as up-regulation of fascin, which was confirmed in tissues by Western blotting and immunohistochemistry. The lack of overlap between these studies was perhaps surprising, even given the use of different gradients for isoelectric focusing. The study of Shi and coworkers identified fewer changes overall and fewer conserved changes. Although the tumors were a more heterogeneous group, there was a significant overlap in samples between the two studies, so this is unlikely to account for the difference. However, both studies identified changes that were validated in tissues.

VHL cell-line pairs generated from established VHL-defective RCC cell lines transfected with vector control or VHL have also been compared by 2-D PAGE to specifically identify VHL-dependent changes in gene expression that accompany tumor formation [42]. Changes in the expression of mitochondrial proteins following VHL transfection suggested that loss of mitochondrial proteins, which is widespread in RCC, may be explained in part by loss of VHL, confirming and extending previously published data [43]. This study also identified VHL-dependent changes in SEPT2 and analysis in tissues suggested that increased expression of this protein was a common event in RCC [42].

Conditioned medium from VHL cell-line pairs has also been profiled to look for differentially shed or secreted biomarkers. Levels of the known HIF targets, insulin-like growth factor binding protein3 (IGFBP3) and plasminogen activator inhibitor-1 (PAI-1), were increased in VHL-defective cells, while clusterin was decreased [44]. Reduced clusterin release by VHL-defective cells was not part of a hypoxic response nor was it restored by type 2C VHL mutants that retain the ability to regulate HIF. Furthermore, analysis of tissue samples by immunohistochemistry showed reduced clusterin expression in VHL-deficient RCCs compared to those with wild-type VHL, and in VHL-defective pheochromocytomas. Clusterin may therefore be a useful marker for HIF-independent VHL function.

Analysis of plasma membrane fractions of established RCC cell lines by 1-D PAGE and MS was carried out in a descriptive study that aimed to produce a catalog of proteins present on the cell surface of tumor cells. One of the proteins identified was the TNF receptor family member, CD70, which was shown to be highly expressed on the cell surface of several RCC cell lines by FACS analysis [45]. Immunohistochemistry showed that CD70 was expressed in 16/20 primary clear cell tumors and 8/11 metastases while normal kidney showed no staining. The potential utility of CD70 as a target for toxin-conjugated antibody therapy was demonstrated using an internalizing anti-CD70 antibody and saponin-conjugated secondary antibody, which produced 50% killing of A498 RCC cells. This data is consistent with other reports showing that CD70 may be useful as a biomarker/target for RCC [46, 47].

Finally, in a study looking for markers of response to interferon-α (IFN)-α, the growth of a bank of RCC cell lines was assessed for an effect of treatment with IFN-α and cell lysates from two cell lines that were sensitive and two that were resistant to the anti-proliferative effects of IFN-α were analyzed by SELDI-TOF MS using a variety of ProteinChip surfaces [48]. This preliminary study highlighted a number of peaks specific for either sensitive or resistant cell lines that now require identification. Validation of the results in further cell lines and in patient material is needed to test their usefulness. This study complements a recent analysis at the mRNA level, which developed a model based on four genes that provided predictive information on the response of cells to IFN-α [49].

A distinct area of proteomics-based research that makes use of cell lines as well as tissues, and should therefore be mentioned here, is that of searching for tumor-associated proteins that elicit an immune response in patients. The most common strategy that has been employed to look for antigens recognized by RCC patient sera, namely 2-D PAGE and Western blotting, has been carried out using cell line extracts in some cases, with studies focusing on heat shock proteins, cytoskeletal family members, and metabolic enzymes [50–52]. This topic is discussed in more detail in chapter 12. Investigation of the MHC ligandome, that is the repertoire of peptides presented at the cell surface by MHC molecules, via immunoaffinity purification of MHC-peptide complexes and MS has also been carried out using RCC cell lines [53]. Thus, cell lines can be used in screens for both B and T cell targets.

These studies illustrate the potential of using cell lines as a model system in proteomic profiling studies to investigate various aspects of RCC; however, this area of research is still in its infancy. Further studies are now needed to build on these preliminary findings, including validating changes in further sample sets and clinical samples to ascertain which are relevant *in vivo* and taking forward candidate biomarkers in more extensive downstream studies to fully test their potential usefulness. Future work is clearly required to extend profiling studies to lower-abundance proteins, which will involve incorporation of sample prefractionation such as purification of particular subcellular compartments or particular subsets of proteins such as glycoproteins and phosphoproteins and using the full capabilities of the technologies available to facilitate improved protein separation. More targeted profiling using antibody arrays is also likely to be adopted to study molecules that fall below the sensitivity of other strategies; such arrays have been used most frequently for the analysis of cytokines with good results [54]. It seems hopeful that protein profiling of cell lines will play a key part in the identification of biomarkers and targets for RCC.

References

1. Lam, J.S., Shvarts, O., Leppert, J.T., Figlin, R.A., and Belldegrun, A.S. (2005) Renal cell carcinoma 2005: new frontiers in staging, prognostication and targeted molecular therapy. *J Urol*, **173**, 1853–1862.
2. Drucker, B.J. (2005) Renal cell carcinoma: current status and future prospects. *Cancer Treat Rev*, **31**, 536–545.
3. Motzer, R.J., Rini, B.I., Bukowski, R.M., Curti, B.D., George, D.J., Hudes, G.R., Redman, B.G., Margolin, K.A., Merchan, J.R., Wilding, G., Ginsberg, M.S., Bacik, J., Kim, S.T., Baum, C.M., and Michaelson, M.D. (2006) Sunitinib in patients with metastatic renal cell carcinoma. *J Am Med Assoc*, **295**, 2516–2524.
4. Motzer, R.J., Michaelson, M.D., Redman, B.G., Hudes, G.R., Wilding, G., Figlin, R.A., Ginsberg, M.S., Kim, S.T., Baum, C.M., DePrimo, S.E., Li, J.Z., Bello, C.L., Theuer, C.P., George, D.J., and Rini, B.I. (2006) Activity of SU11248, a multitargeted inhibitor of vascular endothelial growth factor receptor and platelet-derived growth factor receptor, in patients with metastatic renal cell carcinoma. *J Clin Oncol*, **24**, 16–24.
5. Motzer, R.J., Hutson, T.E., Tomczak, P., Michaelson, M.D., Bukowski, R.M., Rixe, O., Oudard, S., Negrier, S., Szczylik, C., Kim, S.T., Chen, I., Bycott, P.W., Baum, C.M., and Figlin, R.A. (2007) Sunitinib versus interferon alfa in metastatic renal-cell carcinoma. *N Engl J Med*, **356**, 115–124.
6. Ratain, M.J., Eisen, T., Stadler, W.M., Flaherty, K.T., Kaye, S.B., Rosner, G.L., Gore, M., Desai, A.A., Patnaik, A., Xiong, H.Q., Rowinsky, E., Abbruzzese, J.L., Xia, C., Simantov, R., Schwartz, B., and O'Dwyer, P.J. (2006) Phase II placebo-controlled randomized discontinuation trial of sorafenib in patients with metastatic renal cell carcinoma. *J Clin Oncol*, **24**, 2505–2512.
7. Kaelin, W.G. Jr (2007) The von Hippel-Lindau tumor suppressor protein and clear cell renal carcinoma. *Clin Cancer Res*, **13**, 680s–684s.
8. Yin-Goen, Q., Dale, J., Yang, W.L., Phan, J., Moffitt, R., Petros, J.A., Datta, M.W., Amin, M.B., Wang, M.D., and Young, A.N. (2006) Advances in molecular classification of renal neoplasms. *Histol Histopathol*, **21**, 325–339.
9. Okuducu, A.F., Hahne, J.C., Von Deimling, A., and Wernert, N. (2005) Laser-assisted microdissection, techniques and applications in pathology (review). *Int J Mol Med*, **15**, 763–769.
10. Sarto, C., Valsecchi, C., and Mocarelli, P. (2002) Renal cell carcinoma: handling and treatment. *Proteomics*, **2**, 1627–1629.
11. Sarto, C., Marocchi, A., Sanchez, J.C., Giannone, D., Frutiger, S., Golaz, O., Wilkins, M.R., Doro, G.,

Cappellano, F., Hughes, G., Hochstrasser, D.F., and Mocarelli, P. (1997) Renal cell carcinoma and normal kidney protein expression. *Electrophoresis*, **18**, 599–604.

12. Baer, P.C., Nockher, W.A., Haase, W., and Scherberich, J.E. (1997) Isolation of proximal and distal tubule cells from human kidney by immunomagnetic separation. *Kidney Int*, **52**, 1321–1331.
13. Helbert, M.J., Dauwe, S.E., Van der Biest, I., Nouwen, E.J., and De Broe, M.E. (1997) Immunodissection of the human proximal nephron: flow sorting of S1S2S3, S1S2 and S3 proximal tubular cells. *Kidney Int*, **52**, 414–428.
14. Ebert, T., Bander, N.H., Finstad, C.L., Ramsawak, R.D., and Old, L.J. (1990) Establishment and characterization of human renal cancer and normal kidney cell lines. *Cancer Res*, **50**, 5531–5536.
15. Anglard, P., Trahan, E., Liu, S., Latif, F., Merino, M.J., Lerman, M.I., Zbar, B., and Linehan, W.M. (1992) Molecular and cellular characterization of human renal cell carcinoma cell lines. *Cancer Res*, **52**, 348–356.
16. Shin, K.H., Ku, J.L., Kim, W.H., Lee, S.E., Lee, C., Kim, S.W., and Park, J.G. (2000) Establishment and characterization of seven human renal cell carcinoma cell lines. *BJU Int*, **85**, 130–138.
17. Shi, T., Dong, F., Liou, L.S., Duan, Z.H., Novick, A.C., and DiDonato, J.A. (2004) Differential protein profiling in renal-cell carcinoma. *Mol Carcinog*, **40**, 47–61.
18. Perego, R.A., Bianchi, C., Corizzato, M., Eroini, B., Torsello, B., Valsecchi, C., Di Fonzo, A., Cordani, N., Favini, P., Ferrero, S., Pitto, M., Sarto, C., Magni, F., Rocco, F., and Mocarelli, P. (2005) Primary cell cultures arising from normal kidney and renal cell carcinoma retain the proteomic profile of corresponding tissues. *J Proteome Res*, **4**, 1503–1510.
19. Seliger, B., Hohne, A., Knuth, A., Bernhard, H., Meyer, T., Tampe, R., Momburg, F., and Huber, C. (1996) Analysis of the major histocompatibility complex class I antigen presentation machinery in normal and malignant renal cells: evidence for deficiencies associated with transformation and progression. *Cancer Res*, **56**, 1756–1760.
20. Lichtenfels, R., Ackermann, A., Kellner, R., and Seliger, B. (2001) Mapping and expression pattern analysis of key components of the major histocompatibility complex class I antigen processing and presentation pathway in a representative human renal cell carcinoma cell line. *Electrophoresis*, **22**, 1801–1809.
21. Ryan, M.J., Johnson, G., Kirk, J., Fuerstenberg, S.M., Zager, R.A., and Torok-Storb, B. (1994) HK-2: an immortalized proximal tubule epithelial cell line from normal adult human kidney. *Kidney Int*, **45**, 48–57.
22. Racusen, L.C., Monteil, C., Sgrignoli, A., Lucskay, M., Marouillat, S., Rhim, J.G., and Morin, J.P. (1997) Cell lines with extended in vitro growth potential from human renal proximal tubule: characterization, response to inducers, and comparison with established cell lines. *J Lab Clin Med*, **129**, 318–329.
23. Baer, P.C., Tunn, U.W., Nunez, G., Scherberich, J.E., and Geiger, H. (1999) Transdifferentiation of distal but not proximal tubular epithelial cells from human kidney in culture. *Exp Nephrol*, **7**, 306–313.
24. Ornstein, D.K., Gillespie, J.W., Paweletz, C.P., Duray, P.H., Herring, J., Vocke, C.D., Topalian, S.L., Bostwick, D.G., Linehan, W.M., Petricoin, E.F. III, and Emmert-Buck, M.R. (2000) Proteomic analysis of laser capture microdissected human prostate cancer and in vitro prostate cell lines. *Electrophoresis*, **21**, 2235–2242.
25. Celis, A., Rasmussen, H.H., Celis, P., Basse, B., Lauridsen, J.B., Ratz, G., Hein, B., Ostergaard, M., Wolf, H., Orntoft, T., and Celis, J.E. (1999) Short-term culturing of low-grade superficial bladder transitional cell carcinomas leads to changes in the

expression levels of several proteins involved in key cellular activities. *Electrophoresis*, **20**, 355–361.

26. Craven, R.A., Stanley, A.J., Hanrahan, S., Dods, J., Unwin, R., Totty, N., Harnden, P., Eardley, I., Selby, P.J., and Banks, R.E. (2006) Proteomic analysis of primary cell lines identifies protein changes present in renal cell carcinoma. *Proteomics*, **6**, 2853–2864.
27. Gorg, A., Weiss, W., and Dunn, M.J. (2004) Current two-dimensional electrophoresis technology for proteomics. *Proteomics*, **4**, 3665–3685.
28. Yan, J.X., Wait, R., Berkelman, T., Harry, R.A., Westbrook, J.A., Wheeler, C.H., and Dunn, M.J. (2000) A modified silver staining protocol for visualization of proteins compatible with matrix-assisted laser desorption/ionization and electrospray ionization-mass spectrometry. *Electrophoresis*, **21**, 3666–3672.
29. Tonge, R., Shaw, J., Middleton, B., Rowlinson, R., Rayner, S., Young, J., Pognan, F., Hawkins, E., Currie, I., and Davison, M. (2001) Validation and development of fluorescence two-dimensional differential gel electrophoresis proteomics technology. *Proteomics*, **1**, 377–396.
30. Alban, A., David, S.O., Bjorkesten, L., Andersson, C., Sloge, E., Lewis, S., and Currie, I. (2003) A novel experimental design for comparative two-dimensional gel analysis: two-dimensional difference gel electrophoresis incorporating a pooled internal standard. *Proteomics*, **3**, 36–44.
31. Shaw, J., Rowlinson, R., Nickson, J., Stone, T., Sweet, A., Williams, K., and Tonge, R. (2003) Evaluation of saturation labelling two-dimensional difference gel electrophoresis fluorescent dyes. *Proteomics*, **3**, 1181–1195.
32. Wang, Y., Wu, R., Cho, K.R., Shedden, K.A., Barder, T.J., and Lubman, D.M. (2006) Classification of cancer cell lines using an automated two-dimensional liquid mapping method with hierarchical clustering techniques. *Mol Cell Proteomics*, **5**, 43–52.
33. Hurst, R.E., Kyker, K.D., Dozmorov, M.G., Takemori, N., Singh, A., Matsumoto, H., Saban, R., Betgovargez, E., and Simonian, M.H. (2006) Proteome-level display by 2-dimensional chromatography of extracellular matrix-dependent modulation of the phenotype of bladder cancer cells. *Proteome Sci*, **4**, 13.
34. Shiwa, M., Nishimura, Y., Wakatabe, R., Fukawa, A., Arikuni, H., Ota, H., Kato, Y., and Yamori, T. (2003) Rapid discovery and identification of a tissue-specific tumor biomarker from 39 human cancer cell lines using the SELDI ProteinChip platform. *Biochem Biophys Res Commun*, **309**, 18–25.
35. Akashi, T., Nishimura, Y., Wakatabe, R., Shiwa, M., and Yamori, T. (2007) Proteomics-based identification of biomarkers for predicting sensitivity to a PI3-kinase inhibitor in cancer. *Biochem Biophys Res Commun*, **352**, 514–521.
36. Currid, C.A., O'Connor, D.P., Chang, B.D., Gebus, C., Harris, N., Dawson, K.A., Dunn, M.J., Pennington, S.R., Roninson, I.B., and Gallagher, W.M. (2006) Proteomic analysis of factors released from p21-overexpressing tumour cells. *Proteomics*, **6**, 3739–3753.
37. Chen, E.I., Florens, L., Axelrod, F.T., Monosov, E., Barbas, C.F. III, Yates, J.R. III, Felding-Habermann, B., and Smith, J.W. (2005) Maspin alters the carcinoma proteome. *FASEB J*, **19**, 1123–1124.
38. Mann, M. (2006) Functional and quantitative proteomics using SILAC. *Nat Rev Mol Cell Biol*, **7**, 952–958.
39. Liang, X., Zhao, J., Hajivandi, M., Wu, R., Tao, J., Amshey, J.W., and Pope, R.M. (2006) Quantification of membrane and membrane-bound proteins in normal and malignant breast cancer cells isolated from the same patient with primary breast carcinoma. *J Proteome Res*, **5**, 2632–2641.

40. Ross, P.L., Huang, Y.N., Marchese, J.N., Williamson, B., Parker, K., Hattan, S., Khainovski, N., Pillai, S., Dey, S., Daniels, S., Purkayastha, S., Juhasz, P., Martin, S., Bartlet-Jones, M., He, F., Jacobson, A., and Pappin, D.J. (2004) Multiplexed protein quantitation in Saccharomyces cerevisiae using amine-reactive isobaric tagging reagents. *Mol Cell Proteomics*, **3**, 1154–1169.
41. Keshamouni, V.G., Michailidis, G., Grasso, C.S., Anthwal, S., Strahler, J.R., Walker, A., Arenberg, D.A., Reddy, R.C., Akulapalli, S., Thannickal, V.J., Standiford, T.J., Andrews, P.C., and Omenn, G.S. (2006) Differential protein expression profiling by iTRAQ-2DLC-MS/MS of lung cancer cells undergoing epithelial-mesenchymal transition reveals a migratory/invasive phenotype. *J Proteome Res*, **5**, 1143–1154.
42. Craven, R.A., Hanrahan, S., Totty, N., Harnden, P., Stanley, A.J., Maher, E.R., Harris, A.L., Trimble, W.S., Selby, P.J., and Banks, R.E. (2006) Proteomic identification of a role for the von Hippel Lindau tumour suppressor in changes in the expression of mitochondrial proteins and septin 2 in renal cell carcinoma. *Proteomics*, **6**, 3880–3893.
43. Hervouet, E., Demont, J., Pecina, P., Vojtiskova, A., Houstek, J., Simonnet, H., and Godinot, C. (2005) A new role for the von Hippel-Lindau tumor suppressor protein: stimulation of mitochondrial oxidative phosphorylation complex biogenesis. *Carcinogenesis*, **26**, 531–539.
44. Nakamura, E., Abreu-e-Lima, P., Awakura, Y., Inoue, T., Kamoto, T., Ogawa, O., Kotani, H., Manabe, T., Zhang, G.J., Kondo, K., Nose, V., and Kaelin, W.G. Jr (2006) Clusterin is a secreted marker for a hypoxia-inducible factor-independent function of the von Hippel-Lindau tumor suppressor protein. *Am J Pathol*, **168**, 574–584.
45. Adam, P.J., Terrett, J.A., Steers, G., Stockwin, L., Loader, J.A., Fletcher, G.C., Lu, L.S., Leach, B.I., Mason, S., Stamps, A.C., Boyd, R.S., Pezzella, F., Gatter, K.C., and Harris, A.L. (2006) CD70 (TNFSF7) is expressed at high prevalence in renal cell carcinomas and is rapidly internalised on antibody binding. *Br J Cancer*, **95**, 298–306.
46. Diegmann, J., Junker, K., Gerstmayer, B., Bosio, A., Hindermann, W., Rosenhahn, J., and von Eggeling, F. (2005) Identification of CD70 as a diagnostic biomarker for clear cell renal cell carcinoma by gene expression profiling, real-time RT-PCR and immunohistochemistry. *Eur J Cancer*, **41**, 1794–1801.
47. Law, C.L., Gordon, K.A., Toki, B.E., Yamane, A.K., Hering, M.A., Cerveny, C.G., Petroziello, J.M., Ryan, M.C., Smith, L., Simon, R., Sauter, G., Oflazoglu, E., Doronina, S.O., Meyer, D.L., Francisco, J.A., Carter, P., Senter, P.D., Copland, J.A., Wood, C.G., and Wahl, A.F. (2006) Lymphocyte activation antigen CD70 expressed by renal cell carcinoma is a potential therapeutic target for anti-CD70 antibody-drug conjugates. *Cancer Res*, **66**, 2328–2337.
48. Nakamura, K., Yoshikawa, K., Yamada, Y., Saga, S., Aoki, S., Taki, T., Tobiume, M., Shimazui, T., Akaza, H., and Honda, N. (2006) Differential profiling analysis of proteins involved in anti-proliferative effect of interferon-alpha on renal cell carcinoma cell lines by protein biochip technology. *Int J Oncol*, **28**, 965–970.
49. Shimazui, T., Ami, Y., Yoshikawa, K., Uchida, K., Kojima, T., Oikawa, T., Nakamura, K., Honda, N., Hinotsu, S., Miyazaki, J., Kunita, N., and Akaza, H. (2007) Prediction of in vitro response to interferon-alpha in renal cell carcinoma cell lines. *Cancer Sci*, **98**, 529–534.
50. Lichtenfels, R., Kellner, R., Bukur, J., Beck, J., Brenner, W., Ackermann, A., and Seliger, B. (2002) Heat shock protein expression and anti-heat shock protein reactivity in renal cell carcinoma. *Proteomics*, **2**, 561–570.

51. Kellner, R., Lichtenfels, R., Atkins, D., Bukur, J., Ackermann, A., Beck, J., Brenner, W., Melchior, S., and Seliger, B. (2002) Targeting of tumor associated antigens in renal cell carcinoma using proteome-based analysis and their clinical significance. *Proteomics*, **2**, 1743–1751.
52. Lichtenfels, R., Kellner, R., Atkins, D., Bukur, J., Ackermann, A., Beck, J., Brenner, W., Melchior, S., and Seliger, B. (2003) Identification of metabolic enzymes in renal cell carcinoma utilizing PROTEOMEX analyses. *Biochim Biophys Acta*, **1646**, 21–31.
53. Flad, T., Spengler, B., Kalbacher, H., Brossart, P., Baier, D., Kaufmann, R., Bold, P., Metzger, S., Bluggel, M., Meyer, H.E., Kurz, B., and Muller, C.A. (1998) Direct identification of major histocompatibility complex class I-bound tumor-associated peptide antigens of a renal carcinoma cell line by a novel mass spectrometric method. *Cancer Res*, **58**, 5803–5811.
54. Huang, R.P. (2007) An array of possibilities in cancer research using cytokine antibody arrays. *Expert Rev Proteomics*, **4**, 299–308.

12
PROTEOMEX Analysis of Renal Cell Carcinoma

Rudolf Lichtenfels and Barbara Seliger

12.1
Introduction

During the last decades, enormous efforts have been made to increase the knowledge regarding the etiology and the pathogenesis of a number of major diseases including various cancers. However, these advances have not been paralleled by a comparable progress in the treatment results. One important factor is the complexity of malignancies. Almost each tumor consists of many subtypes, which are associated with a distinct clinical behavior and different treatment susceptibilities. The early identification of disease prior to detection of clinical symptoms, the prediction of prognosis as well as the development of efficient therapies remain the major challenge in cancer research. Definitions of the underlying molecular mechanisms of tumors and of disease-associated risk factors are required for the development of innovative and more effective molecular targets for therapy. As an important indicator of the cancer status and/or its progression and thus for the (patho)physiological state of a cell at a specific time, biomarkers represent powerful tools for early detection, diagnosis, prognosis, monitoring the course of disease, risk assessment, disease prediction, and evaluation of treatment modalities. Hence, there is a critical need for the identification of novel biomarkers allowing the design of more effective therapeutic interventions in cancer research.

12.2
Proteome-Based Approaches for Biomarker Discovery and Antigen Identification in Tumors

Recent progress in the development of improved and new proteome-based technologies has opened new routes for the discovery of new cancer-related biomarkers leading to specific pattern recognition in neoplasms. Using quantitative proteomics methods, high-resolution, high-speed, high-throughput, and high-sensitivity mass spectrometry, and protein chips in combination with bioinformatics, putative

Renal and Urinary Proteomics: Methods and Protocols. Edited by Visith Thongboonkerd

ISBN: 978-3-527-31974-9

biomarkers have been identified. These might be able to reliably and accurately diagnose the tumor entity/subtype, and to predict the outcome of a line of cancer treatment, and the efficacy of the treatment. In this context, the identification of putative biomarkers relying on the detection of humoral immune responses in patients directed against disease-associated antigens, in particular, tumor-associated antigens (TAAs), has rekindled interest. This screening concept is based on the findings that cancer cells can elicit immune responses resulting in the activation of both T- and B-lymphocytes recognizing MHC class I and II-restricted TAA, and thus resemble responses to common pathogens.

However, most of these antigens represent self-antigens linking cancer to some degree to autoimmunity. Thus, the isolation and characterization of cancer-specific serum antibodies is of great interest for the subsequent identification of their cognate antigens and might indeed reveal potent biomarkers for the diagnosis of disease, its monitoring, the follow-up of treatment regimen, as well as for the detection of recurrence [1–3]. This approach offers cutting-edge capabilities to accelerate translational research linking discoveries in experimental science to their implementation in the clinical routine.

12.3
The PROTEOMEX Approach

PROTEOMEX, the combination of classical 2-DE-based proteome analysis and serex [4], also known as serological proteome analysis (SERPA) [5] or serological and proteomic evaluation of antibody responses (SPEARs) [6] represents one of the three major serological target identification methods in addition to SEREX (the serological identification of antigens by recombinant expression cloning) [7] and autoantibody-mediated identification of antigens (AMIDA) [8]. The serological screening of tumor samples separated by gel-based proteomic technologies enables the screening for TAAs in the "natural context" and thus offers a great advantage over screenings relying on the profiling of cDNA/mRNA microarrays or even recombinant expressed proteins (SEREX). In addition, it can also address tumor-specific posttranslational modifications (PTMs) of proteins, like phosphorylation, glycosylation, acetylation, methylation, and other modifications that may play crucial roles in cancer and autoimmunity [9–12]. The selection of potential TAA candidates is driven by the distinct immune recognition pattern defined on tumors in comparison to normal specimen using sera from patients and healthy donors. The parallel determination of serum reactivity on corresponding nonmalignant normal tissues and cell cultures in the PROTEOMEX approach allows the exclusion of targets recognized by naturally occurring autoantibodies [13].

12.4
Serum Markers for Cancers Detected by PROTEOMEX

So far, sera from patients with different cancer types or from healthy subjects have been screened individually, allowing the immunodetection of relevant antigens among several thousands of individual proteins separated by 2-DE. These studies have demonstrated that the occurrence of autoantibodies of the IgG isotype is highly conserved between individuals and that these autoantibodies are directed against a limited set of antigens independently of the cancer status [14]. The identification of such (auto)antibody signatures detected by proteome-based methods may be used as diagnostic markers or for early disease detection as it has been shown, for example, in non-small cell lung carcinoma [15, 16], breast carcinoma [14, 17], and prostate cancer, respectively [18, 19].

12.5
Identification of Serum Markers in RCC Patients

To date, 15 RCC lesions and corresponding normal kidney tissues, 9 RCC cell lines and 1 cell line derived from normal renal epithelium have been analyzed by PROTEOMEX. The characteristics of the RCC samples are summarized in Table 12.1. From these different RCC samples, 71% represented RCC of the clear cell types, whereas the other 29% were of unknown origin. In total, 50 differentially expressed proteins have been identified by PROTEOMEX from which 21 have been validated by different techniques like real time polymerase chain reaction (RT-PCR), Western blot analysis, and immunohistochemistry. These different proteins were mainly classified as stress proteins, transporter proteins, and proteins involved in cell motility, metabolism, signal transduction, and expression control (Table 12.2).

Table 12.1 Number and grading of RCC samples used for PROTEOMEX analyses.

Grading of tumors	No. of RCC biopsies	No. of RCC cell lines
G1	1	1
G2	3	3
G3	3	2
G4	2	0
n.d.	6	3
Total number	15	9

n.d.: not determined.

Table 12.2 Currently available validation data on selected candidate biomarkers as defined by PROTEOMEX in renal cell carcinoma.

Target validation

Swiss-Prot ID	Entry name	Cellular function	Cellular compartment	Validation method	Reference
P10809	60 kDa heat shock protein	Stress protein	Mitochondrion	WB	[20]
P11021	78 kDa glucose-regulated protein	Stress protein	Endoplasmic reticulum	WB	[20]
P07355	Annexin A2	Proliferation	Multiple compartment	IHC	[21]
P09525	Annexin A4	Signal transduction	Cytoplasm	IHC, RT-PCR, WB	[21–23]
P14625	Endoplasmin	Stress protein	Endoplasmic reticulum	WB	[20]
P30084	Enoyl-CoA hydratase	Metabolism	Mitochondrion	RT-PCR	[24]
P09211	Glutathione S-transferase P	Metabolism	Cytoplasm	WB, IHC	[25]
P08108	Heat shock cognate 70 kDa protein	Stress protein	Cytoplasm	WB	[20]
P04792	Heat shock protein 27	Stress protein	Multiple compartment	WB, qRT-PCR, IHC	[20, 26, 27]
P08238	Heat shock protein HSP 90-beta	Stress protein	Cytoplasm	WB	[20]
P05787	Keratin, type II cytoskeletal 8	Structural protein	Cytoskeleton	WB, IHC	[28]
P26038	Moesin	Cell motility	Cytoskeleton	WB	[29]
P38646	Stress-70 protein	Stress protein	Mitochondrion	WB	[20]
P04179	Superoxide dismutase (Mn)	Redox status	Mitochondrion	qRT-PCR, WB, IHC	[18, 23, 25]
P10599	Thioredoxin	Redox status	Cytoplasm	WB	[25]
Q01995	Transgelin	Proliferation	Cytoplasm	NB, IHC	[5]
P60174	Triosephosphate isomerase	Metabolism	Cytoplasm	WB, IHC	[25]
P06753	Tropomyosin alpha-3 chain	Cell motility	Cytoskeleton	WB	[28]
Q71U36	Tubulin alpha-3 chain	Structural protein	Cytoskeleton	WB	[28]
P09936	Ubiquitin carboxyl-terminal hydrolase L1	Metabolism	Cytoplasm	WB, IHC, RT-PCR	[25, 30]
P08670	Vimentin	Cell motility	Cytoskeleton	IHC, WB	[26, 28]

12.6
Methods and Protocols

The flow chart of the PROTEOMEX approach is shown in Figure 12.1.

12.6.1
Cultivation and Harvesting of RCC Cells

12.6.1.1 Expansion of RCC Cell Lines

- Cultivate established, primary RCC and normal kidney epithelium-derived cell lines in a CO_2 incubator at 5% CO_2 in DMEM supplemented with 10% FCS, 2 mM glutamine, 100 U ml^{-1} penicillin/100 μg ml^{-1} streptomycin. Grow the cells to about 80% confluence, which will yield in 175-cm^2 flasks 5×10^6 to 1×10^7 cells per flask.
- Expand the cell lines in 175-cm^2 flasks until the required cell count is reached by consecutively splitting them during the expansion phase in the ratio 1 : 3. The expansion of normal kidney epithelium cells requires immortalization by SV40LT and/or hTERT transformation. In this case, employ the respective selection media depending on the expression vector used such as, for example, hygromycin B or G418.

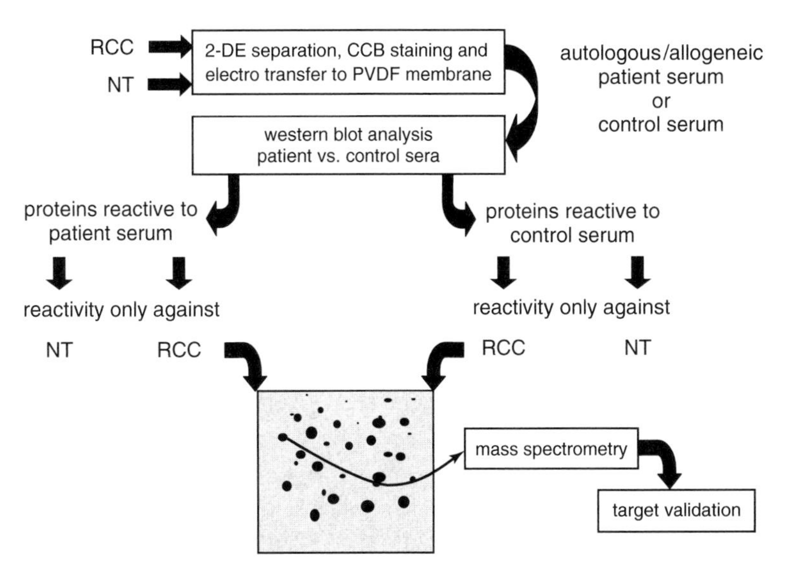

Figure 12.1 Flow chart of the PROTEOMEX approach. RCC next to normal renal epithelium tissue (NT) specimen are subjected to 2-DE and the respective gels stained by colloidal Coomassie blue and subsequently blotted onto PVDF membranes. The resulting blots are further subjected to side-by-side immunoblotting using sera either from patients or from healthy controls. Protein spots showing immunoreactivity restricted to the RCC-derived blots are further subjected to mass spectrometry for target identification. Finally, the resulting candidates have to be validated by different means to qualify as potential biomarkers.

12.6.1.2 Cell Harvesting

- Remove the media, briefly rinse the cell layer with PBS and trypsinize cells with trypsin/EDTA.
- Rinse the remaining flask with 15 ml fresh media for collection of remaining cells, transfer the cell suspension to a sterile 50-ml Falcon tube, and centrifuge the suspension for 5 minutes at 1400 rpm to pellet the cells.
- Discard the supernatant and resuspend the cell pellet in 40 ml PBS.
- Wash the cell pellet at least three times in 40 ml fresh PBS.
- Determine cell number using a Neubauer chamber and adjust cells to 1×10^7 ml^{-1} PBS.
- Aliquot the cell suspensions in sterile previously labeled 1-ml cryo tubes and centrifuge the aliquots for 5 minutes at 1400 rpm.
- Carefully remove the supernatant and store the residual cell pellets (1×10^7 cells) in a liquid nitrogen tank system or in a $-80\,^{\circ}C$ deep freezer until use.

12.6.2
Protein Extraction

- Add 200 μl lysis buffer (7 M urea, 2 M thiourea, 0.2 mM NDSB 256, 4% CHAPS, 1% DTT, 0.5% pharmalytes, and a trace amount of bromophenol blue) to the frozen cell pellet and resuspend the cell pellet on ice by repetitive pipetting.
- Transfer the lysate to a sterile 1.5-ml microreaction tube.
- Sonicate the lysate with a microneedle by pulsing the lysate with two cycles of five pulses each. The pulsing time is set to 0.5 seconds interrupted by a 1-second pause. Keep the sample on ice during the sonication procedure.
- Centrifuge the lysate for 90 minutes at 13 000 rpm at $15\,^{\circ}C$ in a microfuge.
- Transfer the cleared supernatant into a sterile microreaction tube and proceed with the protein determination using a modified Bradford's method [31].

12.6.3
2-DE and Staining

After protein extraction, the recovered proteins were resolved by 2-DE and stained with colloidal Coomassie staining (modified staining based on Neuhoff *et al.* [32]). The colloidal Coomassie staining method established by Neuhoff offers a high sensitivity (about 30 ng per spot) and little background staining, but requires extensive incubation time (>6 hours, the best results are obtained with overnight staining). For subsequent immune detection of the 2-DE-separated proteins via PROTEOMEX analyses, the gels are neither fixed prior to the staining procedure nor is the resulting protein–dye complex stabilized. Destaining is merely achieved by continual rinsing of the gels in destilled water. The resulting staining pattern not only provides the basis for the alignment of immune reactive spots but it is also suited as a quality-control step for the 2-DE separation process.

12.6.4
Two-Dimensional Western Blotting

The transfer of the proteins from prestained gels onto PVDF membranes is performed in a multiple-gel 2-DE unit, which can be easily equipped with a Western blotting insert allowing up to five gel transfers in parallel, or, alternatively, in large tank blotting chambers or semidry blotting devices. Mind that the blotting of prestained gels onto nitrocellulose membranes is not compatible with the subsequent immunodetection approach.

12.6.4.1 Transfer Preparations

- Prepare six sheets of blotting paper and one sheet of PVDF membrane per transfer cassette.
- Equilibrate the prestained and scanned gels for at least 20 minutes with 250 ml fresh transfer buffer (25 mM Tris, 192 mM glycine, 0.1% SDS, 20% methanol) in separate Pyrex glass trays.
- Make sure that the orientation of the gel is correct. The acidic pH range should be on the left side and the dye front should be facing the front of the tray.
- Activate the hydrophobic PVDF membranes by briefly (1–5 seconds) rinsing them in pure methanol using Pyrex glass trays. Make sure that the membranes are fully soaked in the organic solvent.
- Equilibrate the activated membranes for at least 10 minutes in a Pyrex glass tray containing 500 ml fresh transfer buffer.
- Soak the blotting paper with fresh transfer buffer in another Pyrex glass tray.
- In the meantime, remove the running buffer from the Hoefer Dalt vertical system.
- Remove the gel cassette holders from the gel tank unit and replace them with the Western blotting insert.
- Refill the tank unit with 20 l of fresh precooled transfer buffer and set the thermostat unit to 10 °C.

12.6.4.2 Blotting Cassette Assembly

- Use a Pyrex glass tray or a suitable plastic tray for the blot assembly.
- Remove a transfer cassette holding the transfer buffer-soaked sponges from the Hoefer Dalt vertical system tank unit and place the cassette in the assembly tray.
- Open the cassette and place three layers of transfer buffer-soaked blotting paper on the lower sponge mat.
- Mount each layer separately by first soaking the blotting paper in fresh transfer buffer and then placing it air-bubble-free on the supporting sponge mat.
- Use a 10-ml pipette to remove trapped air bubbles.
- Transfer a preequilibrated PVDF membrane to the tray containing the equilibrating gel by slipping it underneath the gel.
- Remove the gel from the tray by using the transfer membrane as a support and place it on the stack of the mounted blotting paper layers.

- Remove air bubbles between the membrane and the gel by rolling a 10-ml pipette along the gel surface.
- Mount three additional layers of presoaked blotting paper sheets air-bubble-free on the gel surface.
- Add the upper sponge mat and close the transfer cassette carefully.
- Transfer the assembled transfer cassette back to the blotting insert, minding the correct orientation of the blot assembly.
- Assemble the remaining transfer cassettes in the same way.
- Prior to the start of the transfer process, eliminate air bubbles trapped in the plastic grids of the transfer cassettes by brief stirring with a pipette. Also remove all air bubbles stuck between the flanking blotting cassettes and the electrodes.
- Place some weight on top of the blotting insert, as it tends to float. This can be achieved by adding one or two empty gel cassettes on the top of the blotting insert.
- Adjust the buffer volume so that the blotting insert is just covered with transfer buffer.
- Close the tank unit and apply up to 50 V constant voltage.
- The blotting is complete if at least 500 V h is reached. Best results are obtained with an overnight run at constant 40 V.

12.6.4.3 Blot Disassembly

- At the end of the run, transfer the cassettes to Pyrex glass trays and disassemble the blotting sandwich starting from the anode-facing side.
- Remove the blotting paper layers until the membrane is reached.
- Note the identity of the sample along with the blotting date with a ball pen at the backside of the membrane.
- Additionally, mark the bands of the separated molecular weight markers at the backside of the membrane.
- Place the wet and carefully labeled membrane one-by-one between transparency foils to prevent their rapid drying and digitize the wet membranes with a scanner to preserve and document the obtained spot pattern.
- Use the scanned image of the wet membrane later on as the primary matrix for the alignment of the resulting immune signal pattern.
- Air-dry the PVDF membranes for (long-term) storage or proceed directly with the immunostaining protocol.

12.6.5 Immunostaining

12.6.5.1 Handling of Serum Specimens

- Collect heparinized blood samples from patients and healthy donors.
- Spin the blood samples for 15 minutes at 3000g.
- Transfer the cell-free supernatants into separate sterile Falcon tubes.
- Aliquot the samples into 0.5–1-ml fractions in sterile cryo-/microreaction tubes.

- Deep freeze the aliquots and store them until use, either in liquid nitrogen or at −80 °C.
- Avoid multiple freeze–thaw cycles.
- Use for the PROTEOMEX analysis either individual sera from the respective panels of collected specimens or pools of sera representing either patients or healthy individuals.
- Alternatively, target-specific antibodies or custom-defined sets of antibody mixtures can be applied to perform PROTEOMEX analyses.

12.6.5.2 Reactivation of Air-Dried Membranes

Whereas freshly transferred blotting membranes can be processed directly following the scanning procedure, air-dried membranes need to be reactivated by brief rinsing in methanol prior to the blocking step.

- Prepare two Pyrex glass trays, one containing methanol, the other containing 1X TBS (14 mM NaCl, 1 mM Tris/HCl, pH 7.4).
- Wet the dry membrane briefly (1–5 seconds) in pure methanol.
- Immediately transfer the reactivated membrane to the TBS-containing tray and equilibrate the membrane for at least 10 minutes.

12.6.5.3 Blocking of Membranes

- Equilibrate the membranes for at least 10 minutes in 1X TBS.
- Incubate the membranes for at least 90 minutes in 100 ml TBSBL (1X TBS, 0.4% Tween-20, 5% skim milk, 10% horse serum) using a rocking platform from now on to ensure that the entire membrane is wetted (no more than back-to-back incubation of two membranes is tolerable).
- Invert the membranes every 30 minutes.
- Wash the membranes for 10 minutes three times in 100 ml of 1X TBS and invert the membranes every 5 minutes.

12.6.5.4 Incubation with the Primary Antibody Mix

- In between, dilute the serum sample in TBSAb (1X TBS, 0.1% Tween-20, 2% skim milk, pH 7.4) (define the adequate ratios in preliminary test runs, common dilutions range from 1 : 50 up to 1 : 1000 depending on the respective serum activities).
- But mind to only compare sera diluted at the same ratios for the analyses.
- Calculate with 20 ml primary antibody mix per freezer bag (no more than back-to-back incubation of two membranes is tolerable).
- Transfer the membrane/s into a freezer bag and add 20 ml primary antibody mix.
- Remove air bubbles by rolling a 10-ml pipette across the freezer bag but avoid spills of the mix.
- Seal the bag with a welting device.
- Incubate the membrane/s in the primary antibody mix overnight at 4 °C.

- Collect the primary antibody mix by draining the freezer bag into a 50-ml Falcon tube (the primary antibody mix can be stored at 4 °C and reused several times if an adequate preserving agent is added).

12.6.5.5 Washing the Membranes

- Transfer the membranes to a Pyrex glass tray and wash for 15 minutes three times in washing buffer (1X TBS, 0.1% Tween-20).
- Invert the membranes once per washing cycle.

12.6.5.6 Incubation with the Secondary Antibody Mix

- In between, dilute the secondary antibody adequately in TBSAb.
- Calculate with 20 ml secondary antibody mix per freezer bag.
- Transfer the membranes into a freezer bag and add 20 ml of the secondary antibody mix.
- Remove air bubbles by rolling a 10-ml pipette across the freezer bag and avoiding spillover.
- Seal the bag with a welting device.
- Incubate the membranes in the secondary antibody mix for at least 90 minutes at room temperature.
- In between, set up the exposition cassettes by fitting two sheets of blotting paper per cassette.
- Align two sheets of blank X-ray films/cover sheets to fit the lower sheet of the blotting paper.
- Mark the outline of the films with a black marker pen to form masks, which will subsequently be used to perform repetitive expositions.
- In between, set up a Falcon tube containing 50 ml of the Lumi-Light Western Blotting Substrate (Roche) by mixing the two components in a 1 : 1 ratio.

12.6.6 Immunodetection

- Remove the membranes from the washing buffer and place them between transparency sheets.
- Upon adding 3–5 ml of substrate across the membranes, transfer the sandwiches to the exposition cassette and align them with the previously designed masks.
- Fix the sandwiches with tape and move on to the dark room.
- Start with short-term exposures (1 minute, 2 minutes, 5 minutes) prior to extended exposures times. The substrate is stable for at least 2 hours.
- Use the masks to align the film sheets under red light illumination and label each film properly with sample, date, and exposure time.
- Finally remove the membranes from the sandwiches and wash them twice for 15 minutes in washing buffer at room temperature to remove the substrate solution.
- Air-dry the membranes or seal them wet in a freezer bag to store them at 4 °C in case the membranes are to be shortly reused.

12.6.7 Data Interpretation and Spot Identification

- Print the scans of the blotted membranes and the corresponding gels on both plain papers and transparencies.
- Align the immune recognition pattern captured on the X-ray films with the matching printouts on the transparencies.
- Use a white light box to improve the matching of weakly expressed spots.
- Mark the immunoreactive spots on the paper printouts using distinct colors for the recognition pattern of the test and the control sera.
- Neglect the spots that are frequently recognized by both types of sera. Select spots that are either recognized exclusively or at least with a highly biased frequency by the test serum.
- Design a spot-picking pattern by matching the differentially recognized spots to the staining pattern of a preparative 2-DE gel run in parallel to the PROTEOMEX gels.
- Cut out the spots of interest from the gel matrix.
- Destain the cut spots (the procedure will vary depending on the staining protocol used, for example, colloidal Coomassie blue or silver nitrate).
- Trypsin-digest the sample within the gel matrix.
- Elute the resulting peptide fragments.
- Subject to mass spectrometric analysis (for example, MALDI-TOF/TOF, ESI-TOF, LTQ-FT).

12.6.8 Target Validation

Finally validate the findings by means of independent approaches such as semiquantitative/real time polymerase chain reaction ((q)RT-PCR), Western blot analysis with target-specific antibodies, or immunohistochemical staining on microtissue arrays using a larger panel of samples. If possible, also include functional assays to validate the PROTEOMEX data.

12.7 Concluding Remarks

PROTEOMEX represents a highly sophisticated analytical approach. However, on the basis of its combinatorial nature linking 2-DE gel-based proteomics and immunodetection on large-scale blotting membranes, it demands significant handling time and asks for expert lab skills. Using tumor and normal tissue representing cell lines (alternatively, even the use of tumor biopsy material is applicable) as a resource, this approach allows the identification of antigens triggering tumor-specific humoral immune responses including potential biomarkers gaining their immunoreactivity due to undergoing PTMs. Furthermore, the PROTEOMEX approach can be adapted to specifically address distinct subcellular compartments such as cytosolic or nuclear fractions next to microsomes or mitochondria. In addition, the overall sample complexity can be considerably reduced by focusing the analysis on more restricted pH gradients. A wide range of commercially available IPG strips can be applied.

On the other hand, the PROTEOMEX approach has also well known limitations. On the basis of the use of the 2-DE gel-based technology for the separation of the proteins, the potential recognition pattern is limited to rather abundant proteins. In addition, quite a number of proteins are as yet underrepresented on classical 2-DE gels, in particular, highly hydrophobic (membrane), high molecular weight (>150 kDa), and low molecular weight (<10 kDa) proteins. This is also true for proteins with extreme isoelectric points (<3 or >11). There is also a chance, in particular, using broad-range pH gradients, that some of the immunoreactive spots contain multiple proteins extending the list of candidates for the subsequent validation process.

Another limitation that is linked to the use of 2-DE gels is that the immune recognition pattern is likely restricted to the detection of predominantly linear antigenic epitopes. Although some of the proteins might partially refold during the transfer from the gel to the blotting membrane, it has to be taken into account that a considerable number of conformational epitopes might be lost. Nevertheless, the sera still give rise to fairly active immune blots.

Owing to the lack of adequate software, the matching of the immune blots to the corresponding membranes and gel scans as well as to the respective preparative gels serving as a resource for the identification of the cognate antigens still presents a challenge. Careful matching has to be performed to reduce the risk of selecting false positive spots.

In addition, by relying on mass spectrometric analysis of proteins for the readout, the identification rate is strictly dependent on the sensitivity and recovery rate for the mass spectrometric approach of choice. Finally, it still presents a major challenge to truly define the exact nature of PTMs with limited sample material.

Acknowledgments

The work was supported by grants from the BMBF 031U-101H201 and PTJ-Bio-031-3376.

References

1. Caron, M., Choquet-Kastylevsky, G., and Joubert-Caron, R. (2007) Cancer immunomics using autoantibody signatures for biomarker discovery. *Mol Cell Proteomics*, **7**, 1115–1122.
2. Cho-Chung, Y.S. (2006) Autoantibody biomarkers in the detection of cancer. *Biochim Biophys Acta*, **1762**, 587–591.
3. Finn, O.J. (2005) Immune response as a biomarker for cancer detection and a lot more. *N Engl J Med*, **353**, 1288–1290.
4. Lichtenfels, R., Ackermann, A., Bukur, J., Melchior, S., Brenner, W., Kellner, R., Huber, Ch., and Seliger, B. (2001) PROTEOMEX: Identification of Proteins by Two-Dimensional Polyacrylamide Gels via Immunoblotting with Immune Sera. International Meeting From Biology to Pathology–The Proteomics

Perspective, British Electrophoresis Society, New York.

5. Klade, C.S., Voss, T., Krystek, E., Ahorn, H. *et al.* (2001) Identification of tumor antigens in renal cell carcinoma by serological proteome analysis. *Proteomics*, **1**, 890–898.
6. Unwin, R.D., Harnden, P., Pappin, D., Rahman, D. *et al.* (2003) Serological and proteomic evaluation of antibody responses in the identification of tumor antigens in renal cell carcinoma. *Proteomics*, **3**, 45–55.
7. Sahin, U., Tureci, O., Schmitt, H., Cochlovius, B. *et al.* (1995) Human neoplasms elicit multiple specific immune responses in the autologous host. *Proc Natl Acad Sci U S A*, **92**, 11810–11813.
8. Gires, O., Munz, M., Schaffrik, M., Kieu, C. *et al.* (2004) Profile identification of disease-associated humoral antigens using AMIDA, a novel proteomics-based technology. *Cell Mol Life Sci*, **61**, 1198–1207.
9. Anderton, S.M. (2004) Post-translational modifications of self antigens: implications for autoimmunity. *Curr Opin Immunol*, **16**, 753–758.
10. Gallego, M. and Virshup, D.M. (2007) Post-translational modifications regulate the ticking of the circadian clock. *Nat Rev Mol Cell Biol*, **8**, 139–148.
11. Kim, K.I. and Baek, S.H. (2006) SUMOylation code in cancer development and metastasis. *Mol Cells*, **22**, 247–253.
12. Krueger, K.E. and Srivastava, S. (2006) Posttranslational protein modifications: current implications for cancer detection, prevention, and therapeutics. *Mol Cell Proteomics*, **5**, 1799–1810.
13. Seliger, B. and Kellner, R. (2002) Design of proteome-based studies in combination with serology for the identification of biomarkers and novel targets. *Proteomics*, **2**, 1641–1651.
14. Caron, M., Choquet-Kastylevsky, G., and Joubert-Caron, R. (2007) Cancer immunomics using autoantibody signatures for biomarker discovery. *Mol Cell Proteomics*, **6**, 1115–1122.
15. Brichory, F.M., Misek, D.E., Yim, A.M., Krause, M.C., Giordano, T.J., Beer, D.G., and Hanash, S.M. (2001) An immune response manifested by the common occurrence of annexins I and II autoantibodies and high circulating levels of IL-6 in lung cancer. *Proc Natl Acad Sci U S A*, **98**, 9824–9829.
16. Zhong, L., Peng, X., Hidalgo, G.E., Doherty, D.E., Stromberg, A.J., and Hirschowitz, E.A. (2004) Identification of circulating antibodies to tumor-associated proteins for combined use as markers of non-small cell lung cancer. *Proteomics*, **4**, 1216–1225.
17. Shin, B.K., Wang, H., and Hanash, S. (2002) Proteomics approaches to uncover the repertoire of circulating biomarkers for breast cancer. *J Mammary Gland Biol Neoplasia*, **7**, 407–413.
18. Wang, X., Yu, J., Sreekumar, A., Varambally, S., Shen, R., Giacherio, D., Mehra, R., Montie, J.E., Pienta, K.J., Sanda, M.G., Kantoff, P.W., Rubin, M.A., Wei, J.T., Ghosh, D., and Chinnaiyan, A.M. (2004) Autoantibody signatures in prostate cancer. *N Engl J Med*, **353**, 1224–1235.
19. Bradford, T.J., Wang, X., and Chinnaiyan, A.M. (2006) Cancer immunomics: using autoantibody signatures in the early detection of prostate cancer. *Urol Oncol*, **24**, 237–242.
20. Lichtenfels, R., Kellner, R., Bukur, J., Beck, J. *et al.* (2002) Heat shock protein expression and anti-heat shock protein reactivity in renal cell carcinoma. *Proteomics*, **2**, 561–570.
21. Unwin, R.D., Craven, R.A., Harnden, P., Hanrahan, S. *et al.* (2003) Proteomic changes in renal cancer and co-ordinate demonstration of both the glycolytic and mitochondrial aspects of the Warburg effect. *Proteomics*, **3**, 1620–1632.
22. Shi, T., Dong, F., Liou, L.S., Duan, Z.H. *et al.* (2004) Differential protein

profiling in renal-cell carcinoma. *Mol Carcinog*, **40**, 47–61.

23. Zimmermann, U., Balabanov, S., Giebel, J., Teller, S. *et al.* (2004) Increased expression and altered location of annexin IV in renal clear cell carcinoma: a possible role in tumour dissemination. *Cancer Lett*, **209**, 111–118.
24. Balabanov, S., Zimmermann, U., Protzel, C., Scharf, C. *et al.* (2001) Tumour-related enzyme alterations in the clear cell type of human renal cell carcinoma identified by two-dimensional gel electrophoresis. *Eur J Biochem*, **268**, 5977–5980.
25. Lichtenfels, R., Kellner, R., Atkins, D., Bukur, J. *et al.* (2003) Identification of metabolic enzymes in renal cell carcinoma utilizing PROTEOMEX analyses. *Biochim Biophys Acta*, **1646**, 21–31.
26. Seliger, B., Menig, M., Lichtenfels, R., Atkins, D. *et al.* (2003) Identification of markers for the selection of patients undergoing renal cell carcinoma-specific immunotherapy. *Proteomics*, **3**, 979–990.
27. Perego, R.A., Bianchi, C., Corizzato, M., Eroini, B. *et al.* (2005) Primary cell cultures arising from normal kidney and renal cell carcinoma retain the proteomic profile of corresponding tissues. *J Proteome Res*, **4**, 1503–1510.
28. Kellner, R., Lichtenfels, R., Atkins, D., Bukur, J. *et al.* (2002) Targeting of tumor associated antigens in renal cell carcinoma using proteome-based analysis and their clinical significance. *Proteomics*, **2**, 1743–1751.
29. Craven, R.A., Stanley, A.J., Hanrahan, S., Dods, J. *et al.* (2006) Proteomic analysis of primary cell lines identifies protein changes present in renal cell carcinoma. *Proteomics*, **6**, 2853–2864.
30. Seliger, B., Fedorushchenko, A., Brenner, W., Ackermann, A. *et al.* (2007) Ubiquitin COOH-terminal hydrolase 1: a biomarker of renal cell carcinoma associated with enhanced tumor cell proliferation and migration. *Clin Cancer Res*, **13**, 27–37.
31. Ramagli, L.S. (1999) Quantifying protein in 2D-PAGE solubilization buffers, in *Methods in Molecular Biology: 2D Proteome Analysis Protocols*, vol. **112** (ed. A.J. Link), Humana Press, Totawa, pp. 99–103.
32. Neuhoff, V., Stamm, R., and Eibl, H. (1985) Highly sensitive colloidal staining in phosphatatic acid. *Electrophoresis*, **6**, 427–448.

13
In Vivo labeling of the Kidney by Means of CyDye DIGE Fluors prior to Proteomic Analysis

Corina Mayrhofer, Sigurd Krieger, Günter Allmaier, and Dontscho Kerjaschki

13.1
Introduction

A central role of the kidney is to filter plasma during the formation of primary urine. The functional units of the kidney consist of the glomeruli that are the source of the plasma ultrafiltrate and its associated tubular system, where the plasma ultrafiltrate is concentrated by reabsorption and the end urine is produced. Filtration takes place through the glomerular capillary wall that is composed of fenestrated endothelial cells, the glomerular basement membrane, and the foot processes of the podocytes with intervening slit diaphragms [1]. This composite structure of the glomerular capillary wall forms the basis of the selective sieving that allows the excretion of potentially hazardous small molecules and water but retaining essential plasma proteins larger in size than albumin [2]. In addition to size [3, 4], the barrier function for macromolecules of the glomerular capillary wall is also selective in terms of shape [5] and charge state [6–8] (Figure 13.1). A number of inherited and acquired diseases lead to dysfunction of the glomerular filtration barrier causing a reduced permselectivity and a pathological increase of proteins in the urine [9].

13.2
Proteomic Approach to Analyze Glomerular Surface Proteins

Nowadays, proteomic tools offer the possibility to analyze a large set of proteins and allow the detection of protein expression changes in association with disease states, thus providing a deeper insight into the underlying pathogenic molecular mechanisms. A proteomics-based approach includes typically the following steps: protein isolation and to a certain extent enrichment and quantitation, separation, and finally mass-spectrometry-based protein identification. Two-dimensional polyacrylamide gel electrophoresis (2-D PAGE) is a widely used technique for separation of proteins and the development of fluorescence difference gel electrophoresis (DIGE) [10, 11] has extended the possibility to analyze protein expression differences. The latter

Renal and Urinary Proteomics: Methods and Protocols. Edited by Visith Thongboonkerd

ISBN: 978-3-527-31974-9

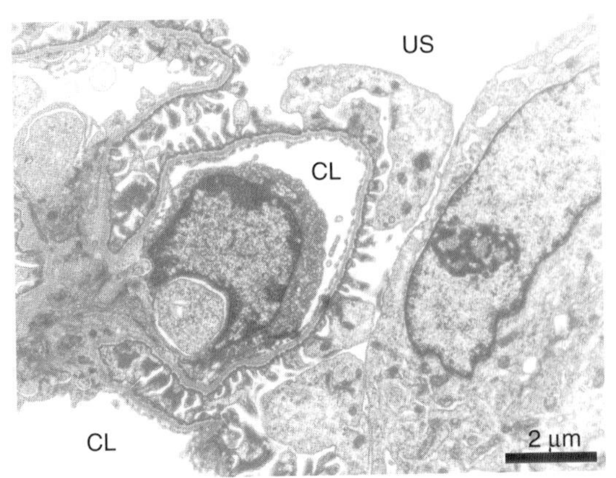

Figure 13.1 Ultrastructural morphology of a glomerular capillary. Transmission electron micrograph of a rat glomerular capillary loop. The endothelium lies closest to the capillary lumen and the visceral epithelial cells, known as *podocytes*, form the outermost layer of the glomerular capillary. Between the endothelium and the epithelium lies the glomerular basement membrane. Abbreviations: (CL) capillary lumen, (US) urinary space.

technique relies on preelectrophoretic labeling of proteins derived from samples obtained at different stages with spectrally resolvable fluorescent dyes and has been used, for example, to identify urinary biomarkers in association with renal diseases [12, 13] or to identify expression differences of proteins isolated from isolated renal structures [14].

A class of proteins of substantial interest are cell surface proteins. These proteins exert key functions in cellular processes, including signaling, adhesion, and interaction with the cytoskeleton and extracellular matrix. However, a problem associated with the analysis of the expression of cell surface proteins stems from their limited abundance. Thus, a fractionation and/or enrichment step is typically required for successful and efficient analysis of this particular protein subset. This can be achieved, for example, by subcellular fractionation [15] or direct targeting of cell surface exposed proteins by surface biotinylation and subsequent affinity enrichment. Surface biotinylation has been applied to cultured cells [16, 17] and further has been used to label proteins accessible in the bloodstream in vivo [18]. Recently, we have established a new strategy to target cell surface exposed proteins [19]. The technique relies on labeling of vital cells with a solution containing CyDye™ DIGE fluors minimal dyes (GE Healthcare, Uppsala, Sweden), which are covalently linked via the ε-amino group of lysine using a *N*-hydroxysuccinamide (NHS) ester. Cell surface exposed proteins that carry primary amino groups and that are accessible to the labeling solution are tagged with the fluorescent dyes and can be directly or after fractionation analyzed by 2-D PAGE. The protocol presented here (see Figure 13.2 for a schematic overview), describing the *in situ* perfusion of rat kidneys, has been applied for labeling glomerular cell surface proteins.

As illustrated in Figure 13.3, perfusion of rat kidneys with the labeling solution containing CyDye DIGE Fluors minimal dyes results in efficient fluorescence

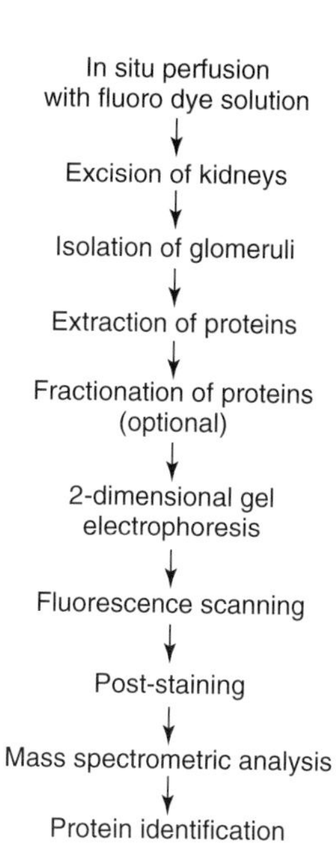

Figure 13.2 Proteomic approach to analyze glomerular surface proteins *in situ*. Schematic representation of the relevant steps of the *in situ* labeling method using CyDye™ DIGE fluors. After perfusion of kidneys and isolation of in situ labeled glomeruli, glomerular proteins are extracted and analyzed by 2-D PAGE. Optionally, a fractionation step can be included to reduce the sample complexity. Fluorescent dye-tagged glomerular proteins are imaged by fluorescence scanning. For subsequent mass spectrometric-based identification, gels are poststained.

labeling of different renal structures, including glomeruli and the brush border of proximal tubuli as well as peritubular capillaries and occurs in the absence of any morphological changes of the glomerular capillary wall and leaves the cells intact as observed by transmission electron microscopy (TEM). For subsequent proteomic analysis of glomerular proteins, fluorescence-labeled glomeruli are isolated by differential sieving (Figure 13.4a,b) and extracted proteins are subjected to 2-D PAGE. Fluorescent dye-tagged cell surface exposed proteins can further be detected and distinguished from other cellular proteins by comparison of the fluorescence pattern (Figure 13.4c) with the poststained 2-D pattern of the same gel (Figure 13.4d).

13.3 Methods and Protocols

13.3.1 Perfusion of Rat Kidneys with a Solution of CyDye DIGE Fluors Minimal Dyes

Sacrified rats are opened through a midline abdominal incision. The dissected aorta is clamped distal to the renal arteries and a flexible tube (inner diameter 0.75 mm, external diameter 1.22 mm) is inserted into the abdominal aorta below the renal arteries. Kidneys are perfused via the renal arteries with 50 ml Hanks' buffered salt solution (HBSS, Gibco BRL, Paisely, Scotland, UK) to remove blood components. Immediately after starting the perfusion the vena cava is opened. The kidneys are then flushed with 10 ml HBSS, pH 8.5 (adjusted with NaOH) containing 7 mM

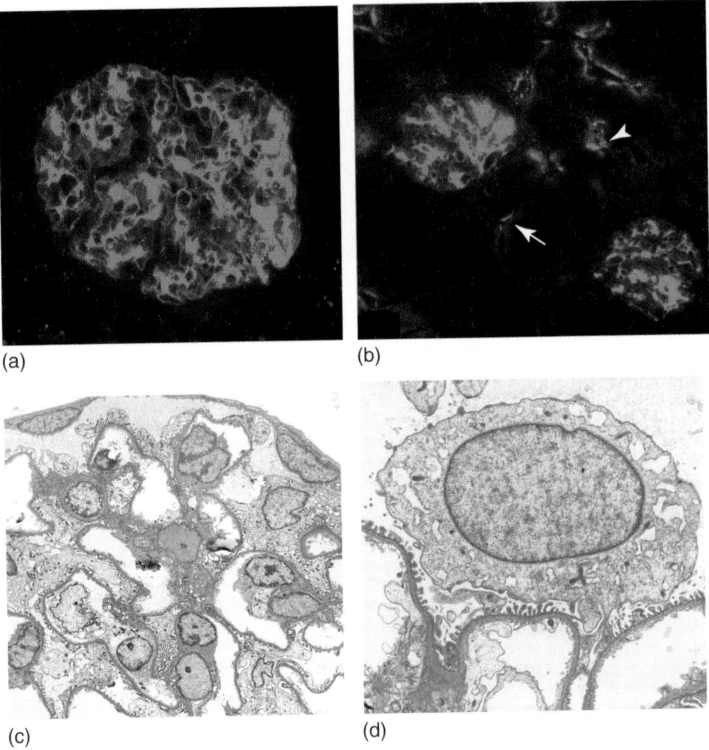

Figure 13.3 Ultrastructural morphology of renal structures after perfusion labeling with 2 nmol CyDye DIGE fluor Cy2. (a, b) Fluorescence micrographs of frozen sections of the kidney showing details of a glomerulus (a) and an overview of renal structures (b) at lower magnification. Note peritubular capillaries (arrows) and the brush border of proximal tubules (arrowheads). The images have been false colored. (c, d) TEM images of a glomerular capillary wall after labeling. (d) Higher magnification points out the detail of a podocyte body and its foot processes (original magnification: a × 400; b × 100; c × 2000; d × 3000) (reproduced with permission from [19]).

DTT (Sigma Aldrich, St. Louis, MO, USA) and 1 M urea. This is followed by perfusion with 5 ml HBSS, pH 8.5 containing 2 nmol fluorescent dyes (CyDye DIGE Fluors minimal dyes, GE Healthcare, Uppsala, Sweden), 7 mM DTT and 1 M urea. Subsequently, the kidneys are rinsed with 50 ml HBSS, pH 7.4 to remove any unbound dye. The perfusion solutions are prepared immediately before use.

13.3.2
Isolation of Glomeruli

Rat glomeruli are separated from other renal structures by graded sieving. After perfusion, kidneys are excised and kept in a plastic dish in ice-cold HBSS, pH 7.4. The kidney cortex is removed using a razor blade and minced into fine pieces (2–3 mm cubes). Minced tissue pieces are then flushed through a 100-μm cell

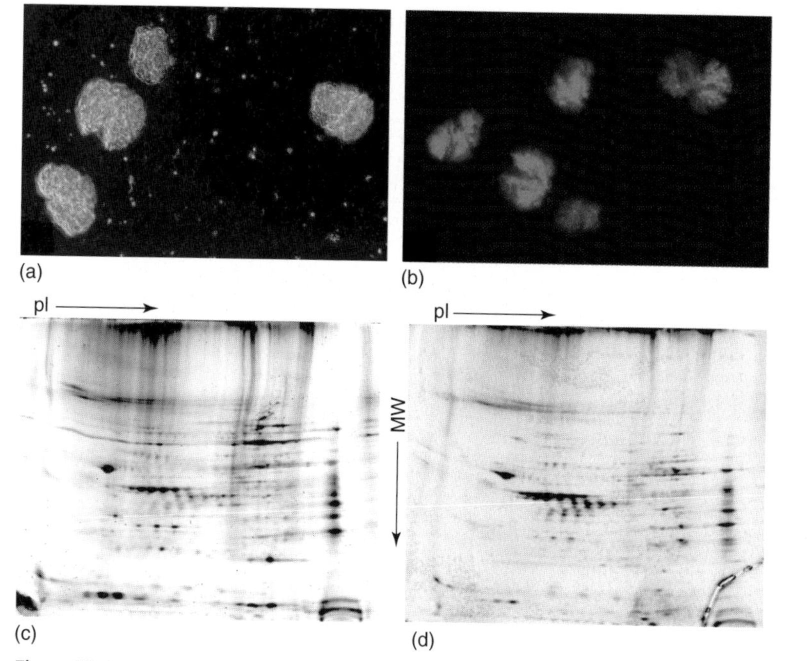

Figure 13.4 Proteomic analysis of glomeruli. (a) Phase-contrast micrograph of glomeruli isolated by sieving. (b) Fluorescence micrograph showing isolated rat glomeruli after perfusion of the kidney with 2 nmol CyDye DIGE fluor Cy2. The image has been false colored. (c, d) 2-D pattern of Triton X-114 fractionated hydrophobic glomerular proteins of Cy5 *in situ* labeled kidneys. The first dimension was performed using a pH 3–10 nonlinear IPG-strip. The second dimension was performed on a 3.6–15% self-made gradient polyacrylamide gel. (c) Fluorescent dye-tagged proteins were visualized by scanning using the Typhoon™ 9400 imager and (d) staining of total protein spots was accomplished by silver staining. (modified with permission from [19]).

strainer (BD Biosciences, Stockholm, Sweden) using a flattened pestle and flushed with HBSS, pH 7.4. The suspension is further filtered through a 70-μm cell strainer (BD Biosciences) and the glomeruli layered on the cell strainer are washed at least six times with HBSS, collected and resuspended in HBSS. All isolation steps are performed on ice or at 4 °C. The purity of the preparation is further assessed by phase-contrast microscopy.

Other techniques to isolate glomeruli from either animal sources or human biopsies include embolization of magnetic beads into glomerular capillaries [20, 21] and laser capture microdissection [22]. More details of glomerular isolation can be found in other chapters of this book.

13.3.3
Analysis of Tissue Samples

Staining efficiency of renal structures is monitored by fluorescence microscopy. Kidney sections are embedded in Tissue-Tek O.C.T. compound (Sakura Finetek,

Zoeterwoude, NL), immersed in cooled isopentane, and snap frozen in liquid nitrogen. The 4-μm cryostat sections are analyzed by fluorescence microscopy using the appropriate wavelength filter.

13.3.4
Proteomic Analysis of Glomerular Cell Proteins after Perfusion Labeling

Glomerular proteins can be extracted and fractionated, for example, using a Triton X-114 phase separation kit (2-D Sample Prep for Membrane Proteins, Pierce, Rockford, IL, USA) [23]. Extracted proteins are further solubilized in 7 M urea, 2 M thiourea, 4% CHAPS, 15 mM DTT, 2% octyl-β-D-glucopyranoside and 0.5% (v/v) immobilized pH gradient (IPG) buffer (GE Healthcare) and can be subjected to 2-D PAGE. We use commercial IPG gels (Immobiline Dry Strips, pH 3–10 nonlinear, GE Healthcare) that are rehydrated overnight with the sample solution. The isoelectric focusing step is performed using an IPGphor (GE Healthcare) unit. For SDS-PAGE, IPG strips are equilibrated with 1% DTT in equilibration buffer containing 7 M urea, 30% (87% v/v) glycerol, 0.05 M Tris–HCl, pH 8.8, 2% SDS for 10 minutes followed by 4% iodoacetamide in equilibration buffer for additional 10 minutes and SDS-PAGE is performed using self-made vertical 3.6–15% polyacrylamide gels. Fluorescence-labeled protein spots are detected using a Typhoon 9400 imager (GE Healthcare). To visualize the proteins for subsequent manual excision, the gels are silver-stained; we recommend the protocol of Blum *et al.* [24]. Protein spots that are visible in both images are then subjected to in-gel enzymatic digestion and subsequent matrix-assisted laser desorption/ionization reflectron time-of-flight and tandem mass spectrometric analysis.

13.4
Notes

We have established this labeling technique to tag surface exposed proteins *in vitro* and also *in situ* using commercially available fluorescent dyes that are suitable to detect expression differences in multiple samples. A limitation of the use of the CyDye DIGE Fluors minimal dyes in general is that only proteins containing free primary amino groups are labeled. In principle, this protocol can be adapted to other organs that are accessible for perfusion. The technique allows sufficient discrimination between surface and intracellular localization and because of the washing steps after the perfusion labeling, only compounds that are fixed in the ultrastructure are detected. Various postseparation staining methods can be applied to visualize the proteins in different ways, for example, for MS-based identification, but one should keep in mind that the sensitivity of fluorescence labeling is very high, and to visualize the tagged proteins, the applied poststain should also be quite sensitive. The technique provides the advantage that proteins are labeled prior to isolation in their native environment and that no further purification step is absolutely necessary before separation by 2-D PAGE.

References

1. Kriz, W., Kretzler, M., Provoost, A.P., and Shirato, I. (1996) Stability and leakiness: opposing challenges to the glomerulus. *Kidney Int*, **49**, 1570–1574.
2. Miner, J.H. (2003) A molecular look at the glomerular barrier. *Nephron Exp Nephrol*, **94**, e119–e122.
3. Brenner, B.M., Hostetter, T.H., and Humes, H.D. (1978) Glomerular permselectivity: barrier function based on discrimination of molecular size and charge. *Am J Physiol*, **234**, F455–F460.
4. Chang, R.L., Ueki, I.F., Troy, J.L., Deen, W.M., Robertson, C.R., and Brenner, B.M. (1975) Permselectivity of the glomerular capillary wall to macromolecules. II. Experimental studies in rats using neutral dextran. *Biophys J*, **15**, 887–906.
5. Bohrer, M.P., Deen, W.M., Robertson, C.R., Troy, J.L., and Brenner, B.M. (1979) Influence of molecular configuration on the passage of macromolecules across the glomerular capillary wall. *J Gen Physiol*, **74**, 583–593.
6. Chang, R.L., Deen, W.M., Robertson, C.R., and Brenner, B.M. (1975) Permselectivity of the glomerular capillary wall: III. Restricted transport of polyanions. *Kidney Int*, **8**, 212–218.
7. Brenner, B.M., Bohrer, M.P., Baylis, C., and Deen, W.M. (1977) Determinants of glomerular permselectivity: insights derived from observations in vivo. *Kidney Int*, **12**, 229–237.
8. Bohrer, M.P., Baylis, C., Humes, H.D., Glassock, R.J., Robertson, C.R., and Brenner, B.M. (1978) Permselectivity of the glomerular capillary wall. Facilitated filtration of circulating polycations. *J Clin Invest*, **61**, 72–78.
9. Tryggvason, K. and Wartiovaara, J. (2001) Molecular basis of glomerular permselectivity. *Curr Opin Nephrol Hypertens*, **10**, 543–549.
10. Alban, A., David, S.O., Bjorkesten, L., Andersson, C., Sloge, E., Lewis, S., and Currie, I. (2003) A novel experimental design for comparative two-dimensional gel analysis: two-dimensional difference gel electrophoresis incorporating a pooled internal standard. *Proteomics*, **3**, 36–44.
11. Unlu, M., Morgan, M.E., and Minden, J.S. (1997) Difference gel electrophoresis: a single gel method for detecting changes in protein extracts. *Electrophoresis*, **18**, 2071–2077.
12. Rao, P.V., Lu, X., Standley, M., Pattee, P., Neelima, G., Girisesh, G., Dakshinamurthy, K.V., Roberts, C.T. Jr, and Nagalla, S.R. (2007) Proteomic identification of urinary biomarkers of diabetic nephropathy. *Diabetes Care*, **30**, 629–637.
13. Sharma, K., Lee, S., Han, S., Lee, S., Francos, B., McCue, P., Wassell, R., Shaw, M.A., and RamachandraRao, S.P. (2005) Two-dimensional fluorescence difference gel electrophoresis analysis of the urine proteome in human diabetic nephropathy. *Proteomics*, **5**, 2648–2655.
14. Curthoys, N.P., Taylor, L., Hoffert, J.D., and Knepper, M.A. (2007) Proteomic analysis of the adaptive response of rat renal proximal tubules to metabolic acidosis. *Am J Physiol Renal Physiol*, **292**, F140–F147.
15. Adam, P.J., Boyd, R., Tyson, K.L., Fletcher, G.C., Stamps, A., Hudson, L., Poyser, H.R., Redpath, N., Griffiths, M., Steers, G., Harris, A.L., Patel, S., Berry, J., Loader, J.A., Townsend, R.R., Daviet, L., Legrain, P., Parekh, R., and Terrett, J.A. (2003) Comprehensive proteomic analysis of breast cancer cell membranes reveals unique proteins with potential roles in clinical cancer. *J Biol Chem*, **278**, 6482–6489.
16. Jang, J.H. and Hanash, S. (2003) Profiling of the cell surface proteome. *Proteomics*, **3**, 1947–1954.
17. Shin, B.K., Wang, H., Yim, A.M., Le Naour, F., Brichory, F., Jang, J.H., Zhao, R., Puravs, E., Tra, J., Michael, C.W., Misek, D.E., and

Hanash, S.M. (2003) Global profiling of the cell surface proteome of cancer cells uncovers an abundance of proteins with chaperone function. *J Biol Chem*, **278**, 7607–7616.

18. Rybak, J.N., Ettorre, A., Kaissling, B., Giavazzi, R., Neri, D., and Elia, G. (2005) In vivo protein biotinylation for identification of organ-specific antigens accessible from the vasculature. *Nat Methods*, **2**, 291–298.
19. Mayrhofer, C., Krieger, S., Allmaier, G., and Kerjaschki, D. (2006) DIGE compatible labelling of surface proteins on vital cells in vitro and in vivo. *Proteomics*, **6**, 579–585.
20. Takemoto, M., Asker, N., Gerhardt, H., Lundkvist, A., Johansson, B.R., Saito, Y., and Betsholtz, C. (2002) A new method for large scale isolation of kidney glomeruli from mice. *Am J Pathol*, **161**, 799–805.
21. Katsuya, K., Yaoita, E., Yoshida, Y., Yamamoto, Y., and Yamamoto, T. (2006) An improved method for primary culture of rat podocytes. *Kidney Int*, **69**, 2101–2106.
22. Sitek, B., Potthoff, S., Schulenborg, T., Stegbauer, J., Vinke, T., Rump, L.C., Meyer, H.E., Vonend, O., and Stuhler, K. (2006) Novel approaches to analyse glomerular proteins from smallest scale murine and human samples using DIGE saturation labelling. *Proteomics*, **6**, 4337–4345.
23. Qoronfleh, M.W., Benton, B., Ignacio, R., and Kaboord, B. (2003) Selective enrichment of membrane proteins by partition phase separation for proteomic studies. *J Biomed Biotechnol*, **2003**, 249–255.
24. Blum, H., Beier, H., and Gross, H.J. (1987) Improved silver staining of plant proteins, RNA and DNA in polyacrylamicle gels. *Electrophoresis*, **8**, 93–99.

Part Two
Urinary Proteomics

14
Simple Methods for Sample Preparation in Gel-Based Urinary Proteomics

Visith Thongboonkerd

14.1
Introduction

Currently, biomarker discovery is the major goal in the *clinical proteomics* arena. Several attempts have been made to search for novel noninvasive biomarkers for the clinical use in diagnostics and prognostics of several human diseases. Among various clinical samples, urine has become one of the most interesting and useful biofluids for such biomarker discovery because of its availability in almost all patients. Moreover, the collection of urine is very simple, without any need for sophisticated or invasive procedures. Urine is originated from plasma filtration through the microstructure in the kidney, namely, *glomerulus.* This intrarenal microstructure can selectively determine the components other than water to pass through it, using an ultrastructure, namely, *glomerular barrier* [1]. This barrier restricts the filtration of proteins from plasma into the urinary space, on the basis of molecular size, electrical charge, and protein structure. Large and negatively charged molecules are less readily filtered than small and neutral or positively charged ones. Proximal renal tubules can then reabsorb filtered proteins and water until <150 mg per day proteins are excreted with waste products as urine ($<1\%$ of ultrafiltrate volume) [2]. Various cells in the glomerulus and nephron can also secrete proteins into the glomerular ultrafiltrate and renal tubular fluid, respectively, in their free forms or in microvesicles (i.e., glomerular-podocytes-derived microvesicles and tubular-cells-derived exosomes) [3–5]. Approximately 70% of proteins in the normal human urinary proteome are kidney originated, whereas the remaining 30% are derived from plasma proteins [6–8]. On the basis of this information and because urine directly bathes the renal glomeruli, tubules, and collecting system, urine is thus an ideal biofluid for biomarker discovery in kidney diseases. Because a portion of urinary proteins is originated from the renal-circulating plasma, urine also offers opportunities to search for biomarkers in nonkidney diseases. Therefore, the discovery of urinary biomarkers is beneficial for the diagnostics and prognostics of not only kidney diseases but also nonkidney diseases.

Renal and Urinary Proteomics: Methods and Protocols. Edited by Visith Thongboonkerd

ISBN: 978-3-527-31974-9

Proteomic technologies, which are commonly used in recent urinary proteomics studies, include two-dimensional polyacrylamide gel electrophoresis (2-D PAGE) [9], two-dimensional difference gel electrophoresis (2-D DIGE) [10, 11], liquid chromatography coupled to tandem mass spectrometry (LC-MS/MS) [12, 13], surface-enhanced laser desorption/ionization time-of-flight mass spectrometry (SELDI-TOF MS) [14], capillary electrophoresis coupled to mass spectrometry (CE-MS) [15, 16], mass spectrometric immunoassays [17, 18], microarrays [19], and microfluidic technology on a chip [20]. Among these, 2-D PAGE remains the most commonly used method for urinary proteome analysis. The major challenge in 2-D gel-based urinary proteomics is that normal human urine has very diluted protein concentration with high salt contents that can interfere with 2-D PAGE analysis, especially during isoelectric focusing (IEF). Moreover, the protein recovery yield remains another challenging issue due to protein loss during sample preparation. It is, therefore, crucial to use an effective protocol to isolate/concentrate urinary proteins and to eliminate interfering substances.

This chapter summarizes simple methods for urinary sample preparation before 2-D gel-based proteome analysis, while details of preparation methods used for other proteomic technologies can be found in chapters 21–26 of this book or elsewhere [9–21].

14.2
Methods and Protocols

14.2.1
Sample Collection, Initial Processing, and Storage

Midstream urine is collected in a clean container. Cells and debris are removed by a centrifugation at 1000g–2000g for 5 minutes. The urinary supernatant is then stored at −80 °C until use.

14.2.2
Precipitation

Precipitation of urinary proteins with different organic solvents/compounds can be performed using similar protocol. Acetic acid, acetone, acetonitrile, ammonium sulfate, chloroform, ethanol, or methanol is added to urine to reach various final concentrations (e.g., 10, 25, 50, 75, or 90%). The mixture is then incubated at 4 °C for 10 minutes and the precipitate is isolated by a centrifugation at 12 000g (at 4 °C for 5 minutes). The supernatant is discarded and the pellet is allowed to air dry and is then resuspended with the solubilizing buffer (7 M urea, 2 M thiourea, 4% CHAPS, 2% (v/v) ampholytes (pH 3–10), 120 mM DTT, and 40 mM Tris-base). The solution is then dialyzed against 18 MΩ cm (dI) water (at 4 °C overnight), lyophilized, and then resuspended with the solubilizing buffer. Protein concentration is measured using Bradford's method, and 50–200 μg proteins are resolved by 2-D PAGE (see Figure 14.1).

14.2.3 Centrifugal Filtration

The sample is filtrated through 10-kDa cut-off (or other cut-off sizes) centrifugal column (Vivaspin; Vivascience AG; Hannover, Germany) by a centrifugation at 12 000*g* and 4 °C until approximately 1/30 of the initial volume remains. The concentrated urine is collected, dialyzed against dI water (at 4 °C overnight), lyophilized, and then resuspended with the solubilizing buffer as aforementioned. Protein concentration is measured using Bradford's method, and 50–200 μg proteins are resolved by 2-D PAGE (see Figure 14.2).

14.2.4 Lyophilization

Urine is lyophilized until dry and then resuspended in the solubilizing buffer. The resuspension is dialyzed against dI water (at 4 °C overnight), lyophilized again, and then resuspended with the solubilizing buffer. Protein concentration is measured using Bradford's method, and 50–200 μg proteins are resolved by 2-D PAGE (see Figure 14.3).

14.2.5 Ultracentrifugation

Ultracentrifugation of urine is performed at 200 000*g* (4 °C for 2 hours) using the Optima LE-80K Ultracentrifuge (Beckman Coulter, Inc., Fullerton, CA). The supernatant is discarded and the remaining thin film of proteins is resuspended with the solubilizing buffer. The resuspension is dialyzed against dI water (at 4 °C overnight), lyophilized, and then resuspended with the solubilizing buffer. Protein concentration is measured using Bradford's method, and 50–200 μg proteins are resolved by 2-D PAGE (see Figure 14.4).

14.2.6 Anionic Batch Adsorption Using DEAE Cellulose Beads

Urine is diluted (1 : 4) with dI water and its pH is adjusted to 7.3. DEAE cellulose beads (DE-52, Whatman Inc.), which have been equilibrated with a buffer containing 0.01 M Tris-HCl and 0.05 M NaCl (pH 7.3), are added to urine and stirred at room temperature for 30 minutes. The slurry is filtered through a sintered glass filter and the cake is washed with 0.01 M Tris-HCl in 0.05 M NaCl (pH 7.3) until the filtrate becomes colorless. The adsorbed proteins are then eluted with 0.01 M Tris-HCl in 0.6 M NaCl (pH 7.3). Thereafter, the eluate is dialyzed against dI water for 24 hours (with two changes) and lyophilized to reduce the sample volume and

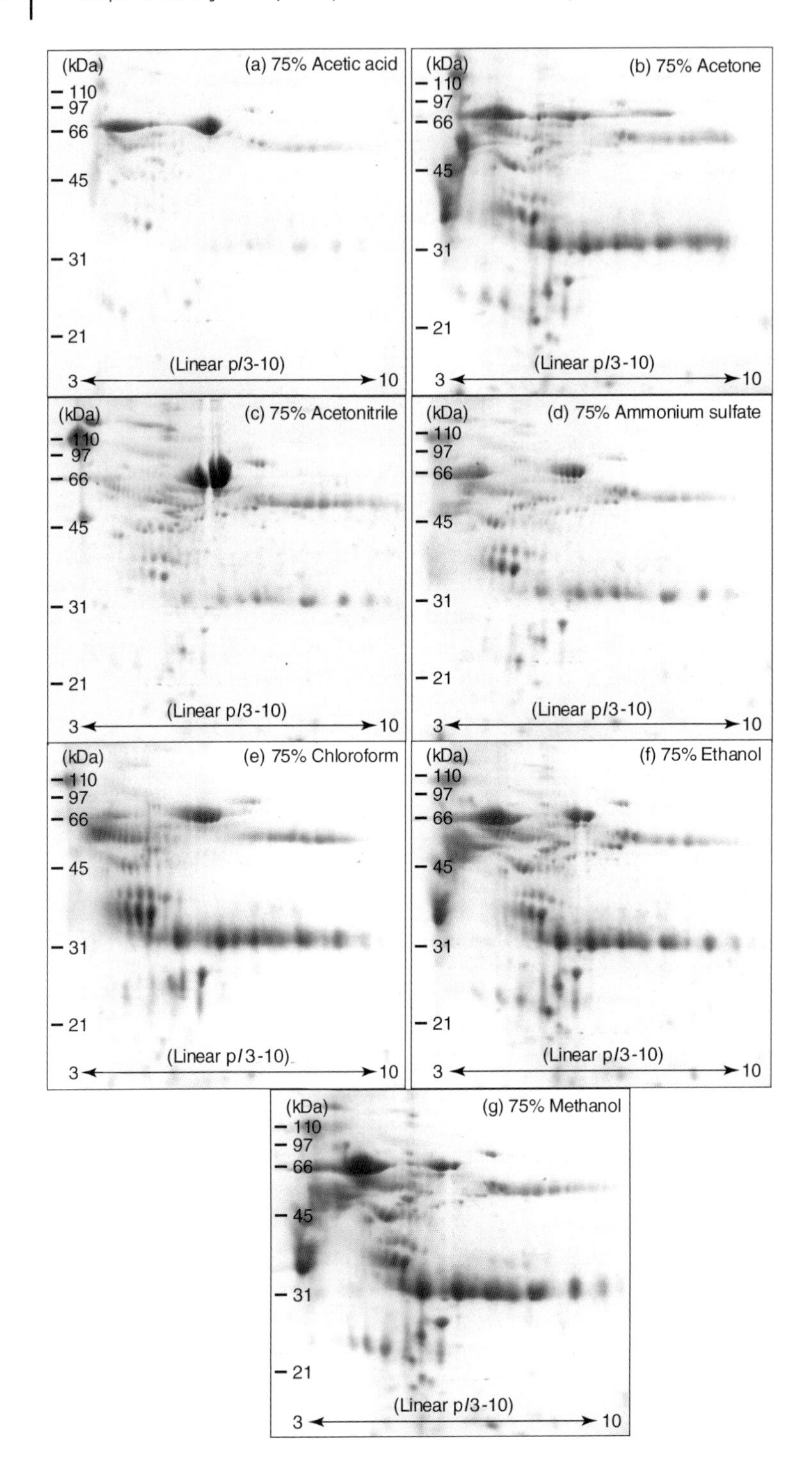
(kDa)
(a) 75% Acetic acid
110
97
66
45
31
21
(Linear pI 3-10)
3
10
(b) 75% Acetone
(c) 75% Acetonitrile
(d) 75% Ammonium sulfate
(e) 75% Chloroform
(f) 75% Ethanol
(g) 75% Methanol

◀ **Figure 14.1** 2-D urinary proteome profiles obtained from precipitation with acetic acid (a), acetone (b), acetonitrile (c), ammonium sulfate (d), chloroform (e), ethanol (f), and methanol (g). Identical concentration (75%) of organic solvents/compounds was used. All these samples were derived from the same urine pool of healthy individuals and equal amount (200 μg) of total protein was loaded in each gel. All gels were stained with Coomassie Brilliant Blue R250. Modified from [22] with permission from the American Chemical Society.

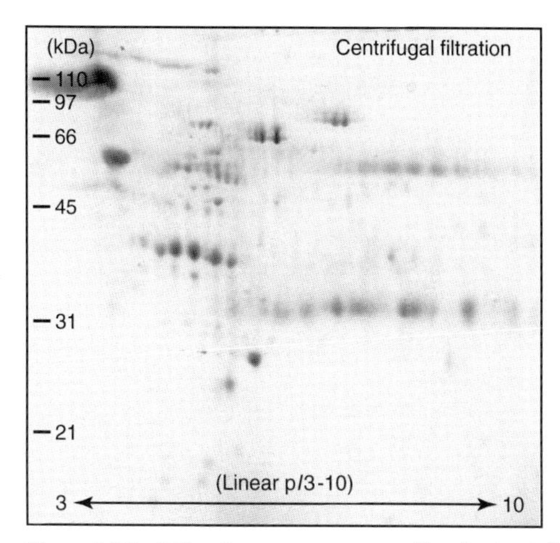

Figure 14.2 2-D urinary proteome profile obtained from centrifugal filtration. The sample was derived from the same urine pool of healthy individuals as those used in other protocols and equal amount (200 μg) of total protein was loaded. The gel was stained with Coomassie Brilliant Blue R250. Modified from [22] with permission from the American Chemical Society.

to concentrate the proteins. Protein concentration is measured using Bradford's method, and 50–200 μg proteins are resolved by 2-D PAGE (see Figure 14.5).

14.2.7
Cationic Batch Adsorption Using SP Sepharose 4 Fast Flow Beads

Urine is diluted (1 : 4) with dI water and its pH is adjusted to 6.0. SP Sepharose 4 Fast Flow beads (GE Healthcare, Uppsala, Sweden), which have been initially equilibrated with 50 mM glacial acetic acid (pH 6.0), are added to urine and stirred at 4 °C for 30 minutes. The urine–beads mixture is filtered through a sintered glass filter and the cake is washed with 50 mM glacial acetic acid (pH 6.0) until the filtrate becomes colorless. The adsorbed proteins are then eluted with a buffer containing 50 mM glacial acetic acid and 1 M NaCl (pH 6.0). Thereafter, the eluate is dialyzed against dI water for 24 hours (with two changes) and then lyophilized to reduce the volume and to concentrate the recovered proteins. Protein concentration is measured using Bradford's method, and 50–200 μg proteins are resolved by 2-D PAGE (See Figure 14.6).

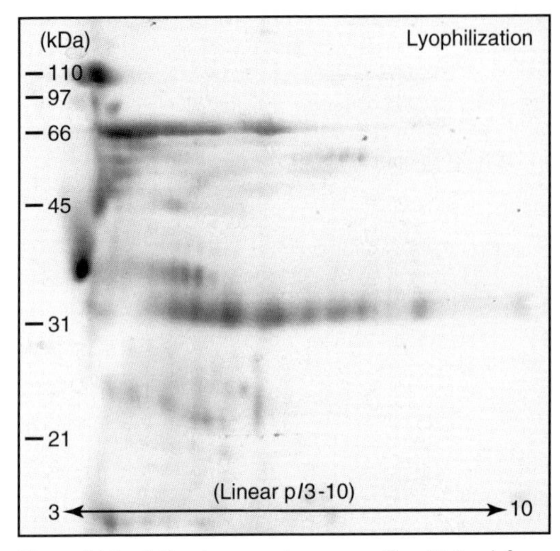

Figure 14.3 2-D urinary proteome profile obtained from lyophilization. The sample was derived from the same urine pool of healthy individuals as those used in other protocols and equal amount (200 μg) of total protein was loaded. The gel was stained with Coomassie Brilliant Blue R250. Modified from [22] with permission from the American Chemical Society.

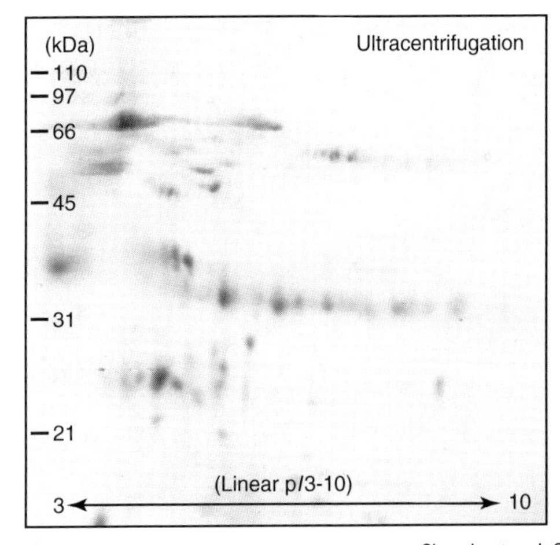

Figure 14.4 2-D urinary proteome profile obtained from ultracentrifugation. The sample was derived from the same urine pool of healthy individuals as those used in other protocols and equal amount (200 μg) of total protein was loaded. The gel was stained with Coomassie Brilliant Blue R250. Modified from [22] with permission from the American Chemical Society.

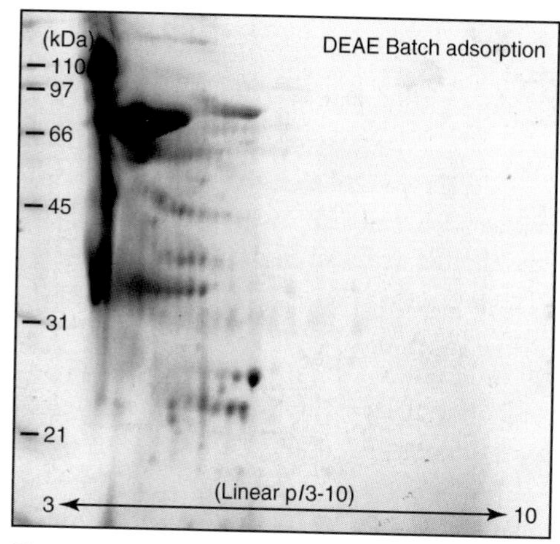

Figure 14.5 2-D urinary proteome profile obtained from anionic batch adsorption using DEAE cellulose beads. The sample was derived from the same urine pool of healthy individuals as those used in other protocols and equal amount (200 μg) of total protein was loaded. The gel was stained with Coomassie Brilliant Blue R250.

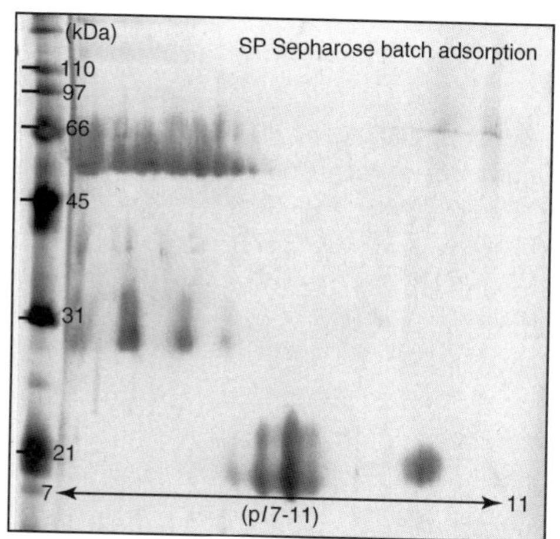

Figure 14.6 2-D urinary proteome profile obtained from cationic batch adsorption using SP Sepharose 4 Fast Flow beads. The sample was derived from the same urine pool of healthy individuals as those used in other protocols. A total of 50 μg of proteins was loaded in a basic-range IPG strip (p*I* 7–11). The resolved protein spots were visualized with EMBL silver staining. Modified from [23] with permission from the American Chemical Society.

14.2.8
2-D PAGE and Staining

Immobiline DryStrip, linear pH 3–10 (or other ranges), 7-cm long (GE Healthcare), is rehydrated overnight with 50–200 μg total protein (*note*: equal loading for comparative or differential proteomics study) that is premixed with rehydration buffer containing 7 M urea, 2 M thiourea, 2% CHAPS, 2% (v/v) ampholytes (pH 3–10), 120 mM DTT, 40 mM Tris-base, and bromophenol blue (to make the final volume of 150 μl per strip). The first dimensional separation (IEF) is performed in Ettan IPGphor II IEF System (GE Healthcare) at 20 °C, using stepwise mode to reach 9083 V h (or greater when the longer IPG strips are used). After completion of the IEF, proteins on the strip are equilibrated with a buffer containing 6 M urea, 130 mM DTT, 30% glycerol, 112 mM Tris-base, 4% SDS, and 0.002% bromophenol blue for 10 minutes, and then with another buffer containing 6 M urea, 135 mM iodoacetamide, 30% glycerol, 112 mM Tris-base, 4% SDS, and 0.002% bromophenol blue for 10 minutes. The IPG strip is then transferred onto 12% acrylamide slab gel ($8 \times 9.5\,cm^2$) and the second dimensional separation is performed in SE260 Mini-Vertical Electrophoresis Unit (GE Healthcare) with the current of 20 μA per gel until the dye front reaches the bottom of the gel. Separated protein spots are then visualized with various dyes (commonly used dyes include Coomassie Brilliant Blue, silver stain, SYPRO Ruby, and Deep Purple).

14.3 Discussion

The IEF of human urinary proteins is generally problematic because of the interference by salts, minerals, and other charged compounds such as lipids and glycosaminoglycans in urine. Lyophilization is not capable of removing these interfering compounds during sample preparation. Therefore, dialysis to remove salts and other interfering substances is required. Precipitation and centrifugal filtration are expected to remove these interfering substances. However, the poor IEF is occasionally observed in precipitated and filtrated urine samples, most likely because the interfering substances remain in the samples. Although the unsatisfactory IEF does not always occur when precipitation and centrifugal filtration protocols are used for isolating or concentrating urinary proteins, routine dialysis to remove salts and other interfering substances is highly recommended during sample preparation for 2-D gel-based urinary proteome analysis.

Various concentrations of organic solvents/compounds can be used for precipitating urinary proteins. It should be borne in mind that the quantity (protein recovery yield) as well as quality (pattern or component of visualized protein spots) of the urinary proteome to be obtained can differ significantly among various precipitation protocols. A previous study [22] has demonstrated that the greatest protein recovery yield (92.99%) is obtained from the sample precipitated with 90% ethanol, whereas precipitation with 10% acetic acid has the least protein recovery yield (1.91%). Theoretically, a protein in the solution tends to precipitate when pH of the solution is at or closed to the isoelectric point (p*I*) of that protein. For precipitation with acetic acid, acetone, acetonitrile, ethanol, and methanol, the higher concentration provides the greater recovery yield. For precipitation with ammonium sulfate and chloroform, the maximal recovery yield is obtained at concentrations of 75 and 50%, respectively.

Using equal amount of total protein loaded and identical concentration of organic solvents/compounds, precipitation with acetonitrile provides the greatest number of visualized protein spots, whereas lyophilized and acetic-precipitated samples have the smallest number of visualized protein spots. It should be noted that precipitation methods may create protein variants as a result of chemically induced modifications; thus, more spots may be observed as compared to nonprecipitation methods. This should be kept in mind when posttranslationally modified proteins are examined. Although precipitation with acetonitrile provides the greatest number of visualized protein spots, it is associated with the low protein recovery yield. Therefore, precipitation with ethanol, methanol, or acetone is recommended as the method of choice for a routine gel-based urinary proteome analysis, because these protocols provide the high protein recovery yield and the large number of protein spots visualized in a 2-D gel, allowing a wide variety of proteins to be examined in a gel. However, it should be emphasized that there is no single perfect sample preparation method that can be used for the analysis of the whole urinary proteome. Combination of more than one protocol may be required in some instances.

References

1. Tryggvason, K. and Pettersson, E. (2003) Causes and consequences of proteinuria: the kidney filtration barrier and progressive renal failure. *J Intern Med*, **254**, 216–224.
2. Moe, O.W., Baum, M., Berry, C.A., and Rector, F.C. (2004) Renal transport of glucose, amino acids, sodium, chloride and water, in *Brenner and Rector's The Kidney* (ed. B.M. Brenner), WB Saunders, Philadelphia, pp. 413–452.
3. Lescuyer, P., Pernin, A., Hainard, A., Bigeire, C., Zimmermann-Ivol, C., Sanchez, J.C., Schifferli, J.A., Hochstrasser, D.F. *et al.* (2008) Proteomic analysis of a podocyte vesicles-enriched fraction from human normal and pathological urine samples. *Proteomics Clin Appl*, **2**, 1008–1018.
4. Zhou, H., Yuen, P.S., Pisitkun, T., Gonzales, P.A., Yasuda, H., Dear, J.W., Gross, P., Knepper, M.A. *et al.* (2006) Collection, storage, preservation, and normalization of human urinary exosomes for biomarker discovery. *Kidney Int*, **69**, 1471–1476.
5. Pisitkun, T., Shen, R.F., and Knepper, M.A. (2004) Identification and proteomic profiling of exosomes in human urine. *Proc Natl Acad Sci U S A*, **101**, 13368–13373.
6. Thongboonkerd, V., McLeish, K.R., Arthur, J.M., and Klein, J.B. (2002) Proteomic analysis of normal human urinary proteins isolated by acetone precipitation or ultracentrifugation. *Kidney Int*, **62**, 1461–1469.
7. Pieper, R., Gatlin, C.L., McGrath, A.M., Makusky, A.J., Mondal, M., Seonarain, M., Field, E., Schatz, C.R. *et al.* (2004) Characterization of the human urinary proteome: a method for high-resolution display of urinary proteins on two-dimensional electrophoresis gels with a yield of nearly 1400 distinct protein spots. *Proteomics*, **4**, 1159–1174.
8. Thongboonkerd, V. and Malasit, P. (2005) Renal and urinary proteomics: current applications and challenges. *Proteomics*, **5**, 1033–1042.
9. Thongboonkerd, V., Klein, J.B., Jevans, A.W., and McLeish, K.R. (2004) Urinary proteomics and biomarker discovery for glomerular

diseases, in *Proteomics in Nephrology* (eds V. Thongboonkerd and J.B. Klein), Karger, Basel, pp. 292–307.

10. Sharma, K., Lee, S., Han, S., Lee, S., Francos, B., McCue, P., Wassell, R., Shaw, M.A. *et al.* (2005) Two-dimensional fluorescence difference gel electrophoresis analysis of the urine proteome in human diabetic nephropathy. *Proteomics*, **5**, 2648–2655.
11. Ngai, H.H., Sit, W.H., Jiang, P.P., Thongboonkerd, V., and Wan, J.M. (2007) Markedly increased urinary preprohaptoglobin and haptoglobin in passive heymann nephritis: a differential proteomics approach. *J Proteome Res*, **6**, 3313–3320.
12. Cutillas, P.R., Chalkley, R.J., Hansen, K.C., Cramer, R., Norden, A.G., Waterfield, M.D., Burlingame, A.L., and Unwin, R.J. (2004) The urinary proteome in Fanconi syndrome implies specificity in the reabsorption of proteins by renal proximal tubule cells. *Am J Physiol Renal Physiol*, **287**, F353–F364.
13. Cutillas, P.R., Norden, A.G., Cramer, R., Burlingame, A.L., and Unwin, R.J. (2004) Urinary proteomics of renal Fanconi syndrome. *Contrib Nephrol*, **141**, 155–169.
14. Nguyen, M.T., Ross, G.F., Dent, C.L., and Devarajan, P. (2005) Early prediction of acute renal injury using urinary proteomics. *Am J Nephrol*, **25**, 318–326.
15. Meier, M., Kaiser, T., Herrmann, A., Knueppel, S., Hillmann, M., Koester, P., Danne, T., Haller, H. *et al.* (2005) Identification of urinary protein pattern in type 1 diabetic adolescents with early diabetic nephropathy by a novel combined proteome analysis. *J Diabetes Complications*, **19**, 223–232.
16. Mischak, H., Kaiser, T., Walden, M., Hillmann, M., Wittke, S., Herrmann, A., Knueppel, S., Haller, H. *et al.* (2004) Proteomic analysis for the assessment of diabetic renal damage in humans. *Clin Sci (Lond)*, **107**, 485–495.
17. Kiernan, U.A., Nedelkov, D., Tubbs, K.A., Niederkofler, E.E., and Nelson, R.W. (2004) Proteomic characterization of novel serum amyloid P component variants from human plasma and urine. *Proteomics*, **4**, 1825–1829.
18. Kiernan, U.A., Tubbs, K.A., Nedelkov, D., Niederkofler, E.E., and Nelson, R.W. (2002) Comparative phenotypic analyses of human plasma and urinary retinol binding protein using mass spectrometric immunoassay. *Biochem Biophys Res Commun*, **297**, 401–405.
19. Liu, B.C., Zhang, L., Lv, L.L., Wang, Y.L., Liu, D.G., and Zhang, X.L. (2006) Application of antibody array technology in the analysis of urinary cytokine profiles in patients with chronic kidney disease. *Am J Nephrol*, **26**, 483–490.
20. Thongboonkerd, V., Songtawee, N., and Sritippayawan, S. (2007) Urinary proteome profiling using microfluidic technology on a chip. *J Proteome Res*, **6**, 2011–2018.
21. Thongboonkerd, V. (2007) Practical points in urinary proteomics. *J Proteome Res*, **6**, 3881–3890.
22. Thongboonkerd, V., Chutipongtanate, S., and Kanlaya, R. (2006) Systematic evaluation of sample preparation methods for gel-based human urinary proteomics: quantity, quality, and variability. *J Proteome Res*, **5**, 183–191.
23. Thongboonkerd, V., Semangoen, T., and Chutipongtanate, S. (2007) Enrichment of the basic/cationic urinary proteome using ion exchange chromatography and batch adsorption. *J Proteome Res*, **6**, 1209–1214.

15
Prefractionation of Urinary Proteins

Annalisa Castagna, Daniela Cecconi, Egisto Boschetti, and Pier Giorgio Righetti

15.1
Introduction

Modern proteome analysis is a very complex science, due mainly to protein heterogeneity. In any proteome, a few proteins dominate the landscape and often obliterate the signal of the rare ones, so that, when the researchers try to detect them, they remain hidden. In addition, proteomes of any origin can be extremely complex, impervious to even the most sophisticated analytical tools. For instance, according to Anderson *et al.* [1, 2], the human plasma should contain most, if not all, human proteins, as well as proteins derived from viruses, bacteria, and fungi. In addition, numerous posttranslationally modified forms of each protein are present, along with, possibly, millions of distinct clonal immunoglobulin sequences. To this intrinsic complexity, one can add the enormous dynamic range, encompassing 8 orders of magnitude and more between the least abundant (e.g., interleukins, at concentrations of $<1\,\mathrm{ng\,ml^{-1}}$) and the most abundant (e.g., albumin, $\approx 50\,\mathrm{mg\,ml^{-1}}$). For these reasons, any scientist working on any proteomic project deserves the title "detective" [3]. A number of published papers have highlighted major limitations of available technologies for proteome investigation. Current approaches are incapable of attaining a complete picture of the proteome, even limited with respect to structural aspects.

Nevertheless, for decades, clinical chemistry research has focused on finding new indicators or markers for disease in any tissue specimen and also in body fluids (plasma, urine, tears, lymph, seminal fluid, milk, saliva, and spinal fluid). This search for biomarkers is particularly appealing in such fluids, since their collection is minimally invasive or, in the case of urine, noninvasive. Yet, even body fluids are not immunized from severe problems that have so far hampered the discovery of novel markers, both plasma and serum exhibit tremendous variations in abundance of individual proteins, typically of the order of 10^{10} or more, with the result that, in any typical two-dimensional (2-D) map, only the high-abundance proteins are displayed [1, 2, 4, 5]. In the case of urine, the problems are further aggravated by its very low protein content necessitating a concentration step of 100- to 1000-fold, coupled with its high salt levels, requiring their concomitant removal before any

Renal and Urinary Proteomics: Methods and Protocols. Edited by Visith Thongboonkerd

ISBN: 978-3-527-31974-9

analytical step [6]. In recent years, considerable research efforts have been devoted to mapping the human urinary proteome, since urine is perhaps the only one that can be collected in a fully noninvasive manner and in large volumes repeatedly over a period of time and for a considerable period of time. Many scientific papers have outlined high-throughput methods for preparing large quantities suitable for 2-D mapping as well as attempts at establishing near-standard proteomic maps for clinical–chemical purposes [7–16].

Although the vast majority of them have exploited 2-D maps, a few reports have also described 1-D and 2-D chromatographic approaches [7, 15, 16]. Moreover, 2-D map analysis has been already exploited in different clinical settings, for example, bladder cancer [17, 18], Bence Jones proteinuria [19, 20], rheumatoid arthritis [21], urinary tract infections, glomerular or nonglomerular diseases [22, 23], chronic exposure to cadmium [24], characterization of urinary apolipoproteins and monitoring adaptive changes in unilateral nephrectomy [25], and searching for novel candidate markers for prostatic cancer [26]. Most of these approaches require a large number of steps for urine preparation before 2-D mapping, such as precipitation with protamine sulfate, removal of glycosaminoglycans, several dialysis steps, lyophilization, gel filtration, immuno-subtraction of the most abundant proteins, and other prefractionation tools [27]. All of these steps render proteomic analysis of human urine quite cumbersome and time consuming, without even taking into consideration the severe sample losses occurring at each sample handling step.

Prefractionation, in all possible variants has been deemed the logical way to go in the attempt to move toward the right direction. As stated by Pedersen *et al.* [28], prefractionation could be a formidable tool for "mining below the tip of the iceberg to find low-abundance and membrane proteins." A wide variety of prefractionation protocols, exploiting all possible variations of chromatographic and electrophoretic procedures, have been described (for reviews, see [29–32]).

It should be appreciated that, in the armamentarium of prefractionation tools available for such complex analysis, no single method has been sufficient to carry out this task. The approach that is gaining momentum, especially in the analysis of biological fluids, is sequential or simultaneous immunoaffinity depletion of the most abundant proteins present in the samples [4]. However, even this approach may not be good enough to gain access to the "deep proteome." Although depletion of the nine most abundant proteins represents the removal of as much as 90% of the overall protein content, the vast number of serum proteins that comprise the remaining 10% remain dilute and the improvement in detecting rare proteins might be quite disappointing. Another major drawback of such immuno-subtraction methods appears to be codepletion. Considering all the drawbacks discussed above, our group has recently proposed a novel method for protein prefractionation and applied it successfully to urinary proteome. This novel approach for capturing and identifying the "hidden proteome" is based on the use of large solid-phase ligand libraries as a multitude of affinity like systems.

15.2 Solid-Phase Ligand Libraries Technology

This technology consists of a combinatorial library of hexapeptides coupled, via a short spacer, on poly(hydroxymethacrylate) beads in such a way so that each bead carries a single peptide structure (Figure 15.1). This is obtained by a modified Merrifield approach [33] and the process is detailed in [34]. The library is constituted of polymeric chromatographic beads; each bead carrying a unique hexapeptide ligand at a local concentration similar to classical affinity sorbents (e.g., about 50 μmol ml^{-1} of swollen beads). The number of different ligands (hence different types of affinity beads) dictated by the combinatorial synthesis approach could reach a few dozens of millions of structures. Since the number of different ligands from the library is significantly larger than the total number of proteins predicated in the initial mixture, theoretically each protein may find its corresponding ligand. The technology of this ligand library is based on the pioneering work of Merrifield on solid-phase synthesis. Using the "split, couple, recombine" method, libraries of potentially billions of different peptide ligands can be created (Figure 15.2). Owing to this tremendous ligand diversity, within the library there is theoretically a ligand, for every protein, antibody, peptide, and so on, present in the starting material. This unique approach is based on a different philosophy in regard to the problem of discovering the hidden proteome: instead of simplifying the complex mixture into fractions, or partitioning away the most abundant species, it captures all the species present in the solution up to the saturation of the solid-phase ligand library. This automatically results in a large dilution of the most abundant species, with a concurrent concentration of the dilute and rare ones depending on the load of initial biological sample. The ligands are represented throughout the porous structure of the beads and can achieve an amount of ≈15 pmol per bead of the same hexapeptide

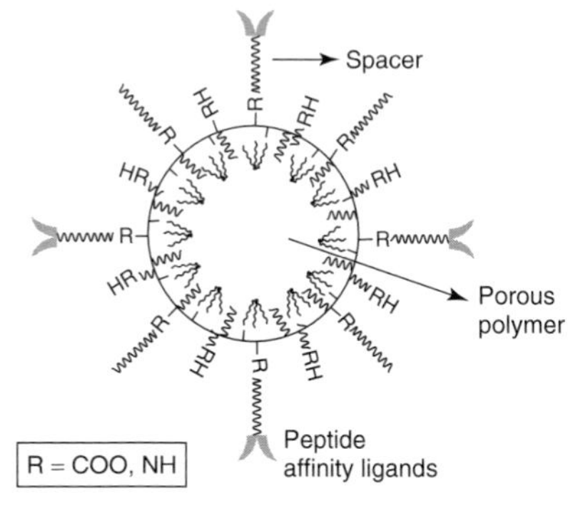

Figure 15.1 Scheme of the architecture of beads on which peptide ligands are attached. The structure shown here consists of an organic porous polymer on the matrix of which peptide ligands are covalently attached. "R" represents the linker that could be either a primary amino group or a carboxyl group. The peptide can be attached via a spacer. From [40] with permission.

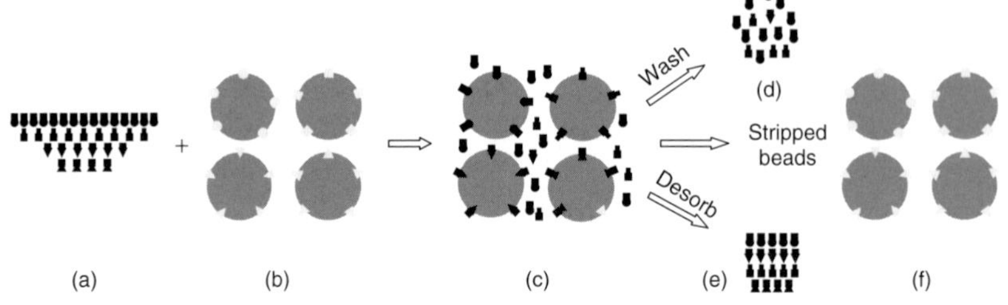

Figure 15.2 Representation of the principle of reducing the concentration difference between species mixed within the same solution. A solution composed of a mixture of four species (a) at different concentrations is loaded on a combinatorial solid-phase ligand library (b) under large overloading conditions. Each bead captures the corresponding protein partner until saturation, leaving the excess in the supernatant (c). A following wash eliminates all the excessive proteins (d). Resulting beads are then treated with appropriate solutions so that adsorbed species are collected (e) and the stripped beads (f) are separated. This latter contains all considered species but at different concentrations compared to the initial sample (a). From [30] with permission.

distributed throughout the core of the pearl. This amounts to a ligand density of $\approx$ 40–60 µmol ml^{-1} bead volume (average bead diameter $\approx$65 µm). As a result of the nonrandomized combinatorial hexapeptide construction, each bead has many copies of a single, unique ligand, and each bead has a different ligand from every other bead. Each bead captures a different dominant protein and coadsorbs a small amount of a very few other species [35]. The principle has been used to identify the hexapeptide ligand structure specific to selected proteins [36, 37]. It should be noted, however, that a given protein can adsorb to more than one peptide ligand structure and a given peptide structure can partner with more than one protein. The ligand structure governs the affinity constant value and can be used as the basis for selecting the interacting ligand for affinity chromatography purification; the affinity constant can also be modulated by varying the pH, ionic strength, temperature, and chemicals used for the preparation of the buffer. When proteins have multiple possibilities for peptide–ligand interaction, they are more enriched than others; this is clearly the case for apolipoproteins from human serum, for example. Lengthening the bait to a heptamer or even an octamer would generate a much larger number of diverse ligands, probably more heterogeneous than all the diverse proteins synthesized by all known living organisms. The use of a hexapeptide ligand to establish an affinity interaction might be considered a rather weak binding event; however, our experience has shown that, in fact, such a complex can have very high affinity and require very strong elution conditions. The hexameric ligands are linked to the organic polymer in such a way as to be stable under typical experimental conditions, such as prolonged incubation at reduced or elevated pH, and ionic strengths and organic solvents used to elicit complex formation with cell/tissue lysates and subsequent elution from the beads. An earlier article outlining the synthesis of the beads and some of their fundamental properties has recently been published [38], together with reviews describing the basic concepts [30, 39, 40].

There are a number of examples of the application of such technology to proteomic analysis so far [3]; in particular, it gave successful results in the analysis of human urine, serum, tissue lysates, such as *Escherichia coli* and *Saccharomyces cerevisiae* extracts, as well as impurity tracking and polishing of recombinant DNA products, especially biopharmaceuticals meant for human consumption. Here, the solid-phase ligand library is applied to the analysis of very dilute urinary proteins that normally escape to regular detection methods [41].

15.3 Application of Peptide Libraries to Urinary Protein Prefractionation

The urine sample, after treatment and concentration, was subjected to prefractionation with protein combinatorial peptide library beads. A total amount of 150 mg of protein was used for the beads prefractionation. After the process of contact, adsorption, and wash, the adsorbed fractions were eluted from the beads using two different solutions, the "TUC" solution, containing urea, thiourea, and 3-[3-cholamidopropyl dimethylammonio]-1-propanesulfonate (CHAPS) and another solution containing 9 M urea and citric acid at pH 3.8. These eluates were then analyzed for protein content, in comparison with the initial unfractionated urine sample, by means of different analytical methods, such as 1-D and 2-D separations, surface-enhanced laser desorption/ionization (SELDI) mass spectrometry, and Fourier transform-ion cyclotron resonance (FT-ICR) mass spectrometry. 2-D gels results are shown in Figure 15.3. The first eluate (TUC) exhibits many more

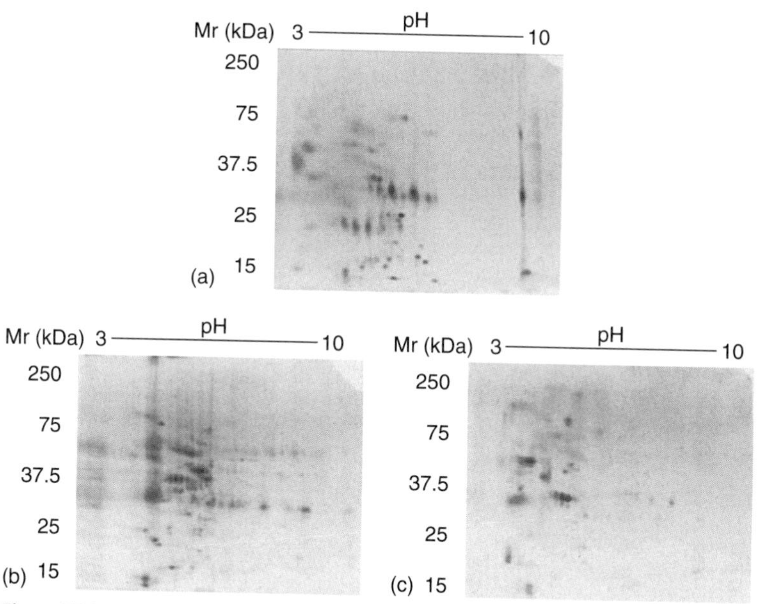

Figure 15.3 Two-dimensional maps of urine. (a) Control; (b) first eluate (TUC); and (c) second eluate (9 M urea, pH 3.8), stained with Sypro Ruby. An equal load of total protein was applied for all samples analyzed (90 μg total protein per gel). From [41] with permission.

spots in the entire pH interval (>300, as counted with the PDQuest) as compared with the unfractionated (control) urine (=100). Interestingly, the second eluate, although displaying a significantly lower number of spots (=120), shows only a limited redundancy (common spots) with the TUC eluate; most of the desorbed proteins are specific to the second elution step. Also, SELDI-MS profiling analysis related to the same protein fractions (Figure 15.4) (as limited to two m/z windows, one covering the 2500–7500 range (a) and the other covers the 7500–20 000 m/z interval (b)) confirms the above data, suggesting a total of at least three times as many signals as compared with the control. Actually, the number of peaks from control urine was 84 as compared to 135 and 146 peaks counted for TUC and acidic urea eluate, respectively. Within the explored interval (2500–20 000 Da), the total number of unique nonredundant peaks revealed by this analysis was 224 or almost three times larger than that of the control. It is, here, interesting to note that the peak count was limited only to proteins that interacted with the immobilized metal affinity chromatography (IMAC)-Cu^{2+} surface. These three samples were then subjected to FT-ICR analysis as described in Section 15.4. The control urine revealed a total of 134 unique gene products, while the TUC eluate allowed the identification of 317 unique protein species and the second eluate of an additional

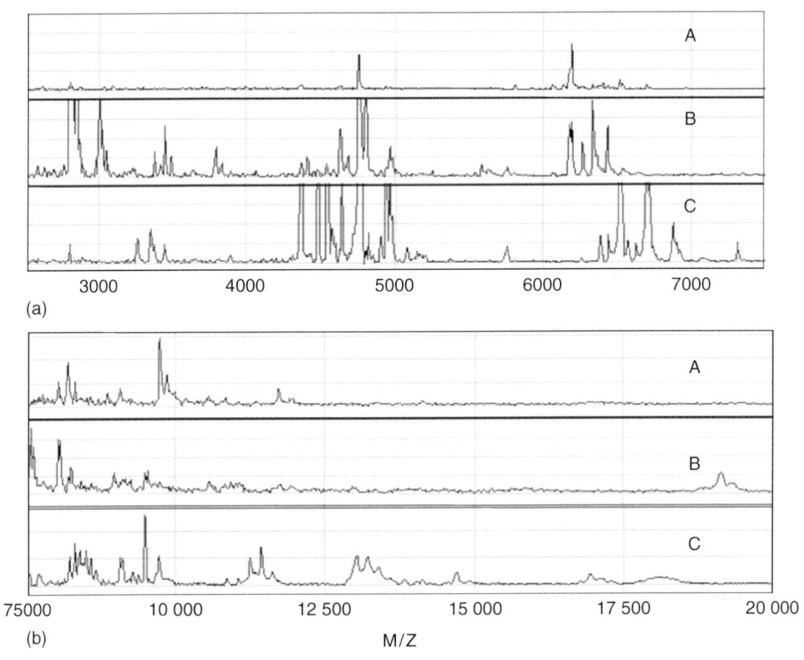

Figure 15.4 Narrow extract of SELDI-MS data obtained using an IMAC-Cu^{2+} ProteinChip Array covering two molecular mass ranges: 2500–7500 Da (a) and 7500–20 000 Da (b). "A" represents the initial crude urine proteins with affinity for IMAC copper ions surface; "B" represents the first elution using TUC buffer; and "C" represents the second elution using 9 M urea, pH 3.8. From [41] with permission.

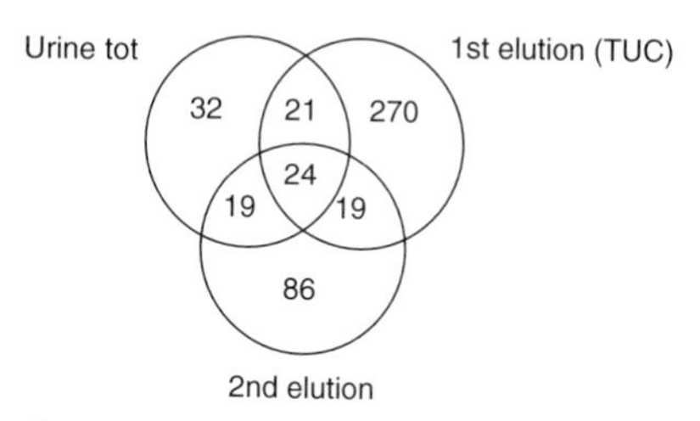

Figure 15.5 Diagram of overlap among the proteins identified in total urine (134 unique proteins), first (317 species), and second (95 species) eluates. From [41] with permission.

95 species. The data obtained have been extensively summarized in our previous report [41], which contains all the identified protein entries, the International Protein Index (IPI) accession number, the total number of peptides identified, the MASCOT score, theoretical Mr, and pI, as resulted from "Compute pI/Mw" tool of Expasy (http://www.expasy.org/tools/pi_tool.html), and additional information about the presence of the identified proteins in the total urine or in the first or second eluate, as well as in the listed references. It is of interest also to calculate which polypeptide chains are unique to each sample, and thus to obtain the degree of redundancy. This is shown schematically in Figure 15.5 that gives the degree of overlap of the various protein species in the different fractions. By eliminating the redundancies and summing up all the species detected, a total count of 383 unique protein species has been obtained in urine. We have additionally represented the above data as the bar graph in Figure 15.6, which gives the increment in species obtained in the sum of the two eluates, as compared with the control, while simultaneously expressing their Mr distribution. The latter is a skewed distribution with

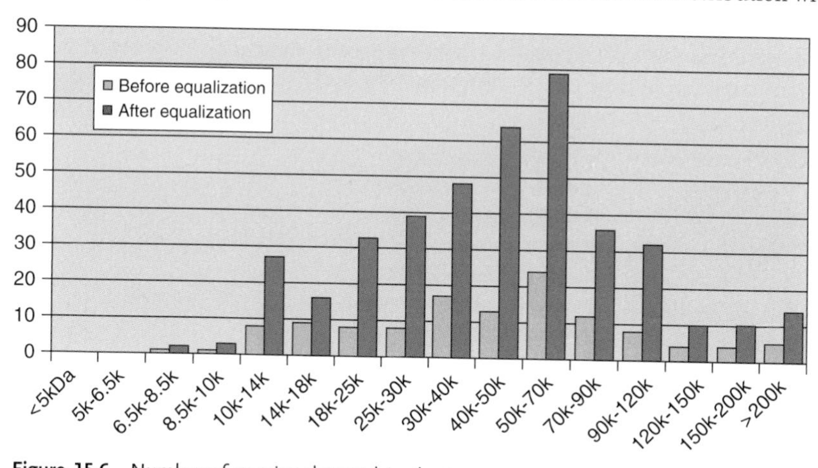

Figure 15.6 Number of species detected in the two combined eluates (black bars) as compared with control (grey bars) urines, as a function of their respective Mr values in the 5 kDa to >200 kDa range. From [41] with permission.

a peak in the Mr 30–50 kDa range, as expected due to the filtering properties of the glomeruli. The significance of these results is highlighted later. The identified proteins were also classified according to their different molecular functions, as summarized in Figure 15.7. Most of the proteins seem to have binding properties, followed by catalytic and signal transducer, respectively. To date, the total detection of proteins in human urine could be summarized as follows: (i) 124 gene products in [42] (results obtained via liquid chromatography-tandem mass spectrometry of peptide fragments); (ii) 103 species in [7] (a sum of data from 2-D maps, 1-D and 2-D liquid chromatography); (iii) 150 unique protein annotations in [8] (as obtained via extensive sample prefractionation and 2-D map analysis); (iv) 113 different proteins in Oh *et al.* [9] (from 2-D gel mapping, followed by peptide mass fingerprinting); (v) 47 unique proteins in [10] (a total sum of those polypeptides found in two fractions, acetone-precipitated and ultracentrifugation collected); and (vi) 48 nonredundant proteins in [15] (but only 22 reported in their Table 1; as obtained by adsorbing urinary proteins into a C18 resin and eluting at progressively higher levels of acetonitrile, from 30 to 70%). To those, one might want to add the 38 different urinary proteins listed in the Danish Centre for Human Genome Research (http://biobase.dk/cgibin/celis). It is intriguing to see what fraction of these proteins is in common among all the reported databases just quoted. The total sum of unique proteins reported in [7–10, 42] is 537. When checked against our data, 132 were found to be in common, thus leaving 251 unique species identified in this study. This leaves out some 269 entries as available only in [7, 10, 43]. Thus, by summing up our data with those already existing in the literature, it would appear that in urine, we can identify about 800 gene products, definitely a huge improvement from the early papers published in the 1980s, when just about a dozen proteins could be identified in 2-D maps, mostly of serum origin, and essentially only via immunoblots [6]. What fraction of the total proteins possibly present in human urine would these 800 proteins represent is difficult to determine at present, but some considerations can be proposed. First of all, 30 proteins, found in the control urine (and representing about 7% of the total proteins detected), were

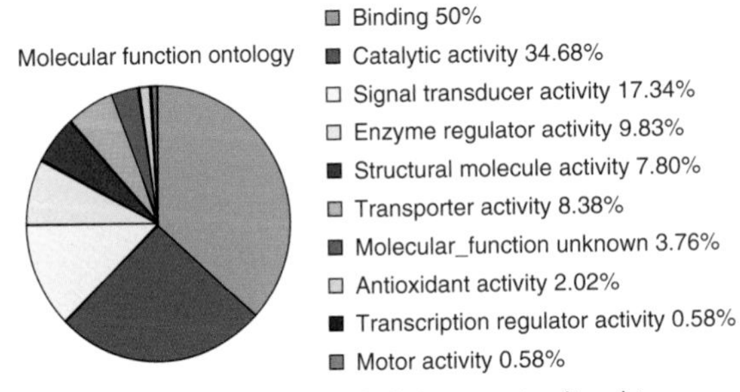

Figure 15.7 Classification of all the proteins listed in Table 1 of [41] according to their molecular function, obtained with the program FatiGO (www.fatigo.org). Note that single proteins may belong to more categories, which explains the total sum being substantially larger than 100%. From [41] with permission.

missing in the two bead eluates; in a previous paper, the loss was estimated to be approximately 2–3% [38]. The capturing mechanism, given the short length of the bait, should be mostly via classical physicochemical interactions among complementary polypeptide chains, such as ion exchange, hydrogen bonds, van der Waals, and hydrophobic interactions [42]. These interactions are present concomitantly under a combinatorial configuration, thus inducing a panel of interactions ranging from rather weak dockings to very strong associations. Weak interactions could potentially lead to instability in capturing some proteins from the sample or their release upon the wash step before the elution. Conversely, very strong interactions may induce difficulties in the desorption of captured species. Given the observed elution behavior, it is clear that a number of interactions appear to be rather strong. TUC alone – pretty drastic denaturing solution – does not release all the species adsorbed. A second elution in 9 M urea under acidic conditions is necessary to release a substantial number of additional proteins, not present in the first eluate. The interaction between urinary proteins and ligands from the library is not just dependent on affinity constants; large proteins may be adsorbed on beads by a multipoint possible interaction, resulting in a stronger apparent association constant. In this situation, elution could become very difficult, explaining that a limited number of proteins may still be adsorbed on beads and hence undetectable by our analytical approaches. Another possible scenario that could represent the limit of this process is the absence among the numerous structures of ligands having an affinity for some proteins; in this specific circumstance, few proteins may escape to the concentration effect. This could also be the case when the dissociation constant has values above the initial concentration of the protein to be captured.

Another reason to explain differences between some of our data and those already reported in the literature (in terms of the types of proteins detected) could be due to the sample processing, which appears to be quite different among the various reports quoted here. For instance, one report [13] has evaluated the total loss of number of spots in a 2-D map upon different sample treatments. Taking as a reference urine that had simply been ultrafiltrated (and exhibited 829 spots in a 2-D map), various sample precipitation procedure resulted in severe losses. For example, trichloroacetic acid (TCA)/acetone precipitation gave only 518 spots, acetone precipitation alone resulted in 467 spots, and acetonitrile (ACN)/trifluoroacetic acid (TFA) precipitation gave the lowest number (393) of spots. Another study [9] reports even more severe losses along the sequence of sample preparation steps. Taken the protein content of the collected urine as 100%, this value diminishes to 94% upon centrifugation and filtration, to 75% upon dialysis, to 27% upon a subsequent lyophilization/dialysis step to end up with only 6% after the last step, precipitation with TCA. The data reported in this study [9] show that species below about 5000 Da are virtually not present; this is probably due to the precipitation of proteins, in a mixture of acetone-methanol before trypsin digestion, which does not allow small proteins to precipitate easily. When making an in-depth observation of SELDI-MS data (in which the sample is processed without preliminary precipitation of proteins) it appears that a large number of peptides, between 1500 and 5000, can be counted. For the untreated urine, the number of species counted here is 43, while it is 90 and 69 from elution with TUC buffer and acidic urea, respectively. If redundancies are withdrawn, the net number of species is as high as 115. This number is presumably even higher in the analysis done with an IMAC-Cu^{2+} biochip; only the protein species interacting with this surface can be detected. Although all these species are clearly present, they are unavoidably ignored when the sample is processed using FT-ICR because of a preliminary protein precipitation operation. According to various protocols adopted by various research groups, different classes of proteins could be lost, resulting into identification of proteins only partly overlapping among the various research groups. Additionally, the presence of a large number of spots in a 2-D map could be misleading. In most cases, such spots represent the strings of polypeptide chains belonging to the same family (isoforms). In other cases, spots scattered through the map could simply be degradation products of a parental protein. A case in point is given in Celis' database [17, 18]; among the 205 spots listed, 45 were identified as albumin or albumin fragments. Similar results (in terms of albumin fragments) have been also reported by Lafitte *et al.* [14]. Even when adopting similar techniques, the data from single groups offer some variations. Thus, in [7], although 1-D liquid chromatography-mass spectrometry (LC-MS) identified only 35 proteins, 2-D LC-MS permitted identification of many more (90 species). Yet, five proteins were detected only in 1-D LC-MS, and not found in a more comprehensive method such as 2-D LC-MS. Even though 2-D maps permit identification of the lowest number of unique species (<30), seven of those are found only via this separation tool and are missing in both 1-D and 2-D-LC-MS [7]. The fact that, in this study, we could see so many more classes of proteins could be due to a minimum of manipulation steps involved in the treatment with combinatorial ligand library, enabling the concentration of the most dilute and rare proteins, while simultaneously cutting down the concentration of the most prominent species. This process, very simple and rapid, is more advantageous because adsorbed proteins can be desorbed sequentially, thus increasing the chance of detecting additional species. A final comment is due to the Mr distribution that is above 100 kDa (for quite a number of proteins), in an apparent contradiction with literature data, which suggest that no protein with Mr >67 kDa could be filtered through the kidney glomeruli. Yet, all the reports cited herein that adopted 2-D maps for separation clearly show polypeptide chains up to 100 kDa and above, so that the presence of such large Mr species seems to be a real phenomenon. For a number of those proteins, though, it is conceivable that only fragments are released into urine and thus their presence is inferred by sequencing them in the FT-ICR mass spectrometer. Nevertheless, for many others, it is clear, as demonstrated by 2-D maps, that large species whose mass may go up to more than 200 kDa are unambiguously present. The present data suggest that the combinatorial ligand libraries could be a novel and effective tool for exploring the hidden, or low-abundance, proteome, that is, very large part of the proteome that has escaped the detection up to the present time, especially when using mass spectrometry for low and medium molecular masses and 2-D mapping protocols for large proteins. The fact of normalizing the concentration of all species present in solution not only has the advantage of revealing the "deep proteome" but also has the added benefit of suppressing, during mass spectrometry analysis, the strong signal due to the very abundant proteins. Notwithstanding, a recent extensive report by Adachi *et al.* [44] demonstrated that using a high-performance FT mass spectrometer combined with pretreatments of urine (albumin depletion, etc.) it is possible to detect about 1500 proteins in human urine. However, this large number of proteins was obtained at the cost of high-performance liquid chromatographic fractionation. Taking into account all the above considerations, we can easily conclude that the urinary proteome is far from being completed and a lot more remain to be discovered.

15.4 Methods and Protocols

15.4.1 Materials

Urea, thiourea, tributylphosphine (TBP), glycine, sodium dodecyl sulfate (SDS), and CHAPS were obtained from Fluka Chemie (Buchs, Switzerland). Bromophenol blue, agarose, and Pharmalytes were from Pharmacia-LKB (Uppsala, Sweden). Acrylamide, *N'*,*N'*-methylenebisacrylamide, ammonium persulfate, *N*,*N*,*N'*,*N'*,tetramethyl-ethylenediamine (TEMED), dithiothreitol (DTT), Sypro Ruby, Silverquest as well as the linear Immobiline dry strips pH gradient 3–10 (7 cm long) were from Bio-Rad Laboratories (Hercules, CA). Ethanol, methanol, glycerol, sodium hydroxide, hydrochloric acid, acetone, and acetic acid were from Merck (Darmstadt, Germany). Molecular marker kit, colloidal Coomassie blue, and 16% acrylamide-tricine gels were from Invitrogen (Carlsbad, CA).

15.4.2 Sample Collection and Preparation

Urines from eight apparently healthy volunteers (age 24–26 years, four males and four females) were collected after informed consent, immediately chilled on ice, and processed as follows. After measuring the volume (200 ml per person), the pH was adjusted to a value of 7.0 with 1 N NaOH, then four "Complete Protease Inhibitor Cocktail" tablets (Roche Diagnostics, Berkeley, USA) were added, to inhibit the activity of endogenous proteases present in the specimen. Urines were centrifuged at 4 °C for 30 minutes at 3000*g* and then filtered through a 0.45 μm Millipore (Millipore, Bedford, MA) filter. The samples were pooled together and concentrated by means of "Centricon Tubes" (Millipore, Bedford, MA) with a Mr cutoff of 3000 Da. Protein concentration of urine was determined using DC Protein Assay (Bio-Rad) by using bovine serum albumin (BSA) as standard. A total volume of 45 ml of sample was obtained, corresponding to 150 mg of protein content.

This material was dialyzed against 25 mM sodium phosphate pH 7.0 and then lyophilized and stored at −20 °C until use.

15.4.3 Sample Treatment with Solid-Phase Ligand Library

The lyophilized urine was solubilized in 22 ml of 25 mM sodium phosphate buffer, pH 7.0. The solution was loaded onto a column (6.6 × 32 mm^2) containing 1 ml of peptide library beads (trade name ProteoMiner and supplied by Bio-Rad Laboratories, Hercules, CA, USA) at a flow rate of 0.24 ml per minute. The column was then washed extensively with 25 mM phosphate buffer, pH 7.0, until the UV absorbance (214 nm) returned to baseline. Absorbed proteins were first eluted with TUC buffer (2.2 M thiourea, 7.7 M urea, 4.4% CHAPS), followed by 9 M urea, pH 3.8 (5% v/v acetic acid) using three column volumes (3 ml) each time. The collected elution fractions were immediately neutralized and frozen.

15.4.4 SDS-PAGE Analysis

Urinary protein samples, before and after treatment, were analyzed by regular SDS-PAGE using a 16% polyacrylamide gel slab from Invitrogen (Carlsbad, CA). The experiments have been conducted under reduced and nonreduced conditions. Protein staining was done by using both Coomassie colloidal blue and silver nitrate (Silverquest Kit).

15.4.5 2-D PAGE Analysis

For proper comparisons, the same protein concentration, that is, 0.6 mg ml^{-1}, was used for all the samples investigated (starting material and the two obtained eluates). The desired volume of each sample (300 µl for the first elution, 820 µl for the second one, and 22.5 µl for the starting material) was subjected to protein precipitation in a cold mixture of acetone and methanol (v/v ratio of 8 : 1) at −20 °C for 2 hours for removing lipids and salts and for regulating the concentration of protein samples, and the solution was then centrifuged at 10 000g for 20 minutes. The pellet was solubilized in the "2-D sample buffer" (7 M urea, 2 M thiourea, 3% CHAPS, 40 mM Tris, 5 mM TBP, and 10 mM acrylamide) and was allowed to be alkylated at room temperature for 90 minutes. To stop the alkylation reaction, 10 mM DTT was added, followed by 0.5% Ampholine and a trace amount of bromophenol blue to the solution. Seven-centimeter-long immobilized pH gradients (IPG) strips (Bio-Rad), pH 3–10, were rehydrated with 150 µl of protein solution for 4 hours. Isoelectric focusing (IEF) was carried out with a Protean IEF Cell (Bio-Rad), at an initial voltage of 1000 V for 15 hours, followed by applying a voltage exponential gradient up to 5000 V until each strip was electrophoresed for 25 kV h. For the second dimension, the IPG strips were equilibrated for 26 minutes in a solution containing 6 M urea,

2% SDS, 20% glycerol, and 375 mM Tris-HCl (pH 8.8) under gentle shaking. The IPG strips were then laid on an 8–18% acrylamide gradient SDS-PAGE with 0.5% agarose in the cathode buffer (192 mM glycine, 0.1% SDS, and Tris to pH 8.3). The electrophoretic run was performed by setting a current of 5 mA per gel for 1 hour, followed by 10 mA per gel for 1 hour and 20 mA per gel until the dye front reached the bottom of the gel. Gels were incubated in a fixing solution containing 40% ethanol and 10% acetic acid for 30 minutes, followed by overnight staining in a ready-to-use Sypro Ruby solution. Destaining was performed in 10% methanol and 7% acetic acid for 1 hour, followed by a rinse of at least 3 hours in pure water. The 2-D gels were scanned with a Versa-Doc image system (Bio-Rad), by fixing the acquisition time at 10 seconds; the gel images were evaluated using PDQuest software (Bio-Rad). After filtering the gel images to remove the background, spots were automatically detected, manually edited, and then counted.

15.4.6
SELDI Analysis

Protein solutions at appropriate concentration, that is, 0.02 mg ml^{-1}, were deposited upon ProteinChip Array surfaces (Bio-Rad Laboratories), using a bioprocessor. Two types of arrays were selected: CM10 (weak cation exchanger) and IMAC 30 (immobilized metal ions affinity capture) loaded with copper ions. Each array contained eight distinct spots over which the adsorption of protein could be performed. After applying the three samples all surfaces were dried and prepared for SELDI-TOF MS analysis by applying 1 µl of matrix solution composed of a saturated solution of sinapinic acid in 50% acetonitrile containing 0.5% trifluoroacetic acid. All arrays were then analyzed by using a Ciphergen PBSIIc ProteinChip Reader. The instrument was used in a positive ion mode, with an ion acceleration potential of 20 kV and a detector gain voltage of 2 kV. The mass range investigated was from 3 to 20 kDa. Laser intensity was set between 200 and 250 U according to the sample tested. The instrument was calibrated with "All-in-1 Protein Standard" mixture (Ciphergen Biosystems Inc.).

15.4.7
Fourier Transform-Ion Cyclotron Resonance (FT-ICR) Mass Spectrometry (MS) Analysis

A 130-µg portion of each sample was precipitated by ethanol with glycogen as carrier and taken up in 8 M urea and 10 mM Tris (pH 8.0). The proteins were reduced using DTT at 37 °C for 30 minutes and subsequently alkylated using iodoacetamide for 20 minutes in the dark at room temperature. The material was then diluted to 2 M urea using 100 mM ammonium bicarbonate and digested with 3 µg trypsin (Proteomics Grade, Sigma) at 37 °C overnight. The peptides were diluted 10-fold using 0.5% acetic acid and desalted using a StageTip with two C18-disks [45]. Peptides were eluted with 50 µl of 80% acetonitrile/0.5% acetic acid and concentrated to below 5 µl using a Speed Vac (Concentrator 5301,

Eppendorf AG, Hamburg, Germany). The volume was adjusted to 5 μl by adding 1% trifluoroacetic acid and analyzed in a single run each on our LC-MS platform. The platform is composed of a LTQ-FT mass spectrometer (ThermoElectron, Bremen, Germany) and an Agilent 1100 binary nano pump (Palo Alto, CA). C18 material (ReproSil-Pur C18-AQ, 3 μm, Dr Maisch GmBH, Ammerbuch-Entringen, Germany) was packed into a spray emitter (75 μm i.d., 8 μm opening, 70 mm length; New Objectives, USA) using an air-pressure pump (Proxeon Biosystems, Odense, Denmark) to prepare an analytical column with a self-assembled particle frit [46]. Mobile phase A consisted of water, 5% acetonitrile, and 0.5% acetic acid and mobile phase B of acetonitrile and 0.5% acetic acid. The peptides were loaded at 700 nl per minute flow rate. The gradient with 300 nl per minute flow rate went, for buffer B, from 0 to 20% in 77 minutes and then in 5 minutes to 80%. The mass spectrometer acquired an MS at a resolution of 50.000 in the FT-ICR cell and eight MS/MS in the linear ion trap section per cycle. Peaks were selected using DTAsupercharge 0.62 (a kind gift from Matthias Mann, Odense, Denmark) with the parameters: precursor-peptide tolerance 0.2, SmartPicking enabled with max depth 8. The peak lists were searched against the IPI human database version 20050411 (http://www.ebi.ac.uk/IPI) using Mascot 2.0 with the parameters: monoisotopic masses, 0.2 Da on MS and 0.5 Da on MS/MS, Q-TOF parameter, carbamidomethylation for cysteine as fixed modification, oxidation on methionine, and N-acetylation as variable modifications, two missed cleavage sites allowed. The significance threshold for the Mascot score was determined to be 18 by searching the data against the IPI database and a reversed IPI database to give a false-positive rate of 5%. Proteins were accepted as identified if their total score was at least 36, resulting in an expected false-positive rate of 0.25% or 1 in 400 (according to this last criterion, the original total of 470 identifications was reduced to only 383 entries, as listed in Table 1 in [34]). Every peptide is assigned to a single protein. Proteins present in the list as different splice variants are matching to separate sets of peptides. Peptides shared between several proteins are only counted for the protein that has overall the most matching peptides. The presented protein list is hence the smallest set of proteins from the database that is needed to explain the presence of the identified peptides.

References

1. Anderson, N.L. and Anderson, N.G. (2002) The human plasma proteome: history, character and diagnostic prospects. *Mol Cell Proteomics*, **1**, 845–867.
2. Anderson, L.N., Polanski, M., Pieper, R., Gatlin, T., Tirumalai, R.S., Conrads, T.P., Veenstra, T.D., Adkins, J.N. *et al.* (2004) The human plasma proteome. *Mol Cell Proteomics*, **3**, 311–326.
3. Righetti, P.G. and Boschetti, E. (2007) Sherlock Holmes and the proteome: a detective story. *FEBS J*, **274**, 897–905.
4. Pieper, R., Gatlin, C.L., Makusky, A.J., Russo, P.S., Schatz, C.R., Miller, S.S., Su, Q., McGrath, A.M. *et al.* (2003) The human serum proteome: display of nearly 3700 chromatographically separated protein spots on two-dimensional electrophoresis

gels and identification of 325 distinct proteins. *Proteomics*, **3**, 1345–1364.

5. Anderson, N.G. and Anderson, N.L. (1982) The human protein index. *Clin Chem*, **28**, 739–748.
6. Edwards, J.J., Tollaksen, S.L., and Anderson, N.G. (1982) Proteins of human urine. III: identification by two-dimensional electrophoretic map positions of some urinary proteins. *Clin Chem*, **28**, 941–948.
7. Pang, J.X., Ginanni, N., Dongre, A.R., Hefta, S.A., and Opitek, G.J. (2002) Biomarker discovery in urine proteomics. *J Proteome Res*, **1**, 161–169.
8. Pieper, R., Gatlin, C.L., McGrath, A.M., Makusky, A.J., Mondal, M., Seonarain, M., Field, E., Schatz, C.R. *et al.* (2004) Characterization of the human urinary proteome: a method for high-resolution display of urinary proteins on 2D electrophoresis gels with a yield of nearly 1400 distinct protein spots. *Proteomics*, **4**, 1159–1174.
9. Oh, J., Pyo, J.H., Jo, E.H., Hwang, S.I., Kang, S.C., Jung, J.H., Park, E.K., Kim, S.Y. *et al.* (2004) Establishment of a near standard 2D human urine proteomic map. *Proteomics*, **4**, 3485–3497.
10. Thongboonkerd, V., McLeish, K.R., Arthur, J.M., and Klein, J.B. (2002) Proteomic analysis of normal human urinary proteins isolated by acetone precipitation or ultracentrifugation. *Kidney Int*, **62**, 1461–1469.
11. Davis, M.T., Spahr, C.S., McGinley, M.D., Robinson, J.H., Bures, E.J., Beierle, J., Mort, J., Yu, W. *et al.* (2001) Towards defining the urinary proteome using liquid-chromatography-tandem mass spectrometry. I: limitations of complex mixtures analyses. *Proteomics*, **1**, 108–117.
12. Schaub, S., Wilkins, J., Weiler, T., Sangster, K., Rush, D., and Nickerson, P. (2004) Urine protein profiling with surface-enhanced laser-desorption/ionization, time-of-flight mass spectrometry. *Kidney Int*, **65**, 323–332.
13. Tantipaiboonwong, P., Sinchaikul, S., Sriyam, S., Phutrakul, S., and Chen, S.T. (2005) Different techniques for urinary protein analysis of normal lung cancer patients. *Proteomics*, **5**, 1140–1149.
14. Lafitte, D., Dussol, B., Andersen, S., Vazi, A., Dupuy, P., Jensen, O.N., Berland, Y., and Verdier, J.M. (2002) Optimized preparation of urine samples for 2D electrophoresis and initial application to patient samples. *Clin Biochem*, **35**, 581–589.
15. Smith, G., Barratt, D., Rowlinson, R., Nickson, J., and Tonge, R. (2005) Development of high-throughput method for preparing human urine for 2D electrophoresis. *Proteomics*, **5**, 2315–2318.
16. Joo, W.A., Lee, D.Y., and Kim, C.W. (2003) Development of an effective sample preparation method for the proteome analysis of body fluids using 2D gel electrophoresis. *Biosci Biotechnol Biochem*, **67**, 1574–1577.
17. Celis, J.E., Wolf, H., and Ostergaard, M. (2000) Bladder squamous cell carcinoma biomarkers derived from proteomics. *Electrophoresis*, **21**, 2115–2121.
18. Rasmussen, H.H., Orntoft, T.F., Wolf, H., and Celis, J.E. (1996) Towards a comprehensive database of proteins from the urines of patients with bladder cancer. *J Urol*, **155**, 2113–2119.
19. Williams, K., Williams, J., and Marshall, T. (1998) Analysis of Bence Jones proteinuria by high-resolution 2D electrophoresis. *Electrophoresis*, **19**, 1828–1835.
20. Marshall, T. and Williams, K.M. (1999) Electrophoretic analysis of Bence Jones proteinuria. *Electrophoresis*, **20**, 1307–1324.
21. Clarck, P.M.S., Kricka, L.J., and Whitehead, T.P. (1980) Pattern of urinary proteins and peptides in patients with rheumatoid arthritis investigated with the Iso Dalt technique. *Clin Chem*, **26**, 201–204.
22. Tracy, R.P., Young, D.S., Hilol, H.D., Cutsforth, G.W., and Wilson,

D.M. (1992) Two-dimensional electrophoresis of urine specimens from patients with renal disease. *Appl Theor Electrophor*, **3**, 5–65.
23. Lapin, A. and Feigl, W. (1991) A practicable 2D electrophoresis of urinary proteins as a useful tool in medical diagnosis. *Electrophoresis*, **12**, 472–478.
24. Marshall, T., Willams, K.M., and Vesterberg, O. (1985) Unconcentrated human urinary proteins analysed by 2D electrophoresis with narrow pH gradients: preliminary findings after occupational exposure to cadmium. *Electrophoresis*, **6**, 47–52.
25. Gomo, Z.A., Henderson, L.O., and Lyrick, J.E. (1988) High-density lipoprotein apolipoproteins in urine. I: characterization in normal subjects and in patients with proteinuria. *Clin Chem*, **34**, 1775–1780.
26. Edwards, J.J., Anderson, N.G., Tollaksen, S.L., von Eschenbach, A.C., and Guevara, J. (1982) Proteins of human urine. II: identification by 2D electrophoresis of a new candidate marker for prostatic cancer. *Clin Chem*, **28**, 160–163.
27. Thongboonkerd, V. and Malasit, P. (2005) Renal and urinary proteomics: current applications and challenges. *Proteomics*, **5**, 1033–1042.
28. Pedersen, S.K., Harry, J.L., Sebastian, L., Baker, J., Traini, M.D., McCarthy, J.T., Manoharan, A., Wilkins, M.R. *et al.* (2003) Unseen proteome: mining below the tip of the iceberg to find low abundance and membrane proteins. *J Proteome Res*, **2**, 303–312.
29. Righetti, P.G., Castagna, A., Herbert, B., Reymond, F., and Rossier, J.S. (2003) Prefractionation techniques in proteome analysis. *Proteomics*, **3**, 1397–1407.
30. Righetti, P.G., Castagna, A., Antonioli, P., and Boschetti, E. (2005) Prefractionation techniques in proteome analysis: the mining tools of the third millennium. *Electrophoresis*, **26**, 297–319.
31. Garbis, S., Lubec, G., and Fountoulakis, M. (2005) Limitations of current proteomics technologies. *J Chromatogr A*, **1077**, 1–18.
32. Vlahou, A. and Fountoulakis, M. (2005) Proteomic approaches in the search for disease biomarkers. *J Chromatogr B*, **814**, 11–19.
33. Merrifield, R.B. (1963) Solid-phase synthesis of peptides. *J Am Chem Soc*, **85**, 2149–2154.
34. Lam, K.S., Salmon, S.E., Hersh, E.M., Hruby, V.J., Kazmierski, W.M., and Knapp, R.J. (1991) A new type of synthetic peptide library for identifying ligand-binding activity. *Nature*, **354**, 82–84.
35. Boschetti, E., Lomas, L., and Righetti, P.G. (2007) Romancing the 'hidden proteome', Anno Domini two zero zero seven. *J Chromatogr A*, **1153**, 277–290.
36. Buettner, J.A., Dadd, C.A., Baumbach, G.A., Masecar, B.L., and Hammond, D.J. (1966) Chemically derived peptides libraries: a new resin and methodology for lead identification. *Int J Pept Protein Res*, **47**, 70–83.
37. Kaufman, D.B., Hentsch, M., Baumbach, G.A., Buettner, J.A., Dadd, C.A., Huang, P.Y., Hammond, D.J., and Carbonnel, R.G. (2002) Affinity purification of fibrinogen using a ligand from a peptide library. *Biotechnol Bioeng*, **77**, 278–289.
38. Thulasiraman, V., Lin, S., Gheorghiu, L., Lathrop, J., Lomas, L., Hammond, D., and Boschetti, E. (2005) Reduction of the concentration difference of proteins in biological liquids using a library of combinatorial ligands. *Electrophoresis*, **26**, 3561–3571.
39. Righetti, P.G., Castagna, A., Antonucci, F., Piubelli, C., Cecconi, D., Campostrini, N., Rustichelli, C., Antonioli, P. *et al.* (2005) Proteome analysis in the clinical chemistry laboratory: myth or reality? *Clin Chim Acta*, **357**, 123–139.
40. Righetti, P.G., Boschetti, E., Lomas, L., and Citterio, A. (2006) Protein equalizer technology: the quest for a 'democratic proteome'. *Proteomics*, **6**, 3980–3992.

41. Castagna, A., Cecconi, D., Sennels, L., Rappsilber, J., Guerrier, L., Fortis, F., Boschetti, E., Lomas, L. *et al.* (2005) Exploring the hidden human urinary proteome via ligand library beads. *J Proteome Res*, **4**, 1917–1930.
42. Van Holde, K.E., Johnson, W.C., and Ho, P.S. (1998) *Principles of Physical Biochemistry*, Prentice Hall, Upper Saddle River.
43. Spahr, C.S., Davis, M.T., McGinley, M.D., Robinson, J.H., Bures, E.J., Beierle, J., Mort, J., Courchesne, P.L. *et al.* (2001) Towards defining the urinary proteome using liquid-chromatography-tandem mass spectrometry. I: profiling an unfractionated tryptic digest. *Proteomics*, **1**, 93–107.
44. Adachi, J., Kumar, C., Zhang, Y., Olsen, J.V., and Mann, M. (2006) The human urinary proteome contains more than 1500 proteins, including a large proportion of membrane proteins. *Genome Biol*, **7**, R80.
45. Rappsilber, J., Ishihama, Y., and Mann, M. (2003) Stop and go extraction tips for matrix-assisted laser desorption/ionization, nanoelectrospray, and LC/MS sample pretreatment in proteomics. *Anal Chem*, **75**, 663–670.
46. Ishihama, Y., Rappsilber, J., Andersen, J.S., and Mann, M. (2002) Microcolumns with self-assembled particle frits for proteomics. *J Chromatogr A*, **979**, 233–2399.

16
Gold Nanoparticle-Assisted Protein Enrichment for Urinary Proteomics

Anne Wang and Shu-Hui Chen

16.1
Introduction

Protein profiling of human urine is a promising approach to discover novel markers associated with many diseases such as bladder and prostate cancers. A large volume of the urine sample can be easily collected by noninvasive means for analysis. The success of the analysis, however, is highly relied on the starting amount of the protein since the sensitivity of most analytical tools such as mass spectrometry (MS) is still rather limited. Therefore, sample clean-up and protein enrichment has been a challenging task for proteomic analysis, especially for samples that contain low amount of protein. Unlike culture cells or other biological fluids, however, the amount of proteins present in the urine is generally low. Protein precipitation by organic reagents such as trichloroacetic acid (TCA) [1], acetone, and ultracentrifugation [2] are commonly used to enrich proteins from crude samples.

Recent developments in nanotechnology have shown that nanosubjects, such as gold (Au) nanoparticles [3–7] and diamond crystals [8], can be applied as a sensitive probe for affinity capture of biomolecules. Because of the large area to mass ratio, the trapping efficiency of nanosubjects is higher than large particles. In this chapter, we show that colloidal Au nanoparticles can be used to aggregate a broad range of proteins in addition to those containing a high percentage of cysteine residues. The Au-precipitated proteins can be digested by trypsins and then directly analyzed by MS for protein identification. Alternatively, the precipitates can be readily dissolved and then pipetted to the gel well to dissociate the aggregated proteins, followed by gel electrophoresis and MS analysis [9].

16.2
Methods and Protocols

16.2.1
Materials

- **DI water:** Barnstead E-pure system with a resistance more than 18.0 mΩ/cm.
- **Au solution:** $HAuCl_4 \cdot 3H_2O$ (0.07625 g; Aldrich, St Louis, MO) is dissolved in 250 ml dI water to form 1 mM HAuCl solution.

Renal and Urinary Proteomics: Methods and Protocols. Edited by Visith Thongboonkerd

ISBN: 978-3-527-31974-9

- **Citrate solution:** Sodium citrate (0.285 g; Sigma, St Louis, MO) is dissolved in 25 ml dI water to form 38.8 mM citrate solution.
- **DTT buffer:** D, L-dithiothreitol (DTT, 0.0154 g; Fluka, Buchs, Switzerland) is dissolved in 100 μl of 100 mM ammonium bicarbonate to form 1 M DTT buffer.
- **IAM buffer:** Iodoacetamide (IAM, 0.00925 g; Fluka, Buchs, Switzerland) is dissolved in 100 μl of 100 mM ammonium bicarbonate to form 500 mM IAM buffer.
- **Loading buffer:** A mixture of 1 g sodium dodecyl sulfate (SDS), 1.2 mg bromophenol blue, 3 ml glycerol, 3.5 ml of 1 M Tris-base (pH = 6.8), and 0.6 ml 2-mercaptoethanol is dissolved in 2.9 ml dI water to yield a final volume of 10 ml solution composed of 10% SDS, 0.175 mM bromophenol, 30% glycerol, 350 mM Tris, and 6% 2-mercaptoethanol.
- **Ammonium bicarbonate buffer:** Ammonium bicarbonate (0.0079 g) is dissolved in 100 μl of 40% ACN solution to form 100 mM ammonium bicarbonate buffer.
- **Urine samples:** Urine is collected from clinical patients and stored at −80 °C before analysis.
- **Trypsin buffer:** Trypsin (20 μg) is dissolved in 200 μl of acetic acid to yield a concentration of 100 ppm and the solution is stored at −20 °C.
- **MALDI matrix:** α-Cyano-4-hydroxycinnamic acid (CHCA, 10 mg) is dissolved in the mixture composed of 500 μl of ethanol and 500 μl of 0.2% trifluoroacetic acid in ACN.

16.2.2 Preparation of Gold Nanoparticles

Au solution is boiled with stirring in a round-bottom flask fitted with a reflux condenser (see Note 1). Citrate solution is added rapidly to Au solution. Remain the solution boiling for another 15 minutes until the color is changed from yellow to brightly red. Cool the solution to the room temperature with continued stirring and stored at 4 °C before use. The particle size determined by UV spectra and TEM is about 13 nm and the concentration is about 11.6 nM.

16.2.3 Urinary Protein Enrichment

The amount of total protein contained in urine sample is determined by Lowry assay according to manufacturers' instructions and the final protein concentration is adjusted to 100 ppm by dilution with dI water. For protein denaturation, a volume of 200 μl of 100 ppm (20 μg protein) urine solution is added with 2 μl of 10 mM DTT buffer and heated at 95 °C for 5 minutes. The solution is immediately added with 22 μl IAM buffer and kept in the dark for 30 minutes.

Before enrichment, a volume of 200 μl urine solution is diluted with dI water to yield 10 ml of 2 ppm protein solution (see Note 2). Mix 690 μl of gold nanoparticle solution, prepared as described above, into the protein solution (see Note 3). The whole mixture is cooled to 4 °C until the precipitation is formed. Centrifuge the solution at a speed of 15 300g at 4 °C for 30 minutes to form the precipitation and wash the precipitates with dI water twice. The color of gold nanoparticle solution would change from brightly red to deep purple when precipitated with urinary proteins. The Au-precipitated proteins can be processed by either direct digestion or gel electrophoresis before MS analysis.

16.2.4
Direct Solution Digestion and MALDI-MS Detection

The Au-precipitated proteins are added with 2 μl of trypsin buffer (see Note 4) in 100 μl of 50 mM ammonium bicarbonate (pH = 8.3) buffer and then incubated at 37 °C overnight. The tryptic digests are added with 0.5 N HCl and MALDI matrix at a volume ratio of 1 : 1 : 2 of peptide/HCl/matrix. The mixture is pipetted onto MALDI plate, dried in vacuum, and then analyzed by MALDI-TOF/MS.

16.2.5
One-Dimensional Gel Electrophoresis and ESI-MS Detection

The Au-precipitated proteins are redissolved with 20 μl dI water. The dissolved solution is mixed with the loading buffer (see Note 5) and then pipetted into the 10–15% SDS-polyacrylamide gel for separation following normal procedures. An applied voltage of 90–120 V at 4 °C for 4–6 hours is recommended. After the separation, protein bands are stained by Coomassie Brilliant Blue dye.

Protein bands are excised from 1-D SDS-PAGE gel, destained twice with 1 ml of ammonium bicarbonate buffer (see Note 6), and then dried under vacuum. After drying, proteins are reduced with 10 mM DTT at 57 °C for 1 hour and alkylated with 50 mM IAM at room temperature for 30 minutes in the dark (see Note 7). Wash the reduced samples with ammonium bicarbonate buffer once and with dI water twice to remove the excess reagents and then dry under vacuum. The tryptic digest is incubated at 37 °C overnight and then is extracted with 50% ACN. The tryptic peptides are ready to be analyzed by LC-MS/MS for protein identification.

16.3
Notes

1. All glass wares used are washed with aqua regia (3 : 1 HNO_3–HCl), rinsed extensively with dI water and dried in an oven.
2. The dilution step is to build up the lower protein concentration situation in large volume. We worked with maximum volume of 15 ml after dilution to make the protein concentration of 1.33 ppm.
3. The volume of gold nanoparticles is estimated according to the molar ratio between the proteins and gold nanoparticles and the suggested molar ratio is 10–100 : 1 of protein/Au nanoparticle.

4. The presence of gold nanoparticles would not cause any interference with the peptide signal in MALDI spectra.
5. The solution color changes soon from deep purple to blue when the loading buffer is added into the precipitates. This indicates that the aggregation is formed, but the solution can still be loaded into the gel well.
6. Wash the protein spots with ammonium bicarbonate buffer until the color is changed from blue to colorless.
7. The volumes of DTT and IAM are added enough to cover the samples. This step can be omitted if the samples are suitable for the following tryptic digestion.
8. We had compared the Au-assisted protein precipitation method with TCA precipitation method; Au-assisted protocol was successful, whereas TCA method showed no enrichment effect for protein samples with large volumes (>2 ml) or with low protein concentrations (4 ppm).

References

1. Sagar, A.J. and Pandit, M.W. (1983) Denaturations studies on bovine pancreatic ribonuclease: effect of trichloroacetic acid. *Biochim Biophys Acta*, **743**, 303–309.
2. Thongboonkerd, V., Mcleish, K.R., Arthur, J.M., and Klein, J.B. (2002) Proteomic analysis of normal human urinary proteins isolated by acetone precipitation or ultracentrifugation. *Kidney Int*, **62**, 1461–1469.
3. Hermanson, G.T. (ed.) (1996) Bioconjugated preparation of colloidal-gold-labeled proteins, in *Bioconjugate Techniques*, Academic Press.
4. Zeng, M. and Huang, X. (2004) Nanoparticles comprising a mixed monolayer for specific bindings with biomolecules. *J Am Chem Soc*, **126**, 12047–12054.
5. Teng, C.-H., Ho, K.-C., Lin, Y.-S., and Chen, Y.-C. (2004) Gold nanoparticles as selective and concentrating probes for samples in MALDI MS analysis. *Anal Chem*, **76**, 4337–4342.
6. Thanh, N.T.K. and Rosenzweig, Z. (2002) Development of an aggregation-based immunoassay for anti-protein A using gold nanoparticles. *Anal Chem*, **74**, 1624–1628.
7. Soto, C.M., Blum, A.S., Wilson, C.D., Lazorcik, J., Kim, M., Gnade, B., and Ratna, B.R. (2004) Separation and recovery of intact gold-virus complex by agarose electrophoresis and electroelution: application to the purification of cowpea mosaic virus and colloidal gold complex. *Electrophoresis*, **25**, 2901–2906.
8. Kong, X.L., Huang, L.C.L., Hsu, C.-M., Chen, W.-H., Han, C.-C., and Chang, H.-C. (2005) High-Affinity capture of proteins by diamond nanoparticles for mass spectrometric analysis. *Anal Chem*, **77**, 259–265.
9. Wang, A., Wu, C.-J., and Chen, S.-H. (2006) Gold nanoparticle-assisted protein enrichment and electroelution for biological samples containing low protein concentrations-A prelude of gel electrophoresis. *J Proteome Res*, **5**, 1488–1492.

17
Enrichment of Human Urinary Proteins Using Solid-Phase Extraction Column Chromatography

Graeme Smith and Robert Tonge

17.1
Introduction

Solid-phase extraction (SPE) has long been used to isolate small molecules, such as pharmaceuticals and narcotics, from urine [1], but it is also possible to enrich proteins from urine using this technology. SPE is an efficient method for isolating and concentrating solutes from a large volume of liquid, where the solutes in question are often in extremely dilute concentrations. The sorbents used for SPE are similar to those used in liquid chromatography, including normal phase, reversed phase, size exclusion, and ion exchange. Advantages of the technique include the following:

- no cross contamination due to disposable cartridges;
- ability to process multiple samples at the same time;
- capable of desalting and concentrating sample;
- inexpensive components required;
- robust system;
- ease of use.

For processing aqueous samples, such as urine and other biofluids, reversed phase is the sorbent of choice. In reversed-phase resins, the bonded phase consists of long chain hydrocarbons, such as octadecyl (C-18) or octyl (C-8), which are chemically bonded to a silica gel sorbent. The chemical nature of resin selected is influenced by the need to achieve efficient capture of analytes from the biological matrix while also allowing subsequent elution of analytes of interest without irreversible binding. For proteins analyzed in a reverse-phase system, the driving force to retention and subsequent elution is hydrophobic attraction for the stationary and mobile phases, respectively. Useful reviews on the choice of supports and bonded phases, albeit for HPLC, can be seen in [2, 3].

Typically, the sorbents consist of 40-μm silica gel with approximately 60-Å pore diameters, but for efficient capture of large protein molecules, larger pore resins are required. The diameter of a typical globular protein is in the region of 4 nm,

Renal and Urinary Proteomics: Methods and Protocols. Edited by Visith Thongboonkerd

ISBN: 978-3-527-31974-9

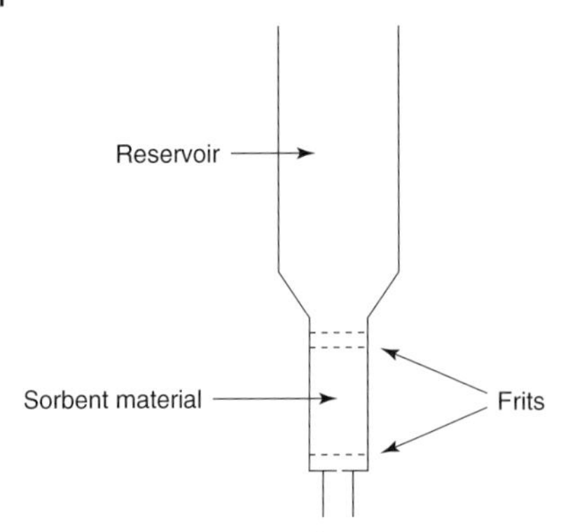

Figure 17.1 SPE cartridge.

or 40 Å, so would struggle to access the pore of a standard SPE resin. When using standard SPE resins, effectively only the surface of the beads is used for protein binding, much reducing the capacity. This is seen in the paper by Pang *et al.* [4], where an effective limit of approximately 40 kDa is seen on 2-D gels when using standard SPE resins.

In our studies, Varian Bond Elut LRC C18 EWP (extra wide pore) cartridges have been used which are based upon standard particle size silica but with 500-Å pores, that allow large molecules to enter the pores and interact with the bonded silica surface. Other large pore resins include Water's Accell Plus QMA and JT Baker's Bakerbond spe Widepore Butyl. For a detailed review of SPE, see [5].

Most SPE cartridges are a "syringe barrel" format, with a casing made of polypropylene. A 20-μm polypropylene frit at the top and bottom retains the sorbent material (Figure 17.1). A number of these cartridges can be run at the same time using a vacuum manifold (Figure 17.2) to apply negative pressure to the system. Each position has a tap so that the flow of liquid over each column can be started or stopped individually, as inevitably the same sample seems to flow over the same type of column at different rates. A trap should be included between the manifold and vacuum source to prevent reagents from contaminating the vacuum source. If only one or two columns are being run, positive pressure can be applied to them individually using a syringe and adapter.

17.2 Methods and Protocols

The method for enrichment of human urinary proteins using SPE is summarized in Diagram 17.1.

17.2.1 Treatment of Urine Prior to SPE

Frozen urine, or that stored at 4 °C, should be allowed to reach room temperature prior to processing, with agitation, to allow any precipitate to resolubilize. This is

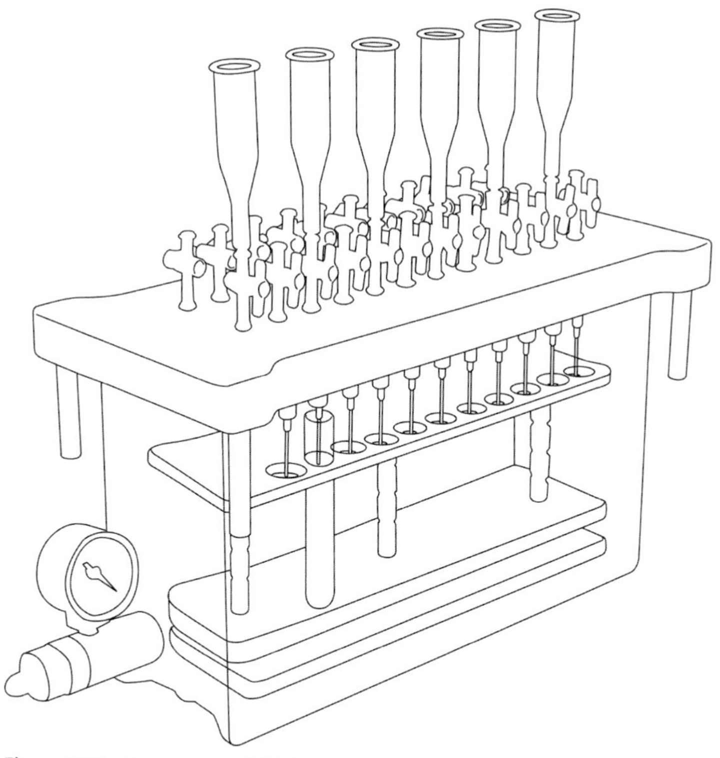

Figure 17.2 Vacuum manifold for running multiple SPE cartridges.

especially important with very concentrated urine samples. In extreme cases, the urine may have to be diluted with water or buffer. Some researchers add TFA to the urine to acidify it, making it as hydrophobic as possible to promote binding, but this has not been found to be necessary with the methods outlined here. If the addition of TFA is desired, the recommended quantity is 1 g to ~100 ml as per [4].

Urine may also contain cell debris and particulates, which can block SPE columns. The most usual method of removing particulate matter from urine is centrifugation, for example, 1500 × *g* for 10 minutes. However, filtration through syringe end filters is the preferred method for relatively small numbers and low volumes of samples. Several changes of filter may be necessary; one filter should process at least 50-ml urine. Luer lock syringes, of 60-ml capacity, are highly recommended to avoid being sprayed with urine if the filter is suddenly blocked!

Filtration through 5-μm syringe end filters (such as Sartorius Minisart) also removes uromodulin, the most abundant protein in normal human urine. This 85-kDa protein, also known as Tamm–Horsfall protein, can make up as much as 50% of the protein content and forms large molecular mass plaques that can easily block SPE columns, so it must be removed prior to analysis. By SDS-PAGE, it appears that no other component is removed in significant quantity by filtration (Figure 17.3). However, with large volumes of samples, centrifugation may be

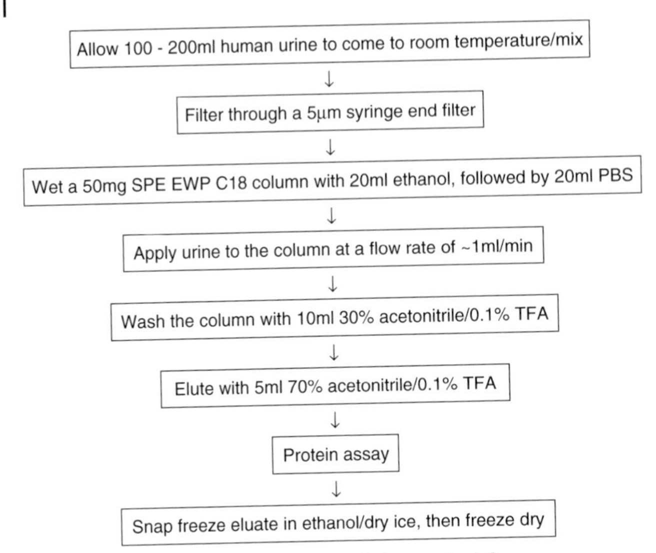

Diagram 17.1 Schematic summary of the method for enrichment of human urinary proteins using SPE column chromatography.

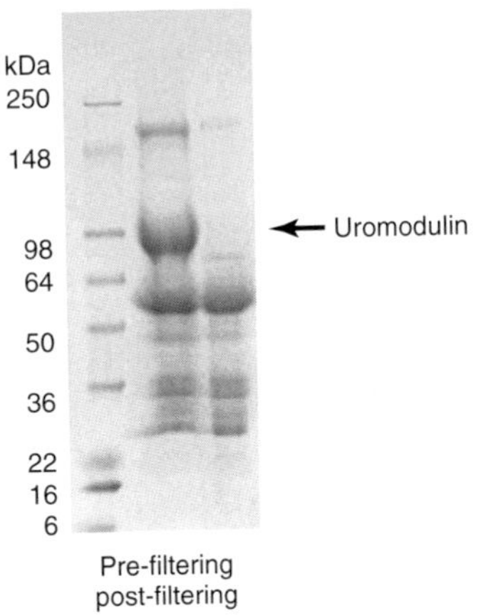

Figure 17.3 SDS-PAGE gel of urine, before and after filtration.

a more convenient processing method. Removing all the supernatant without disturbing the small semisolid pellet can be difficult, and hence, the preference for filtration.

It is prudent to protein assay the urine both pre- and post-filtering. The relatively low protein content of urine (usually 10–100 $\mu g\,ml^{-1}$) will be below the detection range of many protein assays. The Bio-Rad Bradford microassay [6] has a linear range of 1–10 $\mu g\,ml^{-1}$ and so it may be suitable for dilute urine samples, or for samples diluted to this range. The assay uses 800 μl of urine per assay point, that is, 1.6 ml when performing the assay in duplicate. A "rough and ready" assay based upon the Bradford assay that may be suitable for a quick comparison of urine samples is to mix 1 ml of diluted Bio-Rad reagent with 20 μl of urine; the absorbance reading at 595 nm versus water blank gives an absorbance approximately equal to the protein concentration in milligrams per milliliter.

GE Healthcare's 2-D Quant assay [5] is also compatible with urine (linear range of 0–50 μg protein), though it is limited by the number of samples that can be processed simultaneously and the longer processing time.

17.2.2
Cartridge Size

Although not thoroughly understood, it is thought that underutilizing the capacity of the cartridge may promote binding of unknown interfering small molecules, and thus it is recommended that a cartridge of suitable capacity is used.

Recommended cartridge sizes are as follows:

- 100–200 ml of urine: 500-mg cartridge
- 40–100 ml of urine: 200-mg cartridge
- 20–40 ml of urine: 100-mg cartridge
- 10–20 ml of urine: 50-mg cartridge.

17.2.3
Conditioning the SPE Resin

The dry media is conditioned with water-miscible organic solvent to wet the packing material and solvate the functional groups of the sorbent. This also removes the air present in the column and fills the void space with solvent. Methanol is usually stated as the preferred solvent, but it has been found that ethanol also works as a wetting agent and is a less hazardous alternative. Once the column has been wetted with the organic solvent, it is followed by an aqueous buffer, such as PBS, that the separation is to be run in. Ten times the volume of the packing material of both the organic and aqueous solvent are passed over the column by gravity feed alone, that is, 10 ml for 500-mg column, which has a bed volume of approximately 1 ml. The conditioning solvents, along with the sample and wash in subsequent steps, can be collected into separate tubes in the vacuum manifold or into a common waste container.

The column must not be allowed to dry between conditioning and sample addition; the flow must be stopped with the liquid level just short of the top frit. If the column runs dry, it should be reconditioned with organic solvent.

17.2.4
Adding the Sample

The urine is applied to the cartridge at a flow rate of approximately 1 ml min^{-1}, controlled by adjusting the vacuum pressure. This concentrates the dilute solutes onto the resin. The volume of urine to be processed is usually larger than the

Figure 17.4 Drawing up a large volume of urine from a separate reservoir.

reservoir capacity of the cartridge, so funnels or adaptor caps may be used and the cartridges do not need to be constantly topped up. Tubing can be run from the adaptor caps to a separate container of urine, which is drawn up by the negative pressure (Figure 17.4). It is imperative that the cartridge does not run dry and the flow should be stopped just before the fluid level reaches the top frit.

17.2.5
Washing the Column

The column needs to be washed to remove salt and other contaminants. The recommended wash volume is 10 times the volume of the resin, that is, 10 ml for a 500-mg column. This step is performed at a flow rate similar to that for sample loading (1 ml min^{-1}). Again, the cartridge should not be allowed to run dry before elution.

The column can simply be washed with 10% acetonitrile, 0.1% TFA. However, in our experience, this leaves behind some unknown contaminating ionic substance(s), which can interfere with isoelectric focusing (IEF), causing IEF strips to arc and char. This is particularly an issue when large amounts of sample have been loaded, for instance, for preparatory-scale 2-D gels. When run on the second dimension, these heavily loaded gels show evidence of incomplete focusing (Figure 17.5). Increasing the amount of organic solvent in the wash to 30% (still containing 0.1% TFA) was found to overcome this problem (Figure 17.5). If running SDS-PAGE or analytical 2-D gels (less than 100-mg protein), a lower organic solvent concentration in the wash may be acceptable.

Obviously, increasing the amount of organic solvent in the wash step may remove other substances, which are soluble in low organic solvent; no protein processing method is 100% efficient. As with all proteomics methods, the researcher must be aware that they are just looking at a subset of the whole proteome. It can be seen

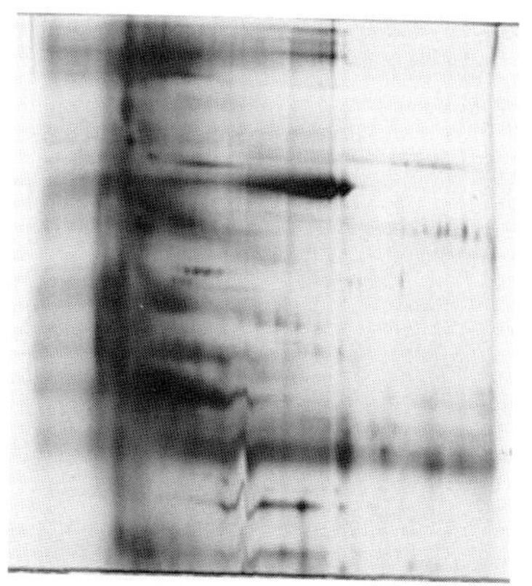

10% acetonitrile wash, 70% acetonitrile elution

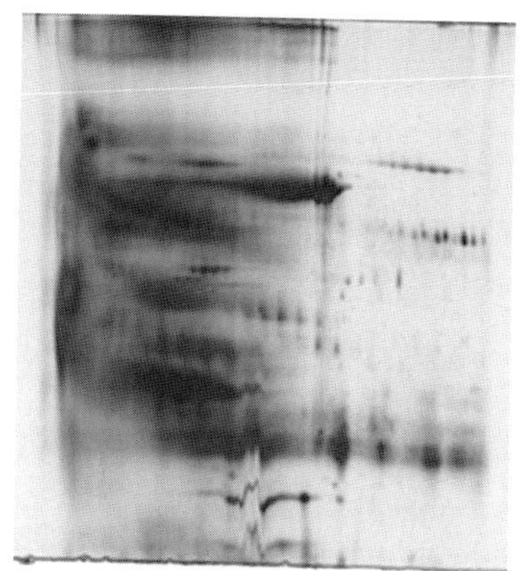

20% acetonitrile wash, 70% acetonitrile elution

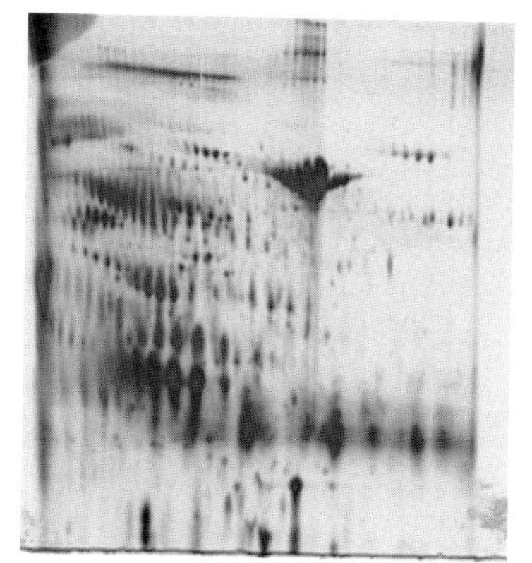

30% acetonitrile wash, 70% acetonitrile elution

Figure 17.5 2-D gels of SPE-prepared urine, washed with increasing amounts of organic solvent (all solutions contain 0.1% TFA).

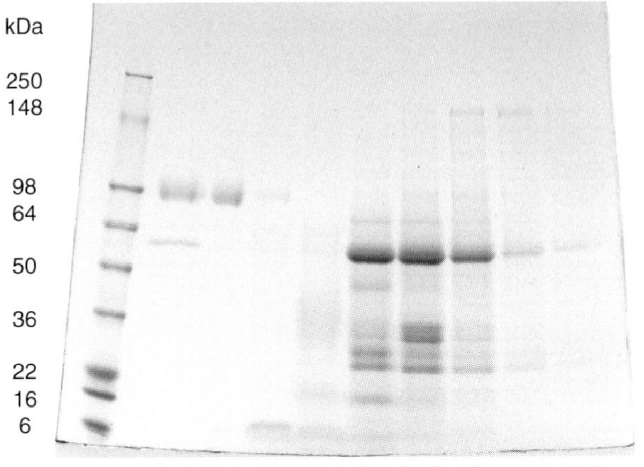

Figure 17.6 Increasing acetonitrile concentration for elution of urinary proteins from SPE cartridge (all including 0.1% TFA).

in Figure 17.6 that acetonitrile washes up to 30% only elute uromodulin and very small molecular mass proteins, not typically observed by 2-D PAGE.

17.2.6 Elution

The proteins of interest are eluted from the column with 70% acetonitrile, 0.1% TFA, which disrupts the bonding between the analyte and the sorbent. This is done at a low flow rate, under gravity alone. The liquid remaining in the cartridge void space can then be recovered by the use of a gentle vacuum. Five times the resin volume is a sufficient volume to elute the bound protein, that is, 5 ml for a 500-mg cartridge. The eluates are collected into separate clean tubes in the vacuum manifold. To assess the recovery, the sample can be protein assayed at this stage with a suitable assay (see earlier discussion of protein assays).

To transfer the sample into a more suitable buffer, the sample is hard frozen in a dry ice/ethanol bath and desiccated in a freeze drier. The resulting desiccant can then be resuspended into a buffer of choice, such as Laemmli buffer for SDS-PAGE, or 2-D sample buffer for 2-D analysis.

17.2.7 Example of 2-D PAGE of SPE Processed Material

Using these methods, it has been possible to produce 2-D gels of urine (Figure 17.7) from which it has been possible to identify >500 individual spots and to identify >50 different proteins by mass spectrometric analysis. For more complete details of the method development of the process, see [7].

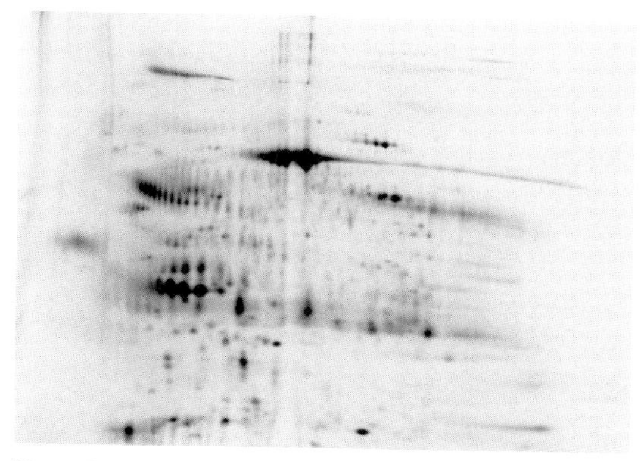

Figure 17.7 2-D gel image of 150 µg urinary proteins recovered from an SPE cartridge elution (equivalent to 20 ml starting urine) run on an 8–16% 2D-PAGE gel.

17.2.8 Animal Urine

Animal urine, such as that obtained from rat and mouse, is generally much more concentrated than human urine and has a different composition of proteins. The most abundant protein in rat urine, Major Urinary Protein, can make up more than 90% of the proteome. The methods detailed in this chapter may provide a starting point for processing animal urine but would require further development.

Acknowledgments

The authors thank Derek Barratt for his help in developing the methods outlined in this chapter.

References

1. Platoff, G.E. Jr and Gere, J.A. (1991) Solid phase extraction of abused drugs from urine. *Forensic Sci Rev*, **3**, 117–133.
2. Esser, U. and Unger, K.K. (1991) Reversed-phase packings for the separation of peptides and proteins by means of gradient elution high-performance liquid chromatography, in *High-Performance Liquid Chromatography of Peptides and Proteins: Separation, Analysis, and Conformation* (eds C.T. Mant and R.S. Hodges), CRC Press, Boca Raton.
3. Meyer, V.R. (2004) Columns and stationary phases, *Practical High-Performance Liquid Chromatography*, Wiley-VCH Verlag GmbH, Weinheim.
4. Pang, J.X., Ginanni, N., Dongre, A.R., Hefta, S.A., and Opiteck, G.J. (2002) Biomarker discovery in urine by proteomics. *J Proteome Res*, **1**, 161–169.
5. Thurman, E.M. and Mills, M.S. (1998) *Solid-Phase Extraction – Principles and Practice*, Wiley-Interscience.

6. Bio-Rad Protein Assay http://www3.bio-rad.com/LifeScience.pdf/Bulletin_9004.pdf (first accessed in 1993).
7. Smith, G.J., Barratt, D., Rowlinson, R., Nickson, N., and Tonge, R. (2005) Development of a high-throughput method for preparing human urine for two-dimensional electrophoresis. *Proteomics*, **5**, 2315–2318.

18
Enrichment and Analysis of Concanavalin A-Captured Urinary Glycoproteins

Linjie Wang and Youhe Gao

18.1
Introduction

Glycosylation, a common and important protein posttranslational modification in eukaryotic organisms, is involved in many biological processes [1], such as cell adhesion [2], signal transduction [3], immune response [4], and inflammatory reactions [5]. More than half of all proteins are thought to be glycoproteins and they alter in both quality and quantity under different conditions in various physiological and pathological states [1].

There are four mechanisms for oligosaccharide-protein attachment [6, 7]: (i) the N-linked glycosylation of asparagine residues, (ii) the O-linked glycosylation of serine and threonine residues, (iii) glycosyl phosphatidyl inositol anchors, and (iv) C-mannosylation of tryptophan residues. Among these, N-linked glycosylation is the main type that exists in body fluid glycoproteins. This specific type of glycosylation has been studied extensively and has been shown to be critical for the membrane localization of mature proteins as well as for the secretion of proteins into the extracellular matrix and body fluids [6].

Urinary proteins, mainly composed of plasma components filtered by renal glomeruli and secretory proteins from kidneys and urogenital tracts, have received much attention in the proteomics field for their simplicity compared to serum as well as their potential in biomarker discovery [8–14]. Many studies have been published in the past few years on urine proteomics [15–19]. One subproteome – the urine glycoproteome – has revealed its importance in this particular field of study. First, many critical proteins, in either normal physiological or pathological processes, are low abundant glycoproteins. By enriching glycoproteins, these critical proteins can then be isolated more easily from other proteins. Enrichment would allow for the earlier detection of their glycosylation alterations under different conditions. Second, albumin is the most abundant protein in urine, particularly when albuminuria occurs in kidney diseases. Under this condition, large amounts of albumin profoundly affect the analysis of the urine proteome and the dynamic range of protein detection will obviously decrease. Since albumin is not N-glycosylated, urine *N*-glycoproteome could deplete it and thus could provide a

Renal and Urinary Proteomics: Methods and Protocols. Edited by Visith Thongboonkerd

ISBN: 978-3-527-31974-9

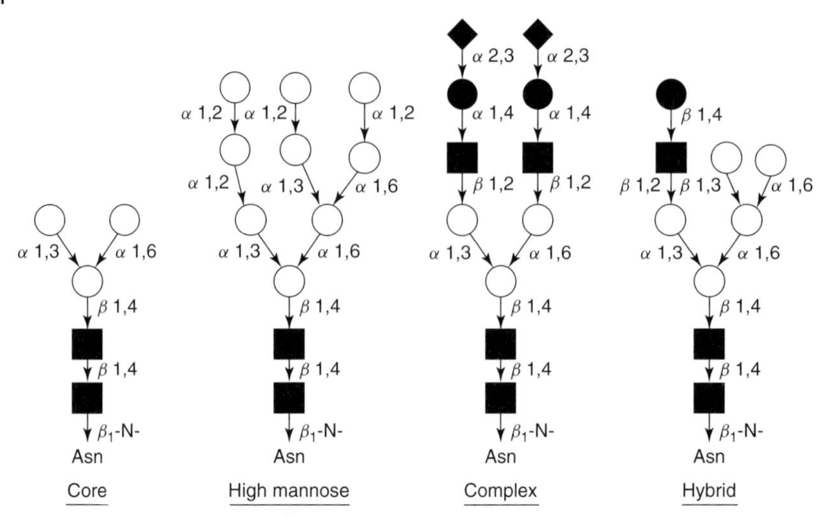

Figure 18.1 Different types of N-linked glycans.

more detailed protein profile in addition to glycosylation information. Therefore, urine *N*-glycoproteome is likely a better choice over proteome itself for albuminuria samples [20].

N-linked glycans are attached to asparagine residues at the consensus sequence of N–X–S/T (where X is any residue except proline) of glycoproteins. Glycans have a common core structure of trimannosyl five sugars and vary in their outer branches. They are classified into three subtypes: high mannose, hybrid, and complex (Figure 18.1) [6]. Since most N-linked glycoproteins are low abundant proteins, enrichment is a necessary procedure before a proteomic analysis. Many methods have been developed to enrich N-linked glycoproteins; the most commonly used method is lectin affinity [21–24].

Lectins are a class of carbohydrate-binding proteins found in plants, bacteria, fungi, and animals, and they can recognize specific oligosaccharide moieties. When the structure of glycans alters, the lectin-binding patterns also change. The most common lectins are listed in Table 18.1 [21].

Among various lectins, Concanavalin A (Con A) is the most widely used for N-linked glycoprotein enrichment, not only because of its broader specificity and higher affinity but also because of the convenience in subsequent analysis by mass spectrometry due to the existence of the specific glycosidase, PNGase F. Con A recognizes α-mannose, which is quite common among N-linked glycans, and it can capture high mannose, hybrid, and biantennary complex types of N-linked glycans with a relatively higher affinity than other types of *N*-glycans [25, 26]. Here, we focus on how to enrich *N*-glycoproteins using Con A. The following sections describe separation methods and analysis of these glycoproteins and their glycosylation site recognition.

Table 18.1 Common lectins and their binding characteristics.

Lectin	Binding sequence	Binding type of glycosylated chain
Concanavalin A (Con A)	α-Mannose	N-type
Wheat germ agglutinin (WGA)	*N*-acetyl-glucosamine (GlcNAc), sialic acid	N-type
Lens culinaris agglutinin (LCA)	Similar to Con A with lower affinity and can bind fucose	N-type
Lectin Jacalin	Galactosyl (β-1,3)-*N*-acetyl-galactosamine (GalNAc)	O-type
Peanut agglutinin (PNA)	Same as Jacalin	O-type

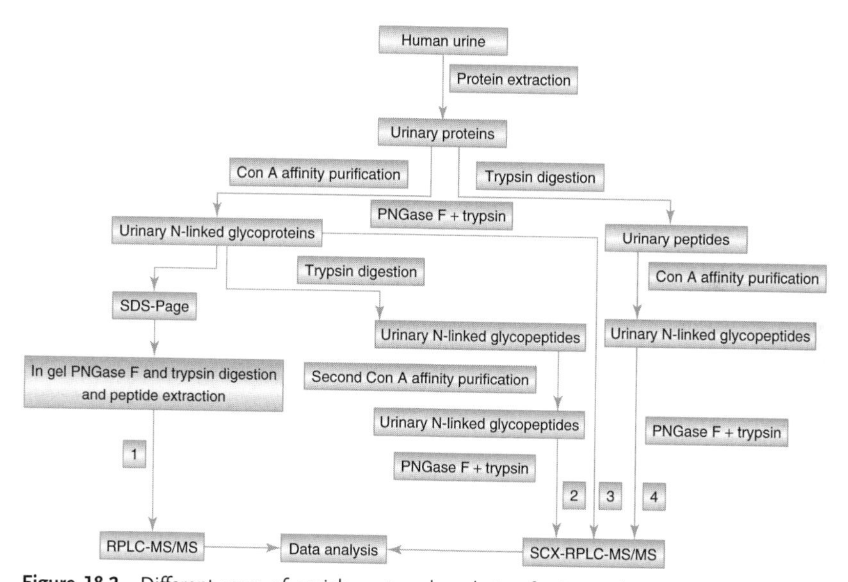

Figure 18.2 Different ways of enrichment and analysis of urinary glycoproteins.

18.2 Methods and Protocols

Sample preparation for total urinary proteins is described in chapter 21. Here, we mainly introduce three ways for the enrichment and analysis of Con A-captured urinary glycoproteins (Figure 18.2).

18.2.1 Step 1: Enrichment of Urinary Glycoproteins or Glycopeptides

18.2.1.1 Sample Preparation

1. Resuspend lyophilized urinary protein pellets in a proper buffer: 25 mM NH_4HCO_3 or Tris-buffered saline (TBS) solution (0.05 M Tris-HCl, pH 7.1, 0.15 M NaCl) containing 1 mM $MnCl_2$ and 1 mM $CaCl_2$.
2. Measure the protein concentration by Bradford assay. The proper protein concentration should be about 1 mg ml^{-1}.

18.2.1.2 Enrichment by Con A Affinity Chromatography

1. Mix the protein/peptide solutions with 2× volume of TBS buffer with a proportional volume of Con A-agarose (sepharose) slurry (according to the product chosen).
2. Incubate the mixture at 4 °C overnight with mild stirring.
3. Centrifuge the mixtures at 1000g for 2 minutes and collect supernatants. These are designated as the flow-throughs.
4. Wash the precipitated Con A beads five times in TBS buffer. The buffer should be removed as much as possible each time to ensure minimal carry-over of unbound contaminants such as albumin.
5. Elute captured proteins by incubating the beads in 1 ml of 500 mM α-D-mannopyranoside for 15 minutes at room temperature. Released proteins are then centrifuged at 1000g for 2 minutes to collect the eluted supernatant fractions, namely, the crude Con A-captured glycoproteins.

18.2.1.3 Protein/Peptide Digestion

1. **Exchange of the elution buffer into enzymatic buffer:** If the Con A affinity chromatography is applied, one can change the 500 mM α-D-mannopyranoside buffer into 25 mM NH_4HCO_3 with 10-kD cut-off devices or just add isovolumic 50 mM NH_4HCO_3 to obtain an alkaline environment for trypsin. If none of the changes are made to urinary proteins, then go to 2.
2. **Reduction:** Add 1 M DTT to a final concentration of 20 mM. Incubate in a 56 °C water bath for 1 hour. Cool the solution to room temperature.
3. **Alkylation:** Add 50 mM iodoacetamide to a final concentration of 50 mM and incubate in the dark at room temperature for 30 minutes.
4. **Deglycosylation of the glycoproteins:** Add endoglycopeptidase PNGase F to 1 U per 100 µg protein and incubate at 37 °C for 1 hour.
5. **Trypsin digestion:** Add sequencing-grade trypsin at a ratio of 1 : 50 (w/w) and incubate in a 37 °C water bath overnight. Add HCl to terminate the reaction.

18.2.2 Step 2: Separation of Urinary Glycoproteins/Glycopeptides

18.2.2.1 SDS-PAGE Followed by Reverse-Phase Liquid Chromatography

1. **SDS-PAGE:** Apply Con A purified glycoproteins. If desired, one can also load samples of total protein content without Con A enrichment, unbound

flow-through fractions, and an agarose control onto SDS-PAGE, with 100 μg protein per lane. Follow SDS-PAGE procedure with Coomassie Brilliant Blue staining.

2. **Decoloration:** Gels of glycoproteins can be cut into several strips according to the protein amount. Each slice is cut into approximately 1 mm^3 pieces. Wash the pieces with deionized water several times. Merge the cleaned gel sections into 50% acetonitrile/25 mM NH_4HCO_3 until decoloration is complete (e.g., no blue stain is observed). Next, dehydrate with 100% acetonitrile.
3. **In-gel digestion:** All slices are in-gel reduced, alkylated, and then digested with PNGase F (0.1 U per section) first, followed by trypsin (200 ng per slice), according to the above-described in Section 2.1.3.
4. **Peptide extraction:** Add 5% trifluoroacetic acid/50% acetonitrile (submerge all the pieces) and incubate at 37 °C for 2–3 hours. Collect the supernatants in Eppendorf tubes. Add 2.5% trifluoroacetic acid/50% acetonitrile to the pieces left over and incubate at 37 °C for 2–3 hours. Collect the supernatants and combine with the previous set of supernatants. Lyophilize extracted peptides and store at −80 °C for further analysis.
5. **Reverse-phase liquid chromatography:** Lyophilized peptides are redissolved in 50 μl HPLC buffer A (0.1% formic acid). Fused silica tubing (150 mm × 170 μm i.d.) packed with C18 5-μm spherical particles with a pore diameter of 300 Å is used to analyze the tryptic peptides. The peptides are eluted with a buffer B (0.1% formic acid, 99.9% acetonitrile; flow rate: 2 μl per minute) gradient from 5 to 30% for a sufficient amount of time.

18.2.2.2 Two-Dimensional Liquid Chromatography (Strong Cation Exchange with Reverse-Phase Liquid Chromatography)

Digested peptides are separated on an SCX capillary column (150 mm × 320 μm i.d.), followed by application to a RP capillary column (150 mm × 170 μm i.d.) online. The ammonium acetate concentrations for a step-gradient elution from the SCX capillary column are as follows: 0 mM, 25 mM, 50 mM, 75 mM, 100 mM, 125 mM, 150 mM, 250 mM, 500 mM, and 1 M. The elution gradient for the RP column is 5–30% buffer B for a sufficient amount of time.

18.2.3 Step 3: Analysis of Glycoproteins/Glycopeptides and Glycosylated Sites

As an example for analyzing these proteins, we choose to use the LCQ-DECA XP^{plus} electrospray ion trap mass spectrometer (Thermo Finnigan, San Jose, CA).

18.2.3.1 Mass Spectrometric Analysis

Eluted peptides from liquid chromatography are analyzed by mass spectrometry. Ions are detected in a survey scan from 400 to 1500 amu (3 μscans), followed by five data-dependent MS/MS scans (5 μscans each, isolation width 3 amu, 35% normalized collision energy, dynamic exclusion for 3 minutes) in a completely automated manner.

18.2.3.2 Analysis of Glycosylated Sites

PNGase F is a glycosidase that cleaves almost all N-linked glycans from their anchoring asparagine residues, removes the entire oligosaccharide, and causes deamidation of the asparagine to an aspartic acid residue, resulting in a 1-Da increase. This mass change can serve as a characteristic tag, cuing to glycan-linked asparagines. If the detection ability of this 1-Da alteration by MS cannot be ascertained, H_2O^{18} can be used in the enzymatic buffer during the PNGase F digestion. This buffer change can lead to a 3-Da alteration that is then easier to detect.

18.2.3.3 Data Processing

All MS/MS spectra are searched using SEQUEST-algorithm-based Bioworks (Thermo Finnigan, San Jose, CA) against the proper database (e.g., IPI human protein database). The searches can be performed with enzyme constraints, a static modification of +57-Da on a cysteine residue and a differential modification of +1-Da/+3-Da on an asparagine residue. The SEQUEST criteria are as follows. (i) DeltaCn score of at least 0.1. (ii) Rsp score of 1. (iii) Xcorr $\geq$ 1.9 for +1 charged peptides, regardless of the end residues; Xcorr $\geq$ 2.2 for +2 charged peptides, with a fully or partially tryptic end; Xcorr $\geq$ 3.0 for +2 charged peptides, regardless to the end residues; and Xcorr $\geq$ 3.75 for +3 charged peptides, without regard to the end residues.

The filtered spectra can be screened with manual validation or the software AMASS (available at http://www.proteomics-cams.com) developed by our laboratory [27]. All the raw data can also be searched against the reversed database in order to estimate false-positive rates of the results, with the same search parameters.

18.3 Pros and Cons

The pros and cons of the above-described enrichment and analysis methods for urinary glycoproteins (as shown in Figure 18.2) are highlighted below.

18.3.1 One-Step Enrichment of Urinary Glycoproteins Followed by SDS-PAGE and Reverse-Phase Liquid Chromatography Separation

18.3.1.1 Pros

1. Electrophoresis can provide direct viewing of the difference between the Con A purified glycoproteins and the untreated samples.
2. Electrophoresis can better separate high abundant glycoproteins, for example, uromodulin, from the lower ones.

18.3.1.2 Cons

1. Multiple manipulations take a longer time to perform and can lead to a significant loss of proteins during the procedures.

2. In-gel digestion and peptide extraction are relatively low efficiency procedures and can introduce contamination into the preparations.

18.3.2 Two-Step Enrichment of Urinary Glycoproteins Followed by SCX-RP Liquid Chromatography Separation

18.3.2.1 Pros

1. Two-step enrichment can maximally eliminate nonspecific contaminants.
2. Peptides that undergo MS analysis are solely glycopeptides, leading to easier identification of glycosylated peptides without nonglycopeptides interference.

18.3.2.2 Cons

1. Multiple manipulations take a longer time to perform and can lead to a significant loss of proteins during the procedures.

18.3.3 One-Step Enrichment of Urinary Glycoproteins Followed by SCX-RP Liquid Chromatography Separation

18.3.3.1 Pros

1. One-step enrichment can reduce the number of procedures.
2. Peptides that undergo MS analysis are solely glycopeptides, leading to easier identification of glycosylated peptides without nonglycopeptides interference.

18.3.3.2 Cons

1. One-step enrichment can introduce nonspecific contaminants.
2. During MS analysis, nonglycopeptides can interfere with the detection of glycopeptides.

18.3.4 Enrichment of Trypsin-Treated Urinary Peptides Followed by Separation by SCX-RP Liquid Chromatography

18.3.4.1 Pros

1. One-step enrichment can reduce the number of procedures.
2. Peptides that undergo MS analysis are solely glycopeptides, leading to easier identification of glycosylated peptides without nonglycopeptides interference.

18.3.4.2 Cons

1. One-step enrichment can introduce nonspecific contaminants.
2. During MS analysis, nonglycopeptides can interfere with the detection of glycopeptides.

This brief introduction shows that every method has its own specific pros and cons. The best-suited method can be chosen for individual use according to the experimental objectives and conditions.

References

1. Haltiwanger, R.S. and Lowe, J.B. (2004) Role of glycosylation in development. *Annu Rev Biochem*, **73**, 491–537.
2. Zhang, Y., Zhao, J.H., Zhang, X.Y., Guo, H.B., Liu, F., and Chen, H.L. (2004) Relations of the type and branch of surface N-glycans to cell adhesion, migration and integrin expressions. *Mol Cell Biochem*, **260**, 137–146.
3. Haltiwanger, R.S. (2002) Regulation of signal transduction pathways in development by glycosylation. *Curr Opin Struct Biol*, **12**, 593–598.
4. Rudd, P.M., Elliott, T., Cresswell, P., Wilson, I.A., and Dwek, R.A. (2001) Glycosylation and the immune system. *Science*, **291**, 2370–2376.
5. Lowe, J.B. (2003) Glycan-dependent leukocyte adhesion and recruitment in inflammation. *Curr Opin Cell Biol*, **15**, 531–538.
6. Durand, G. and Seta, N. (2000) Protein glycosylation and diseases: blood and urinary oligosaccharides as markers for diagnosis and therapeutic monitoring. *Clin Chem*, **46**, 795–805.
7. Furmanek, A. and Hofsteenge, J. (2000) Protein C-mannosylation: facts and questions. *Acta Biochim Pol*, **47**, 781–789.
8. Le Bricon, T., Erlich, D., Bengoufa, D., Dussaucy, M., Garnier, J.P., and Bousquet, B. (1998) Sodium dodecyl sulfate-agarose gel electrophoresis of urinary proteins application to multiple myeloma. *Clin Chem*, **44**, 1191–1197.
9. Grover, P.K. and Resnick, M.I. (1997) High resolution two-dimensional electrophoretic analysis of urinary proteins of patients with prostatic cancer. *Electrophoresis*, **18**, 814–818.
10. Rasmussen, H.H., Orntoft, T.F., Wolf, H., and Celis, J.E. (1996) Towards a comprehensive database of proteins from the urine of patients with bladder cancer. *J Urol*, **155**, 2113–2119.
11. Kageyama, S., Isono, T., Iwaki, H., Wakabayashi, Y., Okada, Y., Kontani, K., Yoshimura, K., Terai, A., Arai, Y., and Yoshiki, T. (2004) Identification by proteomic analysis of calreticulin as a marker for bladder cancer and evaluation of the diagnostic accuracy of its detection in urine. *Clin Chem*, **50**, 857–866.
12. Cutillas, P.R., Norden, A.G., Cramer, R., Burlingame, A.L., and Unwin, R.J. (2003) Detection and analysis of urinary peptides by on-line liquid chromatography and mass spectrometry application to patients with renal Fanconi syndrome. *Clin Sci*, **104**, 483–490.
13. Cutillas, P.R., Chalkley, R.J., Hansen, K.C., Cramer, R., Norden, A.G., Waterfield, M.D., Burlingame, A.L., and Unwin, R.J. (2004) The urinary proteome in Fanconi syndrome implies specificity in the reabsorption of proteins by renal proximal tubule cells. *Am J Physiol Renal Physiol*, **287**, F353–F364.
14. Pang, J.X., Ginanni, N., Dongre, A.R., Hefta, S.A., and Opitek, G.J. (2002) Biomarkers discovery in urine by proteomics. *J Proteome Res*, **1**, 161–169.
15. Thongboonkerd, V., McLeish, K.R., Arthur, J.M., and Klein, J.B. (2002) Proteomic analysis of normal human urinary proteins isolated by acetone precipitation or ultracentrifugation. *Kidney Int*, **62**, 1461–1469.
16. Spahr, C.S., Davis, M.T., McGinley, M.D., Robinson, J.H., Bures, E.J., Beierle, J., Mort, J., Courchesne, P.L., Chen, K., Wahl, R.C., Yu, W., Luethy, R., and Patterson, S.D. (2001) Towards defining the urinary proteome using liquid chromatography-tandem mass spectrometry: I Profiling an unfractionated tryptic digest. *Proteomics*, **1**, 93–107.
17. Davis, M.T., Spahr, C.S., McGinley, M.D., Robinson, J.H., Bures, E.J., Beierle, J., Mort, J., Yu, W., Luethy, R., and Patterson, S.D.

(2001) Towards defining the urinary proteome using liquid chromatography-tandem mass spectrometry: II limitations of complex mixture analyses. *Proteomics*, **1**, 108–117.

18. Pieper, R., Gatlin, C.L., McGrath, A.M., Makusky, A.J., Mondal, M., Seonarain, M., Field, E., Schatz, C.R., Estock, M.A., Ahmed, N., Anderson, N.G., and Steiner, S. (2004) Characterization of the human urinary proteome: a method for high-resolution display of urinary proteins on two-dimensional electrophoresis gels with a yield of nearly 1400 distinct protein spots. *Proteomics*, **4**, 1159–1174.
19. Sun, W., Li, F., Wu, S., Wang, X., Zheng, D., Wang, J., and Gao, Y. (2005) Human urine proteome analysis by three separation approaches. *Proteomics*, **5** (18), 4994–5001.
20. Wang, L., Li, F., Sun, W., Wu, S., Wang, X., Zhang, L., Zheng, D., Wang, J. *et al.* (2006) Concanavalin A-captured glycoproteins in healthy human urine. *Mol Cell Proteomics*, **5** (3), 560–562.
21. Yang, Z. and Hancock, W.S. (2004) Approach to the comprehensive analysis of glycoproteins isolated from human serum using a multi-lectin affinity column. *J Chromatogr A*, **1053**, 79–88.
22. Kaji, H., Saito, H., Yamauchi, Y., Shinkawa, T., Taoka, M., Hirabayashi, J., Kasai, K., Takahashi, N., and Isobe, T. (2003) Lectin affinity capture, isotope-coded tagging and mass spectrometry to identify N-linked glycoproteins. *Nat Biotechnol*, **21**, 667–672.
23. Bunkenborg, J., Pilch, B.J., Podtelejnikov, A.V., and Wisniewski, J.R. (2004) Screening for N-glycosylated proteins by liquid chromatography mass spectrometry. *Proteomics*, **4**, 454–465.
24. Ghosh, D., Krokhin, O., Antonovici, M., Ens, W., Standing, K.G., Beavis, R.C., and Wilkins, J.A. (2004) Lectin affinity as an approach to the proteomic analysis of membrane glycoproteins. *J Proteome Res*, **3**, 841–850.
25. Kristiansen, T.Z., Bunkenborg, J., Gronborg, M., Molina, H., Thuluvath, P.J., Argani, P., Goggins, M.G., Maitra, A., and Pandey, A. (2004) A proteomic analysis of human bile. *Mol Cell Proteomics*, **3**, 715–728.
26. Ohyama, Y., Kasai, K., Nomoto, H., and Inoue, Y. (1985) Frontal affinity chromatography of ovalbumin glycoasparagines on a concanavalin A-sepharose column. A quantitative study of the binding specificity of the lectin. *J Biol Chem*, **260**, 6882–6887.
27. Sun, W., Li, F., Wang, J., Zheng, D., and Gao, Y. (2004) AMASS: software for automatically validating the quality of MS/MS spectrum from SEQUEST results. *Mol Cell Proteomics*, **3**, 1194–1199.

19
Isolation and Enrichment of Urinary Exosomes for Biomarker Discovery

Hua Zhou, Patricia A. Gonzales, Trairak Pisitkun, Robert A. Star, Mark A. Knepper, and Peter S.T. Yuen

19.1
Introduction

Recently, urinary exosomes, a specific fraction of urine, have been recognized as a promising new source to discover urinary biomarkers, which can be the predictors of renal disease that would make them useful in making therapeutic decisions [1, 2]. Pisitkun *et al.* identified 295 proteins in urinary exosomes by LC-MS/MS [3]. Zhou *et al.* demonstrated that 2-D proteomic analysis can provide biomarker candidates for early detection of acute kidney injury [4].

Urinary exosomes are low-density membrane vesicles (<100 nm) that originate from all cell types along the nephron, including glomerular podocytes and tubular epithelial cells [2]. After endocytosis from the plasma membrane, endocytic vesicles, which have an inside-out orientation, can fuse with the outer membrane of a multivesicular body (MVB). The outer membrane of MVB can invaginate, generating internal vesicles that have a right side-out orientation. When the outer membrane of an MVB fuses with the apical plasma membrane of a podocyte or a renal epithelial cell, the vesicles are released into the urinary space as exosomes (Figure 19.1). Urinary exosomes include both membrane proteins and cytosolic proteins enriched in low-density urinary structures [3]. Proteomic analysis of whole urine or urinary sediments may not have sufficient dynamic range to detect changes in the levels of individual proteins associated with kidney diseases, especially low-abundance proteins, but enrichment of urinary exosomes has overcome some of these limitations of proteomic discovery. Pisitkun *et al.* [3] used electron microscopy to demonstrate that vesicles demarcated by lipid bilayers were the predominant component of the low-density fraction of urine. However, fibrils consisting of Tamm–Horsfall protein (THP) (uromodulin) can also be seen, and therefore care must be taken to avoid either entrapment of exosomes in this meshlike structure, and/or contamination of the exosomal fraction with this abundant protein. Indeed, significant losses can occur upon freezing and thawing of urine samples, which is important when using stored samples as a source for urinary exosomes for proteomic discovery. Zhou *et al.* [5] verified

Renal and Urinary Proteomics: Methods and Protocols. Edited by Visith Thongboonkerd

ISBN: 978-3-527-31974-9

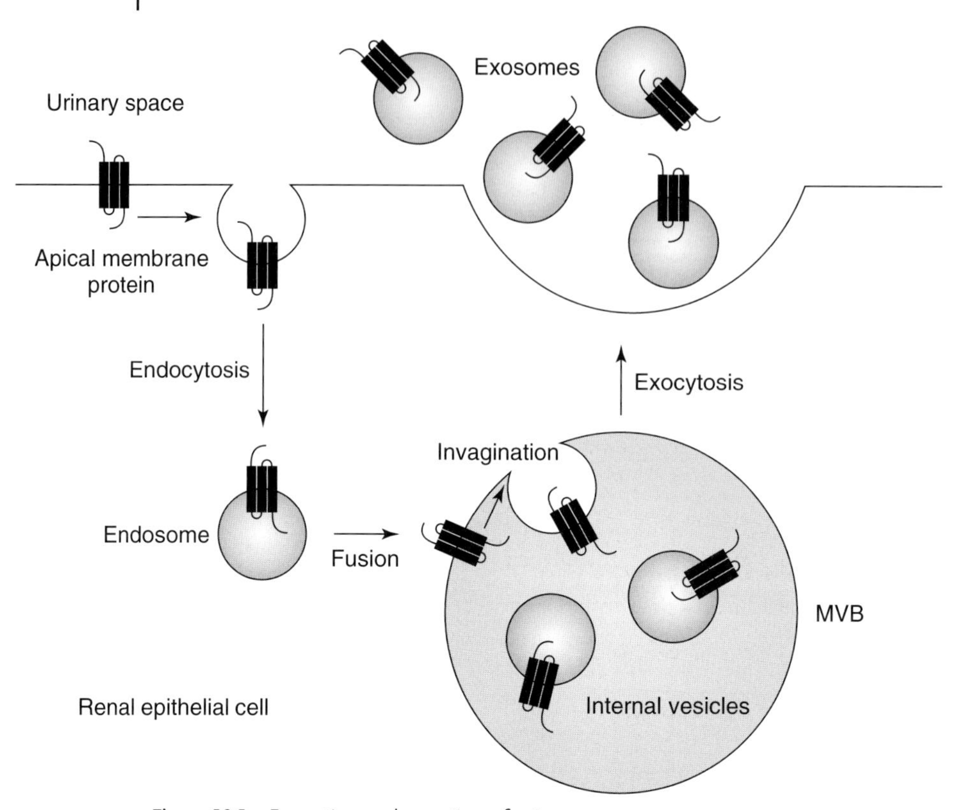

Figure 19.1 Formation and excretion of urinary exosomes. Exosome formation starts from endocytosis of the apical membrane, trafficking to and fusion with a multivesicular body, invagination, then fusion of the outer membrane of the multivesicular body and the apical membrane releases exosomes into the urine; temporal sequence is depicted by arrows. Reprinted with permission from [1].

that −80 °C storage of urine samples and extensive vortexing after thawing are very important to preserve exosomes, as demonstrated by Western blot analysis of exosome-specific biomarkers, such as aquaporin-2 (AQP2, collecting duct), sodium-proton exchanger 3 (NHE-3, proximal tubule), tumor susceptibility gene (TSG101), and apoptosis-linked gene-2 interacting protein X (ALIX) (Figure 19.2). One of the limiting factors for the study of exosomes is the availability of an ultracentrifuge to isolate the low-density membrane fraction, which is sedimented at 200 000*g*. Cheruvanky *et al.* [6] found that a commercially available nanomembrane can also be used to rapidly obtain exosomes using a tabletop centrifuge. In this chapter, we expand upon the methods sections of these three papers and discuss some of the future challenges for studying urinary exosomes and the proteins contained in them. The most current protocol can be found at the UroProt web site http://intramural.niddk.nih.gov/research/uroprot/.

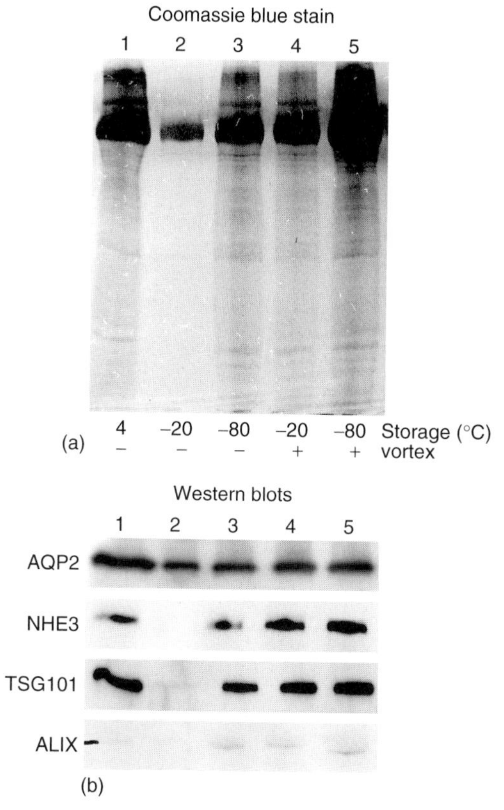

Figure 19.2 Influence of storage and vortexing on urinary exosomes. Urinary exosomal proteins on Coomassie blue-stained gel (a) and Western blots for specific exosome markers (b) from the same urine sample under different storage conditions, with and without vortexing. AQP2, aquaporin2; NHE3, sodium-proton exchanger 3; TSG101, tumor susceptibility gene; and ALIX, apoptosis-linked gene-2 interacting protein X. Reprinted, with permission from [5].

19.2 Methods and Protocols

19.2.1 Preparation of Urine Samples for Exosome Isolation

19.2.1.1 Urine Collection

- If possible, determine the collection time and volume.
- Collect the first or second morning urine, typically a large volume (100–400 ml) for identification of exosomal proteins by proteomic analysis, or a small volume (10 ml) for validation of the identified candidate biomarkers from isolated urinary exosomes.

- First morning urine: first urine after waking (no fluid or fruits after 9 p.m. of the prior evening).
- Second morning urine: Discard the first morning urine. Collect the next voided urine. May have breakfast and undergo regular activity between first and second morning urine.
- Can also use random (spot) urine samples.

Tip: protease inhibitors and antimicrobial agents are recommended in the process of urine collection for exosome isolation. Three kinds of protease inhibitors can be used in the collection.

Homemade Cocktail (Volume per 50 ml of Urine)

- 1.67 ml 100 mM NaN_3.
- 2.5 ml AEBSF (2.75 mg ml^{-1} in dd-H_2O, stable at −20 °C) or 2.5 ml PMSF (2.5 mg ml^{-1} in isopropanol, stable at −20 °C).
- 50 μl Leupeptin (1 mg ml^{-1} in ddH_2O, stable one week at 4 °C, six months at −20 °C).

Protease Inhibitor Cocktail (Sigma P2714) (Volume per 50 ml of Urine)

- Add one bottle of cocktail to 5 ml of dd-H_2O (store at −20 °C).
- Add 625 μl to urine sample (12.5 μl ml^{-1} of urine).

Complete Mini (Roche 11 853 153 001)

- Add one tablet for up to 100 ml of fresh urine.
- Make sure to completely dissolve the tablet by swirling before urine storage.

19.2.1.2 Urine Storage

- Add protease inhibitors in urine samples (see Section 19.2.1.1).
- If possible, process fresh urine samples as soon as possible.
- If the urine samples cannot be processed within 2 hours, the samples should be frozen at −80 °C (*not* at −20 °C) (Figure 19.2).
- Store urine samples in 2–10–50-ml aliquots at −80 °C as soon as possible after collection.

19.2.1.3 Thawing and Handling

- Frozen urine samples should be thawed at room temperature (requires ≈3 hours for a 50-ml urine sample). Avoid prolonged thawing on ice (4 °C).
- While urine is thawing (i.e., is still a mixture of ice and water), extensively and vigorously vortex for 1 minute.
- After the sample completely thaws, vigorously vortex for an additional 30 seconds, and then proceed to differential centrifugation for urinary exosome isolation.
- Insufficient vortexing will result in major loss of urinary exosomes (Figure 19.2).

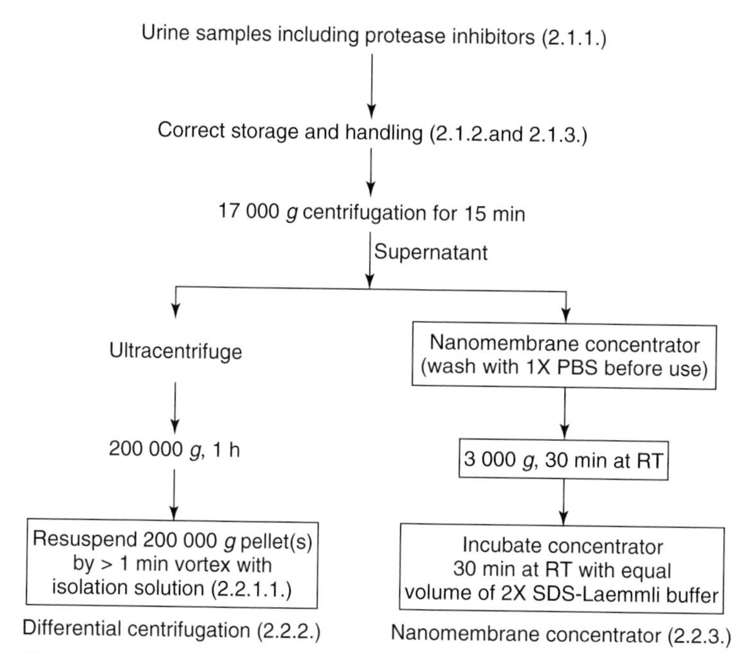

Figure 19.3 Two ways to isolate urinary exosomes. Urinary exosomes can be isolated by ultracentrifugation or nanomembrane concentrator.

19.2.2 Isolation of Exosomes from Urine

The two methods discussed here use differential centrifugation as the means to isolate urinary exosomes from urine. The first step in both methods is a 17 000g centrifugation step that removes whole cells, casts, large membrane fragments, and debris. The resulting supernatant fraction is subjected to a second centrifugation step (i) by ultracentrifugation at 200 000g or (ii) by filtration with a commercially available nanomembrane concentrator at 3000g (Figure 19.3).

19.2.2.1 Solution and Materials for Urinary Exosome Isolation

Isolation Solution (50 ml)

- 10 mM triethanolamine (MW 185.7) 0.093 g
- 250 mM sucrose (MW 342.3) 4.28 g
- dd-H_2O 45 ml
- Adjust pH to 7.6 with 1 N NaOH ~220 µl
- Add dd-H_2O to 50 ml
- Add protease inhibitors on the day of isolation: 2.5-ml AEBSF, 50-µl leupeptin.

Materials

- Tabletop centrifuge with swinging bucket rotor (relative centrifuge force or RCF can reach 3 000g).
- High-speed centrifuge and rotor (RCF can reach 17 000g).
- Ultracentrifuge and rotor (RCF can reach 200 000g).
- Nanomembrane concentrator, Vivaspin 20-PES 100 000 MWCO (Sartorius VS2041)
- Balance tubes, if necessary.

19.2.2.2 **Ultracentrifugation**

Figure 19.3, Left Panel

- Centrifuge urine sample at 17 000g, for 15 minutes at 25 °C (fresh urine samples) or at 4 °C (urine samples stored at −80 °C).
- Transfer 17 000g supernatant to one or more ultracentrifuge tubes, and mark each tube so that the expected location of the pellet on the tube can be found after the spin.
- Ultracentrifuge the supernatant at 200 000g for 1 hour at the given temperature.
- Discard supernatant and resuspend the first tube's pellet (typically light yellow, but not always visible for lower volumes) with isolation solution. Vortex for 30 seconds, and transfer the suspension (as much as possible) to the second tube, resuspend pellet, and repeat for every tube.
- An important practical consideration is how many tubes to use per urine sample versus how many centrifugation runs. Using serial centrifugation runs, as outlined above, maximizes the yield, but larger starting volumes of a 17 000g supernatant fraction would make the number of runs impractical. However, using a large number of ultracentrifuge tubes to minimize the number of runs can dramatically reduce the yield, as the pellets would be distributed in several tubes. Therefore, a compromise between the two extremes of one run/many tubes versus one tube/many runs is most practical.

19.2.2.3 **Nanomembrane Concentrator**

Figure 19.3, Right Panel

- Prewash nanomembrane concentrator with one volume of PBS to remove glycerol and other preservatives and centrifuge at 3000g in a swinging bucket rotor, at 25 °C. The failure to prewash is a frequent reason for variable yields of urinary exosomes.
- Centrifuge the urine samples at 17 000g, for 15 minutes at 25 °C (fresh urine samples) or at 4 °C (urine samples stored at −80 °C).
- Add 20 ml of urine supernatant to nanomembrane concentrator. Centrifuge at 3 000g, at 25 °C for 30 minutes.

- Add an equal volume of 5× SDS-Laemmli buffer to retentate while still in concentrator. Shake at room temperature for 30 minutes.

19.2.3 Proteomic Analysis of Urinary Exosomes

The proteins existing in urinary exosomes can be identified by proteomic analysis including LC-MS/MS after 1-D PAGE [3] and MALDI-TOF-TOF after 2-D DIGE [4]. THP, an abundant protein in urinary exosomes, can obscure low-abundance proteins and decrease the yield of identified proteins. The depletion of THP is required for proteomic analysis before urinary exosomes are separated by gel [3, 4].

19.2.3.1 Solutions for Proteomic Analysis of Urinary Exosomes

Isolation Solution (Same as Section Isolation Solution (50 ml), for 1-D PAGE)

5X SDS-Laemmli (for 1-D PAGE and Validation by Western Blot Analysis)

Lysis Buffer 100 ml (for 2-D DIGE)

• 15 mM Tris	181.7 mg
• 7 M urea	42.0 g
• 2 M thiourea	15.22 g
• 4% CHAPS	4 g
• Dilute HCl adjust pH to	8.5
• Add dd-H_2O to	100 ml

19.2.3.2 Depletion of Tamm–Horsfall Protein

- Resuspend urinary exosomes in isolation solution.
- Add reducing agent DTT to make the final concentration of 200 mg ml^{-1}.
- Incubate at 95 °C for 2 minutes.
- Bring the volume of sample up to an appropriate level with isolation solution, and then repeat ultracentrifugation (200 000g for 1 hour) or use nanomembrane concentrator (3 000g for 30 minutes).
- Solubilize urinary exosomal proteins by Lysis buffer for 2-D DIGE or use isolation solution for 1-D PAGE.

19.2.3.3 Determination of Urinary Exosomal Protein Concentration

Measurement of Protein Concentration

- If exosomes are in isolation solution for 1-D SDS/PAGE, measure protein concentration by bicinchoninic acid (BCA) protein assay kit (PIERCE).

- If exosomes are in Lysis buffer for 2-D DIGE, measure protein concentration by 2-D quant protein assay kit (Amersham).

Optimal Protein Concentration

- 1-D SDS/PAGE: Concentration $\sim$1–2 μg μl^{-1}
- 2-D DIGE: Concentration $\sim$5 μg μl^{-1}

19.2.3.4 LC-MS/MS and MALDI-TOF-TOF

Guidelines for Amounts of Starting Material Needed for Proteomic Analysis of Urinary Exosomal Proteins (Figure 19.4) Although the amount of protein needed for each proteomic technique depends on a number of variables, we provide a general outline of what protein amounts are needed. Around 300 μg of urinary exosomal proteins is required for protein identification by LC-MS/MS. For more sensitive LC-MS/MS techniques, such as LTQ, only $\sim$150 μg of exosomal proteins is required for protein identification. For protein identification by 2-D DIGE/MALDI-TOF-TOF, 500–1000 μg of exosomal proteins is required.

Normalization of Gel Loading In general, the important question is whether a potential biomarker protein is altered in a diseased state relative to normal controls. Since water excretion is highly variable and would dilute exosomes to a variable degree, a simple comparison of the concentration of the potential biomarker, that is, amount per unit volume, is inappropriate. Hence, it is necessary to measure either the rate of excretion of the candidate marker protein (where time is the normalizing variable) or to normalize sample concentration by dividing by the concentration of something that does not change with time. In clinical studies, endogenous creatinine concentration is commonly used for the latter purpose. However, endogenous creatinine concentration can vary substantially among subjects and, consequently, is considered as a relatively crude normalizing variable. Nevertheless, when a potential biomarker is expected to change by very large amounts, for example, an order of magnitude or more, urinary creatinine concentration provides a sufficient means of normalization. Another potential choice is to normalize by total protein concentration. However, if a relatively abundant protein varies considerably between patients and controls, the result will be an artifactual change in the proposed marker. In one study, there was a large variability of urinary protein content, which derived from prostatic proteins. This problem was overcome by using time normalization, pooling samples within each group so that the average total protein content was similar [7]. Another option is to normalize the urinary concentration of a biomarker protein by the concentration of exosomal marker proteins such as TSG101 or Alix. This presumes, however, that total exosome excretion rate is invariant among subjects, a possibility that has not been tested. In the final analysis, normalization by time (measurement of the amount of the biomarker protein per unit time) is superior if it can be completed.

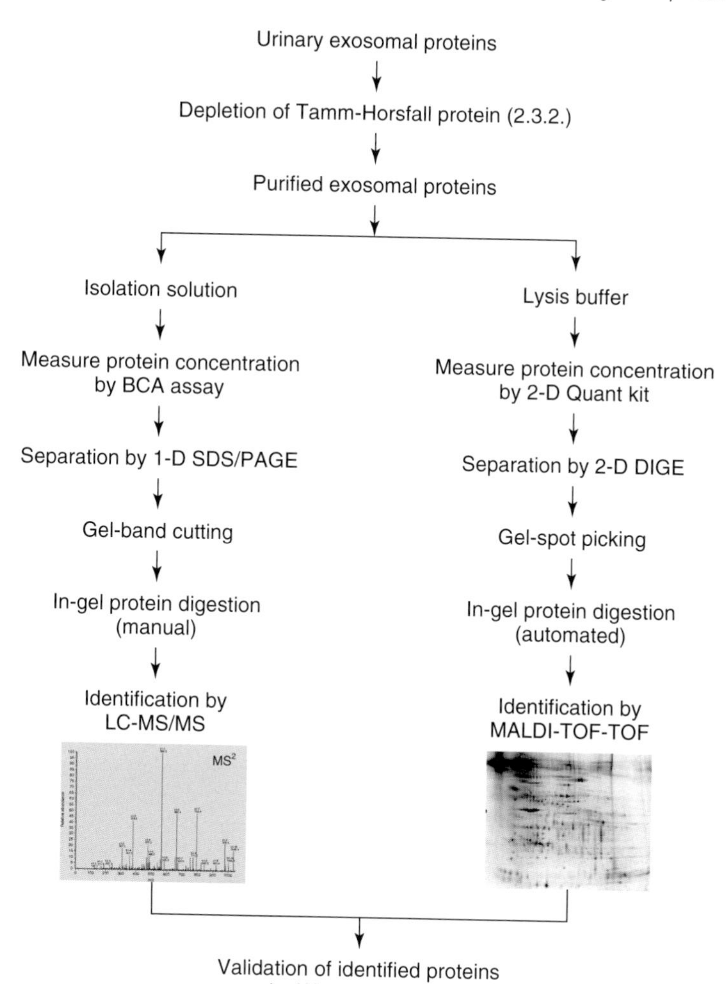

Figure 19.4 Urinary exosomes for proteomic analysis. MALDI-TOF-TOF after 2-D DIGE and LC-MS/MS after 1D SDS/PAGE can be used to identify proteins in urinary exosomes.

19.3
Limitations and Future Directions of Urinary Exosome Proteomics

Ultracentrifugation is currently the standard method to purify urinary exosomes, but the expense and low capacity of ultracentrifuges, both in terms of volume and number of samples, limit its use. In many cases, especially in clinical laboratories, ultracentrifuges are not available. Nanomembrane concentrators are commercially available and simplify the isolation procedure, but recovery of exosomal proteins is not uniform from one protein to another, which can pose quantitative problems for proteomics discovery experiments. Further, nanomembrane concentrators can retain nonexosomal proteins that are present as aggregates or high-molecular-weight complexes.

Proteomic analysis with 2-D gel electrophoresis, particularly using the DIGE method, is good for quantification of changes between control and disease groups, but many membrane proteins do not have sufficient solubility in lysis buffer to enter the first dimension, making it difficult to analyze the major protein component of urinary exosomes [4]. 1-D SDS/PAGE, followed by LC-MS/MS has the capacity of high-throughput for protein identification but it is difficult and time consuming to quantify changes between normal and diseased groups.

Because of the limitations of 1-D SDS/PAGE followed by LC-MS/MS and 2-D DIGE followed by MALDI-TOF-TOF are different; they are complementary methods, each of which may uncover different candidates. The crucial step after completing the discovery process is validation, which generally requires some form of immunoassay. In order to take advantage of the rich information that is readily accessible in exosomes, assays are needed for membrane proteins that are quantitative and high-throughput, so that they can be adopted into clinical diagnostic laboratories. Two promising avenues are modifications of the ELISA format and adaptations of flow cytometry. Despite the challenges in both discovery and validation, urinary exosomes have a largely untapped potential as a source of biomarkers that can reveal the health status of cells along the entire length of the nephron.

References

1. Hoorn, E.J., Pisitkun, T., Zietse, R., Gross, P., Frokiaer, J., Wang, N.S., Gonzales, P.A., Star, R.A. *et al.* (2005) Prospects for urinary proteomics: exosomes as a source of urinary biomarkers. *Nephrology (Carlton)*, **10**, 283–290.
2. Pisitkun, T., Johnstone, R., and Knepper, M.A. (2006) Discovery of urinary biomarkers. *Mol Cell Proteomics*, **5**, 1760–1771.
3. Pisitkun, T., Shen, R.F., and Knepper, M.A. (2004) Identification and proteomic profiling of exosomes in human urine. *Proc Natl Acad Sci U S A*, **101**, 13368–13373.
4. Zhou, H., Pisitkun, T., Aponte, A., Yuen, P.S., Hoffert, J.D., Yasuda, H., Hu, X., Chawla, L. *et al.* (2006) Exosomal Fetuin-A identified by proteomics: a novel urinary biomarker for detecting acute kidney injury. *Kidney Int*, **70**, 1847–1857.
5. Zhou, H., Yuen, P.S., Pisitkun, T., Gonzales, P.A., Yasuda, H., Dear, J.W., Gross, P., Knepper, M.A. *et al.* (2006) Collection, storage, preservation, and normalization of human urinary exosomes for biomarker discovery. *Kidney Int*, **69**, 1471–1476.
6. Cheruvanky, A., Zhou, H., Pisitkun, T., Kopp, J.B., Knepper, M.A., Yuen, P.S., and Star, R.A. (2007) Rapid isolation of urinary exosomal biomarkers using a nanomembrane ultrafiltration concentrator. *Am J Physiol Renal Physiol*, **292**, 1657–1661.
7. Holly, M.K., Dear, J.W., Hu, X., Schechter, A.N., Gladwin, M.T., Hewitt, S.M., Yuen, P.S. and Star, R.A. (2006) Biomarker and drug-target discovery using proteomics in a new rat model of sepsis-induced acute renal failure. *Kidney Int*, **70**, 496–506.

20
2-D Difference In-Gel Electrophoresis (2-D DIGE) and Urinary Proteomics

Satish P. RamachandraRao, Michael A. Shaw, and Kumar Sharma

20.1
Introduction

Applications of urinary proteomics to address questions related to renal pathophysiology will be fundamental in identification of novel pathways and biomarkers. Two-dimensional difference in-gel electrophoresis (2-D DIGE) has been shown to reliably quantify differences between individual protein expression levels in urine proteomes in normal health vis-à-vis clinical conditions. In this chapter, we discuss the promise that 2-D DIGE proteomics holds for urine protein studies, in particular for detecting biomarkers for kidney diseases, and its methods.

20.2
Proteomics and 2-D PAGE

The recent biomedical research has witnessed the development of a series of omes: genome (the DNA sequence of an organism), transcriptome (the messenger RNA (mRNA) expressed at a given time in a cell), and the proteome (the protein equivalent). Whereas the first of these is relatively inert, the other two are highly dynamic and vary greatly according to endogenous and exogenous conditions [1]. Likewise, the technology used to address questions related to DNA and RNA such as PCR [2], shotgun sequencing [3], and fluorescently tagged nucleic acid sequencing are straightforward [4], owing to the fairly homogeneous nature of DNA and RNA. The amenability to automation of the protocols used for studying them makes their analysis almost a routine task. The physicochemical diversity of proteins on the other hand, coupled with the lack of an amplifying method analogous to PCR, makes it an almost impossible venture to develop a singular universal separation method applicable to all proteins.

Renal and Urinary Proteomics: Methods and Protocols. Edited by Visith Thongboonkerd

ISBN: 978-3-527-31974-9

A classic method to assess the proteome is via two-dimensional polyacrylamide gel electrophoresis (2-D PAGE) [5, 6]. Proteins are separated in the first dimension according to their isoelectric point by a phenomenon termed isoelectric focusing (IEF). Proteins are amphoteric molecules carrying positive, negative, or zero net charge, depending on the pH of their surroundings. The isoelectric point designated pI is the specific pH at which the net charge of the protein is zero. During IEF, the proteins migrate to a point in the gel where the pH causes the net charge on the protein to become zero. In the second dimension of 2-D PAGE, proteins are separated according to their mobility in a porous gel; when the porosity of the gel is a constant, under most circumstances, the distance traversed by the protein is inversely proportional to its size, and hence proteins already separated in the first dimension by pI will get separated by size in the second dimension. The 2-D PAGE not only allows resolution of more than 10 000 proteins but also can differentiate among many posttranslationally modified forms of a protein. Upon visualization, when a protein spot in the disease proteome shifts position or changes in density when compared to its normal counterpart, it is considered to play a putative role in disease manifestation/progression. Depending on its hierarchy in the biochemical pathway and its interactions with other proteins, its role in disease could be further elucidated subsequent to bioinformatic analyses.

Though first described in 1975 [5, 6], the 2-D PAGE technology has not been fully utilized to its potential until recently. A combination of two developments has changed this scenario. Primarily, the increased reproducibility of 2-D proteomes that resulted from the introduction of commercial immobilized pH gradient (IPG) first dimension gels [7] made very accurate comparison between two gel images possible. Furthermore, the flexibility of range and linearity of pH allowed for an intelligent choice of IPG strips to help determine unwanted and interfering proteins in the preliminary stages of analysis. Secondly, the technological development of mass spectrometry in an unprecedented manner and the development of the protein databases increased the protein identification throughput. Coupled together, these two giant leaps in the history of protein analyses paved the way for discovery of novel biochemical pathways with an extraordinary speed.

More recently, the technological breakthrough of fluorescently labeling two different populations (such as disease and normal samples) of proteins, and resolving them by 2-D PAGE on a single gel subsequent to combining the two types of proteins, termed eliminates gel–gel variation inherent to 2-D PAGE. First described by Unlu *et al.* [8], 2-D DIGE has also taken sensitivity of detection of proteins to higher levels. In cases where more than three samples need to be compared, the internal standard will tie the multiple gels into a single experiment. In the following sections of this chapter, our discussion is focused on fundamental questions of how urine proteome is a good information resource in regard to kidney function and lends itself to 2-D DIGE studies, the advantages that 2-D DIGE holds over conventional 2-D PAGE, its inherent limitations to the purpose of novel biomarker discovery, and some of the strategies to get around them.

20.3
First of All, Why Study Urine Proteome?

Human urine plays a central role in clinical diagnostics. Systemic alterations in metabolic and biochemical profiles that affect urinary protein excretion can be studied using urine as the starting point [9]. It is an information-rich biofluid whose collection per se is completely noninvasive and almost effortless. It is available in most patients and has a number of advantages as a fluid for biomarker investigation. Since urine collection is a noninvasive procedure, patients/subjects are more open to its collection compared to blood or solid tissue. This ease of collection also enables multiple samples to be collected from a single patient/subject over a period of time to monitor temporal phenomena, such as disease progression or regression [10]. Follow-up studies aiming to look at long-term effects of a drug are rendered less difficult, largely due to this reason. Although whole urine sample analysis may be highly variable and contain substantial amounts of blood proteins [11], the advantages of using urine for biomarker investigation outweigh the limitations such as the one mentioned above. A major disadvantage of using serum/plasma as a biomarker source is the wide range of the relative abundance of proteins, which is shown to span 9 orders of magnitude [12]. General protein abundance concerns to the extent of serum/plasma are not presented regularly in urine proteomics. This, coupled with the ease with which urine can be collected during disease progression/regression over a period of time, makes urine an ideal source of biomarkers.

Under normal physiological conditions, protein amounts in the range of 1–10 mg per voiding are excreted with the urinary fluid [9]. Restrictions are imposed on passage of plasma proteins in the molecular size (M_r) range above ~40 kDa during the filtration process in the glomeruli. The three-component renal glomerular filtration apparatus composed of (i) the fenestrated endothelium, (ii) the glomerular basement membrane (GBM), and (iii) the epithelial cells with characteristic foot processes known as is responsible for this restriction in filtration [13]. Proximal renal tubules not only reabsorb proteins below 40 kDa from the glomerular filtrate but also degrade them. Diseases that adversely affect the function of glomeruli and the tubules, diabetes being the foremost among them, cause excess losses of proteins in the urine, resulting in proteinuria [14, 15]. Other factors affecting glomerular function include renal malignancies, inflammation, and nephrotoxic agents [15]. A number of proteins have been measured in serum to evaluate functional failure of glomeruli (such as albumin, cystatin C, α-1 microglobulin, and β-2 microglobulin) [16], or in urine, where increased excretion of albumin, microalbuminuria, has been observed [17–19]. We have shown that TGF β-1, a multipotent tissue scarring cytokine, is elevated in the diabetic urine [20]. Cystatin C, α-1 microglobulin, and β-2 microglobulin are a few other low M_r proteins bearing direct correlation to disturbed tubular function. The urinary protein content thus serves as a reliable index of the renal performance, and changes in kidney and urogenital tract functions are detected in urine proteome [21]. Diabetic nephropathy (DN) is the leading cause of end-stage renal disease in the United

States [22] accounting for nearly 50% of all new cases [23]. The annual medical cost of treatment of diabetic patients with end-stage renal disease is expected to be ~30 billion US dollars over the next decade [24–27]. It is therefore imperative to identify the predictors of DN risk early in the course of the disease. In view of the foregoing, studying urine proteomes from DN patients and contrasting them with those from the healthy age- and gender-matched volunteers using proteomic analysis is a good starting point for biomarker discovery. For the purpose of this chapter, however, we focus on the role of 2-D DIGE in urine proteomic analysis and its potential to lead to biomarker discovery.

20.4
Initial Considerations: Important Methods for Human Urine Collection

Optimal handling of urine is of utmost importance. It is likely that many proteins may be metabolized by proteases in urine if not frozen rapidly or collected with protease inhibitors. There also appears to be distinct differences in urine proteome pattern from normal females versus normal males. Therefore, it would also be appropriate to separately analyze urine proteomes based on gender. As urine protein concentration could vary widely in random collections, variability of urine proteome would be minimized with timed collections. Table 20.1 summarizes different methods of regularly used urine collection procedures, and their respective advantages/disadvantages.

20.5
Concentrating Urine Proteins and Removal of Salts and Other Interfering Compounds

Protein solubility is a function of three main factors, namely hydropathy, pI and M_r, when temperature and salt concentration are constant [40]. Urine is generally composed of many macromolecules such as carbohydrates, lipids, and nucleic acids in addition to proteins. Urine constitution actively reflects the metabolite consumption of the individual. Salt concentration in urine from each patient is different; hence, the extent to which proteins will be soluble in these samples also will be different. There is additional heterogeneity in the sample from each individual with regard to the physical health, gender, the vitamin richness of food intake, dilution due to water consumption during the night previous to the morning of sample collection, and so on. One option is to use the concentration of protein in unfractionated urine to load gels, although many control urine samples are dilute and may require additional processing to yield enough protein for further investigations. This is a critically important factor to be considered during sample preparation prior to their resolution by 2-D DIGE.

Protein precipitation provides the simplest, fastest, and most cost-effective method for concentrating urinary proteins. It is also the most commonly used

Table 20.1 Methods for urine collection.

Type of urine collection	Features/advantages/disadvantages	References
Random spot collection	Suitable for creatinine clearance estimation	[28, 29]
24-h urine collection	More representative of physiologic status, but has the disadvantage of specimen collection from patients in a nonsupervised manner, thus may reduce consistency of results	[30, 31]
First void collection	Usually hypertonic due to dehydration by kidney overnight; the sample may present risk of bacterial contamination	[32]
Collection by bladder catheterization	Urethral catheterization is generally performed only under extraordinary circumstances, as either a therapeutic or a diagnostic procedure. Therapeutically, catheters are placed to decompress the bladder in patients with acute or chronic urinary retention. In addition, catheters are placed to facilitate bladder irrigation in patients with gross hematuria. Diagnostically, urinary catheters are placed to obtain an uncontaminated urine sample for microbiologic testing, to measure urinary output in critically ill patients or during surgical procedures, or to measure postvoid residuals	[33–38]
Midstream collection	Initial stream serves to flush contaminating cells and microbes from the outer urethra; hence, contamination and false positives are avoidable using midstream urine	[39]

method to remove contaminants and other interfering substances from urine samples. Table 20.2 summarizes some of the routinely practiced methods of removal of contaminants from protein preparation. Specifically, nonproteinaceous impurities including high concentrations of salts, metabolic wastes, lipids, and some small molecules decrease the quality/resolution of the 2-D gel and are routinely removed before analysis. Several protocols have been developed to improve the quality of the urinary proteome maps. Though acetone precipitation method is standardized and has been giving consistently good recovery for urine protein in our laboratory [41], ethyl alcohol precipitation, which results in better yields [42], can also be used. Important factors that need to be considered when ethanol precipitation is attempted for urine protein preparation are the presence of contaminants such as lipids and carbohydrates that tend to contribute to spurious values. Protein quantification methods featuring protocols to remove lipids and carbohydrates will help

Table 20.2 Methods for removal of contaminants from urine proteins.

Method	Comments/remarks	References
Acetone precipitation	Effectively removes most lipids; generally results in protein loss anywhere between 20 and 50%; not suitable for very small quantities of protein	[40, 43]
Ethanol precipitation	Removes most contaminants, handling losses are minimal, usually between 10 and 20%; softer method than acetone precipitation	[42]
Methanol chloroform precipitation	Time-tested method for nonproteinaceous contaminant removal, protein recovery ranges anywhere between 65 and 80% of the starting material. Good for even minute quantities of protein, and can be easily scaled up	[44]
Salting out by ammonium sulfate precipitation	In the presence of high salt concentrations, proteins tend to aggregate and precipitate out of solution. Many potential contaminants (e.g., nucleic acids) will remain in solution. This method is a good starting point for salting out some of the urinary proteins, but suffers from the drawback that it needs additional ammonium sulfate removal step usually by dialysis prior to subsequent analyses; clean protein is not as easily available as in the above methods	[45–47]
TCA precipitation	Though trichloroacetic acid (TCA) is a very effective protein precipitating agent, proteins may be difficult to resolubilize. Extended exposure to this low pH solution is shown to degrade or modify protein	[48–50]

to further purify the protein fraction and may yield identification of low-abundance proteins.

20.6 Further Considerations: Why Remove Abundant Proteins?

One of the most revisited problems in proteomics is the dynamic expression range or the relative protein abundance found in the biofluids. A few proteins dominate the landscape and often obliterate the signal of the low-abundant proteins, irrespective of the power of resolution of 2-D DIGE [51]. This may lead to spurious results; often, the rare protein that may play a significant role in disease goes undetected. If the identity of the abundant proteins is known, they can be removed prior to the analysis of the rest of the proteome to improve the detection of the relatively less abundant proteins. Historically, albumin and IgG were the first proteins to be removed as they accounted for nearly three-fourths of the total proteome [52–54]. Presently, we routinely remove up to 12 most abundant proteins that account for up to 85–90% of the total proteome [55, 56]. This enriches the sample with the rare and less-abundant proteins, and simplifies the analytical protocol subsequent to protein separation.

20.7 2-D DIGE Analysis to Screen Urine Proteomes for Candidate Biomarkers

The principal advantage of 2-D DIGE is that up to three samples being compared are run on the same gel, thus completely eliminating gel–gel variation issues. In experiments involving greater number of samples, the internal standard serves to link multiple gels into one single experiment. The concept of internal standard can be likened to nothing more than a common denominator, against which many different numerator values are computed and compared. Although the singular internal standard does not contribute anything of biological value, technically it serves as a very good purpose to unite multiple-gel experiments. Furthermore, 2-D DIGE not only enables visualization of more than 1000 protein spots from up to three different samples (e.g., disease, normal, and internal standard) simultaneously in a single gel but also quantifies the up-regulation or down-regulation of a disease-sample protein spot as compared to its normal-sample counterpart. A limitation of 2-D DIGE is that, since it is a gel-based platform, its protein-resolving ability is limited by physical factors. Proteins between 10 and 100 kDa are usually reasonably resolved by 2-D DIGE, but proteins of higher size are limited by their capacity to enter the first-dimension gel. Although most proteins encountered in physiology fall in the 10–100 kDa category, disease conditions typically characterized by protein cleavage resulting in peptides of less than 5 kDa virtually go either undetected or are lost due to extensive handling.

20.8 Experimental Design of DIGE

There are two sets of Cy dyes available, namely, minimal and saturation dyes. For the purpose of this review, we confine our discussions to minimal dyes. An example is a typical experiment in which individual proteins from urine samples of five DN patients are compared with the proteins from urine samples of age-matched nondiabetic healthy volunteers. DIGE methodology is applied to the above sample set as follows. Three spectrally resolvable dyes, namely, Cy2, Cy3, and Cy5, are available, and hence up to three different samples can be resolved per gel. Since these dyes are size- and charge-matched, any protein labeled with Cy2 from one source will move to the same spot as its counterpart from a second source labeled with Cy3, which in turn will move to the same spot as its counterpart from yet another source labeled with Cy5. Owing to the spectral resolvability of the dyes,

Table 20.3 CyDye labeling in 2-D DIGE study.

	Gel 1	Gel 2	Gel 3	Gel 4	Gel 5
Cy 2	Internal standard	Internal standard	Internal standard	Internal standard	Internal standard
Cy 3	Patient 1	Normal	Patient 3	Normal	Patient 5
Cy 5	Normal	Patient 2	Normal	Patient 4	Normal

the protein labeled with different dyes will fluoresce to different extents at different wavelengths, thus accurately quantifying the differences in expression levels of proteins from different sources. When there are more than three samples being investigated such as in the following example, to enable comparison of all samples using only three dyes, the concept of internal standard is developed. As shown in Table 20.3, five patient samples can be compared with the normals. The internal standard is prepared by mixing together equal amounts of each sample in the experiment and including this mixture on each gel. Though five gels are run, by virtue of the internal standard, the data of all five patients are combined into a single experiment. The internal control serves as a common denominator against which the protein expression levels for individual patients are compared.

The comparisons can be expanded to as many samples as required, thus resulting in DIGE multiplexes. A notable feature of the fluors is that dye swapping is possible for samples being compared, ensuring against any dye-specific bias in labeling a particular protein. It is clear from the foregoing that when it becomes necessary, any one of the cell contents from Table 20.3 can be replaced by technical replicates or biological replicates, and the individual protein levels can be compared against the internal standard for statistical considerations. Other advantages of CyDye DIGE fluor labels include their pH insensitivity due to which the fluorescent signal of the dye or the hydrodynamic property of the protein remains unaffected over a wide pH range used during first-dimension separation (IEF). This also ensures equivalent migration in SDS gels. In addition, since the fluor dyes are highly sensitive, bright, and photostable, there is a minimal loss of signal during labeling, separation, and scanning.

20.9 DIGE Components and Associated Considerations

20.9.1 Fluorophores

In addition to the above-mentioned properties of the fluorophore dyes, their sensitivity offers subnanogram level detection of proteins, which practically is as good as silver staining for protein detection. Moreover, the CyDye platform shows linearity over 4 orders of magnitude of dynamic range, which makes it far more desirable than other methods that do not show linearity over more than 2 orders of magnitude.

20.9.1.1 Sample Labeling with Minimal Dyes and Subsequent Multiplexing

It is important that primary amines (e.g., ampholytes) and thiols (e.g., DTT) are excluded from the sample until after labeling with the dyes has been completed. For best results, the pH of the labeling reaction at 8.5 and a protein concentration of 5 mg ml^{-1} are shown to be optimal. The dimethylformamide (DMF) used to reconstitute the fluors should be high-quality anhydrous (>99.8% pure). It must not become contaminated with water (<0.005% H_2O); molecular sieves will help keep the DMF in an anhydrous condition. A concentration of 1 nmol µl^{-1} of fluor (e.g., add 25 µl DMF to 25 nmol fluor) is desirable for an efficient labeling reaction. In our experience, the concentrated stock solution is stable for six to eight weeks when stored at −20 °C. To label proteins, the concentrated stock solutions need to be diluted to a working fluor concentration of 400 pmol µl^{-1} using DMF (e.g., add 2 µl of concentrated stock fluor to 3 µl of DMF); and the protein : fluor ratio is generally maintained at 50 µg protein to 400 pmol fluor. A volume of 1 µl of working fluor solution (400 pmol µl^{-1}) is added to a volume of sample containing 50 µg of protein, mixed thoroughly by vortexing, and collected by centrifugation at the bottom of the tube. This reaction mixture is incubated on ice for 30 minutes in the dark. To quench the labeling reaction, 1 µl of 10 mM lysine is added to the reaction mixture, mixed well, and left on ice for 10 minutes in the dark. A pooled internal standard should be created from all of the samples. This needs to be sufficient for inclusion in every gel; it is customary to label the internal standard with Cy2 for uniformity of analyses. As shown in Table 20.3, Cy 3 and Cy 5 can be swapped for labeling the proteins that are under investigation. The labeled sample can either be processed immediately or stored for up to three months at −70 °C in the dark.

The reaction kinetics of the protein labeling protocol outlined in the previous sections are standardized such that only 1–2% of the total protein gets labeled, leaving approximately 98% of the total protein free; this ensures unhindered mass spectral analysis of the protein in the subsequent stages of protein identification. After labeling with different dyes, the protein samples are combined for multiplexing according to the experimental design, the sample mixture is then diluted further in equal volumes of sample buffer prior to resolving the individual proteins by 2-D PAGE, and left on ice for 10 minutes. The sample buffer contains 7 M urea, 2 M thiourea, 2% CHAPS (w/v), 2% IPG buffer or Pharmalyte (v/v) for IEF, 2% DTT (w/v).

20.9.2 2-D PAGE

The differentially labeled proteins are resolved by IEF in the first dimension and SDS-PAGE in the second dimension. Some of the ground-level considerations while performing 2-D PAGE are as follows. The presence of a pH gradient is critical to the IEF technique. Under the influence of an electric field in this pH gradient, a protein with a net positive charge will move toward the negative electrode (cathode), becoming progressively less positively charged as it moves through the pH gradient until it reaches its pI. Exactly the opposite will happen with a protein with a net negative charge: it will migrate toward the anode and become progressively less negatively charged until it reaches zero net charge. If a protein should diffuse away from its pI, it immediately gains charge and migrates back. This is the *focusing* effect of IEF, which concentrates proteins at their pIs and allows proteins to be separated on the basis of very small charge differences. The recently available commercial IPG strips feature covalently immobilized pH gradients that cannot drift. Owing to this, the protein separation is not only true to its actual pI but also the results obtained are perfectly reproducible, have superior resolution, and the observed differences are the effects due to biological variations. Biological phenomena are studied without interference from technical drawbacks, which was the bane of the tube gels using carrier ampholyte-generated pH gradients – the predecessors of IPG strips. Additionally, conducting high-quality IEF/PAGE is also achieved by prefractionation of proteins and/or using zoom IPG strips. A series of narrow-range IPG strips that have overlapping pI ranges will allow high-class resolution of protein of choice, helping to "zoom in" only on the pI of the protein of interest. Thus, an intelligent choice of IPG strips whereby only the proteins of interest are focused, leaving out the rest, can be incorporated into the protocol such that the protein under investigation is resolved very clearly in the 2-D PAGE. In this manner, the serious limitation of 2-D DIGE that it is good only for abundant proteins can be effectively dealt with. Depletion of highly abundant proteins is another mechanism by virtue of which the detectability of relatively less-abundant proteins can be enhanced multiplefold.

The batch-to-batch variation of the second-dimension resolution by SDS-PAGE can be solved by conducting simultaneous gel runs in multiple formats of 6 or 12 gels. This not only increases the reproducibility of results between gels but also reduces the differences in the intensity of fluor dyes if gels were run at different times. The SDS-PAGE gels in the 2-D format can be either poured manually using the glass plates or obtained commercially in the plastic-backed supplies. One of the drawbacks of the commercially available plastic-backed gels is that the plastic support may interfere with the scanning of one of the fluor dye channels, which typically results in high background masking the protein signal. As soon as the gel run is complete, they should be scanned for all three dyes, before fixing.

20.9.3
Fluorescent Scanner

Typhoon variable mode imager is a highly sensitive instrument that meets the specific needs of 2-D DIGE. All three fluor dyes, namely Cy2, Cy3, and Cy5, are detected with exceptional signal-to-noise ratio and a linear protein concentration response over 4 orders of magnitude [57]. The wide dynamic range provided by CyDye 2-D DIGE Fluor minimal dyes, in combination with the Typhoon Variable Mode Imager, enables production of data that is quantitative and reproducible, and offers a sensitivity down to 125 pg protein, compared to approximately 5 ng for silver staining [58]. Guides used for gel-alignment on the scanner enable correct positioning of gels on the scanner. This simplifies gel handling and reduces hands-on time. Two large-format gels can be scanned simultaneously, and the file outputs are separated automatically in a format that is compatible with DeCyder 2-D Differential Analysis Software.

20.9.4 DeCyder Software

Unless otherwise mentioned, our discussion with regard to the 2-D image analysis software is confined to the DeCyder Software developed by Amersham Biosciences and designed specifically for use with the 2-D DIGE technology.

20.9.4.1 DIA – Single Gel: Spot Detection and Normalization of Fluorophore Channels

Gel images are converted from the 2D Image Master software and opened in DeCyder differential in-gel analysis (DIA). All the spots on the gel are then automatically detected. Dust particles that have been detected are removed using a spot filter function and the remaining spots are manually confirmed. DIA is generally used for image development of a single gel having resolved proteins labeled with three distinct fluor dyes, and hence three images per gel. For each protein spot in each of these three images on the gel, DeCyder assigns a spot volume value, which directly reflects the quantity of the spot. Thus, there are three volume ratios for a single spot in Gel 1 featured in Table 20.3. The volume ratios of normal sample/internal standard and patient 1 sample/internal standard are computed by DeCyder DIA. The ratios between these two, in turn, will be used to compute the up- or down-regulation of the protein spot in patient. This process is called normalization. Normalization is necessary to allow a DeCyder DIA experiment to demonstrate changes in abundance for single proteins. Normalization uses relatively robust statistical techniques to find the set of proteins that should be considered for further analysis. It is notable at this point that though this process cancels out the internal standard value, when comparing the spot volume ratios among multiple gels, the concept of internal standard assumes significance. The protein data created thus are exported into XML toolbox from which point onward statistical analysis can be performed. Subsequently, a protein spot pick list is also generated and exported in a text file format.

20.9.4.2 BVA – Multiple Gels: Matching the Spots and Quantitation

When two or more gels are under consideration, the biological variation analysis or BVA is used for protein abundance analysis. DeCyder BVA analyzes multiple sets of spot maps detected in DIGE images by the DeCyder DIA software. Spots are automatically matched between the different spot maps and statistical analysis can then be performed using built-in functions. Protein spot images are generated in DIA. The matched protein spots corresponding to their counterparts in the other gels are then manually inspected, and a list of matched spot data is generated. The gel having maximum number of detectable protein spots is considered the primary (master) gel and the ones having lesser number of protein spots are secondary gels.

20.9.4.3 **Statistics**

For the purpose of statistical analysis, a collection of spot maps that, for the purposes of the experiment, cannot be broken down into further subgroups is considered a group in BVA. For example, three replica gels from one sample are considered one group, and three gels of three different samples treated with exactly the same experimental conditions can also be considered as a group. As many different groups can be generated with DeCyder BVA from the point of view of statistical analysis, the different types of statistical analyses include the following:

- Statistical analysis between two groups or two populations of groups can be performed using Student's *t*-test and calculating average ratio.
- Statistical analysis between all groups can be performed by one-way ANOVA.
- Statistical analysis between groups or populations of groups in an experimental design where there are two dependent factors (e.g., in a time-dose study of a drug in diabetics) can be performed by two-way ANOVA. The internal and mutual effect of the two factors can then be analyzed. Note that Student's *t*-test and the ANOVA measures can be performed in a paired experiment, where data points match between groups.

20.9.5 Pick List Generation

BVA export has a function by virtue of which an XML file with practically all data from the experiment can be generated. Information from this file can be imported using the DeCyder XML Toolbox to a central database or simply for viewing. Subsequent to statistical analysis, the most significantly dysregulated protein spots (e.g., value set at $p \leq 0.005$) will be considered proteins of interest, and can be picked. BVA also exports a tabbed text file holding the coordinates for the picked spots and the picking references. This file can be used directly by the Robotic Spot Picker. Picking lists can be exported either from the Spot Map currently assigned Pick or the Primary Spot Map.

20.9.6 Robotic Spot Processing

The robotic spot picker works on the 2-D gels based on the information provided by the pick list. It is computer-controlled using reference markers as the primary guides and the X and Y coordinates of the spots as the secondary guides and picks out precisely the gel area that purportedly contains the protein spots of interest. One of the advantages of using robotic spot pickers is that while manual spot picking requires the visualization of the individual protein spots robotic spot picking uses geometric *X*, *Y* coordinates and hence the extra step of staining and destaining of the gel is obviated. The picked gel plugs are processed in 96-well format. Robotic spot picker usually has attached features such as in-gel trypsinization of the picked protein spots, automated transference of the peptides released into the liquid bathing the gel plugs into a second 96-well plate, and automated drying of the solvent, so that peptides are in solid state in the second 96-well plate. These are then subjected to protein identification via mass spectrometry by either peptide mass fingerprinting or SEQUEST analysis.

Recent urinary proteomics studies performed by employing 2-D DIGE are summarized in Table 20.4. Some of the problems encountered in 2-D DIGE are the sensitivity of the fluor dyes that can detect the dust spots as false-positive protein spots; the presence of multiple proteins in a single spot; proteins undergoing post-translational modifications which cannot be identified as single protein since they all move as different spots; keratin contamination due to careless handling of the specimen; and protein fragmentation due to time-dependent protein degradation. Each of these problems can be solved by diligent work, for example, dust spots can

Table 20.4 2-D DIGE in urine proteomics.

Nature of the expression profiling	Major finding of the study	References
IgA nephropathy patients urine proteome vs. normal urine proteome	α_1-Microglobulin level in urine of IgA nephropathy patients was much lower compared to that of diabetic nephropathy or even normal urine	[59]
Type 2 diabetic patients urine proteome vs. normal urine proteome	Seven proteins were progressively up-regulated with increasing albuminuria; whereas four proteins exhibited progressive down-regulation. Majority of these 11 proteins were glycoproteins by nature.	[60]
Acute renal failure (sepsis-induced) rat model urine proteome vs. normal rat urine proteome	DIGE-enabled comparison and detection of changes in the excretion of urinary proteins correlated with the presence of sepsis-induced acute renal failure (ARF), which in turn led to the finding that inhibition of brush-border enzymes with actinonin could prevent sepsis-induced ARF	[61]
Diabetic nephropathy patients urine proteome vs. normal urine proteome	Of the 99 spots significantly regulated in the urine proteome of the diabetic samples, 63 were up-regulated and 36 down-regulated. One spot, identified as α_1-antitrypsin was consistently up-regulated by 19-fold across individuals in the diabetic group. Proteomic data for this protein was corroborated by ELISA and immunohistochemical analyses	[41]

be manually filtered out; zoom IPG strips in the first dimension in conjunction with large-format gels in second dimension can increase the power of resolution, thus minimizing the number of proteins in a single spot. Careful handling and quick processing of proteins, wearing gloves throughout, and frequently changing gloves can alleviate the contamination and degradation problems.

20.10 Conclusion

Several recent studies have shown that the urinary proteome contains a much greater variety of proteins than previously recognized [62, 63], some of which may be crucial biomarkers. Analysis of changes in relative abundances of urinary proteins has been shown to be informative not only with regard to changes in tubular and glomerular physiology but also to reflect global changes in renal physiology/morphology. Ideally, the sensitivity and selectivity of the techniques used to study urinary proteomes should enable the investigator to use the information-rich urine to unravel novel kidney disease biomarkers. Especially in the wake of recent and rapid developments in the technology of fluorescent labeling and tandem mass spectrometry, urinary proteomics has received significant attention and growth. 2-D DIGE-based proteomic analysis of urine is a reliable and robust platform for the discovery of biomarkers for kidney diseases alongside other gel-based and gel-free proteomic methodologies.

References

1. James, P. (1997) Of genomes and proteomes. *Biochem Biophys Res Commun*, **231**, 1–6.
2. Saiki, R.K., Scharf, S., Faloona, F., Mullis, K.B., Horn, G.T., Erlich, H.A., and Arnheim, N. (1985) Enzymatic amplification of beta-globin genomic sequences and restriction site analysis for diagnosis of sickle cell anemia. *Science*, **230**, 1350–1354.
3. Adams, M.D., Kelley, J.M., Gocayne, J.D., Dubnick, M., Polymeropoulos, M.H., Xiao, H., Merril, C.R., Wu, A., Olde, B., Moreno, R.F. *et al.* (1991) Complementary DNA sequencing: expressed sequence tags and human genome project. *Science*, **252**, 1651–1656.
4. Scherer, J.R., Kheterpal, I., Radhakrishnan, A., Ja, W.W., and Mathies, R.A. (1999) Ultra-high throughput rotary capillary array electrophoresis scanner for fluorescent DNA sequencing and analysis. *Electrophoresis*, **20**, 1508–1517.
5. O'Farrell, P.H. (1975) High resolution two-dimensional electrophoresis of proteins. *J Biol Chem*, **250**, 4007–4021.
6. Scheele, G.A. (1975) Two-dimensional gel analysis of soluble proteins. Characterization of guinea pig exocrine pancreatic proteins. *J Biol Chem*, **250**, 5375–5385.
7. Bjellqvist, B., Ek, K., Righetti, P.G., Gianazza, E., Gorg, A., Westermeier, R., and Postel, W. (1982) Isoelectric focusing in immobilized pH gradients: principle, methodology and some applications. *J Biochem Biophys Methods*, **6**, 317–339.
8. Unlu, M., Morgan, M.E., and Minden, J.S. (1997) Difference gel electrophoresis: a single gel method

for detecting changes in protein extracts. *Electrophoresis*, **18**, 2071–2077.

9. Pieper, R., Gatlin, C.L., McGrath, A.M., Makusky, A.J., Mondal, M., Seonarain, M., Field, E., Schatz, C.R., Estock, M.A., Ahmed, N., Anderson, N.G., and Steiner, S. (2004) Characterization of the human urinary proteome: a method for high-resolution display of urinary proteins on two-dimensional electrophoresis gels with a yield of nearly 1400 distinct protein spots. *Proteomics*, **4**, 1159–1174.
10. Smith, G., Barratt, D., Rowlinson, R., Nickson, J., and Tonge, R. (2005) Development of a high-throughput method for preparing human urine for two-dimensional electrophoresis. *Proteomics*, **5**, 2315–2318.
11. Saito, M., Kimoto, M., Araki, T., Shimada, Y., Fujii, R., Oofusa, K., Hide, M., Usui, T., and Yoshizato, K. (2005) Proteome analysis of gelatin-bound urinary proteins from patients with bladder cancers. *Eur Urol*, **48**, 865–871.
12. Corthals, G.L., Wasinger, V.C., Hochstrasser, D.F., and Sanchez, J.C. (2000) The dynamic range of protein expression: a challenge for proteomic research. *Electrophoresis*, **21**, 1104–1115.
13. Kalluri, R. (2006) Proteinuria with and without renal glomerular podocyte effacement. *J Am Soc Nephrol*, **17**, 2383–2389.
14. Marshall, T. and Williams, K.M. (1998) Clinical analysis of human urinary proteins using high resolution electrophoretic methods. *Electrophoresis*, **19**, 1752–1770.
15. Waller, K.V., Ward, K.M., Mahan, J.D., and Wismatt, D.K. (1989) Current concepts in proteinuria. *Clin Chem*, **35**, 755–765.
16. Delanghe, J. (1997) Use of specific urinary proteins as diagnostic markers for renal disease. *Acta Clin Belg*, **52**, 148–153.
17. Shihabi, Z.K., Konen, J.C., and O'Connor, M.L. (1991) Albuminuria vs urinary total protein for detecting chronic renal disorders. *Clin Chem*, **37**, 621–624.
18. Watts, N.B. (1991) Albuminuria and diabetic nephropathy: an evolving story. *Clin Chem*, **37**, 2027–2028.
19. Yudkin, J.S., Forrest, R.D., and Jackson, C.A. (1988) Microalbuminuria as predictor of vascular disease in non-diabetic subjects. Islington Diabetes Survey. *Lancet*, **2**, 530–533.
20. Sharma, K., Ziyadeh, F.N., Alzahabi, B., McGowan, T.A., Kapoor, S., Kurnik, B.R., Kurnik, P.B., and Weisberg, L.S. (1997) Increased renal production of transforming growth factor-beta1 in patients with type II diabetes. *Diabetes*, **46**, 854–859.
21. Sun, W., Li, F., Wu, S., Wang, X., Zheng, D., Wang, J., and Gao, Y. (2005) Human urine proteome analysis by three separation approaches. *Proteomics*, **5**, 4994–5001.
22. Centers for Disease Control and Prevention (CDC) (2004) State-specific trends in chronic kidney failure – United States, 1990-2001. *MMWR Morb Mortal Wkly Rep*, **53**, 918–920.
23. Ewens, K.G., George, R.A., Sharma, K., Ziyadeh, F.N., and Spielman, R.S. (2005) Assessment of 115 candidate genes for diabetic nephropathy by transmission/disequilibrium test. *Diabetes*, **54**, 3305–3318.
24. American Society of Health-System Pharmacists (2003) ASHP therapeutic position statement on strict glycemic control in patients with diabetes. *Am J Health Syst Pharm* **60**, 2357–2362.
25. Benjamin, S.M., Valdez, R., Geiss, L.S., Rolka, D.B., and Narayan, K.M. (2003) Estimated number of adults with prediabetes in the US in 2000: opportunities for prevention. *Diabetes Care*, **26**, 645–649.
26. Breyer, M.D. (2007) Diabetic nephropathy: introduction. *Semin Nephrol*, **27**, 129.
27. Greenwald, A. (2006) Current nutritional treatments of obesity. *Adv Psychosom Med*, **27**, 24–41.

28. Babazono, T., Takahashi, C., and Iwamoto, Y. (2004) Definition of microalbuminuria in first-morning and random spot urine in diabetic patients. *Diabetes Care*, **27**, 1838–1839.
29. RPAH Biochemistry Department (2002) Trace and Toxic Elements–Urine Collection Procedures, http://www.cs.nsw.gov.au/csls/handbook/FactSheetView.asp?Number=41.
30. Mushnick, R., (2006) 24-hour Urine Protein, http://www.nlm.nih.gov/medlineplus/ency/article/003622.htm.
31. Bottini, P.V., Ribeiro Alves, M.A., and Garlipp, C.R. (2002) Electrophoretic pattern of concentrated urine: comparison between 24-hour collection and random samples. *Am J Kidney Dis*, **39**, E2.
32. Molitch, M.E., DeFronzo, R.A., Franz, M.J., Keane, W.F., Mogensen, C.E., and Parving, H.H. (2003) Diabetic nephropathy. *Diabetes Care*, **26** (Suppl 1), S94–S98.
33. Schneider, R. (2004) Urologic procedures, in *Clinical Procedures in Emergency Medicine*, 4th edn (eds J.R. Roberts and J.R. Hedges), W.B. Saunders, Philadelphia, pp. 1086–1098.
34. Koraitim, M.M. (1999) Pelvic fracture urethral injuries: the unresolved controversy. *J Urol*, **161**, 1433–1441.
35. Cancio, L.C., Sabanegh, E.S. Jr, and Thompson, I.M. (1993) Managing the Foley catheter. *Am Fam Physician*, **48**, 829–836.
36. Saint, S. and Lipsky, B.A. (1999) Preventing catheter-related bacteriuria: should we? Can we? How? *Arch Intern Med*, **159**, 800–808.
37. Daneshmand, S., Youssefzadeh, D., and Skinner, E.C. (2002) Review of techniques to remove a Foley catheter when the balloon does not deflate. *Urology*, **59**, 127–129.
38. Canes, D. (2006) Male urethral catheterization. *N Engl J Med*, **355**, 1178–1179, author reply 1179.
39. Elley, K. and Semple, M. (1999) Collecting a mid-stream specimen of urine. *Nurs Times*, **95** (Suppl 1-2).
40. Thongboonkerd, V., McLeish, K.R., Arthur, J.M., and Klein, J.B. (2002) Proteomic analysis of normal human urinary proteins isolated by acetone precipitation or ultracentrifugation. *Kidney Int*, **62**, 1461–1469.
41. Sharma, K., Lee, S., Han, S., Lee, S., Francos, B., McCue, P., Wassell, R., Shaw, M.A., and RamachandraRao, S.P. (2005) Two-dimensional fluorescence difference gel electrophoresis analysis of the urine proteome in human diabetic nephropathy. *Proteomics*, **5**, 2648–2655.
42. Thongboonkerd, V., Chutipongtanate, S., and Kanlaya, R. (2006) Systematic evaluation of sample preparation methods for gel-based human urinary proteomics: quantity, quality, and variability. *J Proteome Res*, **5**, 183–191.
43. Barritault, D., Expert-Bezancon, A., Guerin, M.F., and Hayes, D. (1976) The use of acetone precipitation in the isolation of ribosomal proteins. *Eur J Biochem*, **63**, 131–135.
44. Wessel, D. and Flugge, U.I. (1984) A method for the quantitative recovery of protein in dilute solution in the presence of detergents and lipids. *Anal Biochem*, **138**, 141–143.
45. Carmel, R. (1974) A rapid ammonium sulfate precipitation technic for separating serum vitamin B12-binding proteins. Method and applications. *Am J Clin Pathol*, **62**, 367–372.
46. Doskeland, S.O. and Ogreid, D. (1988) Ammonium sulfate precipitation assay for the study of cyclic nucleotide binding to proteins. *Methods Enzymol*, **159**, 147–150.
47. Verhoeven, G., Heyns, W., and De Moor, P. (1975) Ammonium sulfate precipitation as a tool for the study of androgen receptor proteins in rat prostate and mouse kidney. *Steroids*, **26**, 149–167.
48. Lakatos, A. and Jobst, K. (1992) The precipitation of histone-proteins with trichloroacetic acid. *Clin Chim Acta*, **205**, 153–154.
49. Wright, A.P., Bruns, M., and Hartley, B.S. (1989) Extraction and rapid

inactivation of proteins from Saccharomyces cerevisiae by trichloroacetic acid precipitation. *Yeast*, **5**, 51–53.

50. Yeang, H.Y., Yusof, F., and Abdullah, L. (1995) Precipitation of Hevea brasiliensis latex proteins with trichloroacetic acid and phosphotungstic acid in preparation for the Lowry protein assay. *Anal Biochem*, **226**, 35–43.

51. Righetti, P.G., Boschetti, E., Lomas, L., and Citterio, A. (2006) Protein equalizer technology: the quest for a "democratic proteome". *Proteomics*, **6**, 3980–3992.

52. Fountoulakis, M., Juranville, J.F., Jiang, L., Avila, D., Roder, D., Jakob, P., Berndt, P., Evers, S., and Langen, H. (2004) Depletion of the high-abundance plasma proteins. *Amino Acids*, **27**, 249–259.

53. Greenough, C., Jenkins, R.E., Kitteringham, N.R., Pirmohamed, M., Park, B.K., and Pennington, S.R. (2004) A method for the rapid depletion of albumin and immunoglobulin from human plasma. *Proteomics*, **4**, 3107–3111.

54. Hinerfeld, D., Innamorati, D., Pirro, J., and Tam, S.W. (2004) Serum/Plasma depletion with chicken immunoglobulin Y antibodies for proteomic analysis from multiple Mammalian species. *J Biomol Tech*, **15**, 184–190.

55. Echan, L.A., Tang, H.Y., Ali-Khan, N., Lee, K., and Speicher, D.W. (2005) Depletion of multiple high-abundance proteins improves protein profiling capacities of human serum and plasma. *Proteomics*, **5**, 3292–3303.

56. Gong, Y., Li, X., Yang, B., Ying, W., Li, D., Zhang, Y., Dai, S., Cai, Y., Wang, J., He, F., and Qian, X. (2006) Different immunoaffinity fractionation strategies to characterize the human plasma proteome. *J Proteome Res*, **5**, 1379–1387.

57. Yan, J.X., Devenish, A.T., Wait, R., Stone, T., Lewis, S., and Fowler, S. (2002) Fluorescence two-dimensional difference gel electrophoresis and mass spectrometry based proteomic analysis of Escherichia coli. *Proteomics*, **2**, 1682–1698.

58. Gharbi, S., Gaffney, P., Yang, A., Zvelebil, M.J., Cramer, R., Waterfield, M.D., and Timms, J.F. (2002) Evaluation of two-dimensional differential gel electrophoresis for proteomic expression analysis of a model breast cancer cell system. *Mol Cell Proteomics*, **1**, 91–98.

59. Yokota, H., Hiramoto, M., Okada, H., Kanno, Y., Yuri, M., Morita, S., Naitou, M., Ichikawa, A., Katoh, M., and Suzuki, H. (2007) Absence of increased alpha1-microglobulin in IgA nephropathy proteinuria. *Mol Cell Proteomics*, **6**, 738–744.

60. Rao, P.V., Lu, X., Standley, M., Pattee, P., Neelima, G., Girisesh, G., Dakshinamurthy, K.V., Roberts, C.T. Jr, and Nagalla, S.R. (2007) Proteomic identification of urinary biomarkers of diabetic nephropathy. *Diabetes Care*, **30**, 629–637.

61. Holly, M.K., Dear, J.W., Hu, X., Schechter, A.N., Gladwin, M.T., Hewitt, S.M., Yuen, P.S., and Star, R.A. (2006) Biomarker and drug-target discovery using proteomics in a new rat model of sepsis-induced acute renal failure. *Kidney Int*, **70**, 496–506.

62. Nedelkov, D. and Nelson, R.W. (2001) Analysis of human urine protein biomarkers via biomolecular interaction analysis mass spectrometry. *Am J Kidney Dis*, **38**, 481–487.

63. Hampel, D.J., Sansome, C., Sha, M., Brodsky, S., Lawson, W.E., and Goligorsky, M.S. (2001) Toward proteomics in uroscopy: urinary protein profiles after radiocontrast medium administration. *J Am Soc Nephrol*, **12**, 1026–1035.

21
Liquid Chromatography Coupled to Mass Spectrometry for Analysis of the Urinary Proteome

Wei Sun and Youhe Gao

21.1
Introduction

Liquid chromatography (LC)/mass spectrometry (MS) coupling was introduced at the beginning of the 1990s when its importance was recognized for the discovery of biomolecules [1]. At that time, most LC/MS analyses were used for relatively simple mixtures [2, 3], and these early reports identified proteins by the fragment mass fingerprint. With advances in MS, along with the generation of large quantities of nucleotide sequence information and the development of bioinformatics that could correlate them, large-scale protein identification by LC/MS has become a reality [4, 5]. Generally, protein samples are first enzymatically digested into peptide mixtures, then separated by strong cation exchange (SCX) and reverse phase (RP) tandem capillary column on-line, and finally identified by tandem mass spectra. This approach is known as multidimensional protein identification technology (MudPIT). Because this process can be automated and is compatible with some specific classes of proteins (such as very acidic or basic proteins, excessively large or small proteins, and membrane proteins), it has been widely used in proteomic research since its emergence.

In an alternative to peptide fractionation, protein samples are first separated by SDS-PAGE, and an entire lane of the gel is excised into slices. The proteins in each gel slice are then analyzed separately by LC/MS. Because SDS-PAGE separates proteins by molecular weight, this "GeLC–MS" approach includes the apparent molecular weight of the proteins and provides information about protein processing or modification. Furthermore, the analysis was subdivided into several independent analysis runs, which increase confidence in database identifications and extend the dynamic range of the measurement [6, 7].

The LC/MS approach has been also applied to urinary proteome research. For normal urinary proteome reports, Spahr *et al.* [8, 9] employed one-dimensional liquid chromatography/mass spectrometry (1-D LC/MS) to identify 124 urinary proteins from a lyophilized human urine. Sun *et al.* [10] applied three approaches, 1-D LC/MS, 2-D LC/MS, and 1-DE/1-D LC/MS, to human urine proteome and obtained

Renal and Urinary Proteomics: Methods and Protocols. Edited by Visith Thongboonkerd

ISBN: 978-3-527-31974-9

226 identifications. Castagna *et al.* [11] used beads coated with peptide ligand library to enrich urinary proteins and identified 387 unique gene products by 1-D LC/MS. Recently, Adachi *et al.* [12] used 2-D LC/MS and 1-DE/1-D LC/MS in a urinary proteome analysis and identified more than 1500 urinary proteins. Pisitkun *et al.* [13] identified 295 exosome proteins in human urine by 1-D LC/MS. In an approach that permitted analysis of proteins' posttranslational modifications, Wang *et al.* [14] enriched glycoproteins by concanavalin A affinity purification and used 1-DE/1-D LC/MS and 2-D LC/MS to identify 193 glycoproteins in the urinary proteome.

For disease-related urinary proteome research, Pang *et al.* [15] employed 1-D LC/MS, 2-D LC/MS, and 2-DE/MS to investigate acute inflammation due to a pilonidal abscess and were able to identify inflammation-related biomarkers. Cutillas *et al.* [16] used 2-D LC/MS in the analysis of renal Fanconi syndrome and identified some novel proteins and potentially bioactive peptides in urine collected from these patients. Kemperman *et al.* [17] applied 1-D LC/MS and multivariate statistical analysis to proteinuric patients, and found that the LC/MS profiles of urinary compounds was likely to aid in diagnostics, monitoring of disease activity, and therapy.

As discussed above, LC/MS has provided a very useful approach for urinary proteome research. It is likely that LC/MS will also be of great value in forming new hypotheses about disease mechanisms and the effect of therapeutic interventions. In this chapter, the basic steps for LC/MS analysis are introduced, including urinary protein extraction, protein digestion (in solution and in gel), peptide separation (1-D LC, 2-D LC; online and offline), MS identification, and database searching.

21.2
Methods and Protocols

21.2.1
Urine Preparation

Normal human urine has a dilute protein concentration and a high-salt content, which may interfere with proteomic analysis. Therefore, sample preparation is a crucial step in urine proteomic studies. Some methods can be used to isolate and concentrate urinary proteins, including precipitation, lyophilization, ultracentrifugation, and centrifugal filtration. In this chapter, acetone precipitation is used as an example for the method of protein concentration.

21.2.1.1 Materials

Acetone (−20 °C), 25 mM NH_4HCO_3, 1 mM phenylmethylsulfonyl fluoride (PMSF).

21.2.1.2 Protocol

- Collect the urine sample in conical tube. Add protease inhibitors (PMSF) to avoid proteolysis in the urine.

- Centrifuge the urine sample at 3000g for 1 hour at 4 °C. Collect supernatant and discard pellet (cell debris and nuclei).
- Add an equal volume of acetone (−20 °C) to the supernatants and incubate at 4 °C for 1 hour or overnight.
- Centrifuge the sample at 12 000g for 1 hour at 4 °C. Discard supernatant.
- Resuspend the precipitate in 25 mM NH_4HCO_3.
- Centrifuge the resuspended sample at 12 000g for 30 minutes at 4 °C.
- Collect supernatant and store at −80 °C until use. Discard remaining precipitate.

Notes

- Prior to acetone precipitation, samples can be stored up to 48 hours at 4 °C, and for longer than three days at −80 °C.
- Although other lysis buffers can be used, detergents should be avoided because they create problems in LC elution.

21.2.2 Digestion

For proteomic research, protein identification is based on peptide analysis. The accuracy of protein identification is directly related to the completeness of digestion. Consequently, protein digestion is a very important step in sample preparation. Here, two different protocols are presented (standard and microwave-assisted) that are useful for either in-solution or in-gel digestion as appropriate for different separation approaches. Microwave-assisted protocols can be completed in much less time than standard protocols, and can produce similar (in-solution) or higher (in-gel) digestion efficiencies, as compared to standard protocols (see detailed data in [18]).

21.2.2.1 In-Solution Digestion

Materials 25 mM NH_4HCO_3, 1 M DTT in 25 mM NH_4HCO_3, 1.1 M iodoacetamide in 25 mM NH_4HCO_3, and sequence-grade modified trypsin.

Standard Protocol

- To reduce disulfide bonds, add DTT to the sample to a final concentration of 10 mM and incubate at 56 °C for 1 hour.
- To alkylate thiols, add iodoacetamide solution to a final concentration of 55 mM and incubate at room temperature in the dark for 45 minutes.
- Add sequence-grade modified trypsin (1 : 50) and incubate for 16 hours or overnight at 37 °C.

Note If urea is used to dissolve the precipitate, the final concentration should be diluted to <2 M.

Microwave-Assisted Protocol

- Add DTT solution to a final concentration of 10 mM and incubate at 100 °C for 5 minutes.

- Add sequence-grade modified trypsin (1 : 50) and digest under microwave irradiation (850 W) at room temperature for 1 minute.

21.2.2.2 In-Gel Digestion

For in-gel digestion, the utmost care should be taken to eliminate keratin contaminants from the sample. Therefore, the following precautions should be followed: Wear gloves and sleeve protectors. Use methanol to wipe down the outside of all tubes, as well as the outside and inside of the centrifuge, tube racks, bottles, and so on. Wipe razor blades with methanol. The highest purity buffers and water should be used to prepare reagents.

Materials 25 mM NH_4HCO_3, 50% acetonitrile (ACN) in 25 mM NH_4HCO_3, 50% ACN/5% formic acid (FA) (may be substituted with trifluoroacetic acid or acetic acid) in 25 mM NH_4HCO_3, 100% ACN, 10 mM DTT in 25 mM NH_4HCO_3, 55 mM iodoacetamide in 25 mM NH_4HCO_3, 12.5 ng μl^{-1} sequencing-grade modified trypsin in 25 mM NH_4HCO_3.

Standard Protocol

- Cut the gel slice into small pieces (1 mm^3) and place into 1.5-ml siliconized tubes.
- Add 50% ACN/25 mM NH_4HCO_3 to cover and vortex for 10 minutes. Discard the supernatant. Repeat this step once or twice.
- Dehydrate gels with 100% ACN (enough to cover), vortex 5 minutes, spin down. Repeat once.
- Vacuum dry the gel pieces to complete dryness in Speed Vac.
- Add 10 mM DTT/25 mM NH_4HCO_3 to cover the gel pieces. Vortex and spin down briefly. Incubate at 56 °C for 1 hour.
- Remove supernatant and add 55 mM iodoacetamide/25 mM NH_4HCO_3 to cover the gel pieces. Vortex and spin down briefly. Incubate at room temperature in the dark for 45 minutes.
- Discard supernatant. Dehydrate gels with 100% ACN (enough to cover), vortex 5 minutes, and spin down. Repeat once.
- Vacuum dry the gel pieces to complete dryness in Speed Vac.
- Add trypsin solution to just cover the gel pieces.
- Rehydrate the gel pieces in trypsin solution on ice or at 4 °C for 30 minutes.
- Add 25 mM NH_4HCO_3 to cover the gel pieces.
- Spin briefly and incubate at 37 °C for 16 hours or overnight.
- Transfer the digest solution (aqueous extraction) into a clean 1.5-ml siliconized tube.
- Add 50% ACN/5% FA (enough to cover) to the gel pieces, vortex 30 minutes, sonicate 5 minutes. Transfer the supernatants to the digestion solution. Repeat once.
- Vortex the extracted digests, spin and vacuum dry to complete dryness in Speed Vac.

Microwave-Assisted Protocol

- Perform initial steps (from destaining to addition of trypsin and 25 mM NH_4HCO_3) as described above for the standard protocol.

- Digest the gel under microwave irradiation (850 W) for 5 minutes.
- Transfer the digested solution (aqueous extraction) into a clean 1.5-ml siliconized tube.
- Add 50% ACN/5% FA (enough to cover) to the gel pieces, and place under microwave irradiation (850 W) for 5 minutes. Transfer the supernatants to the digested solution. Repeat once.
- Vortex the extracted digests, spin and vacuum dry to complete dryness in Speed Vac.

Note To be compatible to MS, the digested peptide should be prepared by C18 solid-phase-extraction column or ZipTip.

21.2.3 Liquid Chromatography

21.2.3.1 Analysis Column

Depending on the ionization source of the mass spectrometer, different diameter capillary columns and flow rates can be used. For microspray sources, the diameter of column is usually 100–300 μm and the flow rate is about 1–5 μl min^{-1}, and for nanospray sources, the diameter of the column is usually less than 100 μm and the flow rate is around 100–500 nl min^{-1}.

21.2.3.2 Sample

The peptide sample should be desalted and dissolved in 0.1% FA before loading onto the RP column. To avoid sample loss and to achieve better separations, online C18 trap columns can be used before the RP analysis column.

21.2.3.3 1-D LC Protocol

To couple with electrospray MS, a C18 RP chromatography column is usually used for LC/MS analysis. 1-D LC is usually used for simple samples, such as several known protein mixtures, spots from 2-DE, bands from 1-DE, or fractions from other separation methods (e.g., capillary electrophoresis, free flow electrophoresis, liquid-phase isoelectric focusing, etc.). A protocol for microspray is presented here.

Materials

- Buffer A: 0.1% FA + 99.9% H_2O.
- Buffer B: 0.1% FA + 99.9% ACN.
- C18 RP column: 100 mm × 0.18 mm i.d., 5-μm spherical particles with pore diameter 300 Å, flow rate 2 μl min^{-1} (Thermo Finnigan, San Jose, CA).

Protocol

- Load the sample by buffer A for 10 minutes.
- Wash the column with 5% buffer B in buffer A for 10 minutes.
- Elute the peptide from the column using a 5–35% gradient of buffer B in buffer A for 1 hour.

- Regenerate the column by 95% buffer B in buffer A for 10 minutes.
- Reequilibrate the column by buffer A for 10 minutes.

Notes
- The elution time depends on the complexity of the sample. For spots from 2-DE, a 1-hour elution time can produce good separation. For bands from 1-DE and known protein mixtures, the elution time may be 2–3 hours.
- The times for loading, washing, regenerating, and equilibrating the column depend on the loading volume, system dead volume, flow rate, and ionization source.

21.2.3.4 2-D LC Protocol

SCX coupled to RP is the most popular combination for 2-D LC. Other chromatographic techniques, such as strong anion exchange (SAX) and size exclusion, can also be used as the first dimensional separation. For different combinations of chromatographic separations, the elution program should be optimized for proteome analysis. 2-D LC is usually used for complex samples, such as tissue or cell proteome and body fluids. It can also be used to analyze fractions from other separation approaches to improve isolation.

A protocol for SCX-RP protocol is presented as a model. 2-D LC can be run either online using a step gradient elution or offline using a continuous gradient elution. Each fraction from the SCX column is further analyzed using a RP column. The online approach is convenient for complex sample analysis, but compromises the peptide separation efficiency. The offline approach results in better peptide separation but is relatively labor intensive. The protocol presented here is for a mass spectrometer with a microspray ionization source.

Materials
- Buffer A: 0.1% FA + 99.9% H_2O.
- Buffer B: 0.1% FA + 99.9% ACN.
- Buffer C: 0.1% FA + 5% ACN + 0.5 M ammonia acetate.
- Buffer D: 0.1% FA + 5% ACN + 1 M ammonia acetate.
- Buffer E: 0.1% FA + 25% ACN.
- Buffer F: 0.1% FA + 25% ACN + 0.5 M ammonia acetate.
- Buffer G: 0.1% FA + 25% ACN + 1 M ammonia acetate.
- SCX column: 150 mm × 0.32 mm i.d., 5-μm spherical particles with pore diameter 300 Å, flow rate 2 μl min^{-1} (Thermo Finnigan, San Jose, CA).
- C18 RP column: 100 mm × 0.18 mm i.d., 5-μm spherical particles with pore diameter 300 Å, flow rate 2 μl min^{-1} (Thermo Finnigan, San Jose, CA).

Online Protocol
- Load the sample onto SCX column by buffer A for 15 minutes.
- Elute SCX column using the following step gradient: 0, 5, 10, 15, 20, 25, 30, 50 buffer C and 95% buffer D in buffer A for 15 minutes per step.

- After each SCX elution step, wash the column by 5% buffer B in buffer A for 30 minutes.
- Elute the peptides from the RP column using a 5–35% gradient buffer B in buffer A for 3 hours.
- Regenerate the column by 95% buffer B in buffer A for 15 minutes.
- Reequilibrate the column by buffer A for 15 minutes.

Note Each fraction from SCX column contains high salt and should be completely desalted before RP elution.

Offline Protocol

- Load the sample by buffer E for 10 minutes.
- Wash the column by buffer E for 10 minutes.
- Elute the peptide from the column using a 0–30% continuous gradient of buffer F in buffer E for 1 hour.
- Regenerate the column by 95% buffer G in buffer E for 10 minutes.
- Reequilibrate the column by buffer E for 10 minutes.
- Vacuum dry each fraction from the SCX column to complete dryness.
- Dissolve each fraction in buffer A.
- Analyze each fraction by C18 RP as outlined above in the 1-D LC protocol.

Note The SCX column used for the offline method can be a column having larger diameter such as 2.1 or 4.6 mm.

21.2.4 Mass Spectrometry

For the LC/MS approach, electrospray ionization mass spectrometry (ESI-MS) is usually coupled with LC. Nano-LC fractions can also be collected on the MALDI plate for further analysis. In this chapter, ESI-MS is introduced for LC detector. Although there are some differences among mass analyzers, detectors, and so on, from different companies, basic programs for proteomic analysis are similar. Herein, a program for the LCQ DETA XP+ mass spectrometer from Thermo Finnigan is introduced as an example.

21.2.4.1 Electrospray Ionization Source Setting

- For a microspray source, the metal needle voltage is set at 3.2 kV. The sheath gas consists of nitrogen gas (99.999%) and is set at 10 units.
- For a nanospray source, the spray voltage is set at 1.8 kV and no sheath gas is necessary.

21.2.4.2 Mass Spectrometer Setting

Scan Events The instrument is set to acquire a full MS scan between 400 and 1500 m/z (3 μscans per scan) followed by full MS/MS scans (5 μscans per scan) between 400 and 2000 m/z of the top three ions from the preceding MS scan.

Precursor Ion Abundance The most abundant ion above a threshold value of 1×10^5 is selected for MS/MS analysis using an isolation window 3 m/z wide.

Collision Energy Relative collision energy for collision-induced dissociation is set to 35% with a 30-ms activation time.

Dynamic Exclusion

- Repeat count = 2 (the instrument will perform MS/MS twice on the same ion before the m/z value is added to the exclusion list).
- Exclusion duration = 3.0 minutes (m/z values will stay on the dynamic exclusion list for 3 minutes).
- Exclusion m/z width = 3.

21.2.5 Database Searching

21.2.5.1 Positive Database Searching

The MS/MS spectra are searched against the International Protein Index human database from the European Bioinformatics Institute web site (www.ebi.ac.uk/IPI/) using SEQUEST software. The following criteria are set for database searching:

- Peptide molecular weight ranges from 400 to 4500 Da.
- The minimum of acceptable total ion current required for precursor ion fragmentation is 1×10^5, and the minimum number of fragment ions is 50.
- Tryptic cleavages only at Lys or Arg and up to two missed internal cleavage sites in a peptide are allowed.
- Precursor charge states of +1, +2, and +3 are allowed.
- The maximal precursor mass tolerance is 1.4 Da.
- Modification
  - Differential modification: differential modification can be set for different samples: +80 on STY (phosphorylation), +16 on M (oxidation).
  - Static modification: +57 on C (alkylation).

21.2.5.2 False Positive Rate Evaluation

For large-scale proteomic data, to evaluate the false positive rate, a decoy database (random or reverse) should be searched using the same MS/MS spectra. The false positive rate is calculated by dividing matched MS/MS spectra numbers obtained from the decoy database under the same filter score by matched MS/MS spectra numbers obtained from the real database.

References

1. Ishida, M., Tsuda, Y., Onuki, Y., Itoh, T., Shimada, H., and Yamada, H. (1994) Determination of carbenicillin epimers in plasma and urine with high-performance liquid chromatography. *J Chromatogr*, **652** (1), 43–49.
2. Hsieh, Y.L., Wang, H., Elicone, C., Mark, J., Martin, S.A., and Regnier, F. (1996) Automated analytical

system for the examination of protein primary structure. *Anal Chem*, **68** (3), 455–462.

3. Tomlinson, A.J., Jameson, S., and Naylor, S. (1996) Strategy for isolating and sequencing biologically derived MHC class I peptides. *J Chromatogr A*, **744** (1-2), 273–278.
4. Link, A.J., Eng, J., Schieltz, D.M., Carmack, E., Mize, G.J., Morris, D.R., Garvik, B.M., and Yates, J.R. III (1999) Direct analysis of protein complexes using mass spectrometry. *Nat Biotechnol*, **17** (7), 676–682.
5. Washburn, M.P., Wolters, D., and Yates, J.R. III (2001) Large-scale analysis of the yeast proteome by multidimensional protein identification technology. *Nat Biotechnol*, **19** (3), 242–247.
6. Schirle, M., Heurtier, M.A., and Kuster, B. (2003) Profiling core proteomes of human cell lines by one-dimensional PAGE and liquid chromatography-tandem mass spectrometry. *Mol Cell Proteomics*, **2** (12), 1297–1305.
7. Steen, H. and Mann, M. (2004) The ABC's (and XYZ's) of peptide sequencing. *Nat Rev Mol Cell Biol*, **5** (9), 699–711.
8. Spahr, C.S., Davis, M.T., McGinley, M.D., Robinson, J.H., Bures, E.J., Beierle, J., Mort, J., Courchesne, P.L. *et al.* (2001) Towards defining the urinary proteome using liquid chromatography-tandem mass spectrometry. I. Profiling an unfractionated tryptic digest. *Proteomics*, **1** (1), 93–107.
9. Davis, M.T., Spahr, C.S., McGinley, M.D., Robinson, J.H., Bures, E.J., Beierle, J., Mort, J., Yu, W. *et al.* (2001) Towards defining the urinary proteome using liquid chromatography-tandem mass spectrometry. II. Limitations of complex mixture analyses. *Proteomics*, **1** (1), 108–117.
10. Sun, W., Li, F., Wu, S., Wang, X., Zheng, D., Wang, J., and Gao, Y. (2005) Human urine proteome analysis by three separation approaches. *Proteomics*, **5** (18), 4994–5001.
11. Castagna, A., Cecconi, D., Sennels, L., Rappsilber, J., Guerrier, L., Fortis, F., Boschetti, E., Lomas, L. *et al.* (2005) Exploring the hidden human urinary proteome via ligand library beads. *J Proteome Res*, **4** (6), 1917–1930.
12. Adachi, J., Kumar, C., Zhang, Y., Olsen, J.V., and Mann, M. (2006) The human urinary proteome contains more than 1500 proteins, including a large proportion of membrane proteins. *Genome Biol*, **7** (9), R80.
13. Pisitkun, T., Shen, R.F., and Knepper, M.A. (2004) Identification and proteomic profiling of exosomes in human urine. *Proc Natl Acad Sci U S A*, **101** (36), 13368–13373.
14. Wang, L., Li, F., Sun, W., Wu, S., Wang, X., Zhang, L., Zheng, D., Wang, J. *et al.* (2006) Concanavalin A-captured glycoproteins in healthy human urine. *Mol Cell Proteomics*, **5** (3), 560–562.
15. Pang, J.X., Ginanni, N., Dongre, A.R., Hefta, S.A., and Opitek, G.J. (2002) Biomarker discovery in urine by proteomics. *J Proteome Res*, **1** (2), 161–169.
16. Cutillas, P.R., Norden, A.G., Cramer, R., Burlingame, A.L., and Unwin, R.J. (2003) Detection and analysis of urinary peptides by on-line liquid chromatography and mass spectrometry: application to patients with renal Fanconi syndrome. *Clin Sci (Lond)*, **104** (5), 483–490.
17. Kemperman, R.F., Horvatovich, P.L., Hoekman, B., Reijmers, T.H., Muskiet, F.A., and Bischoff, R. (2007) Comparative urine analysis by liquid chromatography-mass spectrometry and multivariate statistics: method development, evaluation, and application to proteinuria. *J Proteome Res*, **6** (1), 194–206.
18. Sun, W., Gao, S., Wang, L., Chen, Y., Wu, S., Wang, X., Zheng, D., and Gao, Y. (2006) Microwave-assisted protein preparation and enzymatic digestion in proteomics. *Mol Cell Proteomics*, **5** (4), 769–776.

22
Comparative Urine Analysis by Liquid Chromatography-Mass Spectrometry: Data Processing and Multivariate Statistics

Peter L. Horvatovich, Ramses F.J. Kemperman, and Rainer Bischoff

22.1
Introduction

Urine is a complex body fluid that contains the metabolic end products of all major biological molecules in the body. The composition of urine is thus quite variable, because it is not subject to the same strict homeostatic control as blood. On the other hand, large quantities of urine can be easily sampled by noninvasive methods and it is therefore widely used in the diagnosis of, for example, metabolic disorders.

Most proteins and peptides in urine are water soluble, which facilitates their analysis. However, urine contains also vesicles, so-called urinary exosomes, which contain proteins that are membrane associated [1–3]. Identification of proteins from urinary exosomes may provide another approach toward discovery of disease biomarkers.

This chapter focuses on the comparative analysis of the low-molecular-weight (LMW) part of the urinary proteome comprising peptides, small soluble proteins, and protein fragments. Urine has served as a rich source of bioactive peptides and small proteins, which can be isolated on a preparative scale due to the availability of large volumes of urine or hemofiltrate [4–6]. The comparative analysis of urine has also been used as a tool to discover novel biomarkers of disease, although many findings require further validation [7–16]. A number of separation technologies have been combined with mass spectrometry (MS) to cover as large a part as possible of the urinary proteome including polyacrylamide gel electrophoresis in the presence of sodium dodecyl sulfate (SDS-PAGE), reverse-phase liquid chromatography (RPLC), two-dimensional polyacrylamide gel electrophoresis (2-D PAGE), and capillary electrophoresis (CE) [17–22]. Sample preparation for urinary proteomics is critical and may entail protein precipitation with acetone, ultracentrifugation [22], solid-phase extraction [17], or enrichment of LMW proteins and peptides by restricted access chromatography [23, 24].

The comparative profiling of urine by any of the above-mentioned hyphenated methodologies generates very large sets of data that need to be processed and statistically analyzed. Notably, methods combining high-resolution separation methods such as high-performance liquid chromatography (HPLC) or CE with MS can

Renal and Urinary Proteomics: Methods and Protocols. Edited by Visith Thongboonkerd

ISBN: 978-3-527-31974-9

easily result in 100 MB- to GB-size datasets per analysis. Processing includes the alignment of chromatograms in the time domain [25–30], the discrimination of peaks from background noise and spikes [31], the normalization of intensities, and the generation of a common peak matrix from all analyzed samples [17, 32]. The quality of this matrix is critical for the subsequent statistical analysis.

There are a number of statistical approaches to compare datasets, but all of them will have to deal with the high dimensionality low sample size (HDSS) problem that is common for most "omics" approaches [33, 34]. For the initial discovery phase, it is often helpful to group samples according to available clinical information (e.g., prostate cancer patients versus controls with benign prostate hyperplasia) that is closely related to the purpose for which a biomarker is sought. This means that the samples can be "preclassified," which allows to search for compounds that contribute to discrimination between the groups. An initial set of samples, a so-called "training-set," is generally used to build a statistical model that discriminates the groups on the basis of the obtained measurements. However, owing to biological and analytical variation, discrimination may occur purely by chance. Therefore, it is necessary to assess the risk of discriminating the groups based on chance alone by validating the statistical model, for example, using cross-validation strategies and permutation tests. This is particularly critical in cases where the number of measured variables exceeds the number of measured samples by a large factor, which is ordinarily the case in comparative analysis of LC-MS data.

If the statistical model turns out to be robust (low cross-validation error; permutation tests show that discrimination is significantly different from random), it has to be further validated using an independent set of samples, a so-called "test-set." For this, it is most appropriate to perform a new round of analyses with samples that were not used to build the original statistical model. On the basis of the obtained results, it should be possible to assess the sensitivity (true-positive rate) and specificity (true-negative rate) of the discriminatory model and the underlying biomarker candidates.

Following the discovery phase, it is often necessary to investigate the biological role of a biomarker candidate in a given disease to find out whether it is causally related to the disease mechanism or rather a general disease marker (e.g., an inflammatory marker related to the response of the organism to the disease). To this end, it is necessary to identify the discriminatory markers (generally by MS/MS) and to annotate them based on existing knowledge of the disease. Furthermore, the development of analytical assays with higher throughput is required for validation in a sufficiently large, relevant population to define their value for later clinical applications [35–38].

In this chapter, we focus on the use of LC-MS for the analysis of urine samples. Emphasis is placed on method validation and on ways to assess the "discriminatory power" of a given method based on the analysis of samples that were "spiked" with known peptides at nanomolar concentrations. The challenge of data preprocessing and the statistical analysis of high-dimensionality datasets obtained from a limited number of analyzed samples are presented on a number of selected examples.

22.2
Methodological Aspects of Comparative Urine Analysis

22.2.1
Quantitative Profiling and Evaluation of the Profiling Method

Quantitative profiling of complex biological samples, such as urine, without the need for labeling or spiking with internal standards, is feasible as long as there is linearity of signal versus concentration and a high degree of reproducibility of sample processing and the LC-MS platform [39]. It is thus important to apply standard validation parameters to assess the quality of the developed methodology, such as reproducibility, linearity, sensitivity, and limit of detection (LOD). It is also important to consider pre and postanalytical normalization to improve comparability between samples, groups of samples, and finally the obtained data sets.

To detect changes that are primarily related to the pathophysiological process of interest, the change in concentration of a candidate biomarker or a panel of biomarkers should ideally exceed the biological variation (homeostasis) in both univariate and multivariate comparisons. The biological variation should be determined in a sufficiently large set of apparently healthy individuals or in a set of individuals having another disease, from which the disease of interest is to be discriminated. The latter is especially relevant if the sensitivity and specificity of a new biomarker (panel) are compared with the sensitivity and specificity of an existing marker. An often used rule of thumb is that the analytical coefficient of variation, that is, imprecision (relative standard deviation; RSD), should be equal to or preferably less than half of the within-subject coefficient of variation [40]. Bias, that is, inaccuracy, should preferably be less than one-fourth of the biological RSD, which is the combination of the average within- and between-subject RSD [40]. Therefore, the variation that is introduced throughout the analytical process should be minimized. Biological variance should be considered when designing a biomarker discovery study. A longitudinal study design, where patients (and controls) are sampled repeatedly over a given time period, is well suited to reduce the between-subject biological variability. Analytical and biological variance are best studied using an experimental design and evaluated by multivariate statistics [41].

Performance characteristics, such as the lower (and upper) LOD, intra- and interassay variation or linearity, as well as information regarding the biological variation are also needed for power calculations when setting up biomarker discovery projects [42]. Despite the lack of clear guidelines for power calculations in LC-MS-driven biomarker research, as a consequence of the large number of variables that are measured, the proteomics community is urged to implement power assessments in future study designs [42]. Power calculation methods for microarray-based studies are emerging and may serve as guidelines for proteomics [43]. Predefined sample numbers with stated powers of detecting a change of a specific magnitude improve the quality of a study and facilitate interpretation of the results. These recommendations also apply for urine-based biomarker research.

The following sections deal with practical aspects of urinary proteomics such as sample preparation, normalization of the amount of urine that is analyzed, analytical method development, and the data processing workflow. Examples are provided to illustrate the evaluation of the profiling method in a multivariate manner with an application to hospitalized patients with and without proteinuria.

22.2.2 Sample Preparation

Systematic evaluation of different sample preparation protocols for gel-based human urinary proteomics showed that a high percentage of organic solvent improves protein recovery through precipitation and produces a larger number of unique protein spots after 2-D PAGE [44]. LC-MS analysis of urine has mostly concentrated on the LMW portion of the proteome present in the supernatant after protein precipitation [17] or following restricted access chromatographic enrichment [23]. The time of sampling of urine is an important matter in itself. First morning urine contains the largest amount of protein per volume but shows only a small number of protein spots in 2-D PAGE [44]. Peptide patterns differ significantly between first and second morning urine [45]. This suggests that 24-hour urine is more appropriate and better reflects endogenous metabolic processes. However, collection of 24-hour urine samples puts a significant burden on patients and healthy volunteers. For our studies, we have generally used first-void midstream morning urine samples and acid-based protein precipitation. Analysis of the effect of acid-based protein precipitation on the recovery of LMW urinary compounds is important for matters of (relative) quantitative profiling. Other matters of concern in urine sample pretreatment are storage time, storage temperature, number of freeze–thaw cycles, and contamination by bacteria and blood [45]. It is possible to assess the importance of such factors by means of an experimental design strategy [41]. Protocols for sample collection and preparation should be thoroughly evaluated to anticipate possible adverse effects of preanalytical variables on downstream results.

Depending on whether 24-hour urine or portion urine samples are required, study participants should store samples during collection at 4 °C and after collection at −20 °C. Furthermore, additives, such as chlorhexidine or acid, can be used to prevent bacterial growth and proteolysis. However, addition of preservatives may also complicate sample analysis or may even render some analytes more vulnerable to storage artifacts. Long-term storage of urine at temperatures below −25 °C and preferably at −80 °C improves analyte stability [46, 47]. Overnight acid-based protein precipitation after one freeze–thaw cycle was used in our case as a mild way to remove proteins. However, precipitation of urinary proteins can result in coprecipitation of LMW analytes. Uromodulin ("Tamm–Horsfall protein"), for example, is a major protein in healthy urine that can bind several LMW proteins and peptides and is able to form fibrils that sediment under certain conditions [48]. Precipitated proteins and sediment are removed by centrifugation and/or filtration. To remove (protein-)precipitate, we have used centrifugation (5 minutes at 1500g and 4 °C). Subsequently, we adjusted the composition of the supernatant to match the loading/washing solution to prevent breakthrough of LMW compounds of interest on the trap column during LC-MS analysis. Care should also be taken to prevent adsorption of compounds of interest to the wall of the vial, which may be tested with suitable internal standards [49].

To assess analytical variation, a pool of urine samples, prepared by mixing equal volumes of urine from seven apparently healthy adults (M/F : 4/3), was processed and analyzed repeatedly by LC-MS. To assess biological variation, individual urine samples from six adults (M/F : 3/3) were processed and analyzed in the same way. A higher number of replicates and individuals is, of course, preferred, because it will provide a more reliable estimation of the analytical and biological variation.

22.2.3
Normalization of Injected Amount

In label-free LC-MS profiling studies, it is important to normalize the amount of urine that is injected to avoid column overloading or ion suppression. We normalized the injected volume (6.6–95 μl) of urine to be equivalent to 50 nmol of urinary creatinine, because creatinine is considered to correct for concentration differences and to be a marker of urinary clearance [50]. This amount of creatinine was chosen based on the capacity of the LC-system and the dynamic range of the mass spectrometer. Next to using creatinine, we also opted for a multicompound normalization strategy, because this seemed better suited to the wide range of different compounds present in urine. Moreover, the creatinine concentration is affected by gender, age, muscle mass, diet, exercise, and so on [50]. To this end, we calculated the area under the curve (AUC) of the UV-chromatogram at 214 nm (AUC_{214}) for each sample within a sample set. The assumption that the AUC_{214} is based on a large number of UV-absorbing compounds did not prove to be completely valid, because one peak in the UV-chromatogram accounted for approximately 44% of the AUC_{214} [17]. However, by comparing the RSD of the AUC of the total ion chromatogram (TIC) (AUC_{TIC}) between creatinine- and AUC_{214}-normalized urine samples of 10 children, we observed a lower RSD (%) for data normalized to AUC_{214} than for data normalized to creatinine (10.1 and 38.4%, respectively).

Normalizing the injected amount to AUC_{214} reduces the variation and improves comparability of the acquired data. However, this approach may also result in normalizing differences between samples due to the fact that UV-absorbing compounds are excreted at an increased rate under pathological conditions. Estimation of the glomerular filtration rate and urinary clearance of endogenous products, using concentrations of serum and urinary creatinine and serum cystatin C, provides valuable information to decide whether to use a single-compound (e.g., urinary creatinine concentration, total protein content, density) or a multicompound (e.g., AUC_{214} and AUC_{TIC}) normalization strategy. It is not yet clear which strategy least affects potentially interesting differences related to biomarkers. It is also possible to normalize after data acquisition at the data processing stage by global or local normalization. Global normalization uses a single constant normalization factor for each chromatogram, for example, by calculating the AUC of the TIC- or UV-trace. Local normalization refers to cases where only a subset of features is used at a time, that is, the use of different subsets for different parts of the data [51]. One example is the NOMIS (normalization using optimal selection of multiple internal standards) method, that utilizes information from multiple internal standard compounds to find optimal normalization factors for each individual molecular species [52].

22.2.4 Analytical System Setup

The LC-MS analyses described within this chapter were performed on the instrumental setup described in detail by Kemperman *et al.* [17]. Briefly, urine samples were desalted on an Atlantis dC_{18} precolumn (Waters) and back flushed from the precolumn onto a temperature-controlled Atlantis dC_{18} analytical column (1.0×150 mm, 3-μm particles, 30-nm pores). Compounds were separated in 90 minutes at a flow rate of 50 μl min^{-1} during which the percentage of solvent B (0.1% formic acid (FA) in acetonitrile (ACN)) in solvent A (0.1% FA in ultrapure H_2O) was increased from 5.0 to 43.6% (0.43% per minute). Following gradient elution, both columns were washed with 85% B for 5 minutes and equilibrated with 5% B for 10 minutes prior to the next injection. Compounds were separated using an 1100 series capillary HPLC system equipped with a cooled autosampler (4 °C) detected with a UV detector ($\lambda = 214$ nm) and by electrospray ionization (positive-mode) mass spectrometry (SL ion trap mass spectrometer; Agilent Technologies). Spectra were saved in centroid mode.

Dilutions of a standard peptide solution (seven 3–8 amino acid peptides: VYV, GYYPT, YPFPG, DRVYIHPF, YGGFL, YGGWL, YPFPGPI; with concentrations ranging from 0.10 to 0.29 mmol l^{-1}) were injected onto the precolumn prior to injection of urine samples to evaluate the sensitivity of the profiling method (LOD: 5.7–21 nmol l^{-1} or 54–204 fmol on column depending on the peptide), its within-day (2.9–19% RSD) and between-day (4.8–19% RSD) analytical variation for peak areas, linearity ($p < 0.001$ for lack-of-fit test; R^2: 0.918–0.999; assessed at eight concentrations ranging from 0 to 500 nmol l^{-1}), and retention time (standard deviation (SD) ≤ 0.52 minutes).

22.2.5 Data Analysis

22.2.5.1 Data Processing

For processing and multivariate statistical analysis, the original Bruker Daltoniks LC-MS data files are converted into ASCII-format with the Bruker Data Analysis software (3.2 build 121). For further data analysis, Matlab (version 7.0.0.19920, Mathworks, Natick, MA, USA) and the PLS toolbox (version 3.5.2, Eigenvector Research Inc., Wenatchee, WA, USA) are used. Matlab is a flexible and user-friendly

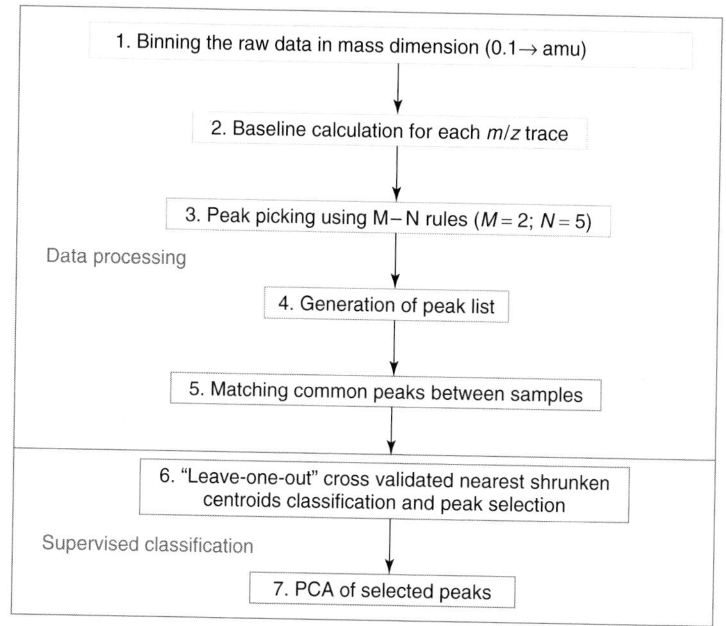

Figure 22.1 Schematic overview over the data processing and statistical analysis process adapted to data obtained by LC-MS using an ion trap mass analyzer. Abbreviations: amu, atomic mass unit; m/z, mass-to-charge ratio; M−N rule, see Radulovic *et al.* [53]; PCA, principal component analysis.

interpreted language. However, compared to C++, it requires more computing power and has limitations in memory management.

Figure 22.1 shows the main steps of the data processing and statistical analysis workflow. At first, all intensities between rounded mass to charge ratio (m/z) values are summed up (binned), reducing data resolution from 0.1 to 1 amu according to Radulovic *et al.* [53]. Binning reduces the amount of data by roughly a factor 10 and partially corrects for slight shifts in m/z values that can arise as a result of ion-trap overfilling during peak elution. For signal filtering and background reduction, data are first smoothed using a moving average filter (three-scan header width (approximately 0.08 minutes), two cycles). In the second step, 30% of the trimmed mean is calculated as baseline for each m/z trace. When this background is lower than 1000 counts, the background value is fixed at 1000 in order to eliminate m/z traces with a low baseline. This baseline is used in step 3 for discrimination between signal and noise (peak-picking) by multiplying the baseline by M (set to 2) to set the threshold. A peak is assigned when the intensity exceeds the threshold (M times the local baseline) for N (set to 5) consecutive points [53]. Peaks are combined into peak lists containing the m/z value of the bin, the mean retention time of the three data points with the highest intensities, and the highest intensity minus the local, calculated baseline.

To obtain optimal settings for *M* and *N*, a similar approach as described by Radulovic *et al.* [53] was applied to two blank LC-MS runs and two LC-MS runs of the pooled urine sample. Settings at which the ratio between the number of peaks in the pooled sample and in the blank sample was the highest and at which a minimal number of peaks was extracted from the noise in the blank chromatogram were used (*M* was allowed to vary between 1.5 and 4, and *N* between 4 and 8).

Time alignment of the chromatograms was not necessary in the presented datasets, because the median SD of the retention time of the standard peptide peaks was only 0.37 minutes (range : 0.23–0.52) but it is often required in studies of larger or more diverse sample sets. One-dimensional peak matching in different samples was performed by using a sliding windows technique in which the same m/z bins are evaluated for peaks proximate in time (step size 0.1 minute; search window 1.0 minute; maximal accepted SD of retention time of all peaks within a group of matched peaks 0.75 minutes). Missing peak allocation was done by extracting the local intensity minus the background in given m/z bins at the mean retention time of the selected peaks. The generated final combined peak matrix, created from the peak matrices of the individual samples, consisted of a peak(*row*)–sample(*column*)–intensity(*value*) matrix. This final peak matrix was used for multivariate statistical analysis. All data processing work was done on a personal computer equipped with a +3600 MHz AMD processor and 4 GB of RAM.

22.2.5.2 **Classification and Multivariate Statistical Methods**

To select the most discriminating peaks, the nearest shrunken centroid (NSC) classification algorithm [54, 55] is applied. NSC regularizes data whereby class-specific centroids are "shrunk" toward the overall (nonclass-specific) centroid, which has the effect of eliminating the influence of peaks showing weak discrimination. This step reduces the number of variables and consequently increases the sample-to-variable ratio [34] retaining peaks that are most relevant for discrimination of the predefined classes [56]. The optimal shrinkage value is chosen using a leave-one-out cross-validation (LOOCV) scheme to reach the lowest classification error in order to minimize the risk of overfitting. In LOOCV, one observation per class is omitted and the remaining data set is used to construct a new classification model, which is in turn used to classify the omitted sample. This procedure is repeated until each sample has been omitted once and the classification errors for a given shrinkage are summed up. In case the same cross-validation error occurs at multiple shrinkage values, it is best to use the highest shrinkage value, because this will result in the lowest number of selected variables (peaks). Variables selected at the optimal shrinkage value are used for construction of the final classification model. The selected peaks are subsequently analyzed and visualized by plotting the first two principal components obtained after principal component analysis (PCA) [41, 57].

22.2.5.3 **Future Perspectives**

All steps of the described data processing and statistical analysis procedure can still be improved in order to extract more information from the raw data. This includes more accurate quantification, better peak matching, improved variable selection, and classification methods that are better adapted to the HDSS problem and finally more accurate statistical validation strategies.

Data reduction inevitably leads to information loss and data processing without the need for data reduction might improve the accuracy of quantification. Application of a noise reduction filter based on two-dimensional Gaussian smoothing might lead to an improved signal-to-noise ratio and facilitate peak picking. Peak-picking algorithms considering peaks as three-dimensional objects and using all available data points for quantification with the capability to resolve overlapping peaks will likely lead to further improved quantification and to an increased concentration dynamic range, because minor peaks can also be assigned accurately. Charge-state determination in combination with the summing of peak volumes of the same compound at different charge states and deconvolution of isotope clusters will simplify the extracted peak list by reducing its redundancy. In examples shown in this chapter, the time shifts between different chromatograms are small and do not require any correction. However, we have often observed larger retention time shifts in sets of chromatograms necessitating application of automated time alignment algorithms. On the basis of our experience, accurate time alignment of complex datasets containing high biological variance is challenging and requires further developments.

A wider range of classification and variable selection methods (e.g., Support Vector Machines, PCA-discriminant analysis) should be evaluated and their reliability in finding discriminating peaks must be assessed at various sample-to-variable ratios. Improved statistical validation is another important issue. LOOCV is used to optimize one classification parameter (shrinkage); however, double cross validation strategies allow parameter optimization using the classification error as a readout without any bias.

Time series analyses using the same patient to minimize biological variability due to genetic differences and differences in lifestyle might result in more detailed information related to the goal of the study (e.g., disease-relevant biomarkers, treatment-related biomarkers; early predictive biomarkers). The statistical analysis of such longitudinal datasets requires dedicated algorithms, many of which are still under active development [58, 59].

22.3 Evaluation of Method Performance by Multivariate Statistics and Application to Proteinuria

22.3.1 Multivariate Evaluation of Analytical and Biological Variation

The effect of the analytical variation (Figure 22.2a) is evaluated using visualization by PCA of all peaks of the final peak matrix containing data from both blank urine and urine spiked with peptides at nanomolar concentrations. Clearly, the PCA plot

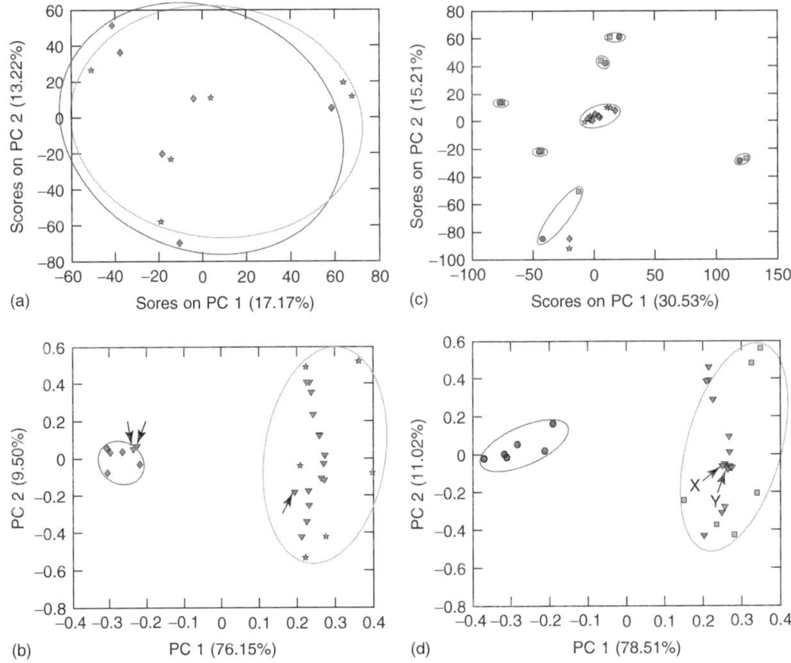

Figure 22.2 Evaluation of analytical and biological variation emphasizing the need for variable selection. To assess the analytical variation only, urine from seven healthy volunteers (M/F: 4/3) was pooled and one set of samples was spiked with 26–74 nM standard peptides (see Section 22.2.4). Blank and spiked samples were analyzed six times each. The combined analytical and biological variation was assessed by analyzing urine from six apparently healthy adults. (a) The plot containing the two largest principal components (PC1, PC2) after principal component (PC) analysis using the complete matrix of 10 029 peaks shows overlap between repetitions of the blank (♦) and the spiked (★ 2–10 times higher than the limit of detection of standard peptides) urine samples indicating that the analytical variation in the data sets precludes differentiation between spiked and blank samples. Together PC 1 and PC 2 explain 30.39% of the variation in the data. (b) The PC biplot depicts the same blank (♦) and spiked (★) urine samples as in (a) after variable selection by the nearest shrunken centroid algorithm, which selected the 17 most discriminating peaks (▼) at a leave-one-out cross-validation error of 0. Arrows indicate peaks that do not belong to the added peptides and that have been incorrectly selected as being discriminatory. In contrast to Figure (a), PC 1 and PC 2 now explain 85.65% of the variation in the data. Ellipses circle groups of samples belonging to the same class (i.e., blank and spiked). Classification is complete in this case. (c) The PC analysis plot using the complete peak matrix of 14 234 peaks shows overlap between the blank (•) and spiked (■; 2–10 times higher than the limit of detection of standard peptides) urine samples of six apparently healthy adults, and six repetitive analyses of a blank ($n = 6$; ♦) and a spiked ($n = 6$; ★) pooled urine sample. Colocalization (red ellipses) of the blank and spiked samples from the same individual (for five out of six individuals) and the pool suggests biological variation to be the main determinant of the observed overall variation. PC analysis using the complete peak matrix provides insufficient discriminatory power to detect the variation caused by spiking. Together PC 1 and PC 2 explain 45.74% of the variation within the data. (d) The PC biplot depicts the same blank (•) and spiked (■) samples from six healthy individuals as in (c) together with the 16 most discriminating peaks (▼) selected by the nearest shrunken centroid classification method at a leave-one-out cross-validation error of 0. After selection, PC 1 and PC 2 explain 89.53% of the variation between blank (•) and spiked (■) samples. Classification is again complete in this case. Two peaks (X and Y) could not be related to the added peptides.

shows overlap between blank and spiked urine samples, emphasizing the need to apply a supervised classification algorithm, to select discriminatory elements prior to PCA (Figure 22.2b). Figure 22.2c shows clustering (ellipses) of blank and spiked urine samples from the same individual when all variables are used for PCA indicating that biological variation is much larger than analytical variation. Figure 22.2d further shows that clear discrimination between spiked and nonspiked samples is possible in the presence of biological variation after variable selection using the NSC algorithm. Of the NSC selected discriminatory peaks, 80–90% can be related to the spiked peptides. The higher number of discriminatory peaks relative to the number of spiked peptides is due to the fact that data were not deconvoluted with respect to charge state and isotopic distribution. Interestingly, two peaks (arrows in Figure 22.2d) cannot be related to the spiked peptides and these peaks also fail to have significantly different peak areas and peak height in a univariate comparison at $\alpha = 0.05$. The notion that biological variation is the main contributor to variability in the data is emphasized by evaluation of the histograms of the RSD of all peaks in the respective peak matrices, which show a significantly ($p < 2.23 \times 10^{-308}$) lower median RSD for all peaks in the peak matrix from the LC-MS data set comprising only analytical variation (Figure 22.3, filled line) compared to all peaks in the peak matrix from the LC-MS data set comprising both biological and analytical variation (Figure 22.3, striped line). This is a convenient way to visualize the relative contributions of analytical and biological variation to the data.

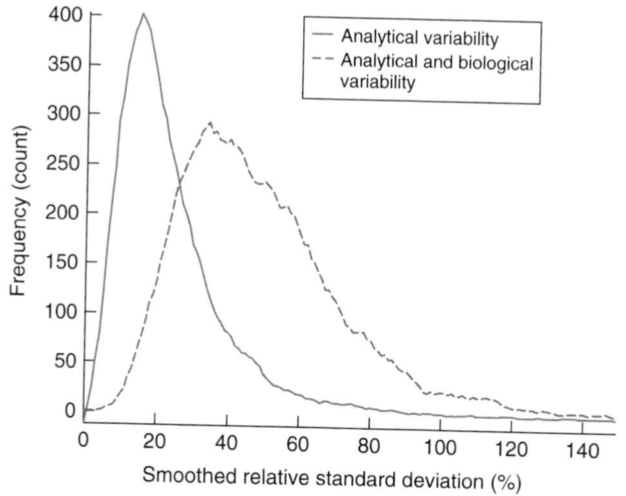

Figure 22.3 Comparison of the contribution of analytical and biological variation to the overall variability. Histograms of the relative standard deviation of all peaks in the final peak matrix from repetitive analyses ($n = 6$) of the pooled urine sample (filled line), which is representative of the analytical variation, and from analysis of urine samples from six healthy individuals (striped line), which is representative of the analytical and the biological variation combined. The median RSD for the analytical variation is lower ($p < 2.23 \times 10^{-308}$) than the median RSD for the analytical and biological variation.

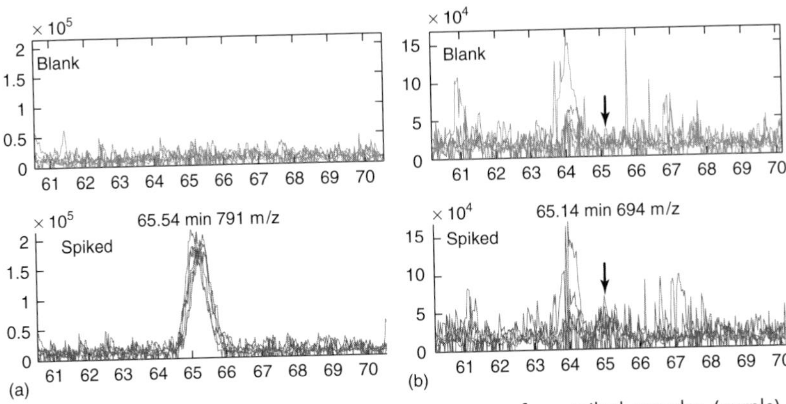

Figure 22.4 Extracted ion chromatograms from 2 of the 17 discriminatory peaks obtained by nearest shrunken centroid classification and variable selection of blank (n = 6) and spiked (n = 6) pooled urine samples (see Figure 22.2a,b). Upper traces (green) are from blank urine samples, while lower traces are from spiked samples (purple). (a) Discriminatory peak related to peptide YPFPGPI (791 m/z, 65.54 minutes). (b) Artifact of data processing appearing as a falsely detected discriminatory peak (694 m/z, 65.14 minutes) coeluting with YPFPGPI.

Figure 22.4 shows the extracted ion chromatograms (EICs) for a discriminatory peak that is related to a spiked peptide YPFPGPI (a) and for another discriminatory peak with a binned m/z of 694 (b) that could not be related to any of the spiked peptides. The latter peak eluted at the same time as YPFPGPI suggesting a peak-picking and/or aberrant baseline calculation effect to have caused this faulty assignment.

22.3.2 Application to Proteinuria

Application of the developed methodology to samples from hospitalized patients with and without proteinuria resulted in the selection of a large number (n = 92) of discriminatory peaks (Figure 22.5a). Manual curation of this list for charge-state and isotopic distribution and univariate comparison of the remaining peaks reduced the number of discriminatory peaks by 40%, emphasizing the need for integrating charge-state deconvolution and deisotoping in the data processing and analysis workflow. Univariate comparison showed that only few peaks had significantly different peak areas, which underscores the need for a multivariate approach. Identification of some biomarker candidates by LC-MS/MS showed that some of these discriminatory peptides were derived from albumin, whereas other identified albumin-derived peptides were, interestingly, nondiscriminatory. Using the list of discriminatory peaks, it was possible to classify the urine samples from apparently healthy adults and the pooled urine sample (as shown in Figure 22.2c) as originating from patients without proteinuria (Figure 22.5b). Remarkably, the healthy adults clustered separately from hospitalized patients with a history of renal disease but without overt proteinuria at the time of sampling, suggesting that a

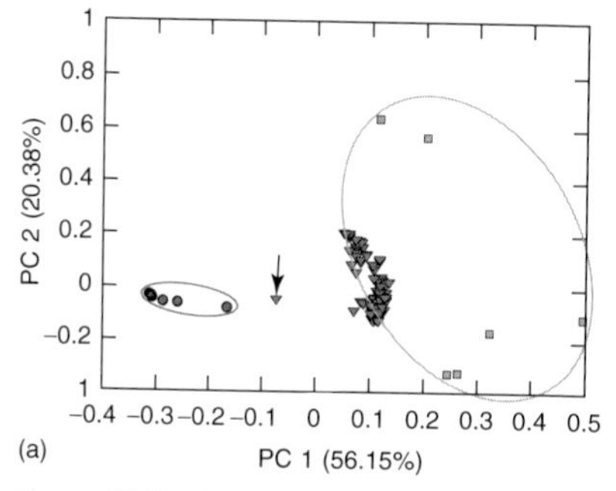

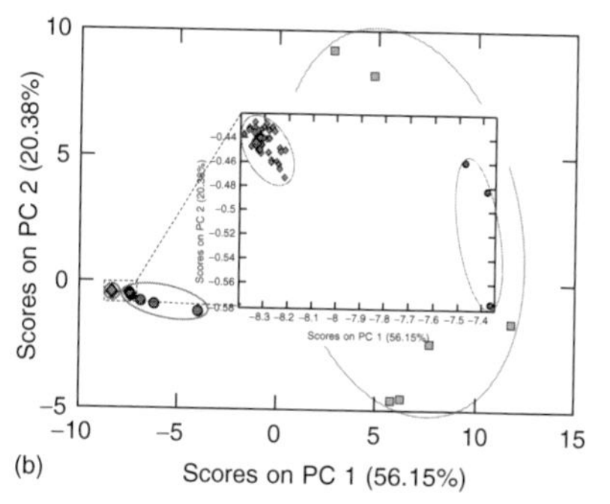

Figure 22.5 Comparison of urine samples from proteinuric and nonproteinuric patients with renal disorders. The PC biplot (a) shows the urine samples from six patients with proteinuria (■; $\geq 1\,\mathrm{g\,l^{-1}}$ total protein concentration in a random portion of urine) as compared to six patients without proteinuria (•; $\leq 0.1\,\mathrm{g\,l^{-1}}$). All patients had a medical history of renal morbidity at time of sampling but exhibited differences in their urinary protein content. 92 peaks (▼) out of 11 867 were selected for their discriminatory properties by the nearest shrunken centroid classification method at a leave-one-out cross-validation error of 0. The arrow indicates an "upregulated" peak in the nonproteinuria group. Using this model, classification of 24 samples from apparently healthy individuals (★; all samples depicted in Figure 22.2) resulted in a dense cluster (purple ellipse) that was clearly separated from the nonproteinuric patients (b) implying large differences in the composition of urine between healthy individuals and hospitalized patients with renal disorders with or without proteinuria. The enlarged area in panel b shows that some patients without overt proteinuria show a trend in PC 1 placing them at various levels toward overt proteinuria.

right-shift along PC 1 is associated with progressing renal disease. The possibility to use this methodology to detect trends is worth further investigation.

22.4 Conclusions and Outlook

Urine is an easily accessible body fluid that contains potentially useful information for diagnosis of disease and for follow-up of therapeutic interventions. Urine is particularly interesting in cases of metabolic disorders and of disorders affecting the urogenitary tract. While urine can be obtained in large quantities by noninvasive measures, its composition can vary widely leading to considerable intra and interpersonal variation. LC-MS is a hyphenated analytical method particularly suited to analyze complex samples such as urine. Data generated by LC-MS define compounds in terms of retention time, m/z, and intensity. Using the supernatant of an acid-precipitated urine sample, where high-molecular weight proteins have been removed, as starting material delivers a peak matrix of 10 000–15 000 features for each LC-MS run. Comparing samples using all of these features is not meaningful, because no discrimination is achieved even in cases where peptides have been added to the urine at 2- to 10-times the LOD (low to medium nanomolar concentration range). Variable selection is therefore of utmost importance to find discriminating features that allow for correct classification of sample groups in spite of significant biological variation. This is shown with spiked versus blank samples but also in the case of patients with renal disorders with or without proteinuria. Variable selection requires, however, that groups of samples can be preclassified based on available upfront information.

This chapter outlines one possible approach to the analysis of urine by LC-MS and it provides one path through the maze of data processing and statistical data analysis. Further developments to improve the performance of chromatographic separation media and mass spectrometers as well as data processing and statistics algorithms will allow to extract more information from urine samples but it can never overcome the inherent variability of urine as a body fluid. These sophisticated analytical approaches, need therefore go hand in hand with more controlled sampling, sample handling and sample preparation approaches as well as with careful selection of patients and controls to obtain clinically meaningful results.

Acknowledgments

The authors would like to thank the members of the Analytical Biochemistry group, especially Berend Hoekman, as well as all members of collaborating groups for their support and enthusiasm. Works in the Analytical Biochemistry group are financially supported by the following grants: Dutch Cancer Fund-KWF (RUG 2004-3165), Netherlands Proteomics Centre (Bsik 3015), Netherlands Bioinformatics Centre (BioRange 2.2.3), European Union (P-Mark, LSHC-CT-2004-503011), and the Dutch Technology Foundation (GPC6150 and TAC7047).

References

1. Pisitkun, T., Shen, R.F., and Knepper, M.A. (2004) Identification and proteomic profiling of exosomes in human urine. *Proc Natl Acad Sci U S A*, **101**, 13368–13373.
2. Pisitkun, T., Johnstone, R., and Knepper, M.A. (2006) Discovery of urinary biomarkers. *Mol Cell Proteomics*, **5**, 1760–1771.
3. Zhou, H., Pisitkun, T., Aponte, A., Yuen, P.S., Hoffert, J.D., Yasuda, H., Hu, X., Chawla, L., Shen, R.F., Knepper, M.A., and Star, R.A. (2006) Exosomal fetuin-A identified by proteomics: a novel urinary biomarker for detecting acute kidney injury. *Kidney Int*, **70**, 1847–1857.
4. Schulz Knappe, P., Raida, M., Meyer, M., Quellhorst, E.A., and Forssmann, W.G. (1996) Systematic isolation of circulating human peptides: the concept of peptide trapping. *Eur J Med Res*, **1**, 223–236.
5. Schulz Knappe, P., Schrader, M., Standker, L., Richter, R., Hess, R., Jurgens, M., and Forssmann, W.G. (1997) Peptide bank generated by large-scale preparation of circulating human peptides. *J Chromatogr A*, **776**, 125–132.
6. Raida, M., Schulz-Knappe, P., Heine, G., and Forssmann, W.G. (1999) Liquid chromatography and electrospray mass spectrometric mapping of peptides from human plasma filtrate. *J Am Soc Mass Spectrom*, **10**, 45–54.
7. Fliser, D., Novak, J., Thongboonkerd, V., Argiles, A., Jankowski, V., Girolami, M.A.,

Jankowski, J., and Mischak, H. (2007) Advances in urinary proteome analysis and biomarker discovery. *J Am Soc Nephrol*, **18**, 1057–1071.

8. Roelofsen, H., Alvarez-Llamas, G., Schepers, M., Landman, K., and Vonk, R.J. (2007) Proteomics profiling of urine with surface enhanced laser desorption/ionization time of flight mass spectrometry. *Proteome Sci*, **5**, 1–9.
9. Nemirovskiy, O.V., Dufield, D.R., Sunyer, T., Aggarwal, P., Welsch, D.J., and Mathews, W.R. (2007) Discovery and development of a type II collagen neoepitope (TIINE) biomarker for matrix metalloproteinase activity: from in vitro to in vivo. *Anal Biochem*, **361**, 93–101.
10. Lokeshwar, V.B. and Selzer, M.G. (2006) Urinary bladder tumor markers. *Urol Oncol*, **24**, 528–537.
11. Ngai, H.H., Sit, W.H., Jiang, P.P., Xu, R.J., Wan, J.M., and Thongboonkerd, V. (2006) Serial changes in urinary proteome profile of membranous nephropathy: implications for pathophysiology and biomarker discovery. *J Proteome Res*, **5**, 3038–3047.
12. Decramer, S., Wittke, S., Mischak, H., Zurbig, P., Walden, M., Bouissou, F., Bascands, J.L., and Schanstra, J.P. (2006) Predicting the clinical outcome of congenital unilateral ureteropelvic junction obstruction in newborn by urinary proteome analysis. *Nat Med*, **12**, 398–400.
13. Theodorescu, D., Wittke, S., Ross, M.M., Walden, M., Conaway, M., Just, I., Mischak, H., and Frierson, H.F. (2006) Discovery and validation of new protein biomarkers for urothelial cancer: a prospective analysis. *Lancet Oncol*, **7**, 230–240.
14. van Gils, M.P., Stenman, U.H., Schalken, J.A., Schroder, F.H., Luider, T.M., Lilja, H., Bjartell, A., Hamdy, F.C., Pettersson, K.S., Bischoff, R., Takalo, H., Nilsson, O., Mulders, P.F., and Bangma, C.H. (2005) Innovations in serum and urine markers in prostate cancer current European research in the P-mark project. *Eur Urol*, **48**, 1031–1041.
15. Pang, J.X., Ginanni, N., Dongre, A.R., Hefta, S.A., and Opitek, G.J. (2002) Biomarker discovery in urine by proteomics. *J Proteome Res*, **1**, 161–169.
16. Celis, J.E., Wolf, H., and Ostergaard, M. (2000) Bladder squamous cell carcinoma biomarkers derived from proteomics. *Electrophoresis*, **21**, 2115–2121.
17. Kemperman, R.F., Horvatovich, P.L., Hoekman, B., Reijmers, T.H., Muskiet, F.A., and Bischoff, R. (2007) Comparative urine analysis by liquid chromatography-mass spectrometry and multivariate statistics: method development, evaluation, and application to proteinuria. *J Proteome Res*, **6**, 194–206.
18. Adachi, J., Kumar, C., Zhang, Y., Olsen, J.V., and Mann, M. (2006) The human urinary proteome contains more than 1500 proteins including a large proportion of membranes proteins. *Genome Biol*, **7**, R80.
19. Fliser, D., Wittke, S., and Mischak, H. (2005) Capillary electrophoresis coupled to mass spectrometry for clinical diagnostic purposes. *Electrophoresis*, **26**, 2708–2716.
20. Pieper, R., Gatlin, C.L., McGrath, A.M., Makusky, A.J., Mondal, M., Seonarain, M., Field, E., Schatz, C.R., Estock, M.A., Ahmed, N., Anderson, N.G., Steiner, S., and Pieper, R. (2004) Characterization of the human urinary proteome: a method for high-resolution display of urinary proteins on two-dimensional electrophoresis gels with a yield of nearly 1400 distinct protein spots. *Proteomics*, **4**, 1159–1174.
21. Idborg-Bjorkman, H., Edlund, P.O., Kvalheim, O.M., Schuppe-Koistinen, I., and Jacobsson, S.P. (2003) Screening of biomarkers in rat urine using LC/electrospray ionization-MS and two-way data analysis. *Anal Chem*, **75**, 4784–4792.
22. Thongboonkerd, V., McLeish, K.R., Arthur, J.M., and Klein, J.B. (2002)

Proteomic analysis of normal human urinary proteins isolated by acetone precipitation or ultracentrifugation. *Kidney Int*, **62**, 1461–1469.

23. Wagner, K., Miliotis, T., Marko-Varga, G., Bischoff, R., and Unger, K.K. (2002) An automated on-line multidimensional HPLC system for protein and peptide mapping with integrated sample preparation. *Anal Chem*, **74**, 809–820.
24. Machtejevas, E., John, H., Wagner, K., Standker, L., Marko-Varga, G., Forssmann, W.G., Bischoff, R., and Unger, K.K. (2004) Automated multi-dimensional liquid chromatography: sample preparation and identification of peptides from human blood filtrate. *J Chromatogr B*, **803**, 121–130.
25. van Nederkassel, A.M., Daszykowski, M., Eilers, P.H.C., and Vander Heyden, Y. (2006) A comparison of three algorithms for chromatograms alignment. *J Chromatogr A*, **1118**, 199–210.
26. Chen, S.S. and Aebersold, R. (2005) LC-MS solvent composition monitoring and chromatography alignment using mobile phase tracer molecules. *J Chromatogr B*, **829**, 107–114.
27. Pierce, K.M., Wood, L.F., Wright, B.W., and Synovec, R.E. (2005) A comprehensive two-dimensional retention time alignment algorithm to enhance chemometric analysis of comprehensive two-dimensional separation data. *Anal Chem*, **77**, 7735–7743.
28. Eilers, P.H.C. (2004) Parametric time warping. *Anal Chem*, **76**, 404–411.
29. Tomasi, G., Van Den Berg, F., and Andersson, C. (2004) Correlation optimized warping and dynamic time warping as preprocessing methods for chromatographic data. *J Chemom*, **18**, 231–241.
30. Bylund, D., Danielsson, R., Malmquist, G., and Markides, K.E. (2002) Chromatographic alignment by warping and dynamic programming as a pre-processing tool for PARAFAC modelling of liquid chromatography-mass spectrometry data. *J Chromatogr A*, **961**, 237–244.
31. Eilers, P.H.C. (2003) A perfect smoother. *Anal Chem*, **75**, 3631–3636.
32. Horvatovich, P., Govorukhina, N., Reijmers, T.H., van der Zee, A.G.J., Suits, F., and Bischoff, R. (2007) Chip-LC-MS for label-free profiling of human serum. *Electrophoresis*, **28**, 4493–4505.
33. Jaffe, J.D., Mani, D.R., Leptos, K.C., Church, G.M., Gillette, M.A., and Carr, S.A. (2006) PEPPeR, a platform for experimental proteomic pattern recognition. *Mol Cell Proteomics*, **5**, 1927–1941.
34. Listgarten, J. and Emili, A. (2005) Statistical and computational methods for comparative proteomic profiling using liquid chromatography-tandem mass spectrometry. *Mol Cell Proteomics*, **4**, 419–434.
35. Lewin, D.A. and Weiner, M.P. (2004) Molecular biomarkers in drug development. *Drug Discov Today*, **9**, 976–983.
36. Alaiya, A., Al Mohanna, M., and Linder, S. (2005) Clinical cancer proteomics: promises and pitfalls. *J Proteome Res*, **4**, 1213–1222.
37. Liu, B.C. and Ehrlich, J.R. (2006) Proteomics approaches to urologic diseases. *Expert Rev Proteomics*, **3**, 283–296.
38. O'Connell, C.D., Atha, D.H., and Jakupciak, J.P. (2005) Standards for validation of cancer biomarkers. *Cancer Biomarkers*, **1**, 233–239.
39. Wang, W.X., Zhou, H.H., Lin, H., Roy, S., Shaler, T.A., Hill, L.R., Norton, S., Kumar, P., Anderle, M., and Becker, C.H. (2003) Quantification of proteins and metabolites by mass spectrometry without isotopic labeling or spiked standards. *Anal Chem*, **75**, 4818–4826.
40. Fraser, C.G. (2001) *Biological Variation: from Principles to Practice*, AACC Press, Washington, DC, pp. 67–70.
41. Brereton, R.G. (2003) Chemometrics: Data Analysis for the Laboratory and Chemical Plant.

42. Mischak, H., Apweiler, R., Banks, R.E., Conaway, M., Coon, J., Dominczak, A., Ehrich, J.H.H., Fliser, D., Girolami, M., Hernjakob, H., Hochstrasser, D., Jankowski, J., Julian, B.A., Kolch, W., Massy, Z.A., Neusuess, C., Novak, J., Peter, K., Rossing, K., Schanstra, J., Semmes, O.J., Theodorescu, D., Thongboonkerd, V., Weissinger, E.M., Van Eyk, J.E., and Yamamoto, T. (2007) Clinical proteomics: a need to define the field and to begin to set adequate standards. *Proteome Clin Appl*, **1**, 148–156.
43. Liu, P. and Hwang, J.T. (2007) Quick calculation for sample size while controlling false discovery rate with application to microarray analysis. *Bioinformatics*, **23**, 739–746.
44. Thongboonkerd, V., Chutipongtanate, S., and Kanlaya, R. (2006) Systematic evaluation of sample preparation methods for gel-based human urinary proteomics: quantity, quality, and variability. *J Proteome Res*, **5**, 183–191.
45. Fiedler, G.M., Baumann, S., Leichtle, A., Oltmann, A., Kase, J., Thiery, J., and Ceglarek, U. (2007) Standardized peptidome profiling of human urine by magnetic bead separation and matrix-assisted laser desorption/ionization time-of-flight mass spectrometry. *Clin Chem*, **53**, 421–428.
46. Jurgens, M., Appel, A., Heine, G., Neitz, S., Menzel, C., Tammen, H., and Zucht, H.D. (2005) Towards characterization of the human urinary peptidome. *Comb Chem High Throughput Screen*, **8**, 757–765.
47. Lauridsen, M., Hansen, S.H., Jaroszewski, J.W., and Cornett, C. (2007) Human urine as test material in 1H NMR-based metabonomics: recommendations for sample preparation and storage. *Anal Chem*, **79**, 1181–1186.
48. Norden, A.G., Rodriguez-Cutillas, P., and Unwin, R.J. (2007) Clinical urinary peptidomics: learning to walk before we can run. *Clin Chim Acta*, **53**, 375–376.
49. van Midwoud, P.M., Rieux, L., Bischoff, R., Verpoorte, E., and Niederlander, H.A. (2007) Improvement of recovery and repeatability in liquid chromatography-mass spectrometry analysis of peptides. *J Proteome Res*, **6**, 781–791.
50. Burtis, C.A., Ashwood, E.R., and Tietz, N.W. (2005) *Tietz Textbook of Clinical Chemistry*, 4th edn., Saunders, Philadelphia
51. Listgarten, J. and Emili, A. (2005) Statistical and computational methods for comparative proteomic profiling using liquid chromatography-tandem mass spectrometry. *Mol Cell Proteomics*, **4**, 419–434.
52. Sysi-Aho, M., Katajamaa, M., Yetukuri, L., and Oresic, M. (2007) Normalization method for metabolomics data using optimal selection of multiple internal standards. *BMC Bioinformatics*, **8**, 93.
53. Radulovic, D., Jelveh, S., Ryu, S., Hamilton, T.G., Foss, E., Mao, Y.Y., and Emili, A. (2004) Informatics platform for global proteomic profiling and biomarker discovery using liquid chromatography-tandem mass spectrometry. *Mol Cell Proteomics*, **3**, 984–997.
54. Tibshirani, R., Hastie, T., Narasimhan, B., and Chu, G. (2002) Diagnosis of multiple cancer types by shrunken centroids of gene expression. *Proc Natl Acad Sci U S A*, **99**, 6567–6572.
55. Tibshirani, R., Hastie, T., Narasimhan, B., Soltys, S., Shi, G., Koong, A., and Le, Q.T. (2004) Sample classification from protein mass spectrometry, by 'peak probability contrasts'. *Bioinformatics*, **20**, 3034–3044.
56. Wagner, M., Naik, D., and Pothen, A. (2003) Protocols for disease classification from mass spectrometry data. *Proteomics*, **3**, 1692–1698.
57. Hilario, M., Kalousis, A., Muller, M., and Pellegrini, C. (2003) Machine learning approaches to lung cancer prediction from mass spectra. *Proteomics*, **3**, 1716–1719.

58. Jansen, J.J., Hoefsloot, H.C.J., van der Greef, J., Timmerman, M.E., and Smilde, A.K. (2005) Multilevel component analysis of time-resolved metabolic fingerprinting data. *Anal Chim Acta*, **530**, 173–183.
59. Smilde, A.K., Jansen, J.J., Hoefsloot, H.C.J., Lamers, R.J.A.N., van der Greef, J., and Timmerman, M.E. (2005) ANOVA-simultaneous component analysis (ASCA): a new tool for analyzing designed metabolomics data. *Bioinformatics*, **21**, 3043–3048.

23
Surface-Enhanced Laser Desorption/Ionization for Urinary Proteome Analysis

Prasad Devarajan and Gary F. Ross

23.1
Introduction

A critical need in clinical proteomics involves the discovery of validated protein biomarkers for the early diagnosis and prediction of specific disease conditions. In addition, biomarkers are also needed for discerning disease subtypes, identifying the etiology and pathogenesis, predicting the severity (risk stratification for prognostication and to guide therapy), and monitoring the response to interventions. Furthermore, biomarkers may play a critical role in expediting the drug development process. The Critical Path Initiative issued by the FDA in 2004 states that "Additional biomarkers (quantitative measures of biologic effects that provide informative links between mechanism of action and clinical effectiveness) and additional surrogate markers (quantitative measures that can predict effectiveness) are needed to guide product development." The concept of developing a new toolbox for earlier diagnosis of disease states is also prominently featured in the NIH Road Map for biomedical research [1].

Biomarker discovery efforts frequently require screening technologies that are sensitive, reproducible, and have high-throughput capacity. These properties can allow for reliable assessment of large sample numbers, which are often available only in limited volumes. One technology which possesses each of these features is surface-enhanced laser desorption/ionization mass spectrometry (SELDI-MS). SELDI, introduced by Hutchens and Yip [2], combines selective retentate chromatography with time-of-flight (TOF) mass spectrometry. SELDI techniques have been the subject of a rapidly growing literature, featuring biomarker studies in a variety of fields including oncology [3–10], neurology [11–13], toxicology [14–17], infectious disease [18–21], and nephrology [22–33]. This chapter updates the readers on the current SELDI methods and protocols that are especially pertinent to urinary proteome analysis.

SELDI instrumentation and products were initially brought to market in 1997 by Ciphergen Biosystems, Inc (Fremont, CA). In November 2006, Bio-Rad Laboratories (Hercules, CA) purchased the rights to produce and market SELDI instrumentation, products, and services. Table 23.1 reviews the nomenclature of

Renal and Urinary Proteomics: Methods and Protocols. Edited by Visith Thongboonkerd

ISBN: 978-3-527-31974-9

Table 23.1 SELDI-TOF-MS instrumentation and software programs.

Ciphergen Biosystems (1997–2006)	Bio-Rad Laboratories (2006 to present)
Instruments	
ProteinChip System 4000 (PCS4000)	ProteinChip SELDI System
Personal Edition[a]	Personal Edition
Enterprise Edition[b]	Enterprise Edition
Autobiomarker Edition[c]	
Protein Biology Systems (PBS)	
PBS I (1997)	
PBS II (2000)	
PBS IIc (2002)	
Software	
Ciphergen Express Data Manager[d]	ProteinChip Data Manager
ProteinChip Software[e]	
Biomarker Wizard[f]	
Biomarker Pattern Software[g]	ProteinChip Pattern Analysis Software

[a] One chip array loaded manually into sample post.
[b] Autoloader allows processing of 8 × 12 cassettes; up to 14 cassettes can be read at once.
[c] Enterprise version is supplemented with Biomek 3000 (Beckman) for robotic chip processing.
[d] Instrument operation and relational database analysis for ProteinChip Systems and PBS IIc.
[e] Instrument operation and relational database analysis for Protein Biology Systems.
[f] Peak cluster analysis program, incorporated into ProteinChip Data Manager.
[g] Supervised classification software, incorporated into ProteinChip Pattern Analysis Software.

ProteinChip systems and software currently marketed by Bio-Rad and versions previously marketed by Ciphergen.

Unique to the SELDI process are specific chip surfaces, which can bind a subset of proteins present in a given sample. The ProteinChip surfaces contain either chromatographic moieties or preactivated chemistries to which user-defined "bait" molecules can be bound. The various ProteinChip arrays as well as their applications in biomarker discovery profiling and protein characterization are discussed below.

23.2 SELDI Chromatographic Arrays

23.2.1 Normal Phase (ProteinChip NP20 Arrays)

The NP20 chips contain silicon oxide (SiO_2) surface groups capable of electrostatic and dipole – dipole interactions. Hydrophilic and charged amino acid residues on the protein surface can bind, and therefore the NP20 array is relatively nonselective. Buffers with a wide pH range are tolerated. Contaminating salts and detergents can easily be removed by washing with distilled water. These properties allow NP20 arrays to be used in collecting mass calibration spectra using defined peptide or protein standards. The NP20 array is useful for monitoring larger scale chromatographic column eluant fractions collected in purification protocols. In addition, NP20 spectra can be used to assess the quality and integrity of protein and peptide reagents (i.e., purified immunoglobulins).

23.2.2 Anion Exchange (ProteinChip Q10 Arrays)

The Q10 chips contain quarternary ammonium groups, providing a cationic surface that can take part in electrostatic interactions with negatively charged aspartic acid and glutamic acid residues. Binding is highly dependent on buffer pH and ionic strength. In the profiling phase of biomarker discovery, permissive binding conditions (i.e., 50 mM Tris-HCl, pH 8–9) are recommended. However, more stringent binding buffers (i.e., 50 mM sodium acetate, pH 5) can also be useful in selecting for strongly acidic proteins. If the isoelectric point (pI) of a target protein is known, a buffer pH at least one unit greater than that pI is recommended.

23.2.3 Cation Exchange (ProteinChip CM10 Arrays)

The CM10 chips are derivatized with carboxylate groups, thus providing an anionic surface. Electrostatic interactions with positively charged arginine, lysine, and histidine residues can take place. As with the Q10 arrays, binding is dependent on buffer pH and ionic strength. Permissive binding conditions (i.e., 100 mM sodium acetate, pH 4.0) are again useful in initial biomarker discovery studies. More stringent binding buffers (i.e., 50 mM sodium Hepes, pH 7–8) will select for strongly basic proteins. If the pI of a target protein is known, a buffer pH at least one unit lower than that pI is recommended.

23.2.4 Immobilized Metal Affinity Capture (ProteinChip IMAC Arrays)

The immobilized metal affinity capture (IMAC) chips incorporate nitrilotriacetic acid (NTA) groups, which form stable octahedral complexes with polyvalent metal ions (i.e., Cu^{+2}, Ni^{+2}, Zn^{+2}, Fe^{+3}, Ga^{+3}). Once "charged" with a selected metal ion, two free sites on the formed octahedral complex can interact with specific amino acids (e.g., His, Cys, Trp) or phosphorylated amino acids (when using Fe^{+3} or Ga^{+3} as metal). Binding buffers (pH 6–8) with high ionic strength (0.5 M NaCl) are commonly used, promoting higher selectivity. Specific proteins can be eluted with increasing concentrations of competitors (such as imidazole) that can displace the protein from the coordinated metal.

23.2.5
Reverse Phase (ProteinChip H4 and H50 Arrays)

These surfaces contain methylene group chains of varying length: H4-(C_{16}), and H50 (C_6–C_{12}). Proteins bind via partitioning of surface hydrophobic residues (Ala, Val, Ile, Leu, Phe, Trp, Tyr) into the lipophilic chip surface. Two methods of binding can be utilized with the H4 and H50 arrays. In the classical reverse-phase mode, aqueous buffer with low organic content (i.e., 10% acetonitrile) is used for binding. Proteins or peptides can be selectively eluted by raising the organic solvent content of subsequent washes. These surfaces can also be used in a hydrophobic interaction chromatography (HIC) format. Here, buffers having high ionic strength (2 M ammonium sulfate) are used for binding. Surface polar groups are occupied by the high salt content. Surface nonpolar groups again interact with the lipophilic chip solid phase. By reducing the buffer ionic strength, proteins can be selectively rehydrated and eluted from the chip surface.

23.2.6
Preactivated Surfaces (RS100 and PS20 ProteinChip Arrays)

ProteinChip RS100 Arrays are preactivated with carbonyldiimidazole groups, and ProteinChip PS20 Arrays are preactivated with epoxide groups. These two reactive SELDI chip surfaces allow specific, user-defined proteins to be covalently bound via free primary amine groups. The PS20 epoxide group can also link via free sulfhydryl (–SH) groups. Those specific "bait" molecules can then serve as affinity surfaces for selective adsorption of complimentary proteins. Typical applications include immunoassays, receptor–ligand binding studies, and transcription factor analysis. Both surfaces are susceptible to nucleophilic attack, principally by protein amine groups. Therefore, the coupling step must be carried out in amine-free buffers. In addition, free sulfhydryl compounds (i.e., 2-mercaptoethanol, dithiothreitol) also should not be present.

23.2.7
Specialty Arrays (ProteinChip Gold, SEND-ID, and PG20 Arrays)

In the ProteinChip Gold Array, the gold chip surface is nonselective in binding characteristics. When samples are allowed to dry on the gold surface followed by matrix application, the SELDI system is comparable to a conventional MALDI-TOF instrument. In the ProteinChip SEND-ID Array, the SEND-ID chip surface has the alpha-cyano-4-hydroxy cinnamic acid (CHCA) matrix incorporated in adition to C_{18} chemistry. This feature greatly reduces the background noise due to desorption of the matrix and thereby increases the sensitivity in the low-mass region. The SEND-ID chip is commonly used to characterize specific proteolytic enzyme digests. In the ProteinChip PG20 Array, Protein G is coupled to the PS20 surface. The PG20 array can be utilized in the development of specific immunoassays.

23.3 Sample Preparation for SELDI

The SELDI retentive chromatography process has been shown to accommodate a wide variety of sample types including serum, plasma, urine, cerebral spinal fluid, cell and tissue lysates, subcellular fractions, conditioned cell culture media, and laser-capture microdissected tissue. Because specific proteins are retained on the chip surface, the sample does not need to be desalted prior to application, thus facilitating the use of a wide variety of sample extraction and lysis buffers. However, as the SELDI MS system involves positive ion flight, it is recommended that SDS be removed from lysis buffers. Nonionic detergents (i.e., Triton X-100, CHAPS) can be substituted.

Because biological samples often contain a complex assortment of proteins with distinct chemical properties, it can be advantageous to prefractionate samples prior to application to the SELDI chip surface. This approach is especially useful when profiling serum or plasma samples, because highly abundant proteins (albumin, immunoglobulins) can preclude binding of other proteins to the chip surface [34]. Strong anion exchange fractionation of serum using Bio-Rad's Serum fractionation kit and buffers is outlined in Figure 23.1. The separated fractions are then profiled on different chip chemistries, excluding the Q10 anion exchange chip.

An alternative prefractionation technique, which also allows enrichment of lower abundance proteins, involves Bio-Rad Equalizer Beads. The beads possess synthetic peptides produced in a combinatorial manner resulting in a wide variety of affinity surfaces. Equalizer Bead eluates have shown to increase the number of detectable SELDI peaks and SDS-PAGE spots in plasma and urine studies [35, 36].

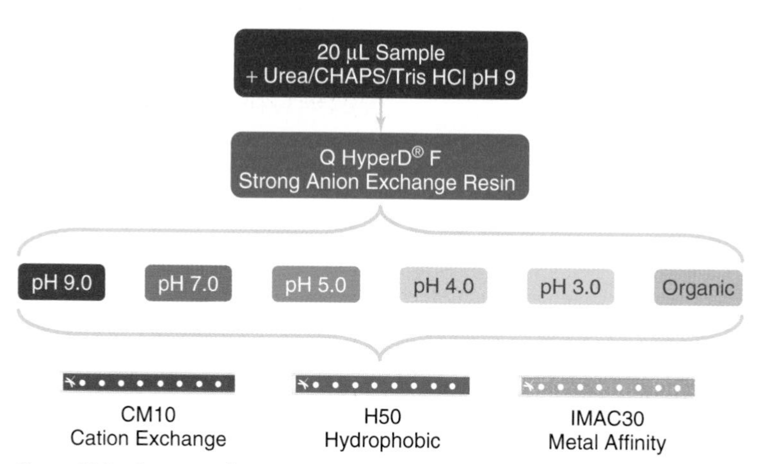

Figure 23.1 Serum prefractionation scheme. Twenty microliters serum is preincubated with 30 μl of 50 mM Tris-HCl, pH 9, 9 M urea, 2% CHAPS. This optional step favors dissociation or protein complexes prior to addition to Q HyperD F anion exchange resin available in filtration plate or spin column formats. Successive elution with buffers of decreasing pH and a final organic solvent wash provide six fractions (Q1–Q6), which can be profiled on different ProteinChip arrays.

Step 3: Complex protein sample is placed on a ProteinChip Array

- Affinity capture – proteins bind to chemical or biological sites on the ProteinChip surface

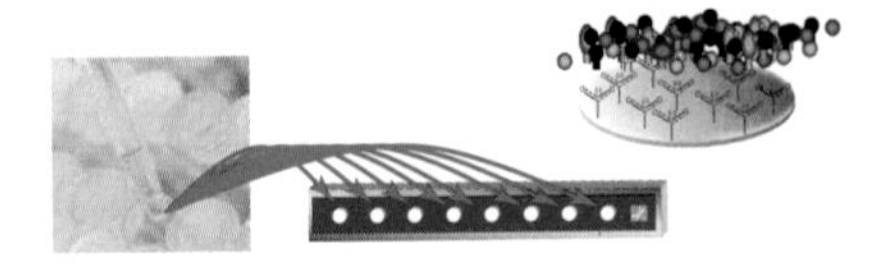

Step 4: Remove unbound proteins

- Wash the ProteinChip with appropriate stringency buffer
- Bound proteins are retained

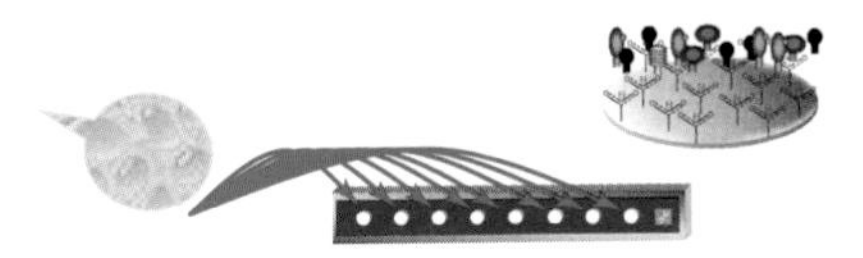

Step 5: Add energy absorbing molecules or "matrix"

- EAM is applied to each spot to facilitate desorption and ionization in the TOF-MS Chip Reader

Figure 23.2 SELDI chip processing. A selected ProteinChip array (step 1) is first preequilibrated in a suitable binding buffer (step 2). In step 3, the sample is applied, typically as a 1 : 10 or 1 : 20 dilution into binding buffer and incubated for 30–60 minutes. On the basis of the selective affinities for the chip surface, a subset of proteins is retained and concentrated on the chip surface. In step 4, loosely bound proteins are removed by washing (3 × 5 minutes) with buffer. In step 5, after a brief low ionic strength wash, energy absorbing molecules (EAMs) are added and allowed to cocrystallize with the retained proteins on the chip surface.

23.4 Basic SELDI Process

A major attribute inherent to the retentate chromatography feature of SELDI is the flexibility of experimental design allowed by the system. This point can be appreciated by an overview of SELDI chip processing (Figure 23.2; steps 3–5 are illustrated).

23.4.1 Selection of Chip Surface

As described above, a variety of chemical and user-designed chip surfaces can be utilized. The chips are sold in an 8 × 12 format. Small sample volumes (5–8 µl) can be applied directly onto the chip surfaces. Alternatively, the chips can be assembled within a Bioprocessor assembly, which includes a fitted reservoir for application of larger sample volumes (up to 400 µl).

23.4.2 Preequilibration Step

For a given chip type, a variety of binding buffers can be utilized. Buffer pH, ionic strength, presence of nonionic detergents (i.e., Triton X-100, CHAPS), or denaturants (urea, GuHCl) can all be varied for experimental optimization.

Typically, the chip arrays are preincubated in a selected binding buffer for two 5-minute periods prior to addition of sample. Depending on the selected chip surface, additional preequilibration steps may be required. IMAC chips are charged with selected metals (e.g., 50 μl of 0.1 M copper sulfate) for 5 minutes. When using copper metal, an additional neutralization wash with 0.1 M sodium acetate, pH 4 is included prior to preincubation with binding buffer. The hydrophobic carbon chains on the H4 and H50 chips are activated by incubation with 50% acetonitrile (or methanol) for 5 minutes prior to preincubation with binding buffer. When using the preactivated RS100 and PS20 arrays to which specific "bait" molecules have been covalently bound, the coupling step is followed by a wash step to remove unbound ligand and then incubation in amine buffer (50 mM Tris-HCl, pH 8, or 0.5 M ethanolamine). This serves to block unreacted sites on the chip surface. Finally, the chips can be preequilibrated in the selected binding buffer.

23.4.3 Binding Step

The sample can then be diluted into the same binding buffer and incubated on the chip surface. The length of time for the binding step can be varied. Thirty to sixty minutes of the binding step are common for biomarker discovery profiling studies. Appropriate binding times will be expected to vary when using ligand-coupled RS100 or PS20 arrays. For example, binding of specific antigens by antibody-coupled arrays may require overnight incubation at 4 °C.

23.4.4 Washing Step

The chip surfaces can then be washed with buffer of the same or increased stringency relative to the binding buffer. Three 5-minutes washes with the same binding buffer are common. However, variable wash buffers can provide valuable information regarding protein – protein interactions as well as in optimization of protein purification strategies (below). Finally, a low ionic strength wash (distilled water or 5 mM sodium HEPES, pH 7) is recommended to remove high concentrations of salt ions. This will help reduce the presence of sodium adducts, which may contribute to the mass heterogeneity of a given peptide/protein.

23.4.5 Application of Energy Absorbing Molecules (EAMs)

Energy absorbing molecules (EAMs) is a generic name for molecules that assist in desorption and ionization of the analyte. These "matrix" molecules are applied in organic solvent, solubilizing many proteins on the array surface. A standard EAM solvent contains 50% acetonitrile and 0.5% TFA. As the EAM solvent evaporates, proteins cocrystallize with the matrix. Commonly used EAM include CHCA) and sinipinic acid (SPA). CHCA is useful in peptide profiling as matrix ions are generally <600 Da. However, production of multiply protonated protein ions can be seen. The matrix ions for SPA can be >1000 Da, with a lesser tendency to form multiply protonated ions. SPA is more commonly used over a wide mass range (2–200 kDa) in biomarker discovery study. The processed chip array is then ready to be loaded into a ProteinChip SELDI TOF-MS for spectra acquisition.

23.5 Data Acquisition

The chip array, containing its complement of bound peptides or proteins, is transferred into the sample post of the ProteinChip SELDI-TOF Reader (Bio-Rad Laboratories markets the PCS4000 edition ProteinChip Reader; an earlier version, protein biology system (PBS)IIc, accounts for a larger number of SELDI instruments in the field and is also supported by Bio-Rad). The sample is then equilibrated with the evacuated flight tube (Figure 23.3). The chip surface can then be impacted by a focused nitrogen laser (337 nm). The energy absorbed facilitates proton transfer from the cocrystallized matrix compound and desorption of the protein from the chip surface. The potential difference across the flight tube draws the positively charged ions to the detector. Impact of protein ions on the detector is converted into an ion current displayed in the form of a chromatographic tracing.

In designing acquisition protocols, several variables can be adjusted to achieve well-resolved peaks with optimal recovery. The laser intensity required to desorb a

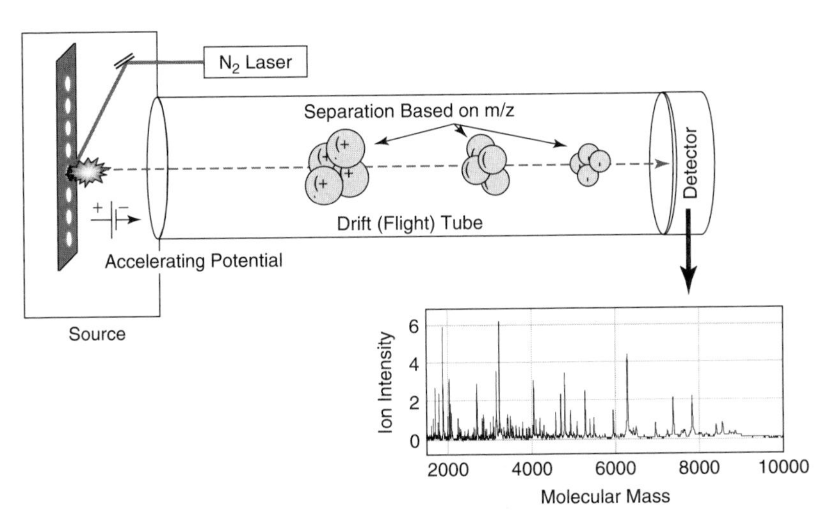

Figure 23.3 SELDI-TOF schematic. The processed ProteinChip array is positioned in line with a 337-nm nitrogen laser. High-energy laser pulses induce proton transfer from EAM, desorption of protein from the chip surface, and entry into the flight tube. The potential difference across the evacuated flight tube promotes the *m/z* dependent drift of H+ ions to the detector. Impact of ions is converted to ion current and displayed chromatographically as a function of the time-of-flight, calibrated to molecular mass.

given protein efficiently increases with the mass of the peptide/protein. However, measures must be taken to control the intensity for detector saturation, signal decay, and to minimize spectral noise particularly evident in the low-mass (<5 kDa) region.

23.5.1
Laser Intensity

The SELDI-MS has variable laser settings, which must be optimized for a given target mass region. The ProteinChip SELDI Enterprise and Personal Systems have the advanced feature that its laser settings are defined in nanojoules (nJ), such that doubling the intensity actually doubles the energy imparted on the chip target. Laser intensities from 500 to 15 000 nJ can be utilized.

23.5.2
Mass Range and Attenuation

The ProteinChip SELDI-MS has detection capabilities extending above 300 kDa. However, spectra with such a wide mass range will have an increasingly large accompanying data file. The mass range should fit the experimental purpose, (i.e., peptide mapping can be done from 0 to 5 kDa). Peak resolution within a given mass range can be optimized by the focus mass setting. This can be especially critical in the case of peptide digest spectra. To guard against detector saturation due to desorbed matrix, the ProteinChip Enterprise and Personal Systems have an adjustable mass attenuation feature, which effectively "deamplifies" the detector until a proscribed time (set by the corresponding mass value) has elapsed. The PBSIIc instrument uses laterally positioned negatively charged deflector plates, which can be activated, transiently, suppressing ion entry into the drift tube. This strategy diminishes the impact of matrix ions onto the detector, also limiting detector saturation. These features diminish the corresponding noise as a consequence of high laser intensity settings.

23.5.3
Spot Partition

The method of EAM application allows uniform dispersion of protein/matrix cocrystals. Laser pulses can sample multiple spot positions without the need for the operator to find the best signal position or "hot spot." The SELDI Enterprise and Personal System's raster laser design has 215 "pixels" to which the laser can be focused and allows for complete spot coverage. The PBSIIc laser makes a focused lateral impact across the width of the spot. The lateral spot positions are on a 0–100 scale. Positions between 15 and 85 are commonly used for quantitative analysis. The uniform sample distribution also allows multiple protocols to be run on the same spot. For example, both low-laser and high-laser protocols can be run sequentially, each directed at different spot partitions.

23.5.4
Pulse Number

The laser repeatedly pulses at an individual spot position for a proscribed number of times before moving to the next position. Typically, 10–15 such pulses can allow for optimal peak detection before significant signal decay might ensue.

By varying the factors listed above, optimized protocols can be designed to profile large sample numbers. Typically, reference or pooled samples are used in the protocol design step. Use of automated, optimized protocols is the essence of the high-throughput capability of the SELDI process. The ProteinChip SELDI Data Manager software can facilitate the collection of hundreds of spectra without further operator input.

23.6
Data Analysis

The raw data obtained in collected spectra is a data average of the multiple pulse transients specified in the acquisition protocol. The data includes TOF increment plotted against ion current, displayed as intensity. Each individual spectrum can include 10^5–10^6 data points. Fung *et al.* [37] provide a detailed review of the biostatistical strategies that can be applied to groups of spectra in search of unique protein biomarkers. A few key points are emphasized here. When considering a set of SELDI spectra to be screened for potential biomarkers, they all must contain the following characteristics: (i) sample type, (ii) chip selection, (iii) matrix selection, and (iv) instrument settings. Collected spectra must then be processed in an identical fashion, including baseline fit and subtraction, noise adjustment, and mass calibration. Spectra can be externally calibrated by application of known m/z peptide or protein standards on an NP20 chip. Standards should be chosen to span the mass range of interest. The standard spectrum should be collected using matrix and instrument settings matching the corresponding sample set. A second polynomial equation relating TOF to the standard m/z values can then be saved and applied to the spectral group being analyzed. Such external calibration results in mass accuracy within 0.1% of true mass (95% confidence).

Each spectrum is then normalized relative to the average total ion current across the group [38]. The normalization step provides a correction for possible pipetting errors during processing of the chip arrays and also allows for identification of spectral outliers, which may need to be reprofiled or eliminated from the data set.

The Bio-Rad ProteinChip Data Manager software facilitates the discovery of distinct protein biomarkers, which may be differentially expressed in a subgroup of spectra. Multiple subgroups can be compared simultaneously. Spectra are subjected to expression difference mapping (EDM) analysis, which creates peak clusters within a prescribed mass window. A minimum signal/noise (S/N) threshold must be met within a prescribed percentage of all spectra to assign a peak cluster at a particular m/z value. Spectra are rescreened using a second, less stringent S/N

threshold. Finally, peak intensity values are estimated using the average m/z value of the existing members of the cluster.

ProteinChip Data Manager then utilizes the Mann – Whitney test (nonparametric equivalent of Student t-test) to calculate p-values revealing the comparative levels of each peak across two defined groups (i.e., control versus diseased). This univariate analysis is an initial step toward the discovery of protein biomarkers, which potentially could differentiate normal versus abnormal phenotypes. ProteinChip Data Manager also provides box-and-whisker and scatter plot visualization of peak intensity distributions. Such scatter plots reveal group mean differences as well as the wide distribution often characteristic of clinical samples. Even though large sample numbers may yield significant p-values, the ability for single markers to discriminate the two groups may still be poor. This points to the need for multivariate analysis of spectra groups. ProteinChip Data Manager also contains two multivariate unsupervised learning analysis methods, namely, principle component analysis and hierarchical clustering, to assess the distribution of the spectral data. These tools allow rapid visualization of spectral outliers, can indicate the presence of subgroups within the data set, and can point to key peak clusters that contribute to the variance across the total sample group.

Bio-Rad Pattern Analysis Software is a supervised learning technique, based on a classification and regression tree method. Peak clusters are assessed for the ability to distinguish defined subgroups. A decision tree is constructed using the most discriminating peak clusters as primary splitting factors. Assignment of unknown samples (spectra) to a specific group is based on comparison to multiple threshold peak intensity values. In the development of potential diagnostic proteomic assays, utilization of multiple biomarkers has resulted in increased sensitivity and specificity values [6].

The high-throughput capacity of SELDI data acquisition allows a large number of independent samples to be used in the construction and cross validation of classification trees using pattern analysis software. By testing a robust data set in the biomarker discovery phase, using either univariate or multivariate analysis programs, markers with a higher degree of validation can be the focus of identification and assay design.

23.7 SELDI-Assisted Purification

The selective binding chemistries present on the ProteinChip array surfaces are identical to chemistries present on solid phase bead (cellulose, dextran, zirconia) supports. By varying the binding and wash conditions on a specific chip array, subsequent mass spectra can give a rapid assay for the retention of a specific biomarker. In addition, the differential binding and elution of neighboring contaminant protein peaks can be monitored. Indeed, monitoring of a specific marker to several chemical surfaces can be done in parallel. This allows for the rapid design of potential purification schemes [39–41]. The basic strategy for the enrichment of a protein biomarker based on SELDI-MS data will implement the following features:

1. The particular chip array to which the biomarker of interest binds best are identified.
2. Alternate binding conditions (i.e., changes in pH) are tested to optimize biomarker retention on the chip surface.
3. Various wash procedures are then screened to establish conditions for elution of the biomarker from the chip array.
4. The conditions established for selective binding and elution are then implemented at a larger scale on a compatible solid phase. For example, Bio-Rad Q HyperD zirconia resins are compatible with the ProteinChip Q10 surface chemistry. Fractions collected from the solid phase resin can be monitored using the nonselective ProteinChip NP20 array.
5. When the best enrichment step is established, a control sample, in which the same biomarker is known to be deficient (as in the discovery phase of the study), is processed in parallel.
6. A second enrichment step can then be carried out using an alternate ProteinChip array.

Ultimately, the specific biomarker needs to be identified. Often, the selective enrichment steps are adequate to allow for subsequent SDS-PAGE isolation of the proteins.

23.8 Protein Identification

Following SDS-PAGE, standard protocols for Coomassie Blue staining can be used. Subsequent protease digestion of selected bands can be achieved using the following approaches: in-gel digestion, passive elution of protein from the gel followed by protease digestion in-solution, and passive elution of protein followed by protease digestion on-chip.

23.8.1 In-Gel Digestion

1. To carry out in-gel protease digestion, SDS must first be removed from the gel. The specific Coomassie-stained gel band (destained in solution containing 50% methanol, 10% acetic acid) should be excised, cut into 1-mm sections and transferred to a 0.5-ml microcentrifuge tube. The corresponding MW region of a control lane (containing the processed "deficient" sample) should also be excised. Alternatively, a protein-free sample lane can be processed in parallel.
2. Add 0.4 ml of 50% methanol/10% acetic acid and incubate with agitation for 1 hour. Replace the solution with fresh 50% methanol, 10% acetic acid; incubate with agitation for an additional 30 minutes. Remove solution.
3. Incubate in 0.4 ml of 50% acetonitrile, 0.1 M ammonium bicarbonate, with agitation for 1 hour. Remove solution.

4. Add 50 µl of 100% acetonitrile. Incubate with agitation for 15 minutes. Remove solution.
5. Dry the gel pieces in a SpeedVac for 15 minutes or until completely dry. Prepare trypsin solution. Sequencing-grade trypsin is dissolved in 10 mM HCl (0.2 $\mu g\,\mu l^{-1}$) = 10X stock stored at −20 or −80 °C. A 1X trypsin solution is made by diluting with nine volumes of 25 mM ammonium bicarbonate, pH 8.0.
6. Add 10 µl of 1X trypsin solution to the dried gel piece(s) in a 0.5-ml microcentrifuge tube. Incubate at room temperature for 15 minutes to rehydrate the gel. Process the "no protein" control tube in the same manner.
7. If necessary, add up to 40 µl of 25 mM ammonium bicarbonate, pH 8.0 to cover the gel. Only add as much as necessary to cover the pieces so that the resultant digest solution is as concentrated as possible. Seal the tube tightly and incubate at 37 °C for upto 16 hours. Shorter incubation times may be adopted. Remove 1-µl aliquots for analysis throughout the digestion period to determine the optimum/minimum digestion time.

23.8.2 Passive Elution and Protease Digestion In-Solution

1. After destaining, transfer the gel piece to a fresh microcentrifuge tube and crush the gel piece at the bottom of the tube with a dissection needle.
2. Passive elution of protein from polyacrylamide gels can be accomplished using either of two extraction solutions. FAPH solution: 50% formic acid, 25% acetonitrile, 15% isopropanol, 10% water (v/v/v/v) is recommended for proteins greater than 10 kDa. ACN/TFA solution: 70% acetonitrile, 30 of 0.2% trifluoroacetic acid (v/v) is effective for proteins less than 10 kDa. Add 40 µl elution solution for an entire protein band or 20 µl for half of a band. The solution should completely cover the gel pieces.
3. Sonicate at room temperature for 30 minutes. Vortex at room temperature for 4 hours. The samples can be stored overnight at 4 °C.
4. Collect the eluant in a fresh microcentrifuge tube and rinse the gel pieces with an equal volume of fresh elution solution. Combine the solutions.
5. Evaporate the eluant in a SpeedVac. Redissolve the eluted protein in 5–10 µl water (for low molecular weight proteins) or 0.5% *n*-octyl glucopyranoside in water (for high molecular weight proteins). Vortex for 15 minutes at room temperature.
6. The samples are ready for analysis. If the protein cannot be recovered after the SpeedVac drying, remove the liquid, wash the vial with a small volume of 20% acetonitrile and examine. If a specific ProteinChip Array was used to detect the protein of interest, the same type of array should be used to analyze the retrieved protein. Binding and washing conditions should be similar to those used originally. Use 1 or 2 µl of recovered protein solution to confirm that the protein of interest was eluted from the gel.
7. The eluted protein can then be incubated with protease (i.e., add an equal volume of 50 mM ammonium bicarbonate, containing 2X trypsin) at 37 °C.

8. The proteolytic digests resulting from these two protocols can then be profiled on several ProteinChip arrays. For NP20, apply 1–2 μl of the digested sample. Let dry. Apply two 0.5-μl aliquots of CHCA (20% saturation) allowing 5 minutes for matrix to dry between each application. For H4 or H50, apply 1–2 μl sample as above. The hydrophobic chains allow for a brief wash step with 2 μl water to reduce salt contaminants. Apply two 0.5-μl aliquots of 20% CHCA. For SEND-ID, dilute the digested sample with an equal volume of 50% acetonitrile, 0.2% TFA. Apply 1–2 μl onto the SEND array and let dry. As CHCA is already incorporated onto the surface, the sample is ready to be inserted into the ProteinChip reader.

23.8.3
Protease Digestion On-Chip

1. Protein, which has been passively eluted from a gel band or purified by alternate methods, can be digested directly on the chip surface. From the initial profiling study, a specific ProteinChip array as well as the best binding – washing conditions can be adopted.
2. Add the eluted protein to the array surface. Rinse with water for 1 minute. Repeat two times for a total of three washes.
3. Prepare a spot with elution solution alone and no protein as a control to identify background signal from protease autolysis. Dry the spots for 5 minutes at 60 °C (heat denaturation of protein before digestion).
4. Add 4–4.5 μl of buffer for digestion (i.e., 25 mM ammonium bicarbonate). Add 0.5–1 μl of protease (1–10 ng μl^{-1}). Protein to protease molar ratio should be at least 50 : 1. Significant excess of protease over the protein will produce ion suppression and nonspecific cleavage.
5. Incubate for 2 hours at optimal temperature. Dry the spot at 60 °C for 2 minutes. Add 1 μl of saturated CHCA solution to 50% acetonitrile/0.05% TFA. Allow the array to dry. Read the array using the ProteinChip Reader.
6. Most accurate m/z estimates (within 0.01% true mass, 95% confidence) can be achieved by internal calibration where peptide standards are mixed with the sample digest and m/z values are derived from the same spectrum. The proper dilution of the peptide standard mixture must be determined. Excess standards may lead to ion suppression and/or masking of specific peptide fragments.
7. The unique proteolytic fragment m/z values can then be searched (using Mascot or ProFound search tools) against protein databases (SwissProt or NCBI) for the identification of the protein biomarker [42, 43].

23.8.4
Direct Sequencing Applications

Sequence analysis of peptides bound to Bio-Rad ProteinChip arrays can be obtained using a ProteinChip Tandem MS Interface [44, 45] as a front-end SELDI ion source for the Applied Biosystem/MDS Sciex QStar Hybrid LC/MS/MS System (Model QStar XL). The ProteinChip Interface uses a 337-nm nitrogen laser with a lensed fiber optic, delivering 150 mJ of energy per pulse at 30 pulses per second. Peptides up to 5 kDa can be selected in Q1 and transmitted to Q2 for collisional induced dissociation, using an applied collisional energy of 50 eV kDa^{-1}. Sample spot scanning is controlled via ProteinChip Interface Control Software (Bio-Rad Laboratories), while data analysis and acquisition are carried out using the QStar System's Analyst software (Applied Biosystems, Inc.). This system is capable of high attomole to low femtomole MS and MS/MS sensitivities. The resulting sequence obtained from spectral data can be submitted to the database mining tools Mascot or ProFound for identification.

23.9 Examples of SELDI-Based Urinary Proteome Analysis

Nguyen *et al.* [23] have employed SELDI-TOF-MS technology to identify urinary biomarker patterns that predict acute renal failure (ARF) in patients undergoing cardiopulmonary bypass (CPB). *ARF* was defined as a 50% or greater increase in serum creatinine. In an initial test set, urine aliquots at baseline ($t = 0$) and 2 hours ($t = 2$ hours) were assigned to control ($n = 15$) or ARF ($n = 15$) groups. The SELDI-TOF-MS analysis of the ARF group at $t = 0$ versus $t = 2$ hours consistently showed a marked enhancement of protein biomarkers with m/z of 6.4, 28, 43, and 66 kDa. The same biomarkers were significantly different when comparing control versus ARF groups at $t = 2$ hours. It should be noted that the serum creatinine in these patients did not increase until day 2–3 after surgery. Scatter plots revealed a dramatic increase in peak intensity of all four novel biomarkers in the ARF group at baseline ($t = 0$) versus 2-hours post CPB, with the area under the curve (AUC) of the receiver-operating characteristic (ROC) curve in the 0.90–0.98 range, indicative of excellent biomarkers [23]. These results have now been confirmed in an unbiased prospective validation set of more than 100 patients (Devarajan *et al.*, unpublished). Thus, the SELDI approach has revealed a distinctive ARF fingerprint comprising of at least four biomarkers that are markedly enhanced within 2 hours of CPB in patients who subsequently developed ARF. An important limitation to this study is that it represents a single center analysis involving only children and young adults with congenital heart disease. A second limitation is the exclusion of patients with preexisting renal insufficiency, diabetes, and nephrotoxin use. These results, therefore, need to be validated in a larger population of susceptible patients. It will also be important to confirm the identity of the four biomarkers uncovered by this study, and to determine their individual and collective robustness for the prediction of ARF.

Several investigators have utilized SELDI-based methods to identify urinary protein peaks that predict acute rejection following kidney transplantation. Clarke *et al.* [46] identified five urinary protein peaks that discriminated between patients who had stable kidney function ($n = 17$) from those who had biopsy-proven acute rejection ($n = 17$). Schaub *et al.* [33] identified three prominent peak clusters in patients with acute rejection episodes confirmed by biopsy, which were largely absent from a stable transplant group ($n = 22$) and in normal controls ($n = 28$). O'Riordan *et al.* [47] similarly identified a cluster of prominent protein peaks that distinguished subjects with biopsy-proven acute rejection ($n = 23$) from transplant recipients with stable graft function ($n = 22$) and healthy volunteers ($n = 20$). These investigators have recently identified two of these polypeptides as beta-defensin-1 and alpha-1-antichymotrypsin, and have demonstrated that the ratio of these peptides in the urine represents a biomarker of acute rejection with an AUC of 0.912 [48]. One limitation of these studies in the transplant population is that they have been cross sectional and they analyzed urine in subjects with known acute rejection. A second limitation is the use of different types of chromatographic surfaces, which perhaps explains the fact that each of the studies identified diagnostic protein patterns of differing characteristics, and not a single protein peak was reported in more than one study. It will be important in the future to design studies that are prospective, initiated soon after transplantation, and utilize standardized approaches to determine whether urinary proteome profiling by SELDI can identify patterns and candidates for the early detection of acute rejection, prior to increases in serum creatinine, and preceding pathological changes.

Jahnukainen *et al.* [49] have compared urine samples from transplant recipients with BK virus-associated nephropathy ($n = 21$), acute rejection ($n = 28$), and stable graft function ($n = 29$) by SELDI-TOF. A distinctive pattern of peaks was demonstrated in the BK virus-associated nephropathy group, which was absent from subjects with acute rejection. Identification of the individual protein peaks and designing robust clinical assays for the identified biomarkers are promising approaches for the future.

Two preliminary cross-sectional SELDI-based studies analyzing the urinary proteome in childhood nephrotic syndrome have been reported. Woroniecki *et al.* [50] compared urine samples from 25 patients with idiopathic nephrotic syndrome and 17 normal controls. The urinary proteome distinguished subjects with nephrotic syndrome and further classified steroid-sensitive versus steroid-resistant cases. A peptide of 4144 Da was found to be the single most important classifier. In a related study, Khurana *et al.* [51] demonstrated the presence of five peaks distinguishing patients with steroid-resistant nephrotic syndrome ($n = 5$) from subjects with steroid-sensitive or steroid-dependent nephrotic syndrome ($n = 19$), or with orthostatic proteinuria ($n = 6$). A peak at m/z of 11 117.4 was the most accurate classifier, and has been identified as β_2-microglobulin, a previously described urinary biomarker of renal tubule damage and steroid-resistant nephrotic syndrome. Prospective studies to identify and establish novel biomarkers that may be predictive of steroid-resistant nephrotic syndrome are warranted.

Two recent SELDI-based studies examining the urinary proteome in systemic lupus erythematosus (SLE) have been reported. Mosley *et al.* [52] were able to discriminate between urine samples from patients with inactive ($n = 49$) and active ($n = 26$) lupus nephritis by the identification of two peptides with m/z of 3340 and

3980 that were enhanced in subjects with active disease. Suzuki *et al.* [53] identified a consistent urinary proteomic signature comprising eight biomarker proteins whose peak intensities were consistently enhanced in subjects with SLE nephritis ($n=32$) when compared with normal controls ($n=10$), SLE patients without nephritis ($n=10$), or patients with juvenile idiopathic arthritis ($n=11$). These biomarkers were strongly correlated with renal disease activity and moderately with renal damage. For the diagnosis of active nephritis, the AUC was >0.90 for the identified biomarkers. Taken together, these reports have identified a urine proteomic signature strongly associated with SLE renal involvement and active SLE nephritis.

23.10 Limitations of the SELDI Technology

Previously quoted problems with calibration and variability of reagents have now been largely resolved by the commercial availability of All-in-one peptide/ protein calibration standards, standardized binding and wash buffers, and standardized protocols for acquisition and normalization of spectra [38]. Concerns about reproducibility have also been largely mitigated by recent assessments of platform reproducibility demonstrating that SELDI-TOF can provide comparable results between various institutions [54, 55].

Inherent to the TOF MS technique is the limited ability to resolve large molecular weight proteins. Both the sensitivity and resolution of specific peaks decrease with increasing mass values. Mass spectrometric techniques, including SELDI, are most sensitive in the low mass range, whereas 2-D PAGE is more sensitive in the high mass range, making them highly complementary techniques. SELDI does not detect equivalent numbers of biomarkers in the high mass region (>50 kDa) as can be detected by 2-D PAGE. The majority of SELDI studies reported to date have identified biomarker candidates in the 3–15 kDa region, a mass range that is more difficult for 2-D PAGE.

Identification of the candidate biomarkers is generally more complex and time consuming on the SELDI system compared to 2-D PAGE. In SDS-PAGE, the spot containing the candidate marker can be excised and digested for subsequent identification on a tandem mass spectrometer. Some other mass spectrometric techniques (iTRAQ, MudPIT, etc.) digest proteins prior to analysis, facilitating direct sequence analysis. Markers that are discovered on SELDI are generally intact proteins or biologically generated fragments. While small candidate markers can usually be identified directly using a tandem MS, most proteins require purification prior to identification.

In the biomarker discovery phase of a study, relative peak intensities across different groups can be calculated. However, such initial data is semiquantitative, as the peak intensities do not correspond to precise (i.e., nanograms per milliliter) concentrations. Protein purification and identification is required to facilitate the preparation of a standard curve for quantitative SELDI-based chromatographic or immunological assays.

23.11 Conclusions

The SELDI-TOF-MS technology has emerged as one of the preferred platforms for urinary protein profiling. This approach allows for rapid high-throughput profiling of multiple urine samples in small quantities, detects low molecular weight biomarkers that are typically missed by other platforms, and even uncovers proteins bound to albumin. The commercial availability of the ProteinChip SELDI System, standardized chromatographic solutions, and standardized protocols for acquisition and normalization of spectra have provided investigators with the ability to rapidly and reproducibly analyze large numbers of crude biological samples and to perform both univariate and multivariate statistical analyses. The limitations of this technology lie in the difficulties in resolving larger peptides, quantifying the peaks, and directly identifying the protein species. This promising approach has already yielded novel protein biomarkers and therapeutic targets for kidney diseases such as ARF, transplant rejection, nephrotic syndrome, and lupus nephritis.

References

1. Zerhouni, E. (2003) The NIH Roadmap. *Science*, **302**, 63–65.
2. Hutchens, T.W. and Yip, T.T. (1993) New desorption strategies for the mass spectrometric analysis of macromolecules. *Rapid Commun Mass Spectrom*, **7**, 576–580.
3. Yip, T.T. and Lomas, L. (2002) SELDI ProteinChip array in oncoproteomic research. *Technol Cancer Res Treat*, **1** (4), 273–280.
4. Wadsworth, J.T., Somers, K.D., Cazares, L.H., Malik, G., Adam, B.L., Stack, B.C., Wright, G.L., and Semmes, O.J. (2004) Serum protein profiles to identify head and neck cancers. *Clin Cancer Res*, **10**, 1625–1632.
5. Wilson, L.L., Tran, L., Morton, D.L., and Hoon, D.S. (2004) Detection of differentially expressed proteins in early-stage melanoma patients using SELDI-TOF mass spectrometry. *Ann N Y Acad Sci*, **1022**, 317–322.
6. Zhang, Z., Bast, R.C. Jr, Yu, Y., Li, J., Sokoll, L.J., Rai, A.J., Rosenzweig, F.M., Cameron, B., Wang, Y.Y., Meng, X.Y., Berchuck, A., van Haaften-Day, C., Hacker, N.F., de Bruijn, H.W.A., van der Zee, H.J.A., Jacobs, I.J., Fung, E.T., and Chan, D.W. (2004) Three biomarkers identified from serum proteomic analysis for the detection of early stage ovarian cancer. *Cancer Res*, **64**, 5882–5890.
7. Semmes, O.J., Feng, Z., Adam, B.L., Banez, L.L., Bigbee, W.L., Campos, D., Cazeras, L., Chan, D.W., Grizzle, W.E., Izbicka, E., Kagan, J., Malik, G., McLarren, D., Moul, J.W., Partin, A., Prasanna, P., Rosenzweig, J., Sokoll, L.J., Srivastava, S., Thompson, I., Welsh, M.J., White, N., Winget, M., Yasul, Y., Zhang, Z., and Zhu, L. (2005) Evaluation of serum protein profiling by surface-enhanced laser desorption/ionization time-of-flight mass spectrometry for the detection of prostate caner: I. Assessment of platform reproducibility. *Clin Chem*, **51**, 102–112.

8. Fung, E.T., Yip, T.T., Lomas, L., Wang, Z., Yip, C., Meng, X.Y., Lin, S., Zhang, F., Zhang, Z., Chan, D.W., and Weinberger, S.R. (2005) Classification of cancer types by measuring variants of host response proteins using SELDI serum assays. *Int J Cancer*, **115**, 783–789.
9. Clarke, C.H., Buckley, J.A., and Fung, E. (2005) SELDI-TOF-MS proteomics of breast cancer. *Clin Chem Lab Med*, **43**, 1314–1320.
10. Li, J., Orland, R., White, C.N., Rosenzweig, J., Zhao, J., Seregni, E., Morelli, D., Yu, Y., Meng, X.Y., Zhang, Z., Davidson, N.E., Fung, E.T., and Chan, D.W. (2005) Independent validation of candidate breast cancer serum biomarkers identified by mass spectrometry. *Clin Chem*, **51**, 2229–2235.
11. Beher, D., Wrigley, J., Owens, A.P., and Shearman, M.S. (2002) Generation of C-terminally truncated amyloid-beta peptides is dependent on gamma-secretase activity. *J Neurochem*, **82**, 563–575.
12. Carrette, O., Demalte, I., Scherl, A., Yalkinoglu, O., Corthals, G., Burkhard, P., Hochstrasser, D.F., and Sanchez, J.C. (2003) A panel of cerebrospinal fluid potential biomarkers for the diagnosis of Alzheimer's disease. *Proteomics*, **3**, 1486–1494.
13. Furuta, M., Shiraishi, T., Okamoto, H., Mineta, T., Tabuchi, K., and Shiwa, M. (2004) Identification of pleiotrophin in conditioned medium secreted from neural stem cells. *Dev Brain Res*, **152**, 189–197.
14. Chu, R., Zhang, W., Lim, H., Yeldandi, A.V., Herring, C., Brumfield, L., Reddy, J.K., and Davison, M. (2002) Profiling of acyl-CoA oxidase-deficient and peroxisome proliferator Wy14, 643-treated mouse liver. *Gene Expr*, **10**, 165–177.
15. Dare, T.O., Davies, H.A., Turton, J.A., Lomas, L., Williams, T.C., and York, M.J. (2002) SELDI detection and identification of urinary parvalbumin-alpha. *Electrophoresis*, **23**, 3241–3251.
16. Yuan, M. and Carmichael, W.W. (2004) Detection and analysis of the cyanobacterial peptide hepatotoxins microcystin and nodularin. *Toxicon*, **44**, 561–570.
17. Vermeulen, R., Lan, Q., Zhang, L., Gunn, L., McCarthy, D., Woodbury, R.L., McGuire, M., Podust, V.N., Li, G., Chatterjee, N., Mu, R., Yin, S., Rothman, N., and Smith, M.T. (2005) Decreased levels of CXC-chemokines in serum of benzene-exposed workers identified by array-based proteomics. *Proc Natl Acad Sci U S A*, **102**, 17041–17046.
18. Papadopoulos, M.C., Abel, P.M., Agranoff, D., Stich, A., Tarelli, E., Bell, B.A., Planche, T., Loosemore, A., Saadoun, S., Wilkins, P., and Krishna, S. (2004) A novel and accurate diagnostic test for human African trypanosomiasis. *Lancet*, **363**, 1358–1363.
19. Gravett, M.G., Novy, M.J., Rosenfeld, R.G., Reddy, A.P., Jacob, T., Turner, M., McCormack, A., Lapidus, J.A., Hitti, J., Eschenbach, D.A., Roberts, C.T., and Nagalla, S.R. (2004) Diagnosis of intra-amniotic infection by proteomic profiling and identification of novel biomarkers. *J Korean Am Med Assoc*, **292**, 462–469.
20. Kang, X., Xu, Y., Wu, X., Liang, Y., Wang, C., Guo, J., Wang, Y., Chen, M., Wu, D., Wang, Y., Bi, S., Qiu, Y., Lu, P., Cheng, J., Xiao, B., Hu, L., Gao, X., Liu, J., Wang, Y., Song, Y., Zhang, L., Suo, F., Chen, T., Huang, Z., Zhao, Y., Lu, H., Pan, C., and Tang, H. (2005) Proteomic fingerprints for potential application to early diagnosis of severe acute respiratory syndrome. *Clin Chem*, **51**, 56–64.
21. Buhimschi, I.A., Christner, R., and Buhimschi, C.S. (2005) Proteomic

biomarker analysis of amniotic fluid for identification of intra-amniotic inflammation. *Br J Obstet Gynaecol*, **112**, 173–181.

22. Schaub, S., Wilkins, J., Weiler, T., Sangster, K., Rush, D., and Nickerson, P. (2004) Urine protein profiling with surface-enhanced laser-desorption/ionization time-of-flight mass spectrometry. *Kidney Int*, **65**, 323–332.
23. Nguyen, M., Ross, G., Dent, C., and Devarajan, P. (2005) Early prediction of acute renal injury using urinary proteomics. *Am J Nephrol*, **25**, 318–326.
24. Thongboonkerd, V. (2004) Proteomics in nephrology: current status and future directions. *Am J Nephrol*, **24**, 360–378.
25. Thongboonkerd, V. (2005) Proteomic analysis of renal diseases: unraveling the pathophysiology and biomarker discovery. *Expert Rev Proteomics*, **2**, 349–366.
26. Vidal, B.C., Bonventre, J.V., and Hsu, S.I.-H. (2005) Towards the application of proteomics in renal disease diagnosis. *Clin Sci*, **109**, 421–430.
27. Pisitkun, T., Johnstone, R., and Knepper, M.A. (2006) Discovery of urinary biomarkers. *Mol Cell Proteomics*, **5**, 1760–1771.
28. O'Riordan, E., Gross, S.S., and Goligorsky, M.S. (2006) Technology insight: renal proteomics – at the crossroads between promise and problems. *Nat Clin Pract Nephrol*, **2**, 445–458.
29. Schaub, S., Wilkins, J.A., Rush, D., and Nickerson, P. (2006) Developing a tool for noninvasive monitoring of renal allografts. *Expert Rev Proteomics*, **3**, 497–509.
30. Hortin, G.L. and Sviridov, D. (2007) Diagnostic potential for urinary proteomics. *Pharmacogenomics*, **8**, 237–255.
31. Gonzalez-Buitrago, J.M., Ferreira, L., and Lorenzo, I. (2007) Urinary proteomics. *Clin Chim Acta*, **375**, 49–56.
32. Fliser, D., Novak, J., Thongboonkerd, V., Argiles, A., Jankowski, V., Girolami, M.A., Jankowski, J., and Mischak, H. (2007) Advances in urinary proteome analysis and biomarker discovery. *J Am Soc Nephrol*, **18**, 1057–1071.
33. Schaub, S., Rush, D., Wilkins, J., Gibson, I.W., Weiler, T., Sangster, K., Nicolle, L., Karpinski, M., Jeffery, J., and Nickerson, P. (2004) Proteome-based detection of urine proteins associated with acute renal allograft rejection. *J Am Soc Nephrol*, **15**, 219–227.
34. Koopman, J., Zhang, Z., White, N., Rosenzweig, J., Fedarko, N., Jagannath, S., Canto, M.I., Yeo, C.J., Chan, D.W., and Goggins, M. (2004) Serum diagnosis of pancreatic adenocarcinoma using surface-enhanced laser desorption and ionization mass spectrometry. *Clin Cancer Res*, **10**, 860–868.
35. Thulasiraman, V., Lin, S., Gheorghiu, L., Lathrop, J., Lomas, L., Hammond, D., and Boschetti, E. (2005) Reduction of the concentration difference of proteins in biological liquids using a library of combinatorial ligands. *Electrophoresis*, **26**, 3561–3571.
36. Castagna, A., Cecconi, D., Sennels, L., Rappsilber, J., Guerrier, L., Fortis, F., Boschetti, E., Lomas, L., and Righetti, P.G. (2006) Exploring the hidden human urinary proteome via ligand library beads. *J Proteome Res*, **4**, 1917–1930.
37. Fung, E.T., Weinberger, S.R., Gavin, E., and Zhang, F. (2005) Bioinformatics approaches in clinical proteomics. *Expert Rev Proteomics*, **2**, 847–862.
38. Fung, E. and Enderwick, C. (2002) ProteinChip clinical proteomics: computational challenges and solutions. *Comput Proteomics*, **32**, S34–S41.
39. Weinberger, S.R., Boschetti, E., Santambien, P., and Brenac, V. (2002) SELDI – a new method

for rapid development of process chromatography conditions. *J Chromatogr A*, **782**, 307–316.

40. Shiloach, J., Santambien, P., Trinh, L., Schapman, A., and Boschetti, E. (2003) Endostatin capture from Pichia pastoris culture in a fluidized bed from on-chip process optimization to application. *J Chromatogr A*, **790**, 327–336.
41. Prahalad, A.K., Hickey, R.J., Huang, J., Hoelz, J., Dobrolecki, L., Murthy, S., Winata, T., and Malkas, L.H. (2006) Serum proteome profiles identifies parathyroid hormone physiologic response. *Proteomics*, **6**, 3482–3493.
42. Thulasiraman, V., McCutchen-Maloney, S.L., Motin, V.L., and Garcia, E. (2001) Detection and identification of virulence factors in Yersinia pestis using SELDI ProteinChip System. *BioTechniques*, **30**, 428–432.
43. Wang, S., Diamond, D.L., Hass, G.M., Sokoloff, R., and Vessella, R.L. (2001) Identification of prostate-specific membrane antigen (PMSA) as the target of monoclonal antibody 107-1A4. *Int J Cancer*, **92**, 871–876.
44. Merchant, M. and Weinberger, S.R. (2000) Recent advancements in surface-enhanced laser desorption-ionization-time of flight-mass spectrometry. *Electrophoresis*, **21**, 1164–1177.
45. Reid, G., Gan, B.S., She, Y.M., Ens, W., Weinberger, S., and Howard, J.C. (2002) Rapid identification of probiotic lactobacillus biosurfactant proteins. *Appl Environ Microbiol*, **68**, 977–980.
46. Clarke, W., Silverman, B.C., Zhang, Z., Chan, D.W., Klein, A.S., and Molmenti, E.P. (2003) Characterization of renal allograft rejection by urinary proteomic analysis. *Ann Surg*, **237**, 660–665.
47. O'Riordan, E., Orlova, T.N., Mei, J., Butt, K., Chander, P.M., Rahman, S., Mya, M., Hu, R., Momin, J., Eng, E.W., Hampel, D.J., Hartman, B., Kretzler, M., Delaney, V., and Goligorsky, M.S. (2004) Bioinformatic analysis of the urinary proteome of acute allograft rejection. *J Am Soc Nephrol*, **15**, 3240–3248.
48. O'Riordan, E., Orlova, T.N., Podust, V.N., Chander, P.N., Yanagi, S., Nakazato, M., Hu, R., Butt, K., Delaney, V., and Goligorsky, M.S. (2007) Characterization of urinary peptide biomarkers of acute rejection I renal allografts. *Am J Transplant*, **7**, 930–940.
49. Jahnukainen, T., Malehorn, D., Sun, M., Lyons-Weiler, J., Bigbee, W., Gupta, G., Shapiro, R., Randhawa, P.S., Pelikan, R., Hauskrecht, M., and Vats, A. (2006) Proteomic analysis of urine in kidney transplant patients with BK virus nephropathy. *J Am Soc Nephrol*, **17**, 3248–3256.
50. Woroniecki, R.P., Orlova, T.N., Mendelev, N., Shatat, I.F., Hailpern, S.M., Kaskel, F.J., Goligorsky, M.S., and O'Riordan, E. (2006) Urinary proteome of steroid-sensitive and steroid-resistant idiopathic nephrotic syndrome syndrome of childhood. *Am J Nephrol*, **26**, 258–267.
51. Khurana, M., Traum, A.Z., Alvado, M., Wells, M.P., Guerrero, M., Grall, F., Libermann, T.A., and Schachter, A.D. (2006) Urine proteomic profiling of pediatric nephrotic syndrome. *Pediatr Nephrol*, **21**, 1257–1265.
52. Mosley, K., Tam, F.W.K., Edwards, R.J., Crozier, J., Pusey, C.D., and Lightstone, L. (2006) Urinary proteomic profiles distinguish between active and inactive lupus nephritis. *Rheumatology*, **45**, 1497–1504.

53. Suzuki, M., Ross, G., Brunner, H., and Devarajan, P. (2006) Identification of signature urinary biomarker patterns in lupus nephritis sub-types. *J Am Soc Nephrol*, **17**, 437A.

54. Rai, A.J., Stemmer, P.M., Zhang, Z., Adam, B.L., Morgan, W.T., Caffrey, R.E., Podust, V.N., Patel, M., Lim, L.Y., Shipulina, N.V., Chan, D.W., Semmes, O.J., and Leung, H.C. (2005) Analysis of Human Proteome Organization Plasma Proteome Project (HUPO PPP) reference specimens using surface enhanced laser desorption/ionization-time of flight (SELDI-TOF) mass spectrometry: multi-institution correlation of spectra and identification of biomarkers. *Proteomics*, **5**, 3467–3474.

55. Semmes, O.J., Feng, Z., Adam, B.L., Banez, L.L., Bigbee, W.L., Campos, D., Cazares, L.H., Chan, D.W., Grizzle, W.E., Izbicka, E., Kagan, J., Malik, G., McLerran, D., Moul, J.W., Partin, A., Prasanna, P., Rosenzweig, J., Sokoll, L.J., Srivastava, S., Thompson, I., Welsh, M.J., White, N., Winget, M., Yasui, Y., Zhang, Z., and Zhu, L. (2005) Evaluation of serum protein profiling by surface-enhanced laser desorption/ionization time-of-flight mass spectrometry for the detection of prostate cancer: I. Assessment of platform reproducibility. *Clin Chem*, **51**, 102–112.

24
Capillary Electrophoresis Coupled to Mass Spectrometry for Urinary Proteome Analysis

Petra Zürbig, Eric Schiffer, and Harald Mischak

24.1
Introduction

Human urine plays a central role in clinical diagnostics. Through the centuries, physicians have examined the urine of their patients to diagnose diseases. Hermogenes wrote about the color and other attributes of urine as indicators of certain diseases [1, 2]. The human urinary proteome has been investigated extensively to analyze disease processes affecting the kidney and the urogenital tract [3–5]. In the case that these organs are damaged or their functions compromised, changes in the urine proteome are plausible due to their role in urine production and excretion. However, urinary proteins not only originate from glomerular filtration, but also from tubular secretion, shedding of epithelial kidney and urinary tract cells, exosome secretion, and semen [6–8]. Thus, in principle, urine is a rich source of biomarkers for a wide range of diseases, as these diseases may alter the urinary proteome profile and/or increase urinary protein excretion (proteinuria) [5, 9, 10]. To realize this potential, large-scale studies are needed to reliably analyze the human urine proteome, quantitatively and with meaningful depth. Efforts have included a variety of techniques, for example, two-dimensional electrophoresis with mass spectrometric and/or immunochemical identification of proteins (2DE-MS) [11, 12], liquid chromatography coupled to mass spectrometry (LC-MS) [13, 14], and surface-enhanced laser desorption/ionization mass spectrometry (SELDI-MS) [15], (Table 24.1).

To date, these urinary proteome analyses have revealed more than 1500 different proteins (see Castagna *et al.* [16], Adachi *et al.* [2], and related articles). In addition, proteomics of urine has been applied in studies on, for example, prostate cancer [17–19], bladder cancer [3, 20–22], diabetic nephropathy [23, 24], chronic renal disease [25–28], detection of transplant-associated complications [29–32], multiple myeloma [33, 34], and heavy metal toxicity impairing kidney function [35]. These results are very encouraging. Unfortunately, the majority of these studies comprise only small numbers of samples (100 samples at best), and mostly only two diagnostic groups, namely, diseased and healthy individuals. Therefore, only a limited number of novel potential biomarkers could be validated. However,

Renal and Urinary Proteomics: Methods and Protocols. Edited by Visith Thongboonkerd

ISBN: 978-3-527-31974-9

Table 24.1 Characteristics, advantages, and disadvantages of various mass spectrometry-based proteomic techniques for clinical applications.

Proteomic method	Characteristics	Advantages	Disadvantages
2-DE-MS	Separation with 2-D electrophoresis (first dimension: isoelectric point, second dimension: molecular weight); protein identification with MS and/or MS/MS	Applicable to large molecules; high resolution	Not applicable to peptides <10 kDa; no automation; time-consuming; quantification difficult; expensive
LC-MS	Separation with LC; analysis of peptide masses with MS	Automation; multidimensional; high sensitivity; MS/MS possibility	Time-consuming; sensitive toward interfering compounds; restricted mass range
SELDI-MS	Separation with different surface chemistries of the arrays; analysis of peptide masses with MS	Easy-to-use-system; automation; low sample volume required	Restricted to selected polypeptides; low-resolution MS; interpretation of data difficult without sequence information
CE-MS	Separation with capillary electrophoresis; analysis of peptide masses with MS	Automation; high sensitivity; fast; low sample volumes required; multidimensional; low cost; MS/MS possibility	Not well suited for larger polypeptides (>20 kDa)

proteomes are highly dynamic and directly react to actual (patho-)physiological situations and environmental influences. This is an invaluable advantage as it reflects the current health state of the organism, but it also poses enormous challenges. The related high degree of heterogeneity suggests that it is crucial to identify panels rather than single markers. A useful method to define such diagnostic panels must therefore sufficiently combine reasonable analysis time with high analytical resolution to enable the analysis of large numbers of samples and the recognition of sufficient features to yield robust diagnostic panels.

Capillary electrophoresis coupled to mass spectrometry (CE-MS) comprises a fast analysis method capable of resolving between 1000 and up to 4000 different peptides per sample within approximately 45 minutes (see also Table 24.1). This approach

Table 24.2 Disease conditions and number of individuals studied, which are to date represented in the human urinary proteome database created by CE-MS.

Disease condition	Number	Medical area
Healthy control	455	Controls
Ureteropelvic junction obstruction	173	Renal disorders
Vasculitides	119	
Diabetic nephropathy	61	
IgA nephropathy	57	
Renal diseases, others	35	
Membranous glomerulonephritis	35	
Kidney stones	39	
Focal segmental glomerulosclerosis	29	
Fanconi syndrome	28	
Systemic lupus erythematosus	25	
Minimal change disease	27	
Henoch–Schoenlein purpura	18	
Renal disease, blinded	197	
Renal transplantation	343	Transplantation
Liver transplantation	113	
Hematopoietic stem cell transplantation	69	
Posttransplant lymphoproliferative disorders	53	
Prostate cancer	182	Oncological disease
Benign prostatic hyperplasia	159	
Kaposi sarcoma	69	
Urothelial cancer	98	
Prostatic intraepithelial neoplasia III	34	
Renal cancer	19	
Prostate cancer, blinded	181	
Urothelial cancer, blinded	172	
Renal cancer, blinded	110	
Diabetes mellitus	271	Others
Coronary artery disease	172	
Alzheimer's disease	126	
Human immunodeficiency virus	49	
Intensive care unit patients	62	
Depression	12	
Hepatitis C	10	
Coronary artery disease, blinded	85	

has been utilized to analyze urine samples from healthy volunteers and patients with a variety of different diseases (Table 24.2). Suitable software solutions facilitate standardized raw data processing including peak detection, charge assignment, calibration, and database deposition. The resulting database containing normal and diseased samples not only describes more then 5000 different polypeptides by mass, CE migration time, and MS signal amplitude, but also represents a comprehensive description of the human urinary proteome itself under various (patho-)physiological conditions.

24.2 CE-MS Methodology

The clinical application of CE-MS demands high reproducibility and comparability of acquired data [36]. Previous studies have demonstrated that, in contrast to blood, urine is stable for several hours at room temperature [37, 38]. This is in part due to the fact that proteolytic degradation by endogenous proteases is essentially complete once urine is voided. Although capillary electrophoresis allows the separation of even crude urine samples, salt and higher molecular weight proteins interfere with separation; hence, it is advantageous to remove these compounds. Ultrafiltration in the presence of urea and sodium dodecylsulfate to eliminate protein–protein and other interactions followed by desalting with a PD-10 column seems to serve this purpose.

For the detection of narrow CE-separated analyte zones, a fast and sensitive mass spectrometer is required. Modern electrospray ionization time-of-flight mass spectrometers (ESI-TOF-MS) provide a resolution >10 000 and a mass accuracy <10 ppm, suggesting that CE-ESI-TOF-MS is a well-suited setup. Each individual CE-MS analysis consists of about 1500 mass spectra. The essential information that needs to be extracted is the identity and quantity of detected polypeptides. The MosaiquesVisu (www.proteomiques.com) software identifies, deconvolutes, and quantifies peaks of an entire CE-MS run within 5 minutes with an error rate generally below 2% [39]. Key to the comparative examination of samples is the ability to reliably retrieve identical polypeptides in consecutive samples. To this end, CE migration time and precursor mass are used to assign the identity of a peptide, enhancing the resolution of the analysis by utilizing two independent and reproducible parameters. These parameters are normalized using peptides that are found with high frequency as internal standards [28, 40]. Finally, a list of unambiguously identified and standardized peptides of a given sample enables digital compilation of individual data sets to specific polypeptide panels that are used for biomarker definition (Figure 24.1).

24.3 Urinary Proteome Database

All detected polypeptides have been deposited in a Microsoft SQL database for further analysis. The listed polypeptides are the aggregation of all peptides found in more than 3000 urine samples, containing approximately 1500 polypeptides in each. To eliminate polypeptides that are of low significance, only polypeptides that were found in more than 20% in at least one diagnostic group were included. By applying these limits, approximately 5000 unique polypeptides characterized

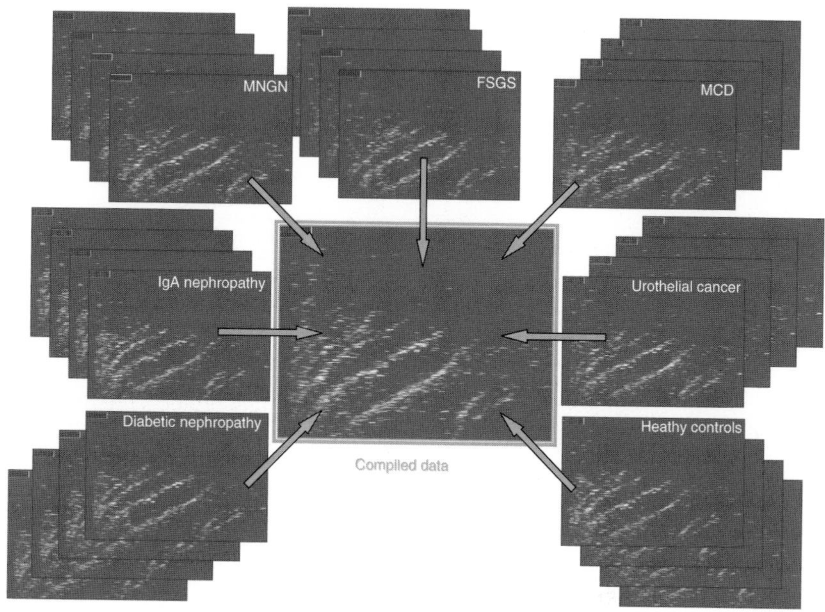

Figure 24.1 Digital data compilation of individual datasets from healthy volunteers as well as from patients with different diseases (diabetic nephropathy, IgA nephropathy, FSGS, MNGN, MCD, urothelial cancer, etc.). Comparison of the different specific panels enables one to pinpoint polypeptides with significant differences in the groups.

by their molecular mass (Da) and normalized CE migration time (minutes) are deposited.

To improve the accurate definition of the polypeptides, we attempted to obtain mass with higher accuracy. For highly accurate mass measurement, several samples have been analyzed using CE coupled online to a Fourier-transform ion cyclotron resonance mass spectrometer (FT-ICR-MS). Owing to the high cost and the lower sensitivity of FT instruments in comparison to TOF instruments, it is not practical to analyze all samples using CE-FT-ICR. Consequently, a set of 20 samples was reanalyzed using CE-FT-ICR. These data were utilized to refine the TOF-MS masses in the human urinary proteome database. As a result, the masses of the listed peptides can be given with a mass deviation of 2 ± 8 ppm (Figure 24.2). This procedure of using FT-ICR-defined calibrants for refinement of TOF-masses combines the high mass accuracy of FT-ICR-MS with the high sensitivity of TOF-MS.

For further validation of polypeptides listed within the proteome database, different MS/MS technologies were applied for polypeptide sequencing. In this context, the direct dependency of CE migration time on the charge density of the analyte represents a valuable key feature of the technology for MS/MS data validation. At the working pH of 2.2, the effective charge of the analyzed polypeptides depends strictly on the number of basic amino acids residues, including the free N-terminus

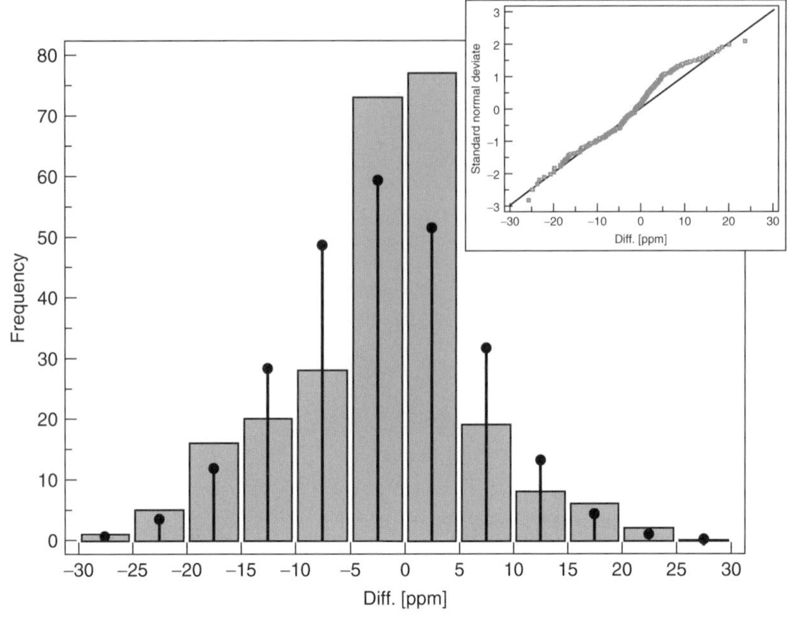

Figure 24.2 The histogram displays the distribution plot of the mass deviation between calculated masses of the known sequences and the measured masses of the database (the black lines depict the theoretic normal distribution). The "Normal plot" which is shown in the small window is a graphical tool to evaluate the normality of the distribution of the data. The horizontal axis of the "Normal plot" shows the mass deviation, and the vertical axis gives the relative frequency in terms of the number of standard deviations from the mean. The mass deviation (squares) almost shows a Normal distribution, because the plot is close to the straight line (gray line).

[41]. Therefore, it is not a prerequisite to use CE-separation for MS/MS sequencing, as the number of basic amino acids in combination with highly accurate masses enables the correlation of the obtained sequence (including number of positively charged residue) to a polypeptide within the database. Most frequently, different fragments of collagens and uromodulin were identified (Figure 24.3). Many of these polypeptides were also found by other research groups such as Adachi *et al.* [2] and Castagna *et al.* [16]. However, some native peptides of proteolytically processed proteins in urine ("top-down" approach [42]) remain undetectable in "bottom-up" approaches. Therefore, top-down approaches appear to be suitable to display the regulated activity of proteases and protease inhibitors [43] by monitoring their substrates. This outweighs the "make or break" challenge of the approach, where fragmentation of some peptides results in elimination of water from an amino acid (asparagine, aspartic acid, glutamine, glutamic acid) [44, 45] or of proline residues caused by partial fragmentation, and the identification of the full sequence becomes uncertain. In this case, typically only MS^n approaches with the use of an ion trap device lead to satisfactory results.

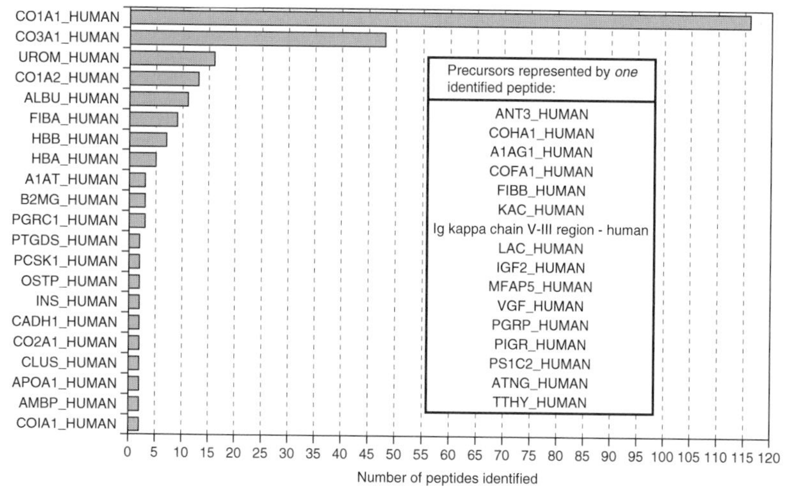

Figure 24.3 Numerical distribution of native peptides with respect to their protein precursors as described by Swiss-Prot entry name. Inset: proteins, from which one native fragment was detected in urine.

Reproducible urine profiling employing the analytical speed of CE-TOF-MS in combination with the high mass accuracy of FT-ICR-based mass calibration, and with peptide sequence data derived from multiple MS/MS platforms, lays the groundwork for the human low-molecular weight urinary proteome in the form of a universal database. This database can be utilized as a tool to validate potential biomarkers for various diseases on a large number of samples. The value of a potentially disease-specific biomarker can be examined by matching it to the database and utilizing the information on those samples currently included within the database. For example, a potential biomarker for cardiovascular disease (CVD), established by examining urine from patients with CVD and comparing with controls, should be evenly distributed in the other controls and disease groups and not be highly overrepresented in a certain oncological disease.

24.4 Biomarkers for Disease

Although CE-MS analysis of urine can identify biomarkers for a variety of diseases of the kidney and urogenital tract [26–28, 46, 47], the variability of polypeptides presents a serious obstacle. Polypeptides in urine change significantly during the day, most likely as a consequence of physical activity, diet, and so on [25, 48]. As a result of this biological variability, reproducibility at the level of single polypeptides is limited. Hence, the clinical usefulness of a single biomarker measurement can only be of low value, even if test accuracy and reproducibility are optimal. In

Table 24.3 CE-MS analysis of urine samples from different indications.

Indications	References
Minimal-change disease	[26, 49]
Focal segmental glomerulosclerosis	[26, 49]
Membranous glomerulonephritis	[26, 27]
IgA nephropathy	[26]
Diabetic nephropathy	[46, 47]
Urothelial cancer	[38]
Prostate cancer	[40]
Ureteropelvic junction obstruction	[50]
Graft-versus-host-disease after allogeneic hematopoietic stem cell transplantation	[29, 51, 52]
Acute tubulointerstitial kidney transplant rejection	[30]
Cardiovascular disease	[53–55]

contrast, a polypeptide panel that consists of an array of well-defined biomarkers is much more robust. Changes in individual analytes will not result in gross changes of the panel. Therefore, we explored whether a common polypeptide panel can be detected in human urine for patients with various disorders (Table 24.3). Diseases affecting urine composition may be directly related to the urogenital tract, but may also be (at first sight) unrelated to it; acute graft-versus-host disease (a-GvHD) and CVD are just two examples of diseases, which may fall into this category. These panels can be used for diagnostic purposes, because each disease produces a specific, characteristic urinary polypeptide panel.

Biomarker patterns based on the urinary proteome analysis are a powerful tool for the diagnosis, stratification, and monitoring of human diseases. These signatures seem to reflect primary pathogenic changes, as well as the reaction of the organism to the disease. Hence, their usefulness extends far beyond the applicability to diseases of the urogenital tract, and may be universally applicable to any disease that produces systemic changes, which is the vast majority of serious illnesses. While genetic analysis can predict the risk of a disease, proteomics can actually determine at which point the risk manifests itself as disease and also permits monitoring of the response to therapeutic measures. Thus, these methods are complementary, but we anticipate that proteomics may have a greater role in personalized medicine. As we now begin to understand the huge individual differences between patients' response to therapy, objective ways to measure these responses will become of prime importance to tailor the therapy to the individual needs of the patient. Here, proteomics has the advantage that monitoring of the therapy success is possible in real time, and adjustments can be made accordingly. This vision is within reach, but its realization entirely depends on the establishment of databases that allow us to quickly compare patients' profiles against healthy controls and other diseased patients in a robust manner.

References

1. Iorio, L. and Avagliano, F. (1999) Observations on the Liber medicine orinalibus by Hermogenes. *Am J Nephrol*, **19**, 185–188.
2. Adachi, J., Kumar, C., Zhang, Y., Olsen, J.V., and Mann, M. (2006) The human urinary proteome contains more than 1500 proteins including a large proportion of membranes proteins. *Genome Biol*, **7**, R80.1–R80.16.
3. Celis, J.E., Rasmussen, H.H., Vorum, H., Madsen, P., Honore, B., Wolf, H., and Orntoft, T.F. (1996) Bladder squamous cell carcinomas express psoriasin and externalize it to the urine. *J Urol*, **155**, 2105–2112.
4. Delanghe, J. (1997) Use of specific urinary proteins as diagnostic markers for renal disease. *Acta Clin Belg*, **52**, 148–153.
5. Marshall, T. and Williams, K. (1996) Two-dimensional electrophoresis of human urinary proteins following concentration by dye precipitation. *Electrophoresis*, **17**, 1265–1272.
6. Oh, J., Pyo, J.H., Jo, E.H., Hwang, S.I., Kang, S.C., Jung, J.H., Park, E.K., Kim, S.Y., Choi, J.Y., and Lim, J. (2004) Establishment of a near-standard two-dimensional human urine proteomic map. *Proteomics*, **4**, 3485–3497.
7. Pieper, R., Gatlin, C.L., McGrath, A.M., Makusky, A.J., Mondal, M., Seonarain, M., Field, E., Schatz, C.R., Estock, M.A., Ahmed, N., Anderson, N.G., and Steiner, S. (2004) Characterization of the human urinary proteome: a method for high-resolution display of urinary proteins on two-dimensional electrophoresis gels with a yield of nearly 1400 distinct protein spots. *Proteomics*, **4**, 1159–1174.
8. Thongboonkerd, V., McLeish, K.R., Arthur, J.M., and Klein, J.B. (2002) Proteomic analysis of normal human urinary proteins isolated by acetone precipitation or ultracentrifugation. *Kidney Int*, **62**, 1461–1469.
9. Shihabi, Z.K., Konen, J.C., and O'Connor, M.L. (1991) Albuminuria vs urinary total protein for detecting chronic renal disorders. *Clin Sci*, **37**, 621–624.
10. Yudkin, J.S., Forrest, R.D., and Jackson, C.A. (1988) Microalbuminuria as predictor of vascular disease in non-diabetic subjects. Islington Diabetes Survey. *Lancet*, **2**, 530–533.
11. Anderson, N.G., Anderson, N.L., and Tollaksen, S.L. (1979) Proteins of human urine. I. Concentration and analysis by two-dimensional electrophoresis. *Clin Chem*, **25**, 1199–1210.
12. Bueler, M.R., Wiederkehr, F., and Vonderschmitt, D.J. (1995) Electrophoretic, chromatographic and immunological studies of human urinary proteins. *Electrophoresis*, **16**, 124–134.
13. Baer, J.C. and Hjelm, M. (1994) Analysis of urinary proteins by liquid chromatography. *Ann Clin Biochem*, **31** (Pt 4), 315–326.
14. Jurgens, M., Appel, A., Heine, G., Neitz, S., Menzel, C., Tammen, H., and Zucht, H.D. (2005) Towards characterization of the human urinary peptidome. *Comb Chem High Throughput Screen*, **8**, 757–765.
15. Cadieux, P.A., Beiko, D.T., Watterson, J.D., Burton, J.P., Howard, J.C., Knudsen, B.E., Gan, B.S., McCormick, J.K., Chambers, A.F., Denstedt, J.D., and Reid, G. (2004) Surface-enhanced laser desorption/ionization-time of flight-mass spectrometry (SELDI-TOF-MS): a new proteomic urinary test for patients with urolithiasis. *J Clin Lab Anal*, **18**, 170–175.
16. Castagna, A., Cecconi, D., Sennels, L., Rappsilber, J., Guerrier, L., Fortis, F., Boschetti, E., Lomas, L., and Righetti, P.G. (2005) Exploring the hidden human urinary proteome via ligand library beads. *J Proteome Res*, **4**, 1917–1930.
17. Edwards, J.J., Anderson, N.G., Tollaksen, S.L., von Eschenbach,

A.C., and Guevara, J. Jr (1982) Proteins of human urine. II. Identification by two-dimensional electrophoresis of a new candidate marker for prostatic cancer. *Clin Chem*, **28**, 160–163.

18. Grover, P.K. and Resnick, M.I. (1997) High resolution two-dimensional electrophoretic analysis of urinary proteins of patients with prostatic cancer. *Electrophoresis*, **18**, 814–818.
19. Rehman, I., Azzouzi, A.R., Catto, J.W., Allen, S., Cross, S.S., Feeley, K., Meuth, M., and Hamdy, F.C. (2004) Proteomic analysis of voided urine after prostatic massage from patients with prostate cancer: a pilot study. *Urology*, **64**, 1238–1243.
20. Irmak, S., Tilki, D., Heukeshoven, J., Oliveira-Ferrer, L., Friedrich, M., Huland, H., and Ergun, S. (2005) Stage-dependent increase of orosomucoid and zinc-alpha2-glycoprotein in urinary bladder cancer. *Proteomics*, **5**, 4296–4304.
21. Rasmussen, H.H., Orntoft, T.F., Wolf, H., and Celis, J.E. (1996) Towards a comprehensive database of proteins from the urine of patients with bladder cancer. *J Urol*, **155**, 2113–2119.
22. Saito, M., Kimoto, M., Araki, T., Shimada, Y., Fujii, R., Oofusa, K., Hide, M., Usui, T., and Yoshizato, K. (2005) Proteome analysis of gelatin-bound urinary proteins from patients with bladder cancers. *Eur Urol*, **48**, 865–871.
23. Sharma, K., Lee, S., Han, S., Lee, S., Francos, B., McCue, P., Wassell, R., Shaw, M.A., and RamachandraRao, S.P. (2005) Two-dimensional fluorescence difference gel electrophoresis analysis of the urine proteome in human diabetic nephropathy. *Proteomics*, **5**, 2648–2655.
24. Thongboonkerd, V., Barati, M.T., McLeish, K.R., Benarafa, C., Remold-O'Donnell, E., Zheng, S., Rovin, B.H., Pierce, W.M., Epstein, P.N., and Klein, J.B. (2004) Alterations in the renal elastin-elastase system in type 1 diabetic nephropathy identified by proteomic analysis. *J Am Soc Nephrol*, **15**, 650–662.
25. Fliser, D., Wittke, S., and Mischak, H. (2005) Capillary electrophoresis coupled to mass spectrometry for clinical diagnostic purposes. *Electrophoresis*, **26**, 2708–2716.
26. Haubitz, M., Wittke, S., Weissinger, E.M., Walden, M., Rupprecht, H.D., Floege, J., Haller, H., and Mischak, H. (2005) Urine protein patterns can serve as diagnostic tools in patients with IgA nephropathy. *Kidney Int*, **67**, 2313–2320.
27. Neuhoff, N., Kaiser, T., Wittke, S., Krebs, R., Pitt, A., Burchard, A., Sundmacher, A., Schlegelberger, B., Kolch, W., and Mischak, H. (2004) Mass spectrometry for the detection of differentially expressed proteins: a comparison of surface-enhanced laser desorption/ionization and capillary electrophoresis/mass spectrometry. *Rapid Commun Mass Spectrom*, **18**, 149–156.
28. Weissinger, E.M., Wittke, S., Kaiser, T., Haller, H., Bartel, S., Krebs, R., Golovko, I., Rupprecht, H.D., Haubitz, M., Hecker, H., Mischak, H., and Fliser, D. (2004) Proteomic patterns established with capillary electrophoresis and mass spectrometry for diagnostic purposes. *Kidney Int*, **65**, 2426–2434.
29. Kaiser, T., Kamal, H., Rank, A., Kolb, H.J., Holler, E., Ganser, A., Hertenstein, B., Mischak, H., and Weissinger, E.M. (2004) Proteomics applied to the clinical follow-up of patients after allogeneic hematopoietic stem cell transplantation. *Blood*, **104**, 340–349.
30. Wittke, S., Haubitz, M., Walden, M., Rohde, F., Schwarz, A., Mengel, M., Mischak, H., Haller, H., and Gwinner, W. (2005) Detection of acute tubulointerstitial rejection by proteomic analysis of

urinary samples in renal transplant recipients. *Am J Transplant*, **5**, 2479–2488.

31. Schaub, S., Rush, D., Wilkins, J., Gibson, I.W., Weiler, T., Sangster, K., Nicolle, L., Karpinski, M., Jeffery, J., and Nickerson, P. (2004) Proteomic-based detection of urine proteins associated with acute renal allograft rejection. *J Am Soc Nephrol*, **15**, 219–227.
32. Schaub, S., Wilkins, J.A., Antonovici, M., Krokhin, O., Weiler, T., Rush, D., and Nickerson, P. (2005) Proteomic-based identification of cleaved urinary beta2-microglobulin as a potential marker for acute tubular injury in renal allografts. *Am J Transplant*, **5**, 729–738.
33. Harrison, H.H. (1991) The "ladder light chain" or "pseudo-oligoclonal" pattern in urinary immunofixation electrophoresis (IFE) studies: a distinctive IFE pattern and an explanatory hypothesis relating it to free polyclonal light chains. *Clin Chem*, **37**, 1559–1564.
34. Tichy, M., Stulik, J., Kovarova, H., Mateja, F., and Urban, P. (1995) Analysis of monoclonal immunoglobulin light chains in urine using two-dimensional electrophoresis. *Neoplasma*, **42**, 31–34.
35. Myrick, J.E., Caudill, S.P., Robinson, M.K., and Hubert, I.L. (1993) Quantitative two-dimensional electrophoretic detection of possible urinary protein biomarkers of occupational exposure to cadmium. *Appl Theor Electrophor*, **3**, 137–146.
36. Mischak, H., Apweiler, R., Banks, R.E., Conaway, M., Coon, J.J., Dominizak, A., Ehrich, J.H., Fliser, D., Girolami, M., Hermjakob, H., Hochstrasser, D.F., Jankowski, V., Julian, B.A., Kolch, W., Massy, Z., Neususs, C., Novak, J., Peter, K., Rossing, K., Schanstra, J.P., Semmes, O.J., Theodorescu, D., Thongboonkerd, V., Weissinger, E.M., Van Eyk, J.E., and Yamamoto, T. (2007) Clinical Proteomics: a need to define the field and to begin to set adequate standards. *PROTEOMICS Clin Appl*, **1**, 148–156.
37. Schaub, S., Wilkins, J., Weiler, T., Sangster, K., Rush, D., and Nickerson, P. (2004) Urine protein profiling with surface-enhanced laser-desorption/ionization time-of-flight mass spectrometry. *Kidney Int*, **65**, 323–332.
38. Theodorescu, D., Wittke, S., Ross, M.M., Walden, M., Conaway, M., Just, I., Mischak, H., and Frierson, H.F. (2006) Discovery and validation of new protein biomarkers for urothelial cancer: a prospective analysis. *Lancet Oncol*, **7**, 230–240.
39. Kolch, W., Neususs, C., Pelzing, M., and Mischak, H. (2005) Capillary electrophoresis-mass spectrometry as a powerful tool in clinical diagnosis and biomarker discovery. *Mass Spectrom Rev*, **24**, 959–977.
40. Theodorescu, D., Fliser, D., Wittke, S., Mischak, H., Krebs, R., Walden, M., Ross, M., Eltze, E., Bettendorf, O., Wulfing, C., and Semjonow, A. (2005) Pilot study of capillary electrophoresis coupled to mass spectrometry as a tool to define potential prostate cancer biomarkers in urine. *Electrophoresis*, **26**, 2797–2808.
41. Zurbig, P., Renfrow, M.B., Schiffer, E., Novak, J., Walden, M., Wittke, S., Just, I., Pelzing, M., Neususs, C., Theodorescu, D., Root, K.E., Ross, M.M., and Mischak, H. (2006) Biomarker discovery by CE-MS enables sequence analysis via MS/MS with platform-independent separation. *Electrophoresis*, **27**, 2111–2125.
42. Pisitkun, T., Johnstone, R., and Knepper, M.A. (2006) Discovery of urinary biomarkers. *Mol Cell Proteomics*, **5**, 1760–1771.

43. Villanueva, J., Shaffer, D.R., Philip, J., Chaparro, C.A., Erdjument-Bromage, H., Olshen, A.B., Fleisher, M., Lilja, H., Brogi, E., Boyd, J., Sanchez-Carbayo, M., Holland, E.C., Cordon-Cardo, C., Scher, H.I., and Tempst, P. (2006) Differential exoprotease activities confer tumor-specific serum peptidome patterns. *J Clin Invest*, **116**, 271–284.
44. Geiger, T. and Clarke, S. (1987) Deamidation, isomerization, and racemization at asparaginyl and aspartyl residues in peptides. Succinimide-linked reactions that contribute to protein degradation. *J Biol Chem*, **262**, 785–794.
45. Stephenson, R.C. and Clarke, S. (1989) Succinimide formation from aspartyl and asparaginyl peptides as a model for the spontaneous degradation of proteins. *J Biol Chem*, **264**, 6164–6170.
46. Mischak, H., Kaiser, T., Walden, M., Hillmann, M., Wittke, S., Herrmann, A., Knueppel, S., Haller, H., and Fliser, D. (2004) Proteomic analysis for the assessment of diabetic renal damage in humans. *Clin Sci (Lond)*, **107**, 485–495.
47. Rossing, K., Mischak, H., Parving, H.H., Christensen, P.K., Walden, M., Hillmann, M., and Kaiser, T. (2005) Impact of diabetic nephropathy and angiotensin II receptor blockade on urinary polypeptide patterns. *Kidney Int*, **68**, 193–205.
48. Sniehotta, M., Schiffer, E., Zurbig, P., Novak, J., and Mischak, H. (2007) CE–a multifunctional application for clinical diagnosis. *Electrophoresis*, **28**, 1407–1417.
49. Wittke, S., Mischak, H., Walden, M., Kolch, W., Radler, T., and Wiedemann, K. (2005) Discovery of biomarkers in human urine and cerebrospinal fluid by capillary electrophoresis coupled to mass spectrometry: towards new diagnostic and therapeutic approaches. *Electrophoresis*, **26**, 1476–1487.
50. Decramer, S., Wittke, S., Mischak, H., Zurbig, P., Walden, M., Bouissou, F., Bascands, J.L., and Schanstra, J.P. (2006) Predicting the clinical outcome of congenital unilateral ureteropelvic junction obstruction in newborn by urinary proteome analysis. *Nat Med*, **12**, 398–400.
51. Weissinger, E.M., Mischak, H., Ganser, A., and Hertenstein, B. (2006) Value of proteomics applied to the follow-up in stem cell transplantation. *Ann Hematol*, **85**, 205–211.
52. Weissinger, E.M., Schiffer, E., Hertenstein, B., Ferrara, J.L., Holler, E., Stadler, M., Kolb, H.J., Zander, A., Zurbig, P., Kellmann, M., and Ganser, A. (2007) Proteomic patterns predict acute graft-versus-host disease after allogeneic hematopoietic stem cell transplantation. *Blood*, **109**, 5511–5519.
53. Fliser, D., Novak, J., Thongboonkerd, V., Argiles, A., Jankowski, V., Girolami, M.A., Jankowski, J., and Mischak, H. (2007) Advances in urinary proteome analysis and biomarker discovery. *J Am Soc Nephrol*, **18**, 1057–1071.
54. von zur Mühlen, C., Schiffer, E., Zürbig, P., Kellmann, M., Brasse, M., Meert, N., Vanholder, R.C., Dominiczak, A.F., Chen, Y.C., Mischak, H., Bode, C., and Peter, K.H. (2009) Evaluation of urine proteome pattern analysis for its potential to reflect coronary artery atherosclerosis in symptomatic patients. *J Proteosome Res*, **6** (1), 335–345.
55. Zimmerli, L.U., Schiffer, E., Kellmann, M., Zürbig, P., Good, D.M, Mouls, L., Pitt, A.R., Coon, J.J., Schmieder, R.E., Peter, K.H., Mischak, H., Kolch, W., Delles, C., and Dominiczak, A.F. (2008) Urinary proteomic biomarkers in coronary artery disease, *Mol Cell Proteomics*, **7** (2), 290–298.

25
Mass Spectrometric Immunoassay in Urinary Proteomics

Urban A. Kiernan, Kemmons A. Tubbs, Eric E. Niederkofler, Dobrin Nedelkov, and Randall W. Nelson

25.1
Introduction

Urine has historically been viewed as an easily accessible biological fluid, rich in biomarkers useful for assessing kidney function as well as for determining the presence of disease in an individual. Because many proteins and peptides cross over from blood through the glomerulus into urine as waste products, disciplines such as proteomics have lately become more interested in better understanding the urine proteome to expand and improve its potential utility in disease diagnostics. Such an improvement extends well beyond the conventional strategy of novel biomarker discovery (classical bottom-up proteomics approaches), and includes the phenotypic characterization of known proteins (i.e., the identification of protein variants) that are associated with known disease states. To perform such succinct analyses requires targeted analyses; therefore, many proteomics approaches have begun to move away from global protein expression methodologies and have gravitated toward candidate-based screens (e.g., SISCAPA, SELDI, mass spectrometric immunoassay (MSIA), etc.).

All these approaches combine affinity targeting followed by mass spectrometric (MS) detection of the captured protein analyte. However, only the MSIA has been repeatedly demonstrated to perform both qualitative and quantitative differentiation of protein analytes, and it is this ability to perform both types of analyses that is being viewed as an essential component to the further successful evolution of proteomics into the clinical and diagnostic realm. Because of its abilities, this chapter focuses on MSIA technology, addressing some of its technical aspects and introducing some of its applications into urinary proteomics.

25.2
Technical Approach

The development of immunoaffinity-based mass spectrometry has the central goal to characterize, in detail, a protein target from a complex biological fluid.

Renal and Urinary Proteomics: Methods and Protocols. Edited by Visith Thongboonkerd

ISBN: 978-3-527-31974-9

This concept toward MS-based protein analysis is reminiscent of preproteomic philosophy, which is captured in the MSIA [1]. This approach is based on the capture of analytes from complex biological samples using immobilized antibodies, followed by their release and analysis using mass spectrometry. The preferred method of analysis is with matrix-assisted laser desorption/ionization time-of-flight mass spectrometry (MALDI-TOF-MS).

This approach has been performed using numerous independent devices, but the introduction of solid supports built into affinity pipette tips, for protein isolation and enrichment prior to MS analysis, is most highly applicable to the examination of human urine. The key aspect of this design is the ability of the functional pipette tip to address large volumes of biological fluid (25 ml have been reported [2]), to compensate for the low concentration of any given protein target found in urine. A general illustration of this approach is provided

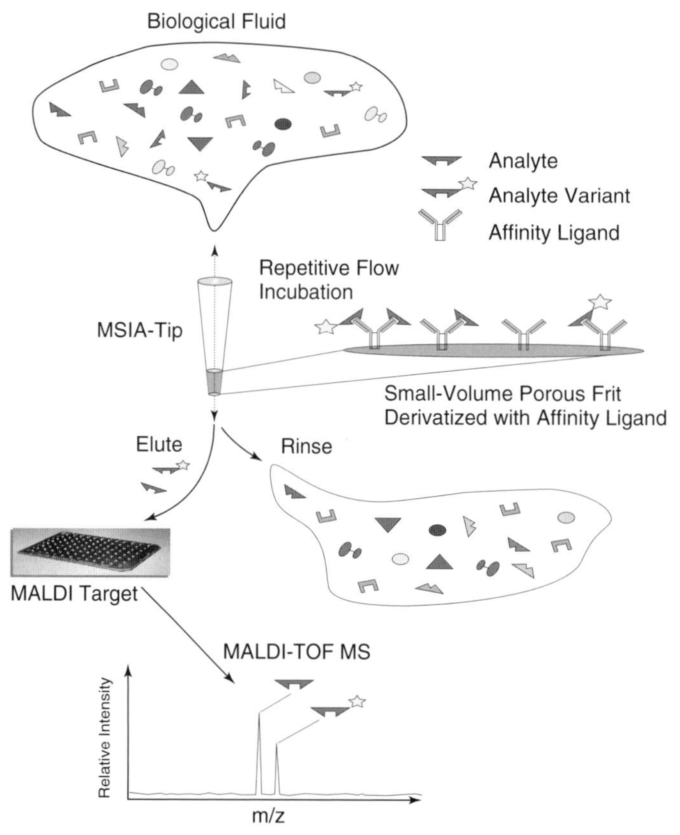

Figure 25.1 Schematic of the MSIA process, in which a biological fluid is repetitively passed through the MSIA-Tip allowing for target protein retrieval, concentration, and purification within the affinity pipette. The retained protein is eluted directly onto a MS target for subsequent interrogated with MADLI-TOF-MS.

in Figure 25.1. The affinity pipette is systematically stepped through the process of target extraction. Incubation of each urine sample consists of 300 cycles (repetitive aspirations and dispenses; 150 μl of sample) through each MSIA-Tip. After incubation, tips are thoroughly rinsed using HBS (10 cycles, 150 μl), doubly distilled water (5 cycles, 150 μl), 20% acetonitrile/2 M ammonium acetate wash (10 cycles, 150 μl), and finally with doubly distilled water (15 cycles, 150 μl). Retained species are eluted by drawing 5 μl of MALDI matrix solution (saturated aqueous solution of sinapic acid (SA) or α-cyano-4-hydroxycinnamic acid (CHCA), in 33% (v/v) acetonitrile, 0.4 % (v/v) trifluoroacetic acid) into each tip and stamping directly onto a 96-well formatted hydrophobic/hydrophilic contrasting MALDI-TOF target [3]. Because the larger sample volumes of urine require a larger number of iterations, extraction-to-elution times are on the order of ~<35 minutes. This approach is also amiable to high-throughput processing. Such analyses have been run in parallel, up to 96 analyses at a time, by equipping a multichannel automated robotics workstation with MSIA tips [4–9]. By doing so, the MSIA process is not disrupted, but enhanced up to 96-fold.

25.3 Applications to Human Urine

25.3.1 Example 1: Qualitative Comparison of Urinary Protein

This first demonstration of urinary MSIA highlights the ability of this approach to differentiate between multiple forms of the same target protein. Phenotypic differentiation has long been known to be associated with disease states and must be accounted for in the next generation of clinical and diagnostic applications.

The target of interest to demonstrate this phenomenon is retinol-binding protein (RBP), a member of the lipocalin family that serves as the primary carrier of retinol (vitamin A) from the liver to peripheral tissues [10]. This small hepatic protein, having molecular weight (MW) 21 065.6, is believed to escape glomerular filtration by noncovalently complexing, in its *holo-* (retinol bound) form, with transthyretin (TTR) [11], whereas free RBP is known to pass through the glomeruli where it is then believed to be catabolized in the tubular cells [12]. Previous studies have shown that individuals with renal diseases known to impair filtering function have an increased concentration of plasma RBP [13]. Evaluations of plasma RBP have also been linked to insulin resistance and Type-2 diabetes [14]. Moreover, the presence of truncated RBP variants, by the loss of one or both C-terminal leucines, may also be markers of renal failure by their increased plasma concentrations, resulting in varied RBP-to-truncated variant ratios [15].

In this study, urine samples (25 ml midstream voids) from five individuals were collected. The urine samples were immediately combined 1 : 1 (v/v) with

2 M ammonium acetate (to adjust the pH to ~6.8–7.2) and 50 µl of protease inhibitor cocktail (AEBSF (100 mM); aprotinin (80 µM); bestatin (5 mM); E-64 (1.5 mM); leupeptin (2 mM); pepstatin A (1 mM) – added to prevent any enzymatic breakdown). Each urine sample was interrogated individually with MSIA Tips derivatized with anti-RBP polyclonal antibody (Cat. # A0040, Dakocytomation, Carpiternia, CA) as described above. Retained target protein was eluted using SA matrix and allowed to air dry. All samples were interrogated on a Linear Bruker Autoflex MALDI-TOF mass spectrometer. Analyses were performed with a 20 kV acceleration voltage, drawout pulse of 1.45 kV, and a 670 ns delay.

The results of the anti-RBP MSIA urine analyses are shown in Figure 25.2, in which Traces A–D were from healthy individuals. All four spectra display a generally consistent phenotypic profile of urinary RBP. Similar to the plasma profiles, wt-RBP and RBP-L were common to the urine of all four healthy subject samples. However, additional signals were observed and corresponded to further C-terminally truncated variants, namely, RBP-LL, RBP-RNLL, and RBP-RSERNLL. In contrast, the profile obtained from the urine of the individual suffering renal dysfunction was less complex (Figure 25.2; Trace E). The profile noticeably lacked the presence of wt-RBP and was dominated by the presence of RBP-L. In addition, a small amount of the RBP-RNLL was also observed.

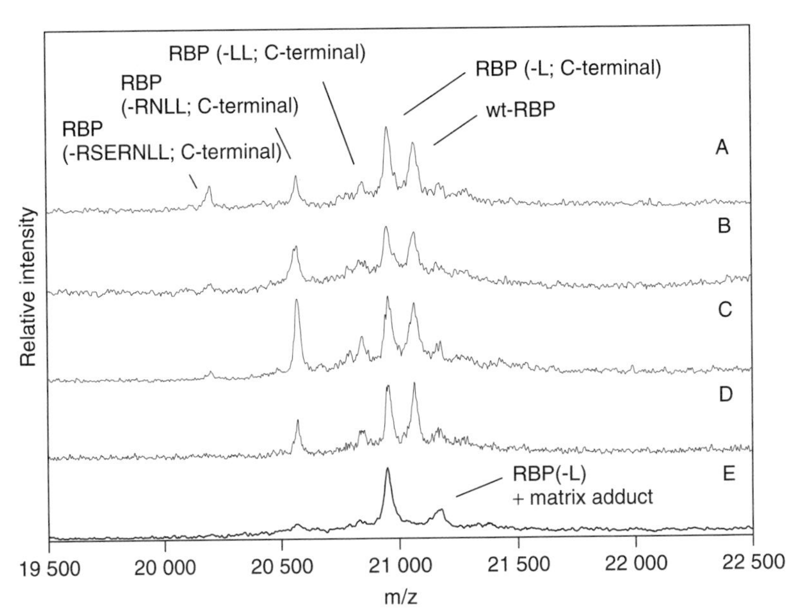

Figure 25.2 Results of the anti-RBP MSIA analysis of urine. Traces A–D were from healthy subjects and display a generally similar phenotypic profile. Wild-type RBP, RBP (-L), RBP (-LL), and RBP (-RNLL) (MW = 20 534) were present in all four mass spectra. Observable signal from RBP (-RSERNLL) (MW = 20 162) was also present in three of the four spectra. However, Trace E was from the individual with kidney impairment and mass spectrum was dominated by RBP-L signal and devoid of wt-RBP.

This example clearly demonstrates how a single protein target is in fact a diverse grouping of multiple targets. More important is the ability to observe how these target protein profiles differentiate between healthy and disease states. It may appear obvious that such processes do exist in human biology; however, there are only a limited number of available technologies that are able to rigorously perform such analyses so quickly and efficiently. This level of molecular differentiation, as achieved with MSIA, is currently lacking in standard clinical and diagnostic platforms, however, with the growing appeal of candidate-based proteomics through immunoaffinity targeting, the need to incorporate such data into mainstream healthcare may not be far off on the horizon.

25.3.2
Example 2: Molecular Characterization of Urinary Protein

Because the goal of this approach is to ascertain the highest content data regarding a single affinity target, a method must be in place to characterize a target at the molecular level in the event that protein variation is observed. This is achieved through the combination of MSIA with Bioreactive Probe (BRP) technologies. The combination of these approaches is described in detail in the following example.

The target selected for this example is human TTR, a small hepatic protein found in serum and cerebral spinal fluid as a homotetramer [16, 17]. Functionally, TTR serves unaccompanied in the transport of thyroid hormones or in complexes with other proteins in the transport of various biologically active compounds. Structurally, wild-type (wt) TTR comprises 127 amino acids and has a MW of 13 762.4. Over 100 point mutations have been cataloged for TTR, with all but 10 potentially leading to severe neurological complications [17]. The majority of mutation-related disorders are caused by amyloid plaques depositing on neurons or tissues, eventually leading to dysfunctions including carpal tunnel syndrome, drussen, and familial amyloid polyneuropathy [18–20].

In this next example, urine samples (25 ml midstream voids) from six participants were individually collected. The urine was collected directly into sterile urine collection cups. The pH for each urine sample was then determined, with native pH ranging from 4.5 to 6.5 for these particular samples. Each sample was then titrated to neutral conditions (pH to ~6.8–7.2) by the addition of a 4 M sodium hydroxide solution and then diluted 1 : 2 using doubly distilled water for immediate analysis.

Each diluted sample was then poured into an individual polyvinyl chloride solution basin prior to analysis. Each sample was individually addressed with an individual MSIA-Tip (Intrinsic Bioprobes, Inc.) derivatized with anti-TTR polyclonal antibody (Cat. # A0002, Dakocytomation, Carpiternia, CA) loaded into a Beckman Multimek 96 robotic workstation for automated processing. It should be noted that even though robotic processing was used in this study, the fundamental processes of MSIA did not change. The same incubation and rinse steps that were previously described still applied. After affinity capture, retained targets were eluted by drawing 5 μl of CHCA MALDI matrix solution into each tip and depositing directly onto a 96-well formatted trypsin immobilized MALDI-TOF target [4] and allowed to air dry. The covalent attachment of modifying enzyme to the surface of the MALDI target results in the production of a BRP. A BRP still functions as a standard MS target, but allows for on-spot chemical/enzymatical modification of a protein target for subsequent molecular characterization. The application of the BRP is discussed in detail later in this section. The duration of sample analysis in this study was ∼ 20 minutes/sample.

Parent MALDI-TOF-MS and data analysis were performed on a linear Bruker Autoflex MALDI-TOF-MS with a full 20.00 kV accelerating potential, a drawout of 1.50 kV and a 250 ns delay. All acquired mass spectra (accumulation of 200 shots) had mass accuracy within 0.01% and resolution of ∼1000 (FWHM), which was sufficient to correctly identify each protein target and the presence of protein variation.

MSIA analysis was successfully performed on all samples, producing mass spectral signatures of urinary TTR from each. These signatures are presented in

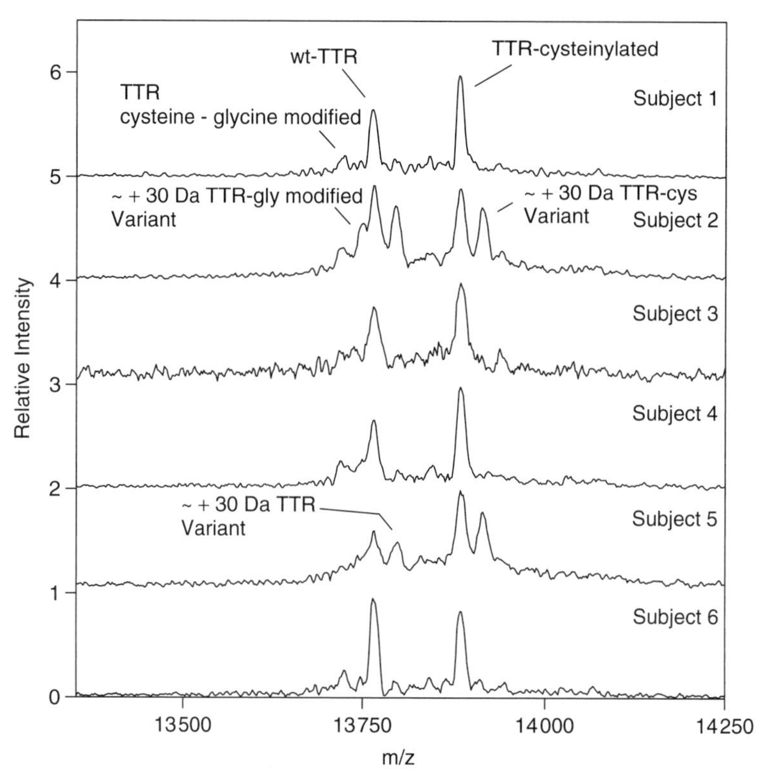

Figure 25.3 Results of the anti-TTR MSIA analysis of urine. Multiple signals for various forms of TTR that are endogenous to human urine were observed. Subjects 2 and 5 showed characteristic signals, peak splitting, suggesting that these individuals had a heterozygous point mutation for TTR.

Figure 25.3. Displayed is the unique TTR MS profile from each individual in the 1 + region of each mass spectrum showing detailed information of the protein target. This close-up clearly displays the presence of several variant forms of TTR and the naturally occurring complexity of each monomer. The "general" MS profile of TTR consists of the wt form ($m/z = 13\,762$) and a naturally occurring cysteinylated variant ($m/z = 13\,881$), both of which were consistently observed in all the samples analyzed. These two basic components of the TTR profile have been widely reported in the literature from both human plasma and urine [2, 4, 21, 22]. Hence, owing to the vast amount of published data surrounding these two forms of TTR, the presence and identity of these observed signals were not in question.

However, other TTR variants were also observed in some of the samples analyzed. These variants include (i) an approximately −46 Da shift observed in four samples, which was consistent with a naturally occurring posttranslational conversion of the cysteine at position 10 to a glycine; (ii) as well as a potential heterozygous point mutation, producing an approximately +30-Da mass shift, observed in two of the samples. From the signals observed, we suspect the presence of a heterozygous point mutation by the observed shifts occurring in the posttranslational modification (PTM) variants as well. Because the presence of two genes would result in the production of two different TTR backbone structures, these gene products would in turn be influenced by the same naturally occurring PTMs. With this, the observation of such peak "splitting" in the posttranslationally modified forms of the protein is a strong indication that multiple genes are being expressed.

To confirm the presence of these less commonly observed variants, more specifically the observed point mutations, these same samples were then proteolytically digested for TTR molecular characterization. This was achieved by activating the trypsin, already present on the MALDI-TOF-MS target, by applying 8 μl of 10 mM Tris solution (pH 9.5) onto each sample spot. Even in the presence of the acidified MALDI matrix, this concentration of buffer was sufficient to render each sample to neutral conditions. Digests were allowed to proceed for 15 minutes at 40 °C in high humidity enclave before termination by the addition of 8 μl of 0.8% TFA. Samples were air dried prior to insertion of the MALDI-TOF target into the mass spectrometer. MS analysis of the mass maps was performed on a Bruker Autoflex II instrument in reflectron mode. The instrument operated with a 19.00 kV full accelerating potential, drawout pulse of 2.05 kV (110 ns delay), 8.40 kV focusing lens, and an ion-mirror voltage of 20.00 kV.

The results of the trypsin molecular characterization resulted in digest of the TTR from all samples analyzed. A representative mass spectrum is shown in Figure 25.4. Signals corresponding to TTR tryptic peptide were labeled accordingly. This specific example showed 100% sequence coverage from the resultant peptides as presented in the coverage map atop the mass spectrum. This sequence coverage ranged from 85 to 100% in all samples analyzed. The figure insets also showed the presence of tryptic peptides that corresponded with the cysteinylated and the cysteinylglycine-modified forms, confirming their presence.

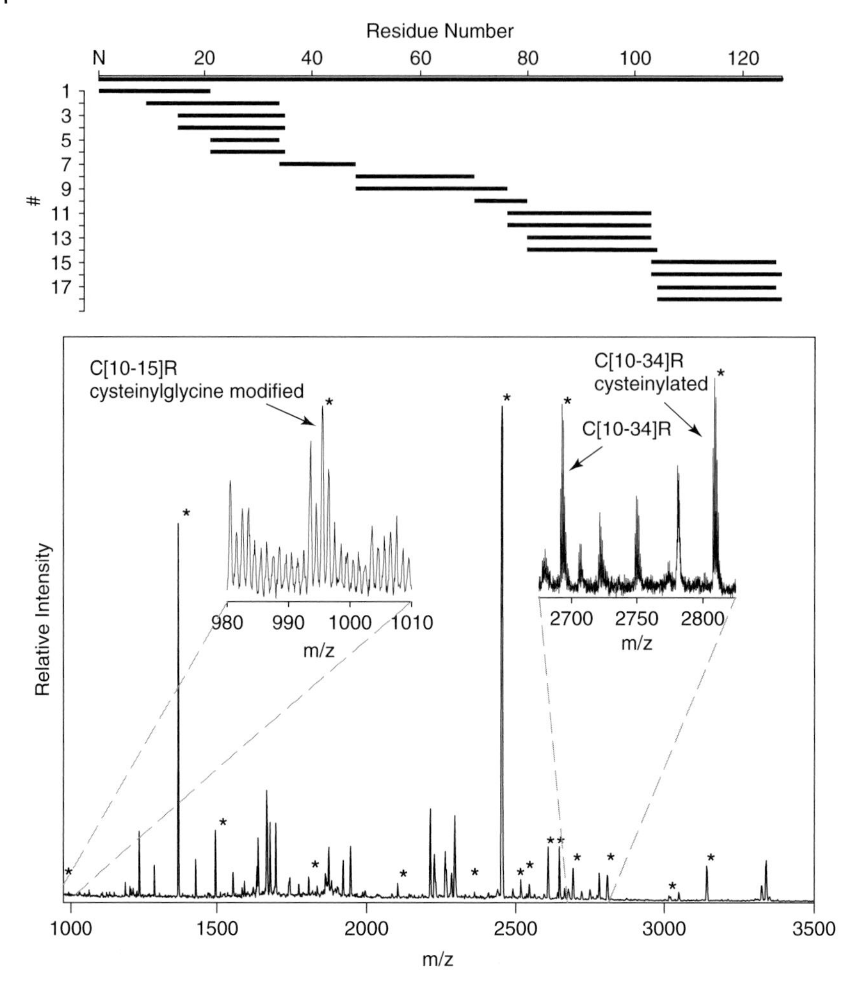

Figure 25.4 Representative mass spectrum of transthyretin trypsin BRP digest. The analysis achieved 100% sequence coverage and was able to identify MS signals for the cysteinylated PTM and the cysteinylglycine-modified forms of TTR.

However, the main reason for performing these molecular characterizations is the observation of potential point mutation in samples 2 and 5. Closer inspection of the digests of both of these samples revealed the presence of a satellite signal with a measured mass shift from a TTR digest peptide of +29.925 Da. This mass shift correlated well with the +30 Da observed in the parent analyses. This is shown in Figure 25.5, in which the *m/z* range 2620–2720 is displayed. Both samples 2 and 5, along with sample 4 (nonmutant control), are presented. This unique satellite peak appeared to originate from the trypsin digest fragment R[104–127]E. With this knowledge, the literature has cited two potential candidates for this point mutation.

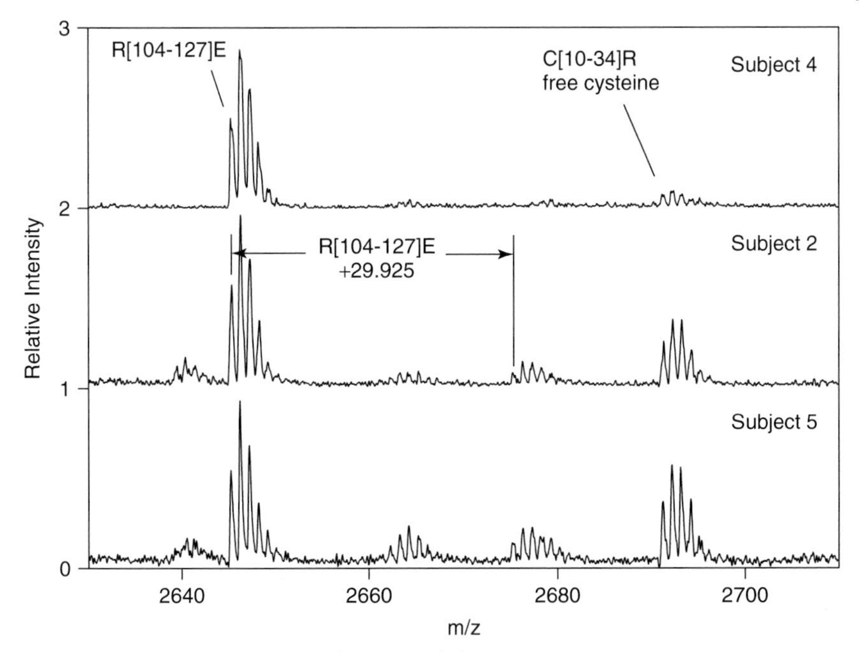

Figure 25.5 Comparative results of MSIA-purified TTR molecular characterization. Subject 4 (top trace) was a nonmutant control. Subjects 2 and 5 (middle and bottom traces) had parent peak splitting by ∼+ 30 Da. Observed satellite signal with observed mass shift of +29.925 Da was correlated with a Thr119Met mutation.

The most likely candidate due to mass accuracy is Thr119Met (theoretical mass shift 29.99281 Da), a nonamyloidogenic mutation found in Portugal and the United States [23].

The results of this molecular characterization demonstrate how MSIA has the ability to pinpoint minute phenotypic differences observed within populations of individuals. The ability of MSIA to qualitatively screen proteins is unprecedented; however, its combination with BRP technology allows for the qualitative screening process to evolve into a rapid protein characterization platform capable of identifying extremely small differences in protein variation. This concept is augmented by the fact that this study was performed on urine proteins, making this analysis completely noninvasive.

25.3.3 Example 3: Quantitation of Urinary Protein

Because the goal of urinary proteomics is to enter clinical and diagnostic application, quantitative measurements are a necessity. This final example demonstrates how MSIA, when used with an internal reference standard, can be used to perform rigorous quantitative measurements.

The target selected for quantitative analysis is human urinary β_2-microglobulin (β_2m), a low molecular mass protein identified as the light chain of the Class I major histocompatibility complex synthesized in all nucleated cells. Upon activation of the immune system, both B- and T-lymphocytes actively release β_2m into circulation where it is later eliminated via glomerular filtration and tubular reabsorption. Serum levels of β_2m have been measured and nominally correlated to a number of ailments. On the other hand, β_2m levels in urine are indicators of glomerular filtration rate and tubular reabsorption [24]. Conventionally, β_2m levels are monitored using a variety of immuno-based assays, including enzyme-linked immunosorbent assays, radioimmunoassays, and particle-enhanced turbidimetry assays. The quantitative dynamic range (spanning β_2m concentrations of $\sim$0.2–20 mg l^{-1}) and the accuracy (1–10%) of these assays are sufficient to cover the normal and elevated levels of β_2m in a variety of biological fluids.

Recently, we have reported on the development of a urine-based MSIA assay targeting on β_2m [25]. Figure 25.6 shows spectra resulting from the MSIA analysis of a human urine sample. For comparison, a MALDI-TOF mass spectrum of whole/unfractionated human urine is shown. The MALDI-TOF spectrum of whole urine shows a number of signals in the peptide region and an absence of signals for β_2m. On the other hand, the results obtained during MSIA were dominated by signals from the β_2m, with few additional signals from nonspecified compounds and of suitable quality upon which to develop a fully quantitative assay. For the

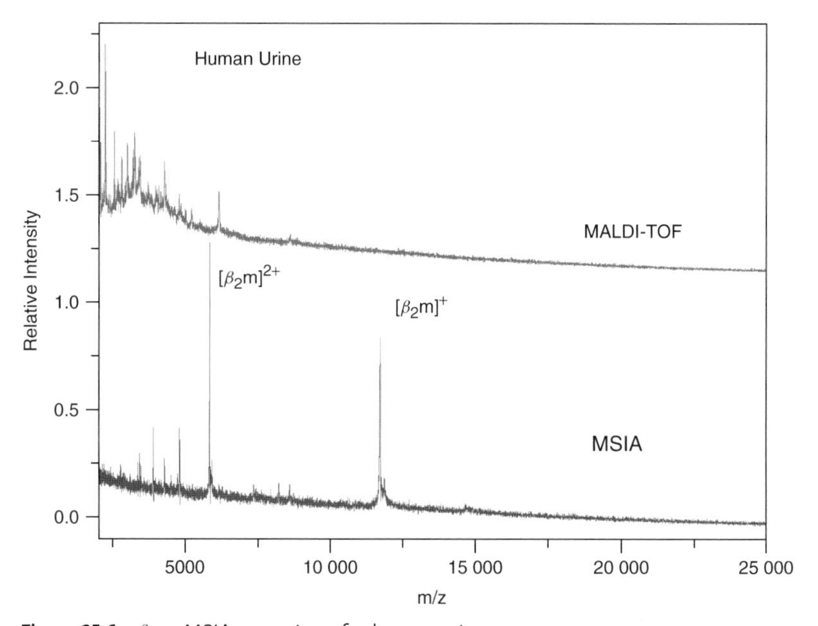

Figure 25.6 β_2m MSIA screening of a human urine sample. Upper trace: direct MALDI-TOF mass spectrum of diluted, unfractioned human urine sample. Bottom trace: mass spectrum of the human urine sample following the MSIA screen.

quantification of β_2m in human urine, internal standards need to be incorporated into the MSIA protocol. Equine β_2m (Eβ_2m) was chosen as an internal reference for quantification because of its high degree of similarity to human β_2m (Hβ_2m) (~75% sequence homology), resolvable mass difference from Hβ_2m ($MW_{E\beta 2m}$ = 11 396.6; $MW_{H\beta 2m}$ = 11 729.7) and because it was easily obtainable. Briefly, horse urine was collected fresh (at a local stable) and treated immediately with protease inhibitor cocktail. Low-solubility compounds were removed from the urine by overnight refrigeration (at 4 °C) followed by centrifugation for 5 minutes at 5000g. The urine was then concentrated 20-fold using a 10-kDa MW cut-off filter, with repetitive HBS and water rinses and with several filter exchanges (4 filters/200 ml urine). Treatment of 200-ml fresh urine resulted in 10 ml of β_2m-enriched horse urine, which served as stock internal reference solution for ~ 100 analyses.

Standards were prepared by step-wise dilution (i.e., × 0.8, 0.6, 0.4, 0.2, and 0.1, in HBS) of a 1.0 mg l^{-1} stock Hβ_2m solution to a concentration of 0.1 mg l^{-1}; the 0.1 mg l^{-1} solution served as stock for an identical step-wise dilution covering the second decade in concentration (0.01–0.1 mg l^{-1}). A blank solution containing no Hβ_2m was also prepared. The samples for MSIA were prepared by mixing 100 µl of each of the Hβ_2m standards with 100 µl of stock horse urine and 200 µl of HBS buffer. MSIA was performed on each sample as described above, resulting in the simultaneous extraction of both Eβ_2m and Hβ_2m. Ten 65-laser-shots MALDI-TOF spectra were taken from each sample, with each spectrum taken from a different location on the target. Care was taken during data acquisition to maintain the ion signals in the upper 50–80% of the *y*-axis range and to avoid driving individual laser shots into saturation. Spectra were normalized to the Eβ_2m signal through baseline integration, and the integral of the Hβ_2m peak was determined. Integrals from the 10 spectra taken for each calibration standard were averaged and the standard deviation calculated. A calibration curve was constructed by plotting the average of the normalized integrals for each standard versus the Hβ_2m concentration. Figure 25.7a shows spectra representing MSIA analyses of Hβ_2m standards in a concentration range of 0.01–1.0 mg l^{-1}. Each spectrum, normalized to the Eβ_2m signal, is one of ten 65-laser shots spectra taken for each calibration point. The calibration curve is shown in Figure 25.7b. Linear regression fitting of the data yielded $Int_{H\beta 2m}/Int_{E\beta 2m}$ = 4.09 (Hβ_2m in mg l^{-1}) + 0.021 (R^2 = 0.983), with a working limit of detection (LOD) of 0.0025 mg l^{-1} (210 pM) and a limit of quantitation (LOQ) of 0.01 mg l^{-1} (850 pM).

For the quantitative MSIA screens, 10 samples were collected from four individuals: female (31 years, pregnant; one sample (F31)), male (30 years; four samples over two days (M30)), male (36 years; two samples over two days (M36)), and male (44 years; three samples over two days (M44)). All of the individuals were in a state of good health when the samples were collected. In preparation for MSIA, 100 µl of each urine sample was mixed with 100 µl of stock horse urine and 200 µl of HBS. Results from the MSIA of the 10 urine samples are shown in Figure 25.8. The bars depict the β_2m concentration determined for each sample. The data for the 10 samples showed remarkable consistency, with an average β_2m concentration of 0.100 ± 0.021 mg l^{-1} (high = 0.127 mg l^{-1}; low = 0.058 mg l^{-1}). An additional

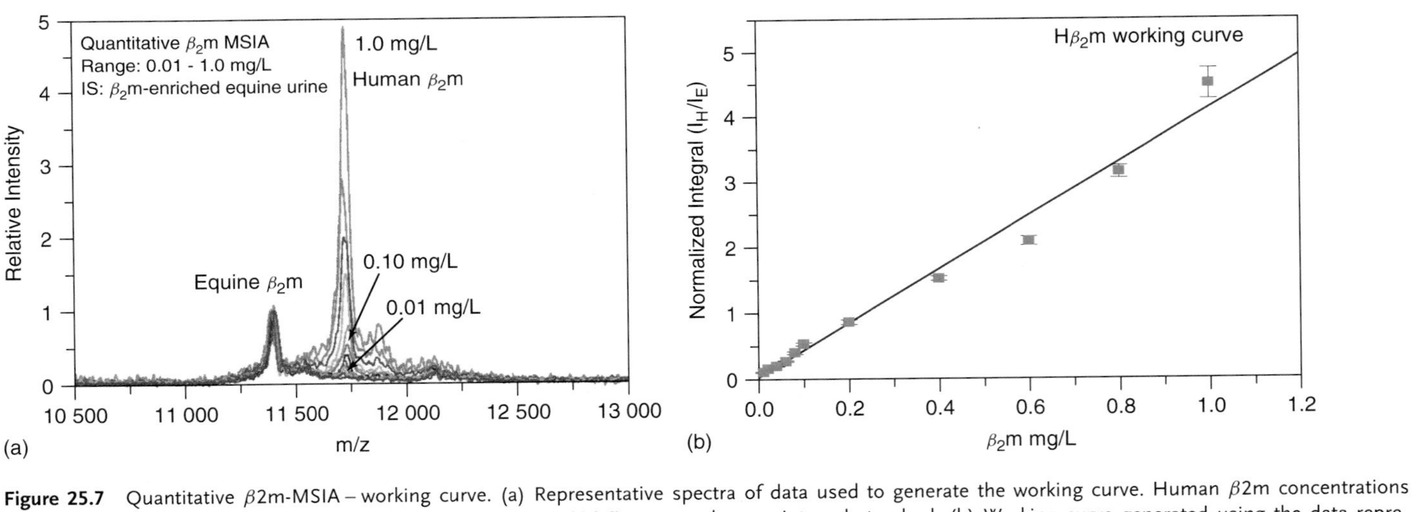

Figure 25.7 Quantitative β2m-MSIA – working curve. (a) Representative spectra of data used to generate the working curve. Human β2m concentrations of 0.01–1.0 mg l^{-1} were investigated. Equine β2m (MW = 11 396.6) was used as an internal standard. (b) Working curve generated using the data represented in (a). The two-decade range was spanned with good linearity ($R2 = 0.983$) and low standard error (~5%). Error bars reflect the standard deviation of 10 repetitive 65-laser shots spectra taken from each sample.

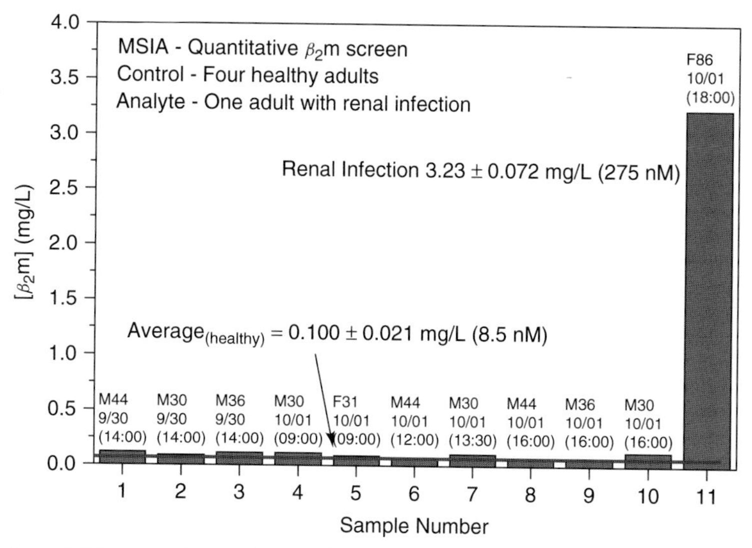

Figure 25.8 Quantitative β2m-MSIA – screening. Human urine samples from five individuals were screened over a period of two days. The average value determined for healthy individuals (10 samples; four individuals (three males; one female) aged 30–44 years) was 0.100 ± 0.021 mg l^{-1}. The level determined for an 86-year-old-female with a recent urinary tract infection indicated a significant increase in β2m concentration (3.23 $\pm$ 0.072 mg l^{-1}).

analysis was performed on a urine sample obtained from an 86-year-old female (F86) who had recently suffered a renal infection. Because of the significantly higher level of β_2m found in this sample, it was necessary to quantitatively dilute the urine by a factor of 10 to keep the β_2m signal inside the dynamic range of the working curve and accurately establish the β2m concentration in F86 (at 3.23 $\pm$ 0.02 mg l^{-1}).

The ability to quantitatively analyze for a given protein target is essential for any proteomics approach to gain a foothold against current clinical and diagnostic platforms. Classical immunoassay technologies are the gold standard for protein measurement. However, mass spectrometry has the ability to quantify with the precision and accuracy needed to perform at a clinical level.

25.4 Summary

The examples presented in this chapter show how MSIA can be used in numerous urinary-based applications. This approach not only meets the current clinical and diagnostic benchmarks of urine analysis with the quantitation of a given protein target, but also provides a new dimension in molecular analysis, through the differentiation of molecular subspecies. It has been suspected that such subspecies exist, but there has not been any readily applicable method for the screening of such targets within populations of individuals. It is clear that these detailed analyses provide more information compared to classical immunoassay approaches. Moreover, as we pioneer the field of proteomics into unknown territory, as with the urine proteome, we are afforded the opportunity to develop new methodologies that set the foundation for future studies and provide an outline for the possibilities in tomorrow's clinical and diagnostic world.

References

1. Nelson, R.W., Krone, J.R., Bieber, A.L., and Williams, P. (1995) Mass-spectrometric immunoassay. *Anal Chem*, **67** (7), 1153–1158.
2. Kiernan, U.A., Tubbs, K.A., Nedelkov, D. *et al.* (2003) Comparative urine protein phenotyping using mass spectrometric immunoassay. *J Proteome Res*, **2** (2), 191–197.
3. Niederkofler, E.E., Tubbs, K.A., Kiernan, U.A., Nedelkov, D., and Nelson, R.W. (2003) Novel mass spectrometric immunoassays for the rapid structural characterization of plasma apolipoproteins. *J Lipid Res*, **44** (3), 630–639.
4. Kiernan, U.A., Tubbs, K.A., Gruber, K. *et al.* (2002) High-throughput protein characterization using mass spectrometric immunoassay. *Anal Biochem*, **301** (1), 49–56.
5. Kiernan, U.A., Nedelkov, D., Tubbs, K.A., Niederkofler, E.E., and Nelson, R.W. (2002) High-throughput analysis of human plasma proteins. *Am Biotechnol Lab*, **20** (3), 26–28.
6. Kiernan, U.A., Addobbati, R., Nedelkov, D., and Nelson, R.W. (2006) Quantitative multiplexed C-reactive protein mass spectrometric immunoassay. *J Proteome Res*, **5** (7), 1682–1687.
7. Nedelkov, D., Tubbs, K.A., Niederkofler, E.E., Kiernan, U.A., and Nelson, R.W. (2004) High-throughput comprehensive analysis of human plasma proteins: a step toward population proteomics. *Anal Chem*, **76** (6), 1733–1737.
8. Nedelkov, D. (2005) Population proteomics: addressing protein diversity in humans. *Expert Rev Proteomics*, **2** (3), 315–324.
9. Niederkofler, E.E., Tubbs, K.A., Gruber, K. *et al.* (2001) Determination of beta-2 microglobulin levels in plasma using a high-throughput mass spectrometric immunoassay system. *Anal Chem*, **73** (14), 3294–3299.
10. Kanai, M., Raz, A., and Goodman, D.S. (1968) Retinol-binding protein: the transport protein for vitamin A in human plasma. *J Clin Invest*, **47** (9), 2025–2044.
11. Naylor, H.M. and Newcomer, M.E. (1999) The structure of human retinol-binding protein (RBP) with its carrier protein transthyretin reveals an interaction with the carboxy terminus of RBP. *Biochemistry*, **38** (9), 2647–2653.
12. Goodman, D.S. (1980) Plasma retinol-binding protein. *Ann N Y Acad Sci*, **348**, 378–390.
13. Beetham, R., Dawnay, A., Landon, J., and Cattell, W.R. (1985) A radioimmunoassay for retinol-binding protein in serum and urine. *Clin Chem*, **31** (8), 1364–1367.
14. Graham, T.E., Yang, Q., Bluher, M. *et al.* (2006) Retinol-binding protein 4 and insulin resistance in lean, obese, and diabetic subjects. *N Engl J Med*, **354** (24), 2552–2563.
15. Jaconi, S., Rose, K., Hughes, G.J., Saurat, J.H., and Siegenthaler, G. (1995) Characterization of two post-translationally processed forms of human serum retinol-binding protein: altered ratios in chronic renal failure. *J Lipid Res*, **36** (6), 1247–1253.
16. Ingenbleek, Y. and Young, V. (1994) Transthyretin (prealbumin) in health and disease: nutritional

implications. *Annu Rev Nutr*, **14**, 495–533.

17. Schreiber, G. and Richardson, S.J. (1997) The evolution of gene expression, structure and function of transthyretin. *Comp Biochem Physiol B Biochem Mol Biol*, **116** (2), 137–160.
18. Damas, A.M. and Saraiva, M.J. (2000) Review: TTR amyloidosis-structural features leading to protein aggregation and their implications on therapeutic strategies. *J Struct Biol*, **130** (2-3), 290–299.
19. Plante-Bordeneuve, V. and Said, G. (2000) Transthyretin related familial amyloid polyneuropathy. *Curr Opin Neurol*, **13** (5), 569–573.
20. Benson, M.D. and Uemichi, T. (1996) Transthyretin amyloidosis. *Amyloid Int J Exp Clin Invest*, **3** (1), 44–56.
21. Schweigert, F.J., Wirth, K., and Raila, J. (2004) Characterization of the microheterogeneity of transthyretin in plasma and urine using SELDI-TOF-MS immunoassay. *Proteome Sci*, **2** (1), 5.
22. Yazaki, M., Tokuda, T., Nakamura, A. *et al.* (2000) Cardiac amyloid in patients with familial amyloid polyneuropathy consists of abundant wild-type transthyretin. *Biochem Biophys Res Commun*, **274** (3), 702–706.
23. Connors, L.H., Lim, A., Prokaeva, T., Roskens, V.A., and Costello, C.E. (2003) Tabulation of human transthyretin (TTR) variants, 2003. *Amyloid*, **10** (3), 160–184.
24. Schardijn, G.H. and Statius van Eps, L.W. (1987) Beta 2-microglobulin: its significance in the evaluation of renal function. *Kidney Int*, **32** (5), 635–641.
25. Tubbs, K.A., Nedelkov, D., and Nelson, R.W. (2001) Detection and quantification of beta-2-microglobulin using mass spectrometric immunoassay. *Anal Biochem*, **289** (1), 26–35.

26
Antibody Microarrays for Urinary Proteome Profiling

Bi-Cheng Liu, Lin-Li Lv, and Lu Zhang

26.1
Introduction

Antibody microarray, in addition to electrophoresis and mass-spectrometry-based technologies, is one of the most promising tools for the screening of complex protein samples. It is characterized by its quantitative ability, high sensitivity, and specificity for a particular set of known proteins and is anticipated to play a crucial role in the studies of protein networks and pathways and development of disease biomarkers. The advent of this technique opens exciting new avenues for urinary proteome profiling, which will accelerate the understanding of kidney disease and promote the identification of urinary proteome patterns to predict chronic kidney disease (CKD) progression or regression. Nevertheless, to succeed, antibody microarrays have to overcome their current limitations regarding technologic aspects and especially data analysis in clinical studies. This chapter aims to introduce this technology and highlights its potential application in investigating renal disease.

26.2
Overview of Antibody Microarray Technology

26.2.1
Antibody Microarray: a Multiplexed Immunoassay Platform for Targeted Proteomics

Proteomics is an active research area that involves identification, characterization, and quantification of proteins in whole cells, tissues, or body fluids. Protein microarray is a miniaturized assay system that contains purified capture reagents or samples immobilized at spatially defined locations and allows for multiplexed measurements. It has become an important proteomics research tool, which is complementary to conventional unbiased separation-based and mass spectrometry-based approaches in proteomic analysis [1]. One of the main advantages of this technology over other proteomic approaches is that it is an

Renal and Urinary Proteomics: Methods and Protocols. Edited by Visith Thongboonkerd

ISBN: 978-3-527-31974-9

intrinsically robust and quantitative system, delivering high throughput, and parallel detection on particular sets of known proteins enabling an exact biological interpretation of the results [2]. Consequently, it is particularly more suited to targeted proteomic investigations compared with conventional unbiased proteomic technologies, which are limited by low throughput and poor sensitivity [3, 4].

A protein microarray comprises many different affinity reagents arrayed at high spatial density as "microspots" on a support surface. Each agent captures its target protein from a complex mixture and the captured proteins are subsequently detected and quantified. The frequently applied capture agents are antibodies because of their exquisite specificity and commercial availability. Accordingly, antibody microarrays form a special subgroup of protein microarrays (Figure 26.1). Historically, the roots of antibody microarrays lie in the development of immunoassays, which is an accepted "gold standard" for single-protein measurement. However, because proteins function within networks, pathways, complexes, and families, multiplexed and parallel measurement is apparently much more logical for biological research. In fact, the feasibility of highly sensitive microarray-based ligand binding assays was already demonstrated in the late 1980s [5]. However, it was not implemented until the late 1990s when the DNA microarray technology progressed and the technical tools for the generation and detection of microarrays became available as standard laboratory equipment. Since then, interest has been shifted to the development of antibody microarrays for the purpose of quantitative targeted proteomics. In 2000, five years after the announcement of the DNA microarray technology, the first

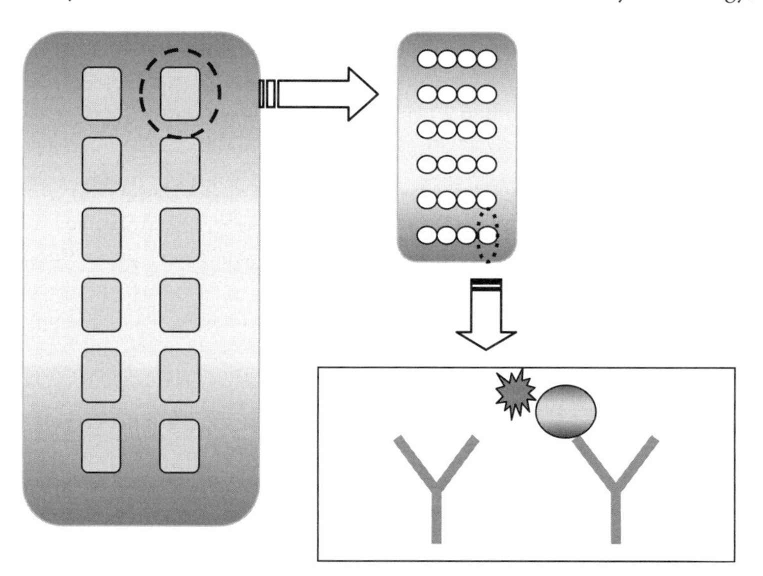

Figure 26.1 A schematic overview of the antibody microarray assay. At the left is an illustration of the platform containing 12 individual arrays. Within each of the arrays, a 4 × 6 configuration of capture antibody is printed at a defined spatial. The antigen–antibody reactions take place at these spots and are detected with a signal generation system.

paper describing the protein array format was published and initiated an exciting new research area [6].

At present, antibody microarray platforms including different cytokines, chemokines, and soluble factors have been commercially developed and the companies and technology providers in the capture array market, such as Schleicher & Schuell (S&S) and BD Clontech, took different approaches to their product with regard to the candidates, substrates, detection systems, and so on. The applications of these products in multiplexed protein measurement of cell culture supernatants, crude serum, and plasma concerning various diseases have been reported [7–10]. More recently, some works indicated that the antibody array, comprising antibodies specific for various posttranslational modifications, is also a potentially powerful approach for profiling protein posttranslational modification events [11]. Despite the rapid development, antibody microarrays are faced with challenges that require a diverse set of specialized tools and tricks, which are described in the following sections [12].

26.2.2
Assays on a Microspot: Principles of Antibody Microarray

The theoretical principle for microarray-based ligand binding assays was proposed by Ekins and coworkers in the late 1980s. He suggested that an ambient analyte assay using "microspots" of antibody on solid supports would provide extremely high sensitivity. This theory has demonstrated that the signal density obtained would increase due to the miniaturized format, the reasons for which are as follows: (i) The binding reaction occurs at the highest possible target concentration; (ii) The capture-detection complex is found only in the small area of the microspot, resulting in a high local signal [13] (Figure 26.2). As for sandwich immunoassay, Saviranta *et al.* [14] performed 24 sandwich microarray assays, which were lastly demonstrated to fulfill the criteria of the "ambient analyte" regime because depletion of analyte molecules from the assay volume was insignificant. However, sensitivities achieved for antibody microarrays are still far from this theoretical limit today. This discrepancy between the experiment and the theory was attributed to a number of causes, which are mentioned in Section 26.3.3 [15].

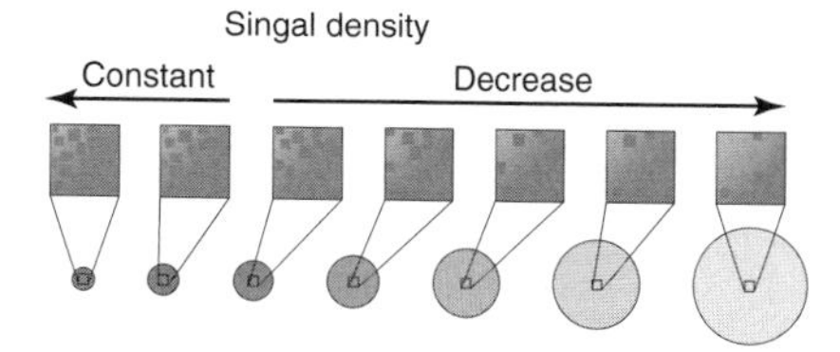

Figure 26.2 Reducing the spot size will decrease the overall signal per microspot but the signal density will increase. Below a certain spot size, the signal density achieves an optimum and will stay approximately constant even in case of further decrease in spot size [13].

26.2.3
Antibody Microarray Assay Formats

Two broad categories of antibody microarray experimental formats have been described: direct labeling, single antibody experiment, and dual antibody, sandwich immunoassay. In the direct labeling method, the proteins to be measured are labeled with a tag to allow detection following incubation on an antibody microarray. In the sandwich immunoassay format, antibody pairs against the different antigenic sites on a specific protein are applied to capture and detect the target protein (Figure 26.3). The pioneer works by Haab and Huang [16, 17], early in the infancy stage of antibody microarray, have demonstrated the feasibility of these two approaches in an array format and both formats have their advantages and disadvantages.

Because only one antibody per target is used in the label-based strategy, the number of antibodies that can be multiplexed is relatively large, making it suitable for large-scale protein profiling compared with sandwich assay. However, covalent attachment of fluorophors decreases the solubility of proteins and can also interfere with the antigen–antibody recognition. Moreover, low-abundance proteins are difficult to observe because of the competitive binding of the more abundant proteins in the sample that subsequently causes high background. In addition, the test and control samples should be labeled with fluorophors and analyzed in reciprocal incubations because of the different extent of labeling [18]. Unlike the single antibody method, the labeled detection antibody is introduced to detect the protein in the sandwich approach, which does not require the proteins to be labeled and thus, in principle, can provide concentration measurements of target proteins

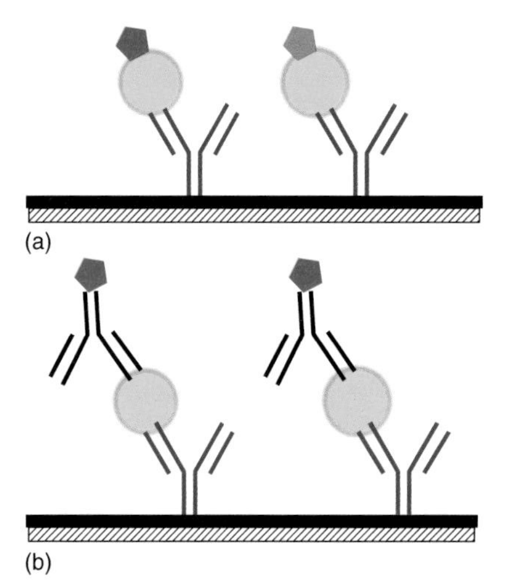

Figure 26.3 Antibody microarray format: (a) Direct labeling assay, in which protein is labeled with Cy3 or Cy5. (b) Sandwich assay, in which a matched pair of antibodies bind the unlabeled analyte followed by a detection system.

in complex biological mixtures. Moreover, low-abundance analytes can be detected through an extra signal amplification system such as rolling circle amplification (RCA) and tyramide signal amplification (TSA), making it an important tool for quantitative screening of low-abundance proteins [19]. However, the requirement of two epitope-specific antibodies increases the cost, time, and complexity of the assay and meanwhile limits the scalability of this format.

26.3 Technical Aspects of Antibody Microarray Production

26.3.1 Choice of Capture Antibody

To construct an antibody microarray platform, the candidates of the capture antibody are decided mainly according to the purpose of its application. To date, most antibody microarrays have been produced with commercially available polyclonal or monoclonal antibodies. Because polyclonal antibodies usually contain multiple epitope specificities, monoclonal (as opposed to polyclonal) antibodies should be the focus of the present research effort in order to generate a reproducible result. The main problem of antibody-mediated analytical protein microarrays is the specificity for most antibodies, which may cross-react with proteins other than the antigens of interest; so cross-reactivity is a particular problem in a multiplexed format and requires careful evaluation [20, 21].

Although the commercially available antibody pairs can be readily applied in sandwich antibody microarray, antibody pairs from different manufacturers for the same antigen with similar sensitivity as ELISA assays perform quite differently in microarrays. Thus, it has become necessary to select optimal pairs for antibody microarrays by comparing all available antibodies and pairs [22].

However, antibody is not a perfect capture agent for protein microarray by its disadvantages per se. The generation of such antibody microarrays is an expensive endeavor because of the labor-intensive nature of monoclonal antibody production. Moreover, limited availability of well-characterized antibodies is a key limitation for both types of antibody microarray approaches. As for large-scale, parallel profiling assays of direct-labeling antibody microarray, proteome-wide sets of antibodies are required. However, for many proteins, there are still no available antibodies. Matched antibody pairs for sandwich immunoassay are much more limited and the possibility of cross-reactivity between detection antibodies increases with additional analytes. Therefore, multiplexed sandwich assays are considered to have a practical size limitation of 30–50 different targets [23]. At present, almost all of these sandwich assays have been developed to study levels of cytokines and growth factors owing to the availability of antibody pairs. Additionally, Paulovich stated that the commercially available antibodies are often not annotated or validated extensively for the specific application required (Proteomic Technologies Reagents Resource Workshop; Chicago; 12–13 December 2005). Antibodies have to be selected for the properties that they are expected to display in a particular application [24]. Consequently, to overcome these limitations, some strategies have been proposed. Companies are expected to distribute small aliquots (more economical) of antibodies that are needed to spot on microarrays. A variety of improvements in antibody production have been made and hold promise for high-throughput antibody production and screening [25–27]. Furthermore, Haab [24] proposed two promising operational models for the antibody validation procedure.

26.3.2
Surfaces and Attachment

The selection of an appropriate surface and immobilization strategy is apparently a key step in antibody array design. Compared with DNA, proteins are prone to denaturation and degradation, so their immobilization on microarray surface is especially tricky [28, 29]. Suitable surfaces should bind capture proteins in a high amount and retain their three-dimensional structure, target-binding function together with low nonspecific binding of proteins to the surface. To meet the requirements of antibody microarrays, several surfaces have been proposed, which can be broken down into two basic categories: two-dimensional (2-D) and three-dimensional (3-D) [30]. The immobilization methods on these surfaces include noncovalent, covalent, and affinity bindings (Table 26.1) [31].

Although several commercial products are available, both for 2-D and for 3-D surfaces, and various slide surfaces have been investigated rather intensively [32], it is difficult to define universal immobilization strategies that do not discriminate between proteins [33]. Functionalized glass surfaces are among the popular products for protein-array construction for their easy operation, great durability, optical properties, and ability to eliminate background fluorescence. However, proteins must remain hydrated to assure the integrity of their structures, an issue not relevant to the production of DNA array. 3-D support is an important optional approach to preserve antibody activity, which allows larger spacing and aqueous surrounding. Mirzabekov's group attached capture agents to modified polyacrylamide gel pads on glass, which were applied to various types of immunoassays, enzymatic reactions, and assays with live cells. The substrate was demonstrated to be compatible with fluorescence and chemiluminescence detection and could maintain the biological properties of most immobilized proteins for at least six months [34–36]. It is commercially available as HydroGel slides from PerkinElmer Life Sciences, with a licensed technology, which has been engineered to possess high probe-loading capacity, low intrinsic fluorescence, and low nonspecific protein binding (www.perkinelmer.com/proteomics). Agarose-coated slides, another kind of 3-D support, are simpler to prepare than polyacrylamide gel surface. They have been shown to be sensitive and inexpensive for protein and DNA microarrays [37–40]. Moreover, it has been reported that

Table 26.1 Attachment strategies for antibody microarray production.

Categories	Surface chemistries
2-D surfaces	
Nonspecific/noncovalent way	PVDF (polyvinylidene fluoride)
	Nitrocellulose
	Poly(L-lysine)
	Calixcrown-5 derivatives
Nonspecific/covalent way	Aldehyde
	Epoxide
	Succinimidyl ester/isothiocyanate
	Photoaffinity reaction
Specific/noncovalent way	Avidin–biotin tag
	Ni-NTA-His tag
	Glutathione-GST tag
	Protein A/G-IgG Fc region
	OligoDNA-oligoDNA
Specific/covalent way	Maleimide
	Thioester
	Glyoxylyl group
3-D surfaces	
	Agarose
	Polyacrylamide
	Gel pad
	PDMS (polydimethylsiloxane) film

Adapted from [31].

specific oriented immobilization consistently produces a significant improvement in performance [41].

Besides, in constructing antibody microarrays, every surface/print mechanism and surface/fluorescent dye combinations require some consideration and optimization. Choosing the right printing buffer, appropriate sample protein concentration, and optimal binding conditions will undoubtedly have a key influence on the immobilization efficiency. A proper additive, such as glycerol, sucrose, or polyvinyl alcohol, is essential for improved performance of antibodies on microarray substrates and long-term stability [42, 43]. Nitrocellulose-coated slides cause light scattering and higher background [44], limiting the use of nitrocellulose substrate for fluorescent detection methods. Finally, as for the quantitative protein measurement, the substrates are expected to be homogeneous with equal amounts of sample attached in a quantitative way to ensure the reliable information of derived concentration.

26.3.3
Signal Detection and Assay Sensitivity

The complexity of any proteome makes all proteome analyses technically challenging in terms of multiplicity, dynamic range, and sensitivity. So far, various signal generation strategies have been employed in antibody arrays, including fluorescence, chemiluminescence, colorimetry, radioactivity, quantum dots, and other nanoparticles [45]. Most of the microarray assay formats rely on fluorescence- or chemiluminescence-based detection methods. Fluorescent dyes are becoming more and more popular because of high sensitivity, availability of multiple colors, ease of use and disposal, availability of labeling method, and relatively simple and low-cost equipment for signal capture and quantitation [46]. Also, several strategies have focused on replacing fluorescence with more powerful label systems, such as quantum dots [47], which are supposed to be a promising addition to established labeling techniques.

Although the theoretical detection limit of a microspot assay has been predicted to be a few femtograms or less, many published antibody microarray systems have reported a detection limit in the nanogram range [48]. It has been difficult to generate detectable signals in the low picogram range probably due to the low stability of proteins, the low affinity of the antibodies applied, as well as the low sensitivity of the detection methods or a high level of the background noise [49]. It is well known that many important biological markers for various diseases are present at very low concentrations in body fluids and the variation in concentrations of biomarkers among different patients can be scattered over as much as 5–6 orders of magnitude as demonstrated in blood and urine [50]. Consequently, from the perspective of clinical application, high-sensitivity detection technologies are obviously necessary for antibody microarray.

A popular approach to enhance detection sensitivity involves the addition of a signal amplification system and one of the most interesting developments is RCA and TSA, which indeed gain a lower detection limit of subpicogram [51, 52]. However, amplification typically requires protein labeling and often means less reproducible techniques with more steps, longer running time, and higher nonspecific signal. Accordingly, more recently, some novel strategies not relevant to amplification have been introduced. Sample format has been reported to be important for designing antibody-based microarrays for screening of low-abundance low-molecular weight protein analytes in directly fluorescence-labeled method [53]. Lately, Kusnezow *et al.* [54] have posed a view about kinetic aspects of antibody microarray, which used to get insufficient attention. They demonstrated that one of the main limitations for the sensitivity of current generation of microspot immunoassays is the strong dependence of antibody microspot kinetics upon mass flux to the spot. Thus, the task to relax the mass-transport limitations should be an important issue in designing the antibody microarrays [54]. An automated assay system integrating a novel microagitation device using surface acoustic wave (SAW) technology has been developed to enhance signal intensity by up to three fold [55]. Knight *et al.* [56] achieved a sensitivity that ranged from 4 to 12 $\mathrm{pg\,ml^{-1}}$ just through careful titration of the antibody-analyte-antibody interactions at the microarray level.

Hence, antibody microarrays for the quantitative analysis of complex proteins require further modifications and careful optimization to overcome the limitations with regard to sensitivity and the continued improvements of methods would undoubtedly promote the application of antibody microarrays for basic and applied biological research.

26.3.4
Technological Challenges of Antibody Microarrays

Antibody microarrays are far more difficult to construct than DNA microarrays and numerous difficulties need to be solved and procedures need to be optimized to establish a reliable protein microarray platform. Compared with a relatively uniform structure of DNA, proteins exist as large complexes for their various biophysical and chemical properties. And posttranslational modification such as acetylation, glycosylation, or phosphorylation is generally involved in protein functions and makes protein analysis more complex in the microarray format [57, 58].

The biggest challenge associated with antibody microarrays is the source of antibodies. A high-density microarray with good performance mainly depends on production of antibodies with high specificity and affinity in a high-throughput manner [59]. Recently, efforts have been undertaken to develop alternative antibody production method such as phage-display techniques and alternative capture agent such as aptamer [60–62]. We may anticipate these great contributions to antibody microarray in the coming few years.

To enter the diagnostic market, a prerequisite for antibody microarray, of course, will be excellent precision with high sensitivity and reliability. It is important to establish reproducible protocols and sample pretreatment strategies that enable quantitative analysis to be reproducible [63]. There are many different factors crucial for the quality and reliability of the final readout of each experiment. Thus common standards for various types of protein microarrays should be agreed upon in order to compare results obtained from different laboratories [64]. Fortunately, several pilot studies have been carried out by the Human Proteome Organization (HUPO: http://www.hupo.org) and the HUPO Standardization Initiative has proposed standards for data exchange (MIAPE).

It is hoped that protein expression microarrays will aid in biomarker discovery, disease diagnosis, monitoring, and prognosis. However, accurately achieving these aims is dependent upon suitable experimental designs, normalization procedures, and appropriate statistical analyses to obtain a reliable result and to expose differential expression patterns [65]. Therefore, improvements in statistical tools and bioinformatics analysis are equally crucial problems to be resolved, especially as the microarray density is increased. Despite all these bottlenecks and challenges within the field of antibody microarray technology, continued improvement will make it a powerful diagnostic tool with accuracy, reliability, and cost-effectiveness in the near future.

26.4
Application of Antibody Microarray to Renal Disease

26.4.1
Analysis of Multiplexed Proteins and Cellular Signaling: a Novel Approach to Investigate Renal Disease

As mentioned above, antibody microarray is characterized by its quantitative ability, high sensitivity, and specificity for a particular set of proteins, which make it uniquely suited for multiplexed protein detection and cellular signaling characterization [66]. The development of disease would definitely be accompanied with changes of certain proteins, which serve as the workhorse of the living cells. Obviously, identification of these disease-associated proteins could promote a better understanding of molecular events involved in disease pathogenesis and eventually encourage the development of novel therapeutic targets [67]. The advent of antibody microarray techniques opens exciting new avenues for simultaneous examination of those proteins, which will lead to better understanding of the protein network and their coregulations in an efficient way [68]. The development of detecting methods for posttranslational modification of intracellular signaling molecules on microarrays would make antibody microarrays invaluable in the study of signaling networks [69, 70]. Several studies using either the commercial antibody microarrays or self-assembled ones have demonstrated the application of antibody microarray technology in studying the proteins from serum, cell cultures, and tissue with regard to various disease states [71–73].

Renal disease is characterized by closely interrelated mechanisms of inflammation, repair, scarring, and atrophy affecting over 20 different intrinsic renal cell types [74]. Over the years, both basic and clinical investigations have identified many proteins involved in pathophysiological processes of kidney diseases among which cytokines have been demonstrated as the key participants in renal injury. Cytokines that are up-regulated during the course of progressive renal disease, include interleukins 1–7, platelet-derived growth factor (PDGF)-A, PDGF-B, tumor necrosis factor (TNF)-α, TNF-β, epithelial growth factor, insulin-like growth factor-1, transforming growth factor (TGF)-α, TGF-β, and so on [75]. Obviously, understanding the precise roles of those complicated cytokine networks is really a tough task. Enzyme-linked immunosorbent assays and Western blot analysis have been commonly used for rapid cytokine determinations for a long time. Unfortunately, insight into the molecular mechanisms that underlie the progression of nephropathy remains limited partly because those conventional research tools have been restricted to single protein or isolated pathways [76]. Understanding these networks at a system level requires technologies to measure both the amounts and activities of multiple proteins in an efficient and accurate manner. Antibody microarray is a potential tool satisfying those requests. Meanwhile, the readily available antibodies of these proteins allow an easy shift to an integrated investigation in the antibody microarray format, which can provide entire information of all members of the pathway [77]. Our previous study has demonstrated that antibody

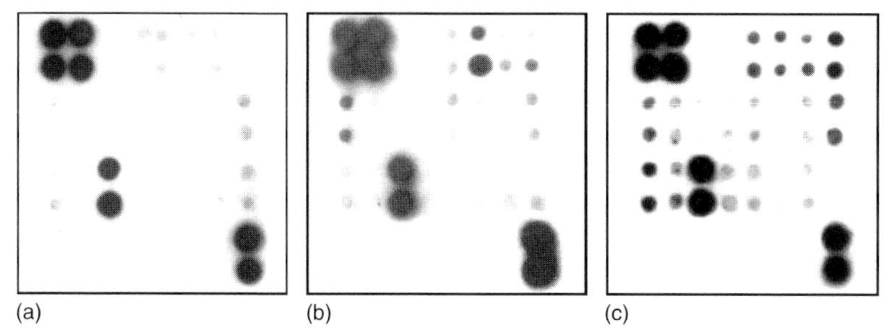

Figure 26.4 Images of Custom Raybio Human Cytokines Array. A total of 20 cytokines were placed on the array. Representative antibody arrays were incubated with urine samples from normal controls (a), CKD without renal failure (b) and CKD with renal failure (c). Four spots in the upper left and two in the lower right corners of the membranes indicate the positive controls.

microarray could serve as a fast, high-throughput, and sensitive tool for detecting the excretion of urinary proteins, revealing a significant change of profiles in patients with CKD compared with normal controls (Figure 26.4) [78]. Moreover, its ability to probe the activities of signaling proteins in a parallel and reliable manner will make antibody microarray powerful in investigating the cellular process of renal disease.

However, the recent proteomic technologies for the study of renal disease have focused on electrophoresis and MS-based platforms, whereas the application of antibody microarray is very limited. This may be partly due to the immaturity of this technology and admittedly its value in this area has still not been explored [79, 80]. Owing to the limited capacity to identify and quantify proteins in complex mixtures for electrophoresis and MS-based technologies, antibody microarray will surely find its crucial place in uncovering the underlying mechanisms for the progression of renal disease with the advancement of this technology [81].

26.4.2
Antibody Microarray for Targeted Urinary Proteome Profiling

The research using antibody microarray currently focuses on a limited number of candidate biomarker proteins, termed as *targeted proteomics*, which requires a prior knowledge about these related proteins [82]. This assay technology provides higher sensitivity, reproducibility, and dynamic range compared with the unbiased separation-based and MS-based approaches. The first step in an antibody microarray study is to define a protein population to be investigated that mainly depends on the purpose of the study. Examination of urine is a particularly appealing focus for kidney disease research because of its accessibility in a large quantity and the abundant information it contains in different pathophysiological states [83–85]. Sandwich microarray measurements can provide quantitative information on a number of analytes in parallel and the single antibody microarray, similar to the two-color mRNA profiling approach, can be used to yield ratiometric information of relative changes of protein abundance [23]. Consequently, both approaches for targeted urinary proteome profiling hold a potential promise in basic and clinical renal research.

26.4.2.1 Handling of Urine Samples

As proteomic technologies move toward higher throughput manner, upstream sample preparation gains emphasis as a potential bottleneck. In the Cambridge Healthtech Institute's Second Annual Conference in 2002, genomic and proteomic sample preparation was chosen as the theme and the quality of protein sample was emphasized for generating accurate and informative data [86]. Urine represents a complex proteome, which harbors soluble proteins derived from glomerular filtration, epithelial cell secretion, and solid phase components [87]. The information from the results of the microarray analysis will certainly be influenced by a number of factors, which make it difficult to make a comparison between different laboratories. Therefore, there is an urgent need for the development of standard protocols for collection, preparation, and storage of urine samples [88].

However, it can be anticipated that the handling of urine sample for cytokine analysis will be similar between antibody microarray and traditional ELISA assay, except only for the complexity resulting from the different detection ranges in a single experiment. After collection, the urine is centrifuged to separate debris and a protease inhibitor cocktail is usually added. Then urine is divided into aliquots and stored at -20 or $-80\,^{\circ}$C for future use. Before cytokine analysis, it is thawed at room temperature. To avoid the effect of urine volume, quantitative analysis of creatinine is usually performed and urinary levels of cytokines can be finally expressed as the ratio of cytokine to urinary creatinine (pg μmol^{-1} creatinine) [89–91]. Depletion of abundant proteins can increase the relative concentration, facilitate the detection of lower-abundance proteins, and become a commonplace or even routine process [92]. And it is typically required to prepare serial dilutions of urine sample to fit the antibody affinity and the corresponding detection dynamic range. Overall, sample collection, preparation, and storage are important aspects and need to be carefully considered as described by Schaub *et al.* [93] and Rogers *et al.* [94].

26.4.2.2 Identification of Biomarkers for Kidney Disease

At present, most renal diseases, particularly diabetes and hypertension, are usually detected at some stage after substantial renal injury based on either the detection of proteinuria or elevation of serum creatinine [95]. Although available therapies can slow down progression to end-stage renal disease (ESRD) in many patients, most of them would inevitably enter ESRD and eventually replacement therapy. Moreover, the difficulties in follow-up progression and determining response to therapy add complexity to clinical practice for nephrologists [96]. Thus, sensitive markers satisfying these desires are desperately needed. Urinary proteome profiling represents an appealing source for biomarker identification and the combinatorial biomarkers obtained are more specific and sensitive compared to each individual biomarker.

In the First International Summit on Kidney Disease Prevention, the consensus document proposed that one of the important laboratory-based studies is to identify urinary proteome patterns using proteomic approaches to predict or detect CKD progression or regression [97]. Actually, some nephrologists have already used electrophoresis and MS-based methods for the discovery of potential disease biomarkers for a number of renal diseases including diabetic glomerulopathy [77], acute renal allograft rejection [98], and lupus nephritis [99], and the list of prognostic biomarkers in patients with kidney diseases is continuously growing [100]. Although DNA microarray has been applied in numerous studies for kidney diseases [101, 102], very limited research employing antibody microarray has been reported so far when such platforms have already shown its potential in the studies of other diseases [103, 104].

A three-stage pipeline is usually described comprising (i) discovery, (ii) verification/validation, and (iii) clinical implementation for the development of biomarkers [105]. The second stage aims to determine the key parameters for a diagnostic test, including normal biovariability, sensitivity, and specificity in relation to target disease in a large population. Therefore, good reproducibility (i.e., coefficients of variation <10%), high sensitivity (i.e., a range from $\mathrm{pg\,ml^{-1}}$ to $\mathrm{mg\,ml^{-1}}$), and quantitative ability are required. Consequently, antibody microarray technology represents the most practical approach for validation in a cost effective and efficient manner [106]. Moreover, the high-density antibody microarray can serve as a discovery tool for biomarker development yielding proteins that show significant differences in expression between controls and patients following validation study [107].

However, there are many challenges encountered when incorporating proteomic study into our diagnostic and therapeutic paradigms [108, 109]. As for antibody microarrays, several new operational issues have to be emphasized, that is, the quality control of the assays, proper data normalization, and the use of diagnostic algorithms to transform microarray data into diagnostic venue [110, 111]. Although development of biomarkers for kidney diseases is still challenging, it is predicted that antibody microarray will play an important role in clinical nephrology in the near future.

References

1. MacBeath, G. (2002) Protein microarrays and proteomics. *Nat Genet*, **32**, 526–532.
2. Barry, R. and Soloviev, M. (2004) Quantitative protein profiling using antibody arrays. *Proteomics*, **4** (12), 3717–3726.
3. Wingren, C. and Borrebaeck, C.A. (2006) Antibody microarrays: current status and key technological advances. *OMICS*, **10** (3), 411–427.
4. Ling, M.M., Ricks, C., and Lea, P. (2007) Multiplexing molecular diagnostics and immunoassays using emerging microarray technologies. *Expert Rev Mol Diagn*, **7** (1), 87–98.
5. Ekins, R., Chu, F., and Biggart, E. (1990) Multispot, multianalyte, immunoassay. *Ann Biol Clin*, **48**, 655–666.
6. MacBeath, G. and Schreiber, S.L. (2000) Printing proteins as microarrays for high-throughput function

determination. *Science*, **289** (5485), 1760–1763.

7. Lia, Y., Schuttea, R.J., Abu-Shakra, A. *et al.* (2005) Protein array method for assessing in vitro biomaterial-induced cytokine expression. *Biomaterials*, **26**, 1081–1085.
8. Sukhanov, S. and Delafontaine, P. (2005) Protein chip-based microarray profiling of oxidized low density lipoprotein-treated cells. *Proteomics*, **5**, 1274–1280.
9. Relucio, K.I., Beernink, H.T., Chen, D. *et al.* (2005) Proteomic analysis of serum cytokine levels in response to highly active antiretroviral therapy (HAART). *J Proteome Res*, **4**, 227–231.
10. Srivastava, M., Eidelman, O., Jozwik, C. *et al.* (2006) Serum proteomic signature for cystic fibrosis using an antibody microarray platform. *Mol Genet Metab*, **87** (4), 303–310.
11. Ivanov, S.S., Chung, A.S., Yuan, Z. *et al.* (2004) Antibodies immobilized as arrays to profile protein post-translational modifications in mammalian cells. *Mol Cell Proteomics*, **3** (8), 788–795.
12. Eisenstein, M. (2006) Protein arrays: growing pains. *Nature*, **444** (7121), 959–962.
13. Templin, M.F., Stoll, D., Schrenk, M. *et al.* (2002) Protein microarray technology. *Trends Biotechnol*, **20** (4), 160–166.
14. Saviranta, P., Okon, R., and Brinker, A. (2004) Evaluating sandwich immunoassays in microarray format in terms of the ambient analyte regime. *Clin Chem*, **50** (10), 1907–1920.
15. Kusnezow, W., Syagailo, Y.V., Ruffer, S. *et al.* (2006) Kinetics of antigen binding to antibody microspots: strong limitation by mass transport to the surface. *Proteomics*, **6** (3), 794–803.
16. Haab, B.B., Dunham, M.J., and Brown, P.O. (2001) Protein microarrays for highly parallel detection and quantitation of specific proteins and antibodies in complex solutions. *Genome Biol*, **2**, RESEARCH0004.
17. Huang, R.P. (2001) Detection of multiple proteins in an antibody-based protein microarray system. *J Immunol Methods*, **255** (1-2), 1–13.
18. Ahn, E.H., Kang, D.K., Chang, S.I. *et al.* (2006) Profiling of differential protein expression in angiogenin-induced HUVECs using antibody-arrayed ProteoChip. *Proteomics*, **6** (4), 1104–1109.
19. Nielsena, U.B. and Geierstanger, B.H. (2004) Multiplexed sandwich assays in microarray format. *J Immunol Methods*, **290**, 107–120.
20. Phizicky, E., Bastiaens, P.I.H., Zhu, H. *et al.* (2003) Protein analysis on a proteomic scale. *Nature*, **422** (13), 208–215.
21. Song, S., Li, B., Wang, L. *et al.* (2007) A cancer protein microarray platform using antibody fragments and its clinical applications. *Mol Biosyst*, **3** (2), 151–158.
22. Shao, W., Zhou, Z., Laroche, I. *et al.* (2003) Optimization of rolling-circle amplified protein microarrays for multiplexed protein profiling. *J Biomed Biotechnol*, **5**, 299–307.
23. Haab, B.B. (2005) Antibody arrays in cancer research. *Mol Cell Proteomics*, **4**, 377–383.
24. Haab, B.B. Paulovich, A.G., Anderson, N.L. *et al.* (2006) A reagent resource to identify proteins and peptides of interest for the cancer community: a workshop report. *Mol Cell Proteomics*, **5** (10), 1996–2007.
25. Wingren, C., Ingvarsson, J., Lindstedt, M. *et al.* (2003) Recombinant antibody microarrys-a viable option? *Nat Biotechnol*, **21**, 223.
26. Chambers, R.S. (2005) High-throughput antibody production. *Curr Opin Chem Biol*, **9** (1), 46–50.
27. Michaud, G.A., Salcius, M., and Zhou, F. (2003) Analyzing antibody specificity with whole proteome microarrays. *Nat Biotechnol*, **21** (12), 1509–1512.
28. Kozarova, A., Petrinac, S., Ali, A. *et al.* (2006) Array of informatics:

applications in modern research. *J Proteome Res*, **5** (5), 1051–1059.
29. Doerr, A. (2005) Protein microarray velcro. *Nat Methods*, **2** (9), 642–643.
30. Angenendt, P. (2005) Progress in protein and antibody microarray technology. *Drug Discov Today*, **10** (7), 503–511.
31. Angenendt, P., Glokler, J., Murphy, D. *et al.* (2002) Toward optimized antibody microarrays: a comparison of current microarray support materials. *Anal Biochem*, **309** (2), 253–260.
32. Guilleaume, B., Buness, A., Schmidt, C. *et al.* (2005) Systematic comparison of surface coatings for protein microarrays. *Proteomics*, **5** (18), 4705–4712.
33. Kusnezow, W., Jacob, A., Walijew, A. *et al.* (2003) Antibody microarrays: an evaluation of production parameters. *Proteomics*, **3** (3), 254–264.
34. Arenkov, P., Kukhtin, A., Gemmell, A. *et al.* (2000) Protein microchips: use for immunoassay and enzymatic reactions. *Anal Biochem*, **278** (2), 123–131.
35. Fesenko, D.O., Nasedkina, T.V., Prokopenko, D.V. *et al.* (2005) Biosensing and monitoring of cell populations using the hydrogel bacterial microchip. *Biosens Bioelectron*, **20** (9), 1860–1865.
36. Rubina, A.Y., Dementieva, E.I., Stomakhin, A.A. *et al.* (2003) Hydrogel-based protein microchips: manufacturing, properties, and applications. *Biotechniques*, **34** (5), 1008–1014, 1016–1020, 1022.
37. Afanassiev, V., Hanemann, V., and Wölfl, S. (2000) Preparation of DNA and protein micro arrays on glass slides coated with an agarose film. *Nucleic Acids Res*, **28**, E66.
38. Jin, S.H., Liu, B.C., Zhang, C.X. *et al.* (2005) Optimizing of the preparation condition of agarose modified glass slide for the production of protein microarray. *Chin J Biotechnol*, **25**, 57–60.
39. Zhang, C.X., Liu, H.P., Tang, Z.M. *et al.* (2003) Cell detection based on protein array using modified glass slides. *Electrophoresis*, **24**, 3279–3283.
40. Lv, L.L., Liu, B.C., Zhang, C.X. *et al.* (2007) Construction of an antibody microarray based on agarose-coated slides. *Electrophoresis*, **28**, 406–413.
41. Vareiro, M.M., Liu, J., Knoll, W. *et al.* (2005) Surface plasmon fluorescence measurements of human chorionic gonadotrophin: role of antibody orientation in obtaining enhanced sensitivity and limit of detection. *Anal Chem*, **77**, 2426–2431.
42. Wu, P. and Grainger, D.W. (2006) Comparison of hydroxylated print additives on antibody microarray performance. *J Proteome Res*, **5** (11), 2956–2965.
43. Wu, P. and Grainger, D.W. (2004) Toward immobilized antibody microarray optimization: print buffer and storage condition comparisons on performance. *Biomed Sci Instrum*, **40**, 243–248.
44. Kukar, T., Eckenrode, S., Gu, Y. *et al.* (2002) Protein microarrays to detect protein-protein interactions using red and green fluorescence proteins. *Anal Biochem*, **306** (1), 50–54.
45. Espina, V., Woodhouse, E.C., Wulfkuhle, J. *et al.* (2004) Protein microarray detection strategies: focus on direct detection technologies. *J Immunol Methods*, **290** (1-2), 121–133.
46. Müller, U.R. (2003) Evolution of high-sensitivity microarray technology. *J Assoc Lab Autom*, **8**, 96–100.
47. Chan, W.C. and Nie, S. (1998) Quantum dot bioconjugates ultrasensitive nonisotopic detection. *Science*, **281** (5385), 2016–2018.
48. Nettikadan, S., Radke, K., Johnson, J. *et al.* (2006) Detection and quantification of protein biomarkers from fewer than 10 cells. *Mol Cell Proteomics*, **5** (5), 895–901.
49. Kusnezow, W., Syagailo, Y.V., Ruffer, S. *et al.* (2006) Kinetics of antigen binding to antibody microspots: strong limitation by mass transport to the surface. *Proteomics*, **6**, 794–803.
50. Drukier, A.K., Ossetrova, N., Schors, E. *et al.* (2005) Ultra-sensitive immunoassays using

multi-photon-detection in diagnostic proteomics of blood. *J Proteome Res*, **4** (6), 2375–2378.
51. Kingsmore, S.F. and Patel, D.D. (2003) Multiplex protein profiling on antibody-based microarrays by rolling circle amplification. *Curr Opin Biotechnol*, **14** (1), 74–81.
52. Woodbury, R.L., Varnum, S.M., and Zangar, R.C. (2002) Elevated HGF levels in sera from breast cancer patients detected using a protein microarray ELISA. *J Proteome Res*, **1** (3), 233–237.
53. Ingvarsson, J., Lindstedt, M., Borrebaeck, C.A. *et al.* (2006) One-step fractionation of complex proteomes enables detection of low abundant analytes using antibody-based microarrays. *J Proteome Res*, **5** (1), 170–176.
54. Kusnezow, W., Syagailo, Y.V., Goychuk, I. *et al.* (2006) Antibody microarrays: the crucial impact of mass transport on assay kinetics and sensitivity. *Expert Rev Mol Diagn*, **6** (1), 111–124.
55. Hartmann, M., Toeglb, A., Kirchner, R. *et al.* (2006) Increasing robustness and sensitivity of protein microarrays through microagitation and automation. *Anal Chim Acta*, **564**, 66–73.
56. Knight, P.R., Sreekumar, A., Siddiqui, J. *et al.* (2004) Development of a sensitive microarray immunoassay and comparison with standard enzyme-linked immunoassay for cytokine analysis. *Shock*, **21** (1), 26–30.
57. Ivanov, S.S., Chung, A.S., Yuan, Z.L. *et al.* (2004) Antibodies immobilized as arrays to profile protein post-translational modifications in mammalian cells. *Mol Cell Proteomics*, **3** (8), 788–795.
58. Delom, F. and Chevet, E. (2006) Phosphoprotein analysis: from proteins to proteomes. *Proteome Sci*, **4**, 15, DOI: 10.1186/1477-5956-4-15.
59. Kozarova, A., Petrinac, S., Ali, A. *et al.* (2006) Array of informatics: applications in modern research. *J Proteome Res*, **5**, 1051–1059.
60. Collett, J.R., Cho, E.J., and Ellington, A.D. (2005) Production and processing of aptamer microarrays. *Methods*, **37**, 4–15.
61. Hanash, S. (2003) Disease proteomics. *Nature*, **422** (13), 226–232.
62. Renberg, B., Nordin, J., Merca, A. *et al.* (2007) Affibody molecules in protein capture microarrays: evaluation of multidomain ligands and different detection formats. *J Proteome Res*, **6** (1), 171–179.
63. Boyle, M.D., Hess, J.L., Nuara, A.A. *et al.* (2006) Application of immunoproteomics to rapid cytokine detection. *Methods*, **38** (4), 342–350.
64. Hultschig, C., Kreutzberger, J., Seitz, H. *et al.* (2006) Recent advances of protein microarrays. *Curr Opin Chem Biol*, **10**, 4–10.
65. Eckel-Passow, J.E., Hoering, A., and Therneau, T.M. (2005) Ghobrial I: experimental design and analysis of antibody microarrays: applying methods from cDNA arrays. *Cancer Res*, **65**, 2985–2989.
66. Espina, V., Mehta, A.I., Winters, M.E. *et al.* (2003) Protein microarrays: molecular profiling technologies for clinical specimens. *Proteomics*, **3**, 2091–2100.
67. Collins, C.D., Purohit, S., Podolsky, R.H. *et al.* (2006) The application of genomic and proteomic technologies in predictive, preventive and personalized medicine. *Vascul Pharmacol*, **45** (5), 258–267.
68. Haab, B.B. (2006) Using array-based competitive and noncompetitive immunoassays. AACR 97th Annual Meeting, Washington, DC, United States, April 1-5, 2006.
69. Nielsen, U.B., Mike, H., Cardone, M.H., Sinskey, A.J. *et al.* (2003) Profiling receptor tyrosine kinase activation by using Ab microarrays. *Proc Natl Acad Sci U S A*, **100** (16), 9330–9335.
70. Gembitsky, D.S., Lawlor, K., Jacovina, A. *et al.* (2004) A prototype antibody microarray platform to monitor changes in protein tyrosine phosphorylation. *Mol Cell Proteomics*, **3** (11), 1102–1118.

71. Srivastava, M., Eidelman, O., Jozwik, C. *et al.* (2006) Serum proteomic signature for cystic Wbrosis using an antibody microarray platform. *Mol Genet Metab*, **87**, 303–310.
72. Sreekumar, A., Nyati, M.K., Varambally, S. *et al.* (2001) Profiling of cancer cells using protein microarrays: discovery of novel radiation-regulated proteins. *Cancer Res*, **61** (20), 7585–7593.
73. Weber, A., Hengge, U.R., Stricker, I. *et al.* (2007) Protein microarrays for the detection of biomarkers in head and neck squamous cell carcinomas. *Hum Pathol*, **38** (2), 228–238.
74. Harris, D.C.H. and Rangan, G.K. (2005) Retardation of kidney failure – applying principles to practice. *Ann Acad Med Singapore*, **34**, 16–23.
75. Wong, W. and Singh, A.K. (2001) Urinary cytokines: clinically useful markers of chronic renal disease progression? *Curr Opin Nephrol Hypertens*, **10** (6), 807–811.
76. Susztak, K., Sharma, K., Schiffer, M. *et al.* (2003) Genomic strategies for diabetic nephropathy. *J Am Soc Nephrol*, **14**, S271–S278.
77. Knepper, M.A. (2002) Proteomics and the kidney. *J Am Soc Nephrol*, **13**, 1398–1408.
78. Liu, B.C., Zhang, L., Lv, L.L. *et al.* (2006) Application of antibody array technology in analysis of urinary cytokine profiles in patients with chronic kidney disease. *Am J Nephrol*, **26** (5), 483–490.
79. Mosley, K., Tam, F.W., Edwards, R.J. *et al.* (2006) Urinary proteomic profiles distinguish between active and inactive lupus nephritis. *Rheumatology*, **45**, 1497–1504.
80. Thongboonkerd, V. and Klein, J.B. (eds) (2004) Overview of proteomics, *Proteomics in Nephrology*, Contributions to Nephrology, vol. **141**, Karger, Basel, pp. 1–10.
81. Aebersold, R., Anderson, L., Caprioli, R. *et al.* (2005) Perspective: a program to improve protein biomarker discovery for cancer. *J Proteome Res*, **4**, 1104–1109.
82. Aderson, L. (2005) Candidate-based proteomics in the search for biomarkers of cardiovascular disease. *J Physiol*, **563** (1), 23–60.
83. Fliser, D., Novak, J., Thongboonkerd, V. *et al.* (2007) Advances in urinary proteome analysis and biomarker discovery. *J Am Soc Nephrol*, **18** (4), 1057–1071.
84. Dihazi, H. and Muller, G.A. (2007) Urinary proteomics: a tool to discover biomarkers of kidney diseases. *Expert Rev Proteomics*, **4** (1), 39–50.
85. Pisitkun, T., Johnstone, R., and Knepper, M.A. (2006) Discovery of urinary biomarkers. *Mol Cell Proteomics*, **5** (10), 1760–1771.
86. Anderson, P., Parallel Sample Preparation for Proteomics Applications. Cambridge Healthtech Institute's Second Annual Conference, Genomic and proteomic sample preparation. Boston, MA, May 2-3, 2002.
87. Gonzalez-Buitrago, J.M., Ferreira, L., and Lorenzo, I. (2007) Urinary proteomics. *Clin Chim Acta*, **375** (1-2), 49–56.
88. O'Riordan, E., Gross, S.S., and Goligorsky, M.S. (2006) Technology Insight: renal proteomics – at the crossroads between promise and problems. *Nat Clin Pract Nephrol*, **2** (8), 445–458.
89. Parikh, C.R., Abraham, E., Ancukiewicz, M. *et al.* (2005) Urine IL-18 is an early diagnostic marker for acute kidney injury and predicts mortality in the intensive care unit. *J Am Soc Nephrol*, **16**, 3046–3052.
90. Kassir, K., Vargas-Shiraishi, O., Zaldivar, F. *et al.* (2001) Cytokine profiles of pediatric patients treated with antibiotics for pyelonephritis: potential therapeutic impact. *Clin Diagn Lab Immunol*, **8** (6), 1060–1063.
91. Aldreda, S., Grantb, M.M., and Griffiths, H.R. (2004) The use of proteomics for the assessment of clinical samples in research. *Clin Biochem*, **37**, 943–952.
92. Steel, L.F., Trotter, M.G., Nakajima, P.B. *et al.* (2003) Efficient and specific removal of albumin from

human serum samples. *Mol Cell Proteomics*, **2** (4), 262–270.

93. Schaub, S., Wilkins, J., Weiler, T. *et al.* (2004) Urine protein profiling with surface-enhanced laser-desorption/ionization time-of-flight mass spectrometry. *Kidney Int*, **65**, 323–332.
94. Rogers, M.A., Clarke, P., Noble, J. *et al.* (2003) Proteomic profiling of urinary proteins in renal cancer by surface enhanced laser desorption ionization and neural-network analysis: identification of key issues affecting potential clinical utility. *Cancer Res*, **63**, 6971–6983.
95. Hostetter, T.H. (2003) Prevention of the development and progression of renal disease. *J Am Soc Nephrol*, **14**, S144–S147.
96. Hewitt, S.M., Dear, J., and Star, R.A. (2004) Discovery of protein biomarkers for renal diseases. *J Am Soc Nephrol*, **15**, 1677–1689.
97. Vidal, B.C., Bonventre, J.V., and I-Hong Hsu, S. (2005) Towards the application of proteomics in renal disease diagnosis. *Clin Sci (Lond)*, **109** (5), 421–430.
98. Ramitez, S.B. (2003) First international summit on kidney disease prevention, 25–27 July 2002: consensus document. *J Am Soc Nephrol*, **14**, S205–S207.
99. Sharma, K., SoHee Lee, S., Han, S. *et al.* (2005) Two-dimensional fluorescence difference gel electrophoresis analysis of the urine proteome in human diabetic nephropathy. *Proteomics*, **5**, 2648–2655.
100. Schaub, S., Rush, D., Wilkins, J. *et al.* (2004) Proteomic-based detection of urine proteins associated with acute renal allograft rejection. *J Am Soc Nephrol*, **15**, 219–227.
101. Mosley, K., Tam, F.W., Edwards, R.J. *et al.* (2006) Urinary proteomic profiles distinguish between active and inactive lupus nephritis. *Rheumatology (Oxford)*, **45** (12), 1497–1504.
102. Zoccali, C. (2005) Biomarkers in chronic kidney disease: utility and issues towards better understanding. *Curr Opin Nephrol Hypertens*, **14**, 532–537.
103. Nagasawa, Y., Takenaka, M., Kaimori, J. *et al.* (2001) Rapid and diverse changes of gene expression in the kidneys of protein-overload proteinuria mice detected by microarray analysis. *Nephrol Dial Transplant*, **16** (5), 923–931.
104. Weber, A., Hengge, U.R., Stricke, I. *et al.* (2007) Protein microarrays for the detection of biomarkers in head and neck squamous cell carcinomas. *Hum Pathol*, **38** (2), 228–238.
105. Mor, G., Visintin, I., Lai, Y. *et al.* (2005) Serum protein markers for early detection of ovarian cancer. *Proc Natl Acad Sci U S A*, **102** (21), 7677–7682.
106. Anderson, N.L. (2005) The roles of multiple proteomic platforms in a pipeline for new diagnostics. *Mol Cell Proteomics*, **4** (10), 1441–1444.
107. Urbanowska, T., Mangialaio, S., Zickler, C. *et al.* (2006) Protein microarray platform for the multiplex analysis of biomarkers in human sera. *J Immunol Methods*, **316** (1-2), 1–7.
108. Azad, N.S., Rasool, N., Annunziata, C.M. *et al.* (2006) Proteomics in clinical trials and practice: present uses and future promise. *Mol Cell Proteomics*, **5** (10), 1819–1829.
109. Wu, T. and Mohan, C. (2007) Proteomics on the diagnostic horizon: lessons from rheumatology. *Am J Med Sci*, **333** (1), 16–25.
110. Master, S.R., Bierl, C., and Kricka, L.J. (2006) Diagnostic challenges for multiplexed protein microarrays. *Drug Discov Today*, **11** (21-22), 1007–1011.
111. Hamelinck, D., Zhou, H., Li, L., Verweij, C. *et al.* (2005) Optimized normalization for antibody microarrays and application to serum-protein profiling. *Mol Cell Proteomics*, **4** (6), 773–784.

27
Proteomic Analysis of Dialysate Fluid and Adsorbed Proteins on Dialysis Membranes during Renal Replacement Therapy

Isao Ishikawa

27.1
Introduction

Dialysis therapy is the most common method of treatment for a number of end-stage renal disease (ESRD) patients. Approximately 1 455 000 of 1 900 000 ESRD patients worldwide have been treated by dialysis for renal replacement therapy [1]. In dialysis therapy, uremic toxins including serum protein components are eliminated by diffusion and convection accompanied by adsorption onto the dialysis membrane [2–4]. Dialysis membranes are not only classified into cellulosic membrane and synthetic polymer membrane but also divided in terms of efficiency into low flux and high flux. On the basis of the cross-sectional structure, synthetic polymer membranes are divided into symmetrical and asymmetrical membranes. The amounts and kinds of protein filtered through the membrane and adsorbed onto the membranes differ among various types of the membranes [2]. As a result, although the death rate of patient does not significantly differ by the kind of dialysis membranes used [5, 6], the incidence of dialysis complications, such as cardiovascular complication and dialysis-related amyloidosis, differ with the type of dialysis membranes used. For various dialysis membranes, proteins that are filtered through the dialysis membranes and/or adsorbed onto the dialysis membrane are only partly known. Although 90 uremic toxins are known at present [7], there is a concern that there may be more uremic toxins than currently known.

In this chapter, recently reported proteomic analyses of polypeptides and proteins in the dialysate fluid as well as those adsorbed onto the dialysis membrane are reviewed. In our study, the same patients were dialyzed using polysulfone membrane and polymethyl methacrylate (PMMA), both of which were classified as the high-flux membranes. Two membranes were compared with regard to the profile of eliminated proteins in the outflow dialysate and proteins adsorbed onto the dialysis membrane using ProteinChip surface-enhanced laser desorption/ionization time-of-flight mass spectrometry (SELDI-TOF-MS) system (Ciphergen Biosystems Inc. Fremont, USA) [7–11] and in-gel digestion followed by tandem MS system.

Renal and Urinary Proteomics: Methods and Protocols. Edited by Visith Thongboonkerd

ISBN: 978-3-527-31974-9

27.2 Literature Review

The polypeptides and proteins in the outflow dialysate and those adsorbed onto the dialysis membrane are mainly intermediate- to low-molecular-weight components (20 000 Da or less). This range of polypeptides and proteins is difficult to detect by the usual two-dimensional (2-D) gel electrophoresis. Thus, ProteinChip SELDI-TOF-MS system effective in the range of 2 000 Da to 20 000–30 000 Da and capillary electrophoresis coupled to mass spectrometry (CE-MS) [12–14] effective in the range of 800–10 800 Da are desired. According to Neuhoff *et al.* [15] who investigated the same series of urine specimens by both these methods, SELDI-TOF-MS is easy to perform and can examine small quantities of specimen, but the spectra detected are limited in number and show lower resolution. Furthermore, additional methods are necessary for protein identification. However, CE-MS is a complex method but detects a larger number of polypeptides. CE-MS can also identify polypeptides sequentially.

27.2.1 Proteomic Analyses of Proteins in the Outflow Dialysate Reported in the Literature

Kaiser *et al.* [14] used CE-MS to compare the dialysis filtrates between the high-flux membrane and the low-flux membrane. There were 2515 polypeptides detected in samples obtained using the high-flux membrane and 1639 polypeptides in those obtained using the low-flux membrane. However, identification of polypeptides was not carried out in this study. Because small polypeptides that were filtered in large amounts by the high-flux membrane were filtered only in small amounts by the low-flux membrane, the high-flux membrane was considered to be more useful for eliminating polypeptides than the low-flux membrane.

Lefler *et al.* [16] examined the filtrate from continuous venovenous hemofiltration (CVVH) in a patient with acute kidney injury. Proteins in the filtrate were separated using 2-D gel electrophoresis combined with reverse-phase liquid chromatography, making the third dimension of separation. Proteins were subsequently identified by peptide mass fingerprinting using MALDI-TOF-MS. Using this method, proteins were successfully identified in 47 of 196 spots. The 47 spots were multiply charged forms of 10 proteins, including albumin, apolipoprotein A-IV, β_2-microglobulin, lithostathine, mannose-binding lectin-associated serine protease 2 associated protein, plasma retinol-binding protein, transferrin, transthyretin, vitamin D-binding protein, and Zn-α_2-glycoprotein.

Ward and Brinkley [17] collected dialysis filtrates from four hemodialysis patients. Protein spots detected by 2-D gel electrophoresis and by in-gel digestion with trypsin were identified by peptide mass fingerprinting using MALDI-TOF-MS. The dialyzer membranes tested were Baxter cellulose triacetate (CT190G), Fresenius polysulfone (F80A), and Gambro high-flux polyarylethersulfone-polyvinylpyrrolidone (Polyflux 17S). Twenty-one proteins including α_1-antitrypsin, β_2-microglobulin, complement factor D, and isoform of retinol-binding protein were identified in the outflow dialysate of hemodialysis patients. The posttranslational modifications of retinol-binding protein, α_1-antitrypsin, β_2-microglobulin, and complement factor D were detected.

Weissinger *et al.* [18] investigated the proteins filtered by dialysis membranes using high-flux polysulfone and low-flux membranes. After protein separation by capillary electrophoresis (CE), polypeptides were identified by electrospray ionization time-of-flight (ESI-TOF) mass spectrometer. In the molecular mass range of 800–10 000 Da, 490 polypeptides were found in the filtrate of normal serum using a low-flux membrane and 544 polypeptides were found using a high-flux membrane. Moreover, 1046 polypeptides by low-flux membrane and 1394 polypeptides by the high-flux membrane were obtained using uremic serum. Many polypeptides of 5 kDa or more were found in the filtrate of uremic serum by high-flux membrane. In uremic ultrafiltrate, the 950.6 and 1291.8 Da polypeptides were identified as a fragment of the salivary proline rich protein and a fragment of α-fibrinogen, respectively.

Molina *et al.* [19] used gel electrophoresis and liquid chromatography coupled to tandem mass spectrometry (LC-MS/MS) to examine filtrate from one patient with acute kidney injury treated by CVVH with polyacrylonitrile (AN69) membrane. Using ultrafiltrate of hemofiltration from which high-molecular-weight serum proteins were removed, the excreted small proteins and polypeptides in the filtrate were investigated. As a result, 70% of proteins in the filtrate consisted of 292 proteins, which were not previously found in normal serum. The molecular weights of more than half of these proteins were less than 40 kDa. Fifty N-terminally acetylated peptides and several identified proteins were found as novel proteins.

Dihazi *et al.* [20] used 2-D gel electrophoresis and Biosystems Voyager-DESTR or Q-TOF Micromass Ultra to resolve and identify proteins in the outflow dialysate. Dialysate from each membrane, Fresenius high-flux polysulfone (FX60), Fresenius high-flux polysulfone (FX80), Asahi high-flux polysulfone (APS650), and Fresenius low-flux polysulfone (F6HPS) used in 28 hemodialysis patients, and continuous ambulatory peritoneal dialysis (CAPD) ultrafiltrate from 10 CAPD patients were analyzed by ProteinChip SELDI-TOF-MS and 2-D gel electrophoresis. The proteins in the outflow dialysate were different largely by the modality such as hemodialysis or CAPD and by the type of the dialysis membrane such as high flux or low flux. The protein peaks obtained by the low-flux membrane were observed in small numbers using ProteinChip SELDI-TOF-MS, whereas 46 different proteins in the outflow dialysate from high-flux membrane and 14 proteins in CAPD dialysate were identified. They concluded that these proteins may be related to complications of long-term dialysis, because they were also involved in the immune system.

27.2.2
Proteomic Analyses of Proteins Adsorbed onto the Dialysis Membrane as Reported in the Literature

Leviski *et al.* [21] investigated proteins adsorbed onto the polyarylethersulfone-polyvinylpyrrolidone (Gambro polyflux) membrane. Proteins were desorbed from membrane by applying 1 N sodium hydroxide for 45 minutes. Thereafter, proteins were separated by 2-D gel electrophoresis and identified by peptide mass fingerprinting. As a result, hemoglobin-α and -β, mannose-binding lectin-associated serine protease-2 related peptide (MAp19), antithrombin-III, calgranulin, and serum amyloid A were identified as the adsorbed proteins on the dialysis membrane. Although the meaning of membrane adsorption of these proteins is unknown, amyloid A is a potent acute phase reactant and MAp19 is important to activate the complement through mannose-binding pathway. They concluded that the membrane adsorption of these proteins may act for the elimination of acute phase reactants and suppression of complement activation.

In a study by Bonomini *et al.* [22], normal blood was circulated in minidialyzers made from ethylene vinyl alcohol (EVAL) and cellulose diacetate (CDA), and then protein adsorption onto EVAL and CDA membranes was studied. These adsorbed proteins were extracted using strong chaotropic solubilization buffer. 2-D gel electrophoresis coupled to both MALDI-TOF-MS and nanoLC-MS/MS identified a number of proteins that were adsorbed onto the CDA membrane. Adsorption of different albumin isoforms was observed for both membranes. Although there were common adsorbed proteins such as apolipoprotein A-IV, asprytyl-tRNA synthetase, proapoprotein A-1, peroxiredoxin 2, transferrin, albumin 2 isoforms, carbonic anhydrase, hemoglobin-β subunit, and hemoglobin-β chain, adsorption of apolipoprotein A-IV and thoredoxine-peroxidase B were adsorbed only onto the CDA membrane and two isoform spots of albumin were found only on the EVAL membrane. They suggested that proteomic analysis is useful to analyze the protein adsorption property of various dialysis membranes.

27.3
Proteomic Analysis of Proteins in the Outflow Dialysate and Proteins Adsorbed onto the Membrane Studied by Our Group

27.3.1
Methods and Protocols

The first set of studies [23] was performed mainly to compare the serum concentrations of β_2-microglobulin and the height of the spectra corresponding to β_2-microglobulin on SELDI-TOF-MS pre and posthemodialysis to determine the validity of SELDI-TOF-MS [9, 11]. The second set of studies [23] was performed to determine differences in the outflow-dialysate proteins and proteins adsorbed onto the membrane between PMMA and polysulfone membranes.

On the basis of the cross-sectional structure, dialysis membranes are divided into symmetrical and asymmetrical membranes. The structure of the polysulfone membrane, which is a representative of the asymmetrical membranes, demonstrates that the inner surface of the hollow fiber in contact with blood has a fine structure, while the external surface has a large foramen structure. However, the structure of the PMMA membrane exploits the stereo complex gained in solvent, consists of isotactic and syndiotactic forms [24], and was chosen as a representative of the symmetrical membrane. This membrane demonstrates an almost uniform symmetrical structure from the inner surface to the external surface of the hollow fibers. The ultrafiltration rates (UFRs) of the polysulfone membrane (TS-1.8UL: Toray Medical Co., Tokyo, Japan) and PMMA membrane (BK-1.8U: Toray Medical Co., Tokyo, Japan) were 34 and 51 $ml^{-1}\,h^{-1}$ mmHg, respectively. The clearance of β_2-microglobulin was 67 ml min^{-1} for polysulfone membrane and 52 ml min^{-1} for PMMA membrane. The amount of protein loss was 0.8 ± 0.3 (mean $\pm$ SD) g of albumin/session using the polysulfone membrane and 0.6 ± 0.5 g of albumin/session using the PMMA membrane.

Routine hemodialysis (4-hours duration, Qb = 200 ml min^{-1} and Qd = 500 ml min^{-1}) was performed using polysulfone and PMMA membranes. Outflow dialysate was accumulated in a 150-l tank, and then 30 ml of the dialysate was desalted. As a preliminary experiment for desorption of the adsorbed proteins onto dialysis membrane, the ratio of detached proteins calculated by all proteins determined with an amino acid analysis was 70, 53, 53, and 22% for 50% acetic acid (acid condition), 0.8 M sodium hydrogen carbonate (basic condition), 8 M urea, and 0.1% SDS, respectively. The greatest detachment of proteins was obtained using 50% acetic acid. However, in the experiment with 50% acetic acid, some dry samples were difficult to dissolve in a solution. It was thought that denaturation of the protein is advanced under a strong acetic acid condition. Therefore, 40% acetic acid was used to detach proteins from the dialysis membrane after hemodialysis. Approximately 40–50% of proteins adsorbed onto the polysulfone or PMMA membrane were desorbed by 10 minute immersion into 40% acetic acid, and the irrigated fluid was collected thereafter. This fluid and the outflow dialysate were concentrated using Spectra/Pro6 filter (membrane cut-off: 1000 Da) and then freeze-dried. The freeze-dried material was resuspended in superpurified water at a 50-fold dilution for outflow dialysate and in 20 mM acetic acid at a 25-fold dilution for desorbed protein fluid. However, it was difficult to confirm that all polypeptides were detached from the membrane as polypeptides and not from the parent proteins during the desorption procedure. Similar spectral patterns of adsorbed proteins from four patients on SELDI-TOF-MS (Figure 27.1) suggested that there was no random fragmentation or aggregation of proteins due to 40% acetic acid.

In our ProteinChip SELDI system [23] (Ciphergen Biosystems Inc. Fremont, USA), the resuspended materials of outflow dialysate and desorbed protein fluid were diluted 10 times and 100 µl of each solution was applied to five different conditions of ProteinChip (negative ion exchange chip Q10 (washed at pH 8 and pH 5), positive ion exchange chip CM10 (washed at pH 4 and pH 7), and metal affinity

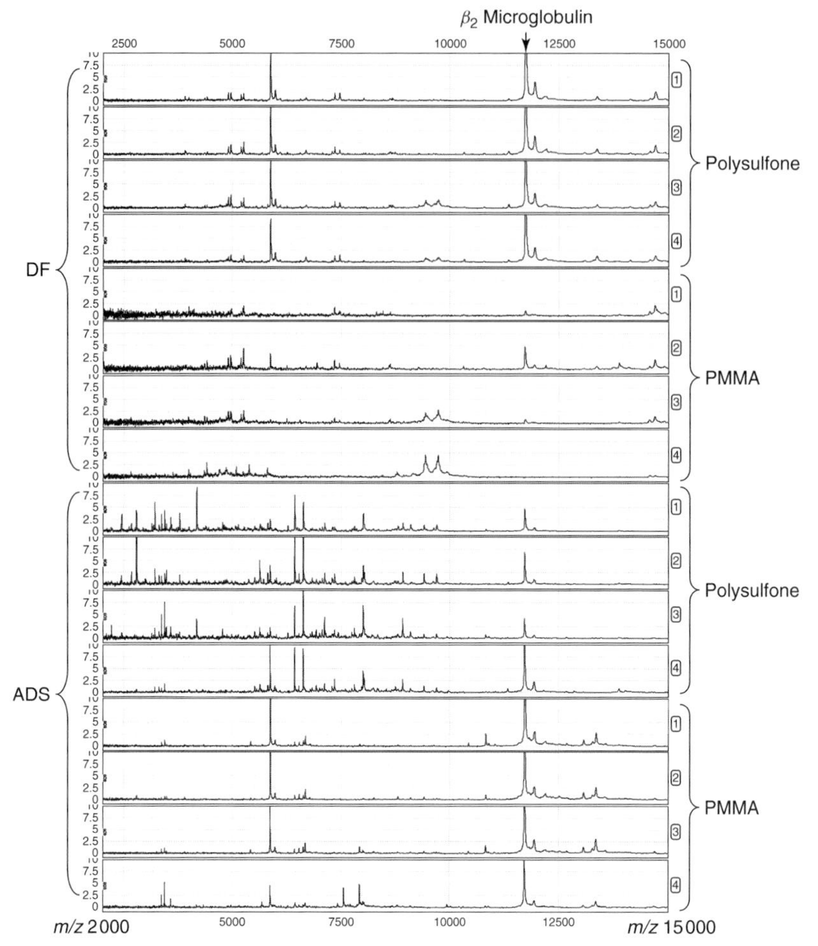

Figure 27.1 SELDI-TOF-MS spectra from metallic affinity ProteinChip (IMAC30) demonstrating differences in polypeptides and proteins filtered and desorbed from polysulfone and PMMA membranes in four patients. Molecular masses were ranged from 2000 to 15 000 Da. DF, proteins in outflow dialysate; ADS, adsorbed proteins onto membrane. Reproduced from [23], by permission of S Karger, AG.

chip IMAC30). The optimized ranges, 2000–15 000 Da and 15 000–120 000 Da, were used. The treated ProteinChips were measured by ProteinChip System IIC (Ciphergen Biosystems Inc. Fremont, USA) and peak clusters were made using Biomarker Wizard software. Thereafter, the peak view and gel view were obtained using ProteinChip software v3.2.1.

The serial dilution (1 : 1, 1 : 2, 1 : 5, 1 : 10, 1 : 50, and 1 : 100) of specimen decreased the heights of the spectra. However, there were no new peaks appearing with serial dilution. Reproducibility of spectra by Ciphergen's SELDI-TOF-MS was evaluated in our previous paper [25] and showed the coefficients of variations of 10.7% for spot-to-spot and 12.6% for chip-to-chip variabilities. The reproducibility of spectra of adsorbed protein by SELDI-TOF-MS in this experiment was calculated and the coefficient of variation for spot-to-spot was 14.3% [23].

Polypeptides (6629 and 6431 Da) were purified by cation exchange chromatography (CM Separose, Amersham Biosciences). Purified proteins were applied to the NP20 ProteinChip (hydrophilic array) and digested by trypsin. Proteins (11 730 Da) were purified by cation exchange chromatography (CM Separose, Amersham Biosciences) and SDS-PAGE. Polyacrylamide gels were stained by Coomassie Blue stain and the band of the selected protein was cut from the polyacrylamide gel. Half of the gel band was extracted by 50% formic acid/25% acetonitrile/15% isopropanol/10% water. The identity of the eluted material was checked by a ProteinChip reader. Then, in-gel digestion by trypsin was conducted on the remaining half. Tryptic digests were analyzed by Q-TOF tandem MS system (Q-TOF tandem mass spectrometer equipped with a PCI1000 ProteinChip Interface). MS/MS data were directly submitted using the MASCOT search engine.

27.3.2 Results

27.3.2.1 The First Set of Studies

The serum spectra (peaks) of eight hemodialysis patients (age = 56.9 ± 9.6 years, duration of dialysis = 230.9 ± 87.0 months), were measured before and after the hemodialysis using a metallic affinity ProteinChip (IMAC30). The relative intensities of the peak (spectra) of mass-to-charge ratio (m/z) 11 730, which represents β_2-microglobulin, were decreased after dialysis both with the polysulfone membrane (from 4.06 ± 0.83 to 1.40 ± 0.90) and with the PMMA membrane (from 3.23 ± 0.73 to 1.61 ± 0.44). Latex agglutination method demonstrated a linear relationship in the range of 1.5–47.0 mg l^{-1} for β_2-microglobulin with the relative intensity of the spectra of m/z 11 730. Relative intensity of spectra in our method was semiquantitative to some degree. The mean reduction rate of β_2-microglobulin during hemodialysis as measured by the conventional latex agglutination method and by the relative intensity (peak) of the m/z 11 730 by ProteinChip SELDI-TOF-MS system was 75.1 ± 6.1% and 67.6 ± 15.4% for polysulfone membrane, and 52.6 ± 7.9 % and 49.7 ± 10.8% for PMMA membrane, respectively.

The comparison of mean relative intensity of β_2-microglobulin in the outflow dialysate of eight patients between hours 0–1 and 3–4 by polysulfone and PMMA membranes were made. Mean relative intensity of m/z 11 730 in the outflow dialysate was 8.4 ± 1.5 at hours 0–1 and decreased to 4.8 ± 1.1 at hours 3–4 ($p < 0.0002$) using polysulfone membrane and was 0.08 ± 0.05 at hours 0–1 and increased to 1.0 ± 0.8 at hours 3–4 ($p < 0.02$) using PMMA membrane, suggesting that the pores of the polysulfone membrane had become clogged and the overflow elimination of m/z 11 730 adsorbed onto PMMA membrane in the dialysate at the end of dialysis session.

27.3.2.2 The Second Set of Studies

Profiles of proteins in the outflow dialysate are shown as DF, and profiles of proteins adsorbed onto the membrane are shown as ADS in Figure 27.1. These profiles were obtained from four patients (age $= 57.5 \pm 7.7$ years, duration of dialysis $= 268 \pm 72$ months, dialyzers with a membrane area of $1.8\,m^2$ were used in three of four patients and that of $2.1\,m^2$ was used in one patient) who were treated with both membranes. Under the same conditions of hemodialysis using polysulfone membrane and PMMA membrane in all four patients, the patterns of the peaks (spectra) on metallic affinity ProteinChip IMAC30 were highly similar, and the reproducibility of our methods appeared quite good. The results for membrane-filtered proteins and membrane adsorbed proteins measured by five different conditions of ProteinChips are shown as gel view (Figure 27.2a,b). The polypeptides and proteins, which were more filtered by polysulfone membrane than PMMA membrane, had molecular weight less than 30 kDa (Figure 27.2a,b).

Thirty-seven proteins with molecular weights larger than m/z 11 730 showed greater filtration through PMMA membrane than through polysulfone membrane. Of these 37 molecular masses, 11 molecular masses with relative intensities greater than 1.0 in the PMMA membrane group were observed [23]. Molecular masses of 34–40, 60, and 79 kDa are seen in outflow dialysate by the PMMA membrane (Figure 27.2b). High-molecular-weight proteins in the outflow dialysate demonstrating a range between m/z 15 000 and 120 000 by IMAC30 were seen with the PMMA membrane, although the band intensities were faint (Figure 27.2b). These findings are compatible with the finding that high-molecular-weight dextran appeared more frequently in the outflow dialysate of PMMA membrane than that of polysulfone membrane in an experiment using dextran [26].

Proteins adsorbed onto membrane obtained by five different conditions of ProteinChips were summarized as follows. Unexpectedly, there were 68 molecular masses of polypeptides that showed greater adsorption onto polysulfone than onto PMMA membrane. Dominant peaks with m/z less than 11 730 were m/z 6431, 6629, 9424, and 9714 indicating a relative intensity greater than 4 [23] (Figure 27.2a). Among these, the two conspicuous peaks, m/z 6629 and 6431, were prominent and identified as apolipoprotein C-I (pI 7.9) and truncated apolipoprotein C-I (pI 8.2), respectively. However, further research is necessary to clarify the clinical significance of apolipoprotein C-I and truncated apolipoprotein C-I. Other peaks have not yet been identified.

There were 149 molecular masses that were adsorbed only onto PMMA or in great quantity onto PMMA membrane than onto polysulfone membrane (Figure 27.2a,b). Dominant peaks were m/z 11 730, 24 319, 27 979, 43 221, 43 401, 51 089, and 51 646, indicating a relative intensity more than 4 [23]. Purification and identification of m/z 11 730 adsorbed onto the PMMA membrane, showed that the peak of m/z 11 730 was the full length of β_2-microglobulin (pI 6.07). The proteins adsorbed mostly onto PMMA membrane were those with high molecular weights in addition to β_2-microglobulin. Many proteins that were adsorbed onto PMMA membrane remain unidentified, although several proteins (β_2-microglobulin, Factor D, Pf-4, lysozyme, somatostatin, and C3a) have been identified by conventional methods

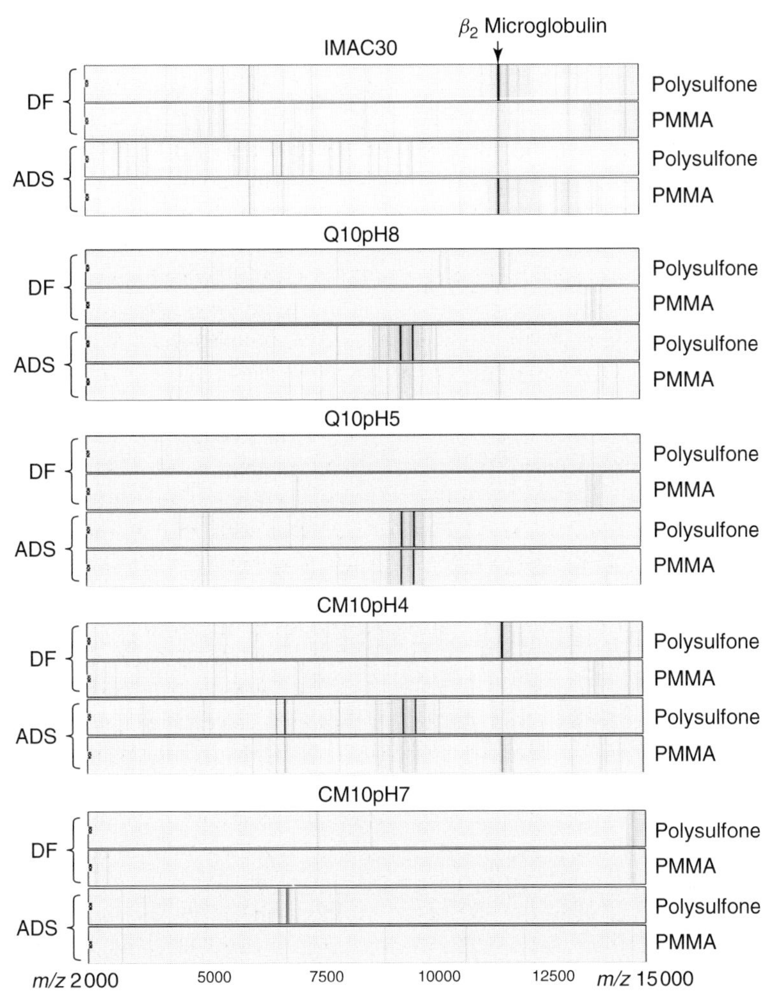

Figure 27.2 SELDI-TOF-MS gel view from ProteinChip IMAC30, Q10 washed at pH 8, Q10 washed at pH 5, CM10 washed at pH 4, and CM10 washed at pH 7 demonstrating differences in adsorbed proteins onto polysulfone and PMMA membranes. Proteins in outflow dialysate (DF) and adsorbed (ADS) proteins onto polysulfone (TS-UL) and PMMA (BK-U) membranes were analyzed at the *m/z* range of 2000–15 000 (a) and 15 000–120 000 (b). Reproduced from [23], by permission of S Karger, AG.

[26]. Identification of more proteins is needed to determine the characteristics of pI and the hydrophobicity of proteins adsorbed to specific membranes, polysulfone, or PMMA.

Although it is important to know which kinds of polypeptides and proteins are filtered through the membrane or adsorbed onto the membrane, it is more important to know how polypeptides and proteins in the serum are removed by hemodialysis, that is, how the relative intensity of the peaks obtained by

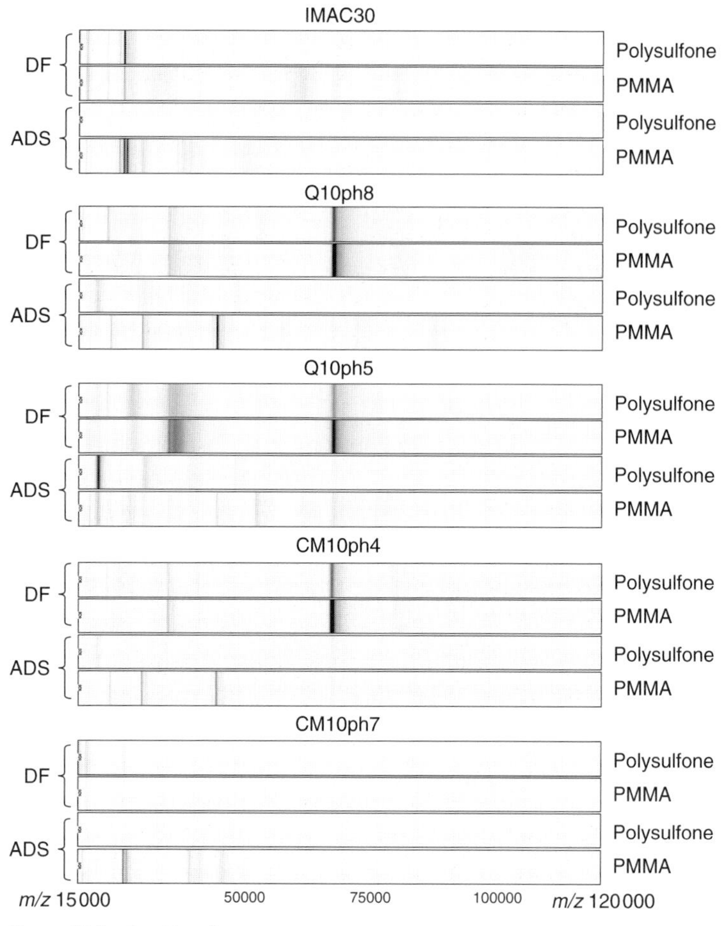

Figure 27.2 *(continued)*

ProteinChip SELDI-TOF-MS system change after hemodialysis. In our study, the rate of reduction in serum proteins was measured according to molecular weight. A similar pattern of reduction was demonstrated between polysulfone and PMMA membranes, but the pattern differed when precise observation was made. It is important to demonstrate whether this subtle change in the reduction rate between the two membranes causes clinically important differences in the prevention of dialysis complications or even patient survival.

27.4 Summary

In the literature review of proteomic studies investigating membrane filtration and adsorption of polypeptides and proteins, only the difference between low-flux and high-flux membranes have received attention to date. Our study suggested that membrane adsorption is an important mechanism for the removal of intermediate-molecular-weight proteins by hemodialysis, using not only PMMA membrane but also polysulfone membrane. Furthermore, the ProteinChip SELDI-TOF-MS system is useful for the analysis of polypeptides and proteins from *m/z* 2000 to a molecular weight less than that of albumin. The first step toward proteomic analysis of dialysate and adsorbed proteins onto dialysis membrane during renal replacement therapy has just been taken. Further study is still needed. If these proteomic analyses are used, not only identification of the accumulating uremic toxins due to uremic condition but also detailed information on polypeptides and proteins filtered through the dialysis membrane and adsorbed onto the dialysis membrane could be obtained.

References

1. Grassmann, A., Gioberge, S., Moeller, S., and Brown, G. (2005) ESRD patients in 2004: global overview of patient numbers, treatment modalities and associated trends. *Nephrol Dial Transplant*, **20**, 2587–2593.
2. Birk, H.W., Kistner, A., Wizemann, V., and Schutterle, G. (1995) Protein adsorption by artificial membrane materials under filtration conditions. *Artif Organs*, **19**, 411–415.
3. Bonomini, M., Fiederling, B., Bucciarelli, T., Manfrini, V., Di Ilio, C., and Albertazzi, A. (1996) A new polymethylmethacrylate membrane for hemodialysis. *Int J Artif Organs*, **19**, 232–239.
4. Clark, W.R., Macias, W.L., Molitoris, B.A., and Wang, N.H. (1994) Membrane adsorption of beta 2-microglobulin: equilibrium and kinetic characterization. *Kidney Int*, **46**, 1140–1146.
5. Eknoyan, G., Beck, G.J., Cheung, A.K., Daugirdas, J.T., Greene, T., Kusek, J.W., Allon, M., Bailey, J. *et al.* (2002) Effect of dialysis dose and membrane flux in maintenance hemodialysis. *N Engl J Med*, **347**, 2010–2019.
6. Grooteman, M.P. and Nube, M.J. (2004) Impact of the type of dialyser on the clinical outcome in chronic haemodialysis patients: does it really matter? *Nephrol Dial Transplant*, **19**, 2965–2970.
7. Vanholder, R., De Smet, R., Glorieux, G., Argiles, A., Baurmeister, U., Brunet, P., Clark, W., Cohen, G. *et al.* (2003) Review on uremic toxins: classification, concentration, and interindividual variability. *Kidney Int*, **63**, 1934–1943.
8. Issaq, H.J., Veenstra, T.D., Conrads, T.P., and Felschow, D. (2002) The SELDI-TOF MS approach to proteomics: protein profiling and biomarker identification. *Biochem Biophys Res Commun*, **292**, 587–592.
9. Lee, S.W., Lee, K.I., and Kim, J.Y. (2005) Revealing urologic diseases by proteomic techniques. *J Chromatogr B Analyt Technol Biomed Life Sci*, **815**, 203–213.
10. Merchant, M. and Weinberger, S.R. (2000) Recent advancements in surface-enhanced laser desorption/ionization-time of flight-mass spectrometry. *Electrophoresis*, **21**, 1164–1177.
11. Seibert, V., Wiesner, A., Buschmann, T., and Meuer, J. (2004) Surface-enhanced laser desorption ionization time-of-flight mass spectrometry (SELDI TOF-MS) and ProteinChip technology in proteomics research. *Pathol Res Pract*, **200**, 83–94.

12. Fliser, D., Novak, J., Thongboonkerd, V., Argiles, A., Jankowski, V., Girolami, M.A., Jankowski, J., and Mischak, H. (2007) Advances in urinary proteome analysis and biomarker discovery. *J Am Soc Nephrol*, **18**, 1057–1071.
13. Hille, J.M., Freed, A.L., and Watzig, H. (2001) Possibilities to improve automation, speed and precision of proteome analysis: a comparison of two-dimensional electrophoresis and alternatives. *Electrophoresis*, **22**, 4035–4052.
14. Kaiser, T., Hermann, A., Kielstein, J.T., Wittke, S., Bartel, S., Krebs, R., Hausadel, F., Hillmann, M. *et al.* (2003) Capillary electrophoresis coupled to mass spectrometry to establish polypeptide patterns in dialysis fluids. *J Chromatogr A*, **1013**, 157–171.
15. Neuhoff, N., Kaiser, T., Wittke, S., Krebs, R., Pitt, A., Burchard, A., Sundmacher, A., Schlegelberger, B. *et al.* (2004) Mass spectrometry for the detection of differentially expressed proteins: a comparison of surface-enhanced laser desorption/ionization and capillary electrophoresis/mass spectrometry. *Rapid Commun Mass Spectrom*, **18**, 149–156.
16. Lefler, D.M., Pafford, R.G., Black, N.A., Raymond, J.R., and Arthur, J.M. (2004) Identification of proteins in slow continuous ultrafiltrate by reversed-phase chromatography and proteomics. *J Proteome Res*, **3**, 1254–1260.
17. Ward, R.A. and Brinkley, K.A. (2004) A proteomic analysis of proteins removed by ultrafiltration during extracorporeal renal replacement therapy. *Contrib Nephrol*, **141**, 280–291.
18. Weissinger, E.M., Kaiser, T., Meert, N., De Smet, R., Walden, M., Mischak, H., and Vanholder, R.C. (2004) Proteomics: a novel tool to unravel the patho-physiology of uraemia. *Nephrol Dial Transplant*, **19**, 3068–3077.
19. Molina, H., Bunkenborg, J., Reddy, G.H., Muthusamy, B., Scheel, P.J., and Pandey, A. (2005) A proteomic analysis of human hemodialysis fluid. *Mol Cell Proteomics*, **4**, 637–650.
20. Dihazi, H., Mueller, C., Mattes, H., Asif, A., and Mueller, G. (2005) Proteomics analysis of dialysates: comparison of different high and low flux dialyzers with peritoneal dialysis. *J Am Soc Nephrol*, **16**, 153A.
21. Levitski, T., Ward, W., Black, N., Arthur, J., and Hamilton, N. (2004) Dialysis membrane protein adsorption: a proteomic approach. *J Am Soc Nephrol*, **15**, 364A.
22. Bonomini, M., Pavone, B., Sirolli, V., Del Buono, F., Di Cesare, M., Del Boccio, P., Amoroso, L., Di Ilio, C. *et al.* (2006) Proteomics characterization of protein adsorption onto hemodialysis membranes. *J Proteome Res*, **5**, 2666–2674.
23. Ishikawa, I., Chikazawa, Y., Sato, K., Nakagawa, M., Imamura, H., Hayama, S., Yamaya, H., Asaka, M. *et al.* (2006) Proteomic analysis of serum, outflow dialysate and adsorbed protein onto dialysis membranes (polysulfone and PMMA) during hemodialysis treatment using SELDI-TOF-MS. *Am J Nephrol*, **26**, 372–380.
24. Sakai, Y. and Tanzawa, H. (1978) Poly(methyl methacrylate) membrane. *J Appl Polymer Sci*, **22**, 1805–1815.
25. Ishikawa, I., Hayama, T., Yoshida, S., Asaka, M., Tomosugi, N., Watanabe, M., Yamato, H., and Sugano, M. (2006) Proteomic analysis of rat plasma by SELDI-TOF-MS under the condition of prevention of progressive adriamycin nephropathy using oral adsorbent AST-120. *Nephron Physiol*, **103**, 125–130.
26. Sugaya, H., Ueno, Y., Yamada, T., and Itagaki, I. (2006) Structure and function of high performance membranes. *Jin-To-Tohseki (Kidney Dial)*, **61** (Suppl), 19–23 (in Japanese).

Index

Renal and Urinary Proteomics: Methods and Protocols. Edited by Visith Thongboonkerd

ISBN: 978-3-527-31974-9